The Nature, Causes, Effects and Mitigation of Climate Change on the Environment

Edited by Stuart A. Harris

Published in London, United Kingdom

IntechOpen

Supporting open minds since 2005

The Nature, Causes, Effects and Mitigation of Climate Change on the Environment
http://dx.doi.org/10.5772/intechopen.94777
Edited by Stuart A. Harris

Contributors
Fidaa Fawwaz Haddad, Stuart Arthur Harris, Deepika Pandey, Emmanuel O. Oladipo, Ojonigu F. Friday Ati, Kayode Aremu, Iliya B. Abaje, Adewale F. Olatunde, Tufa Feyissa Negewo, Arup Kumar Sarma, Moses Zakhele Sithole, Isaac Agholor, Julian Alberto Sabattini, Rafael Alberto Sabattini, Julius Jackson Nyaombo, Terence Epule, Abdelghani Chehbouni, Driss Dhiba, Arthur Martin Kato, Revocatus Twinomuhangi, Adam M. Sebbit, Zulqurnain Khan, Saima Rasheed, Iqra Farooq, Muhammad Bin Mushtaq, Thida Htoo, Firew Bekele Worana, Pawan Thapa, Golam Ali Mozaffari, Muneza Lydie, Lazaro Elibariki Nnko, Inga Nikolaevna Ryumkina, Sergey Ryumkin, Anastasiia Malykhina, Andrei Khanturgaev, Dmitry Ursu, Adebanke L. Adebayo, Yonten Chophel, Amor Escoz-Roldán, Pablo Ángel Meira-Cartea, José Gutiérrez-Pérez, Somala Karthik, M.S. Sai Reddy, Gummudala Yashaswini

First published in London, United Kingdom, 2022 by IntechOpen
IntechOpen is the global imprint of INTECHOPEN LIMITED, registered in England and Wales, registration number: 11086078, 5 Princes Gate Court, London, SW7 2QJ, United Kingdom
Printed in Croatia

British Library Cataloguing-in-Publication Data
A catalogue record for this book is available from the British Library

Additional hard and PDF copies can be obtained from orders@intechopen.com

The Nature, Causes, Effects and Mitigation of Climate Change on the Environment
Edited by Stuart A. Harris
p. cm.
Print ISBN 978-1-83968-611-5
Online ISBN 978-1-83968-612-2
eBook (PDF) ISBN 978-1-83968-613-9

For EU product safety concerns:
IN TECH d.o.o., Prolaz Marije Krucifikse Kozulić 3, 51000 Rijeka, Croatia, info@intechopen.com or visit our website at intechopen.com.

Meet the editor

Stuart Arthur Harris has a BSc (Honours), MSc, and Ph.D. in Geology and a DSc in Geography from the Queen Mary University of London. His experience includes geology, water supply, and engineering. Before becoming Government Soil Surveyor in Guyana, Dr. Harris engaged in soil survey work in Egypt and Iraq for the consulting firm Hunting Technical Services. He also taught in the geography departments of the University of Chicago, Wilfred Laurier University, and the University of Kansas before joining the University of Calgary in 1969. The National Research Council of Canada asked Dr. Harris to study the relationship of climate to permafrost in 1973 and thus he carried out detailed studies of the permafrost landforms in northwest Canada, China, Iceland, the Alps, Poland, Russia, China, Mongolia, New Zealand, and Kazakhstan, and on the Tibetan Plateau. Dr. Harris has published more than 200 papers, books, and reports. In 1996, he received the Nikolai Mihailovich Prjevalsky Medal from the Russian Geographical Society for his research on Alpine permafrost.

Contents

Preface

Climate change is one of the most important problems today. It affects everybody as well as the natural environment in which we live. Recently, the weather has changed causing all kinds of problems for both humans and the current biota. These changes in weather and weather patterns have led to changes in the physical processes occurring in the natural environment, such as frequency and magnitude of floods, droughts, rates of soil erosion, and associated phenomena. Sea level changes, slope stability, thawing of permafrost, and desertification are among the consequences of climate change. Not surprisingly, this has resulted in numerous local and international meetings at which the causes and mitigation of the more problematic effects of climate change have been discussed.

The main idea of global warming was developed by a consortium of mainly European scientists and funded by numerous governments. They argue that the whole world has been warming at an ever-increasing rate since the beginning of the Industrial Revolution, c. 1870 AD. The causes are presumed to be the effects of greenhouse gases and the results are published in numerous reports and updates (e.g., Intergovernmental Panel on Climate Change [IPCC]) [1]. In a previous book [2], the origin and history of the concept of global warming was discussed and it was clear that not all climate scientists agreed. Instead, some thought that climate change could be restricted to regional areas that could become warmer or colder as well as wetter or dryer. Obviously, the high variability of climate at all time scales complicates the situation. Although the IPCC has revised its conclusions many times, not all scientists agree with the results.

The objective of this book is to obtain a picture of the climatic changes occurring around the world during the last 20–50 years together with the resulting changes in land use and the environment. Remarkably few papers have come from countries where the ideas of global warming were developed. Instead, more than 100 potential chapters were offered dealing with the changes in climate in tropical and subtropical countries. These changes did not prove global warming in most of these areas, but rather they described the environmental changes that were due to more moderate changes in climate. One paper from Nigeria analyzed 50 years of data from actual weather stations and concluded that the climate was not changing, but that instead the weather alternated with decades of years with cooler and wetter climates with other periods of warmer weather accompanied by drought.

The book is divided into five sections. Section 1 begins with a chapter discussing some possible causes of the arguments over the concept of global warming. This is followed by a chapter examining the evidence from satellites, which raises doubts about the worldwide application of global warming theory. Other possibilities are explored such as ocean currents and hurricanes piling up heat energy in the North Atlantic region, and periodic deep-water warm, thermo-saline current carrying heat south to the Antarctic seas [3, 4]. This would be counterbalanced by cold surface Antarctic waters flowing north to replace the mass of the warm thermo-saline current. The post-2018 increase in violence of the weather around the Northern Hemisphere appears to have started after a great amount of warm air moved north to the tropical areas instead of going south to Antarctica. Its causes and effects are not currently known. Section 2 provides an example of the response of Spanish schoolchildren to the theory of global warming. Section 3 consists of several chapters discussing the effects of climate change on agriculture and the environment. Chapters describe the land use changes in various tropical and subtropical countries in contrast to the effects of climatic changes in northern Russia. In many cases, the changes involve growing different crops that can cope with the new climatic conditions. Section 4 consists of chapters describing changes in hydrology and appropriate mitigation measures including minimizing the effects of drought. The problems involved include soil erosion, mass wasting, changes in stream courses with using solar panels to improve efficiency of irrigation. Section 5 consists of three chapters dealing with the problems encountered by humans in hot climates. A chapter in this section discusses the problems of survival of bees, which are essential for pollination of plants and crops. Without bees, animals are in trouble since they rely on vegetation as an essential part of the food chain.

Stuart A. Harris
Emeritus Professor,
University of Calgary,
Alberta, Canada

References

[1] IPCC. Special report on the impacts of global warming of 1.5 °C above pre-industrial levels. 2018. Available from: http://www.ipcc.ch/report/sr15/ [Accessed: December 02, 2019]

[2] Harris SA, editor. Global Warming. Croatia: IntechOpen; 2010a. p. 249. DOI: 10.5772/10283

[3] Crowley T. North Atlantic deep water cools the Southern Hemisphere. Paleoceanography. 1992;7:489-497

[4] Broecker WS. Was a change in thermohaline circulation responsible for the Little Ice Age? Proceedings of the National Academy of Sciences. 2000;**97**(4):1339-1342. DOI: 10.1073/pnas.97.4.1339

Section 1

Climate Change: Causes and Processes

Chapter 1

Global Warming and Climate Change (GWCC) Realities

Yonten Chophel

Abstract

The study attempted to investigate the urgency of the global warming and climate change by analyzing the available data from the secondary sources. The document analysis technique was used to examine the available literature. When it comes to the urgency of global warming and climate change, the study showed that there are two schools of thought. One is in support of the motion, claiming that global warming is a real phenomenon triggered by anthropogenic behavior, while the other is opposed to the motion, claiming that global warming and climate change are complicated phenomena, and that forecasting future climates is difficult due to the various players involved, about which climate specialists know little or nothing. Based on document analysis, study infers that there is certain uncertainty about the future of the climate, because climate always changes, and it cannot be certainly affirmed that the climate change is man- made (anthropogenic activities) or is due to natural occurrence. However, it is evident that the global surface temperature, borehole temperature, sea surface temperature, and the sea level is increasing over the years. The study suggests that for the humanity to be certain about their future, treating the global warming and climate change as an act of urgency and working towards prevention and mitigation by limiting the production of greenhouse gases and mindfully consuming the natural resources would be the plausible solution for the larger problem of Global Warming and Climate change.

Keywords: Global warming, Climate change, Global surface temperature, Borehole temperature, Sea surface temperature, Sea level

1. Introduction

Global warming is a disastrous phenomenon compounded by the human activities often drilled by the greed and corporate paradigm of thinking. Its totaling effect in the long run might threaten the survival of mankind and any other species for that matter. Many nations recognize it as serious immediate threat and they forms the associations and organizations to combat against it. However, there are few cynical nations, owing to several reason, economic reason being dominant. But, with passage of time, it has become so apparent that even a rational high school student can comprehend the status and need of action, it is very evident to a layman, that the temperature of the earth, be it in any season is gradually increasing and the people working in the field are awe struck by the crops ability to grow in new higher unnatural altitudes. The lexicon "global warming" might be new to them but they have already felt its consequences.

Climate is very essential concept, owing to dependence of existence of life on it. Living beings survive on food, food is produced by plants and growth of plants depends on climate. So, erratic change in the climate will put billions of lives at stake. Therefore, it is of paramount importance that we discuss, monitor and make sure that the climate for different regions remains at its natural range. Many literatures suggest that global warming caused by the global warming gases emitted enormously by the manufacturing factories and automobiles industries are the cause of erratic climate [1–6]. Global warming if unchecked will influence the climate of world to the point that earth tipping point is reached, where the surviving earth cannot further sustain any lives in it, changing the course of history preceded by mass extinction of species in the world. Owing to the likelihood of paying a bigger price latter due the simple negligence in time, it is duties of all the global citizens to be mindful about it and do come together to find the effective solutions to mitigate the problem that is ever going to be bigger as it gets delayed. It would be grave mistake to decide not to heed on this global immediate call, for it has the ability to either make or break our collective future.

There is two school of thoughts with regard to urgency of the matter, where one school of thought believes that understanding the climate change is complex and it is extremely difficult to predict the future of climate in this complex universe, and are skeptical about the anthropogenic global warming, suggesting that the information that is known about weather and its causes is not enough to predict the future climate and the climate model are far from precision [7]. However, another school of thought, expresses its concern and urgency to save the earth form untimely destruction, advocating the need to limit the production of the global warming gases that enhance the chances and intensity of occurrence of the erratic weather [7, 8]. In this modern era, people spent more time looking into screen than looking into sky for weather. As a responsible global citizen, it is our moral obligation to explore the underlying truth and make informed decisions locally. Thus, this study explored the available literature on the global warming and climate change in terms of its urgency.

2. Literature review

Global warming is the term that was introduced or used for the first time by climatologist Wallace Broecker in his article "*Climate Change: Are we on the Brink of pronounced Global warming?*" Global warming is observable increase in the global temperature of earth (both land and water) and climate change is the effect brought about by the process of warming globally or in general, overall long-term change in our climate, including sea level rising, extreme weather, and ocean acidification. However, the term global warming and climate change are used interchangeably but there is difference in it, global warming is more sensitive and more diverging which results in less advocacy by some subpopulations [9]. Further, Krauss [3] clearly puts it in the context when he quotes Lorenz, "Climate is what you expect; weather is what you get." (p.158). Krauss [3] argues that just because there is an anomalous cold day in Washington, DC does not mean that global warming is not happening. Likewise, just because there is an ultra-hot week in Washington, DC also does not validate global warming.

The solar radiation from the sun is balanced by the thermal radiations reflecting from the earth; this interaction balance and determines the surface temperature of the earth. The incoming solar radiation from the sun is independent but the outgoing thermal radiations depends on the earth's surface temperature and the presence of greenhouse gases, which absorbs some of the thermal radiations. Greenhouse gasses (GHG), such as carbon dioxide, methane, nitrous oxides, water vapor, ozone and chlorofluorocarbon (CFC) are responsible for trapping of heat. For instance, water vapor (40%) is responsible for absorbing majority of thermal radiation from earth, followed by carbon dioxide (30%), methane (20%) and other gases (5%) [7]. So, it indicates that outgoing thermal radiation is mostly absorbed by water vapor and carbon dioxide. Likewise, the change in composition of water vapor due to human activities are negligible, so it implies that the greenhouse gases produced by human activities are likely responsible for the most of the trapping of heat [10, 11].

However, some authors [12] argues that the heat trapping by the CO_2 is not significant and rather it is likely that sun radiations are responsible for the global temperature rise. Again, the debate on whether the sun radiations or GHG is responsible for the global temperature continues, recently, Herring [13] refuted the claim that likeliness of sun radiation as the cause of global warming might not be true. He argues that, it is possible that sun can warm the earth provided that the pattern of the solar intensity increases over the years. Likewise, the sunspot data do indicate that there was a small increase in the amount of sunlight from late 1800s to the mid-1900s which experts estimates that it could have contributed at the most up to 0.1°C of the 1.0°C (1.8°F) of warming observed since the pre-industrial era. However, there has been no significant net change in the sun's energy output from the late 1970s to the present (see **Figure 1**), which is when the most rapid global warming was observed. Further, scientists rule out the significant role of sun in global warming due to the fact that if the sun energy output had intensified then it is logical to expect all the layers of earth's atmosphere to be warmed, which is not the case that is been observed. Rather, satellite and weather balloons observation showed that more warming in the lower atmosphere (troposphere) and cooling in the upper atmosphere (stratosphere) [13]. This pattern of differential warming is what is been expected due to result of increasing GHG trapping heat.

The main source of global warming gases is the burning of the fossil fuels and it is observed that a nation with the abundance of the availability of the fossils fuels tends to depend more on fossil fuels than the nations with low abundance of fossils resources [5]. Global energy consumption with regard to fossil fuels are staggeringly high, and its prediction for the coming years are also projected very high, as reflected in the world energy outlook report of 2012 and 2013. Thereby the production of greenhouse gases will be soared and ultimately it might bring about the increase in the global temperature of the earth. Likewise, the rising population's

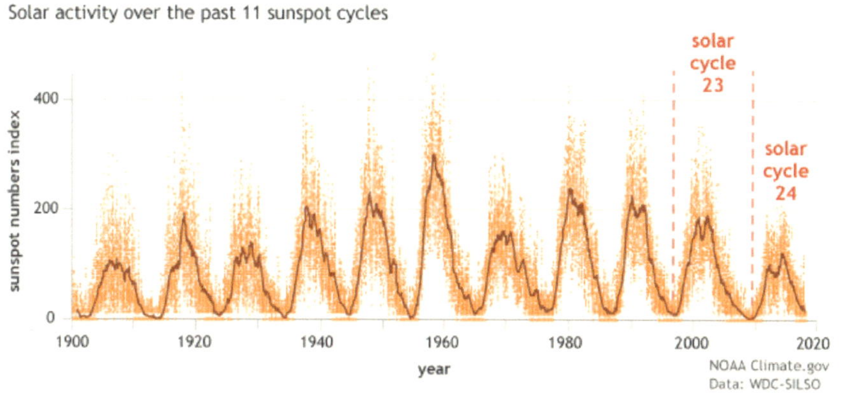

Figure 1.
The peaks and valleys in solar geomagnetic activity since 1900, based on the number of sunspots observed on the face of the sun each day (orange dots). Graph by NOAA Climate.gov, based on data from the WDC-SILSO, Royal Observatory of Belgium (source: Herring [13]).

demand for fuel, as well as the need for economic growth and a higher standard of living, are both factors to consider. Lack of political will and institutional failure in to make and enforce effective environmental policies, as well as a lack of related knowledge and Rampant disinformation, have all hampered action against global warming and reduction of the global warming gases [7]. Furthermore, the global temperature rises correlates with the rising pattern in the consumption of the fossil fuel, indicating the likeliness of the GHG warming the earth [3].

With the increase in the concentration of the greenhouse gases it will act like a thick blanket in the atmosphere where it will absorb the solar radiation. The carbon molecules and the oxygen molecule in the carbon dioxide undergoes vibration like stretching, bending, and this action absorbs the solar radiations [1, 2]. Thereby hindering the reflection of solar radiation by the earth making the earth surface temperature warmer.

Many studies affirms that the GWCC is a worrying thing. For example, a study done by Suonan et al. [14] suggests that due to global warming there will be more than worse phenological alteration in some places like in Tibetan Plateau, more than what is predicated or known. This idea makes us think of worst situation if the temperature of the earth keeps warming up. Moreover, the global sea level has risen by about 8 inches since 1880, and it's projected to rise another 1 to 4 feet by 2100 (as a result of melting of ice), thus the danger of storm surges and high tides would increase the flooding in many regions [8].

Shrinking of glaciers, early breaking up of ice on rivers and lakes, shifting of plant and animal and premature flowering are some of the observable effects global climate change had on environment. Moreover, effects that scientists had predicted in the past are now occurring: loss of sea ice, accelerated sea level rise, more intense heat waves. Scientists have high confidence that global temperatures will continue to rise for decades to come, largely due to greenhouse gases produced by human activities. The Intergovernmental Panel on Climate Change (IPCC) which includes more than 1,300 scientists from the United States and other countries, forecasts a temperature rise of 2.5 to 10 degrees Fahrenheit over the next century [15]. Moreover, the IPCC predicts that increases in global mean temperature of less than 1.8 to 5.4 degrees Fahrenheit (1 to 3 degrees Celsius) above sea levels will produce beneficial impacts in some regions and harmful ones in others. Net annual costs will increase over time as global temperatures increases [8].

However, not everyone believes that the earth's surface temperature is rising, and even if global warming is true, not everyone accepts that human activities are the primary cause. Also, not everyone agrees that climate change is a problem. As a result, critics and deniers of global warming and climate change do not see the need to take steps to delay or reverse these trends [5, 7]. USA disengaging from Kyoto protocol and Paris agreement on climate showed the polarity of attitude of nations and people towards the GWCC issue. Moreover, at number of climate summit, member countries failed to agree to number of "Environmental targets" in recent years (Kyoto, Copenhagen, so on) [5].

Extend to which the idea of global warming and climate change have reached to the mankind largely depends on their exposure to the main stream medias and social media. It is found out that the social media network does plays vital role in spreading the knowledge and awareness of the GWCC and it is also indicated that they understand the concept better when they are expose to those terms in positive or negative light [16]. However, as watchdog it is responsibility of media and social media to uncover the truth, but some studies suggest that newspapers aren't doing much to convince the health impact of GWCC to the public [17]. Moreover, in the study done by Shapiro and Park [18], they found out that people responses to the GWCC in social media particularly their reactions to YouTube video of GWCC are general and shows little or no concern about GWCC, which indicates many people aren't so convince about the reality of GWCC.

On the contrary, the very existence of 175 plus active organization on climate change [19], and 100 plus top websites on climate change [20], indicates the concern and the urgency expressed by the people around the world. Study done by Liu et al. [21] also affirms that many congressmen do believe GWCC as real thing. Further, IPCC asserts that scientific evidence for warming of the climate system is unequivocal, global temperature rise, warming of oceans, shrinking of ice sheets, rising of sea level, acidification of ocean, and declining of Arctic sea ice are some of the events that conveys the story of the happening event so called Global warming and climate change [8]. In addition, study done by Allen and McAleer [22] suggests that the of negative emotions or indifference to global warming might be due to lack of clear logical framework and confusion of short-term variations in localized weather with the long-term global average climate change.

The happenings of global warming can be traced through using the technique of observation and examining the rise in the land surface temperature, borehole temperature profile, sea surface temperature, and sea level [7, 8, 23]. If there are increase in all those four independent parameters, it is the indication that the global warming is occurring [7]. This increase is attributed to the increase in the global warming gases, CO_2 concentrations in the atmosphere have risen from 0.028 percent in pre-industrial 1750 to 0.043 percent today. Until recently, it was thought that stabilizing CO2 levels in the atmosphere at about 0.055 percent by 2035 will be enough to keep global warming below 2°C. However, 3°C is becoming more possible, which will induce wreaking havoc on human colonies, coral reefs, rain forests, and polar ice caps. To keep temperature rises below 2°C, urgent international action is required, which means keeping CO_2 levels below 0.045 percent. Only if governments can negotiate on cooperative national and international action can this become a possibility [23].

Despite the two school of thought on the urgency and status of GWCC, the evidences from the surface temperature, sea level rise, sea surface temperature, and borehole temperature profile indicate that the global warming is happening and it is going to be a major threat in the future, which will be discussed in the result and discussion section in detail.

3. Methodology

The study utilized the qualitative design and uses document analysis approach of the data collected from the existing documented secondary sources. Resources were collected via use of Research for Life search engine, mostly peer-review journals and accredited resources were used in finding the useable data. Data were collected from more than fifteen different sources. Primarily the main sources of the collected data included Journal article titled "Global warming and climate change: Realities, Uncertainties and Measures" by A. P. Alzebeokhai, published in 2009, data from the NASA-Global climate change, data from NOAA (National centers for environmental information: National oceanic and atmospheric administration), and Intergovernmental Panel on Climate Change (IPCC) (latest and updated information as of 2019).

The data collection and analysis are done in line with the research question on "Is Global warming and climate change (GWCC) really occurring, and how urgent it is?" Data from several climate expert organizations such as NASA-Global climate change, IPCC, NOAA and other accredited scientific journals are used for the same.

The data from those separate sources were compared and analyzed. Owing to the lack of primary data, no statistical software was employed for the analysis.

4. Result and discussion

The result from the analysis of four separate sets of observations, including surface temperature measurements, sea surface temperature, sea level changes, and temperature profiles in boreholes, all indicate that the earth's surface temperature is increasing, suggesting that it is warming. Each of these separate sets of observations yields findings that overlap and complement each other, suggesting that the GWCC is a real phenomenon.

4.1 Surface temperature

The surface of the earth has warmed by an average of 1.0°C (1.8°F) in the last 100 years, according to regular measurements of the earth's surface temperature recorded daily from thousands of weather stations around the world, both ashore and stumped. Mean weekly, monthly, and annual temperatures are calculated using daily temperature measurements. As a result, the average annual temperature change can be easily monitored from year to year. The global mean temperature has risen by 0.1°C per decade over the last two decades, with 2005 being the warmest year on record [7]. The effects of large population centers on global mean temperature, referred to as the "urban heat island effect," are calculated and corrected for; however, this accounts for less than 15% of observed global warming. Global warming is not constant across the globe, both in terms of time and space; high latitude regions warm more than low latitude regions [7]. Human activities are estimated to have caused approximately 1.0°C of global warming above pre-industrial levels, with a likely range of 0.8°C to 1.2°C. Global temperature rise is likely to reach 1.5°C between 2030 and 2052, if it continues to increase at the current rate [15]. Moreover, according to the Global climate change report 2018 [23], complied by NOAA, shows that the surface temperature is increasing, the data compared the temperature recorded from 1880 till 2018 which showed that, 2016 is the warmest recorded temperature with 0.95 temperature anomaly degree Celsius, followed by 2015 with 0.91 anomaly degree Celsius, and 2017 with 0.85 anomaly degree Celsius [23].

Additionally, the data from three major compilations based on measured surface temperatures: from GISS (Goddard institute for space studies), HadCRU (global temperature dataset) and NCDC (national oceanic and atmospheric administration) showed upward trend (see **Figure 2**). They have expressed the trend as the temperature difference ("anomaly") with respect to the 1901–2000 average as the baseline [24].

The comparison of the three different dataset form three climate recoding source showed that the temperatures juggle up and down, but the overall trend is upward meaning the globe is warming.

Upon analysing trend through the average of the three datasets over the period 1975–2009 (during which greenhouse gas forcing was the dominant driver of climate change), the following are (see **Figure 3**).

For all three-temperature series, the trend from 1975 to 2009 is about the same (0.17 +/– 0.03 degrees per decade). The error reflects the trend's 95 percent confidence interval, i.e., if the trend analysis were repeated a hundred times on the actual underlying results, the trend will be within the range of 0.14 to 0.20 degrees per decade 95 times out of 100 (see **Figure 3**) [24].

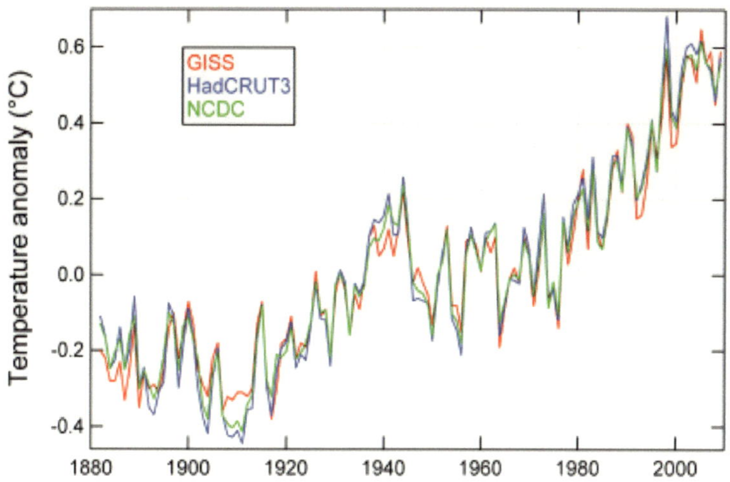

Figure 2.
Comparison of three data set on surface temperature (source: Verheggen [24]).

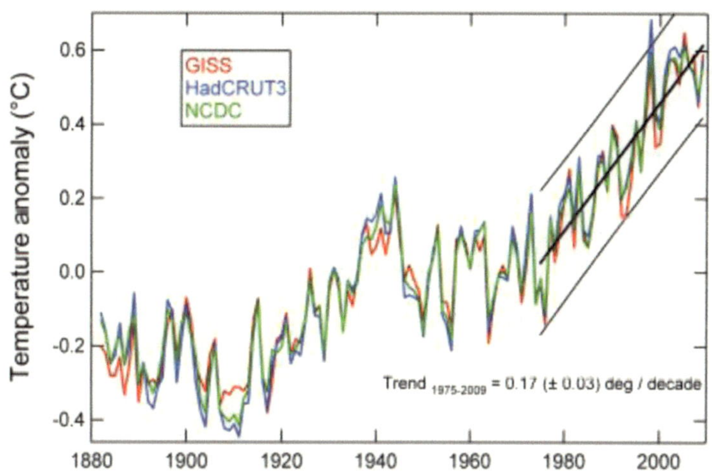

Figure 3.
Showing average temperature from three dataset (source: Verheggen [24]).

The thin black lines (see **Figure 3**) represent the 95% confidence "predictions bands" for the data. Based on the observed variability, 95% of the data are expected to fall within these lines. The observed yearly variability in global temperatures (occasionally exceeding 0.2 degrees) is such that 10 years is too short to discern the underlying long-term trend (0.17 degrees per decade) [24]. Thus, data from all different sources shows and depicts that the surface temperature of the earth has increased over the decade, with different data source agreeing to the value range of 0.1°C increase per decade [24]. This indicates the occurrence of global warming.

4.2 Sea level rise

Another predictor of global warming and climate change comes from a completely different series of findings (the measurements of water level changes). The amount of water in the oceans is rising as a result of thermal expansion of water within the oceans and, as well as due to, melting of glaciers and polar ice as the earth warms.

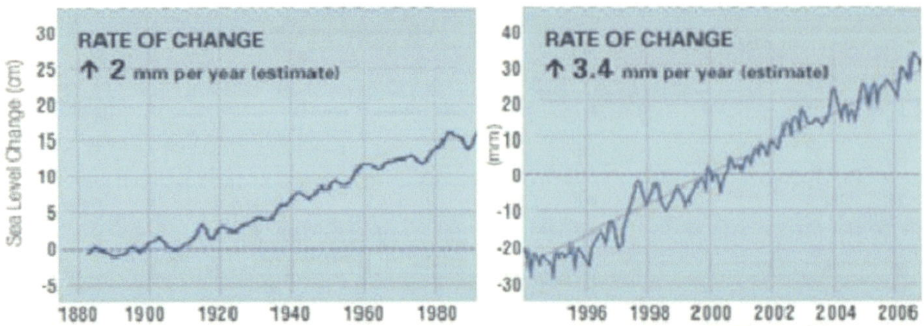

Figure 4.
Mean annual sea level rise associated with the thermal expansion of sea water due to warming and widespread melting of ice sheets (source: Aizebeokhai [7]).

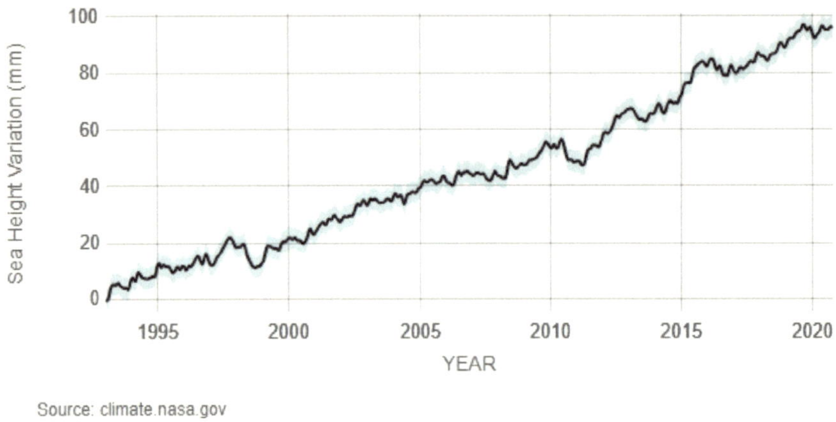

Figure 5.
Trend of increase in sea level (satellite data) (source: NASA-global climate change [25]).

Regular water level observations are taken at various sites, equivalent to temperature measurements; daily water level variations, mostly due to tides and storms, are averaged to obtain mean sea level for a given period of time. **Figure 4** depicts the average annual change in sea level between 1880 and 2008. Over the last century, the average water level has risen by around 18 cm [7]. Between 1880 and 1990, it increased by an estimated 2 mm per year on average (left chart in **Figure 4**) and is now growing at a rate of about 3.4 mm per year (right chart in **Figure 4**). Similar to global temperature changes, sea level changes, aren't constant, so the detailed changes aren't always in line with surface temperature measurements. The thermal expansion of the water column occurs later than the associated change in surface temperature, with ocean currents influencing the timing. Global temperature changes, as well as changes in sea level, are not constant, and the details of these changes are not always in line with surface temperature measurements. The water column's thermal expansion occurs later than the related change in surface temperature; the differences are affected by ocean currents (**Figure 5**) [7].

Furthermore, the latest data from the NASA-Global Climate Change, shows that the trend of sea level is upward and increasing. The increase rate of 3.3 mm per year is recorded [25]. Thus, the data from the existing sources both form past and the recent, indicates the rise in the sea level, agreeing to rise value range of 2 to 3.4 mm per year. Which hints to the occurrence of the global warming and retreat of the ice sheets and glaciers.

4.3 Borehole temperature profile

The thermal history of the earth's subsurface offers a third evidence of global warming and global climate change. The subsurface stores temperature records over time that are related to the prevailing environment at the time. Responsive thermometers are used to calculate temperature profiles with depth in boreholes, caves, and deep mines. Temperature anomalies due to geological features, upward flow of warmth from the earth's interior, heat produced by crustal rocks, and variations in groundwater movement are generally adjusted for. Surface temperature oscillations propagate downward with depth, with shorter duration fluctuations attenuating more than longer period fluctuations. As a result, only long-term fluctuations in temperature penetrate great depths, with seasonal changes penetrating around 15 m until the signals fade. Century-long variations, in contrast to seasonal variations, can be observed to depths of about 150 m, and millennial cycles can be observed to depths of 500 m or more. These depths are easily attained by low-cost drilling. The subsurface serves as a selective filter, eliminating short-term temperature fluctuations and maintaining excellent records of global warming and, as a result, climate change (**Figure 6**) [7, 26].

The temperature profiles suggest substantial warming in the last century from 0.6°C in southeast Utah to more than 2.0°C in Alaska. Curves are arbitrarily offset for display purpose (see **Figure 6**). The temperature profiles of boreholes spread across a length of about 500 km of northern Alaska shows anomalous warming of 2 to 5°C in the upper 100 to 150 m of the permafrost and rocks [7]. Similarly, borehole temperature profiles in eastern Canada shows a less rapid warming of about 1.0°C. A warming of about 0.5 and 1.0°C were observed in Nebraska sites and Utah sites, respectively (see **Figure 6**). These results indicate that geothermal data mimicked the geographic variations of warming observed in weather station data. Baseline temperatures from previous centuries are often inferred from geothermal evidence, enabling researchers to date the start of the industrial revolution to this century and thus determine the effect of industrialization on global warming and climate change [7]. As a result of the data from the borehole temperature profile from various places, the temperature rises in the range of 0.5 to 5°C at 100 to 150 m. This hints to the likelihood of global warming.

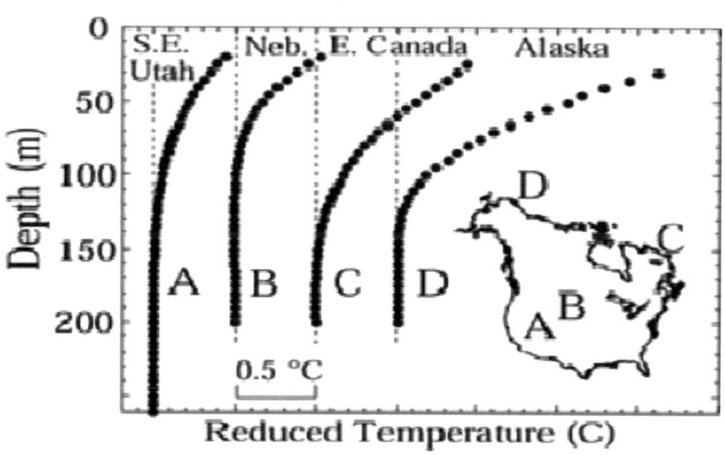

Figure 6.
Borehole temperature profile from sites in North America showing warmer temperatures within near-surface depths of 100–150 meters (source: Aizebeokhai [7]).

4.4 Sea surface temperature

The thermometer measurements made on water samples taken by merchant and navy ships as they sailed the world's oceans date back to about 1850, and constitute the instrumental record of natural processes within the oceans. The data is best for parts of the oceans along major trading routes, and it's understandable that they are scarcer further back in time. These readings, like those from land-based meteorological observation posts, must be gridded to provide a global average sea surface temperature. Since the oceans cover about 75% of the earth's surface, the sea surface temperature record is close to global temperature records, as one would expect (**Figure 7**).

The two records are identical, but the SST (sea surface temperature) varies across a narrower spectrum than the land surface temperature, and the land temperature are subjected to dramatic swings. This disparity is primarily due to the oceans' higher heat potential than the air (it takes a long time to heat and cool the oceans).

The measurement and data collected from hundreds of buoys stationed across the ocean at the depth range of about 2000 m collected over the years as early as 1955, showed that not just the surface of the oceans but the whole upper half of the ocean is gradually warming. Over the past 50 years average of 0.1 to 0.2°C was recorded. So, while the whole ocean has absorbed a huge amount of heat, its overall temperature has changed little. Nevertheless, the very surface of the ocean has warmed almost as much as the rest of earth's surface [26]. Thus, form the past data and the present data it shows that the oceans are warming up slowly which almost resonates with the increase in land surface temperature, indicating the occurrence of global warming.

Analysis of four dimension (indictors) indicates the happening of global warming. The connection between the global warming and climate change is well documented. For instance, according to IPCC [27] "Changes in many extreme weather

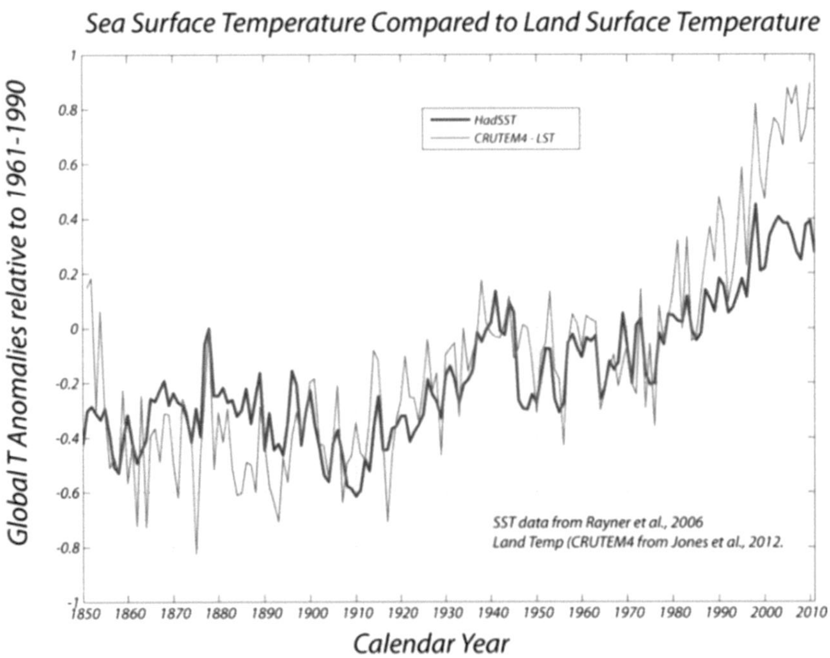

Figure 7.
Comparison of Sea surface temperature and land surface temperature (source: Brawlower and Bice [26]).

and climate events have been observed since about 1950. Some of these changes have been linked to human influences, including a decrease in cold temperature extremes, an increase in warm temperature extremes, an increase in extreme high sea levels and an increase in the number of heavy precipitation events in a number of regions" (p.7). Moreover, IPCC [15] predicts differences in mean temperature in most land and ocean regions, hot extremes in most inhabited regions, heavy precipitation in several regions, and the probability of drought and precipitation deficits in some regions, caused by the phenomenon of global warming.

5. Conclusion

Earth is facing the global warming. Scientifically, there is no longer any doubt that the surface temperature of land, sea surface temperature, and sea levels are increasing. The logical reason of this increasing trend is far beyond the occurrence of natural cause. Despite the different propaganda between the global warming and climate change denier and the advocates, it is of paramount importance to heed for the interest of the humanity and survival of humankind. There is still considerable confusion regarding the exact timing and scale of global warming, as well as the impacts that would result from it, although this is also due as much to human responses to the issue as it is to scientific uncertainties. Growing temperatures, changes in rainfall levels and seasonality; increased frequency of severe weather events such as droughts, floods, and hurricanes; sea level rise; melting of polar ice and glaciers are only a few of the consequences that can be expected. Desertification, loss of tropical forests and coral reefs, decreases in agricultural production, extinction of species, water scarcity, increasing natural disaster losses, and the spread of tropical diseases are likely to be among the ecological and human effects of these changes. Whether the severity of these impacts results in only a slight deterioration in environmental quality and social well-being, or a truly catastrophic collapse that leads to famine, mass displacement, and resource wars, will be determined by how we behave in the coming decades, as well as the probability of an unanticipated response within the climate system to rising temperatures and greenhouse gas emissions.

The polarity of the views with regard to the urgency is demystified by a study in China [28]. The authors reported that on the global scale the average public concern about the GWCC among Chinese citizens are relatively low, further analysis revealed that youth and women with greater post-materialist values had more concern about GWCC than that of their counterparts. Likewise, citizens from provinces with higher economic dependency on carbon-intensive industries were found to have less concern about GWCC than people from provinces with lower carbon dependency. Their study revealed the underlining motive of the GWCC deniers and skeptics. Perhaps, the polarity in views between two school of thought has much to do with national or regional benefit than the actual truth.

Skeptics of global warming are not very convinced of the fact that the change in the climate and the extreme climate, periodically experienced, are the result of the effect of global warming, enraged by the production of greenhouse gases like carbon dioxide. However, they are also equally not able to disregard the voluminous literature which suggests that the climate change is primarily linked with the global warming. Understanding the climate change in totality is sophisticated, owing to the number of players involved in it, which aren't yet fully understood by the experts themselves.

However, the existing literature suggests based on the evidences of increase in the *surface temperature, sea surface temperature, borehole temperature, and sea level*, that it would be wise on the part of humanity to act to minimize the global warming – to which burning of fossil fuels is main culprit. Cause of global warming is well documented and known by majority of the people, what is apparently failing, is in the action to minimize the production of global warming gases and excessive natural resources consumption. Humans are most intelligent species, yet most skeptical and cynical creature who ever walked the earth, choice are-to scum to habit of cynicism, or accept the fact for real and act to maintain the balance in nature for the sake of humanity.

Author details

Yonten Chophel
Daga Central School, Dagana, Bhutan

*Address all correspondence to: yanphel7@gmail.com

References

[1] Agarwal SK. Global warming and climate change. New Delhi: APH Publishing Corporation; 2011.

[2] Shrivastava A. Global Warming. New Delhi: APH Publishing Corporation; 2009.

[3] Krauss LM. The physics of climate change. London: Head of Zeus Ltd; 2021.

[4] Budzianowski W. Time delay of global warming. International Journal of Global Warming. 2011;3(3):289.

[5] Johnsson F, Kjärstad J, Rootzén J. The threat to climate change mitigation posed by the abundance of fossil fuels. Climate Policy. 2018;19(2):258-274.

[6] Henderson RM, Reinert SA, Dekhtyar P, Migdal A. Climate Change In 2017: Implications for Business [Internet]. Hbs.edu. 2017 [cited 20 March 2019]. Available from: https://www.hbs.edu/environment/Documents/Climate_Change_2017.pdf

[7] Aizebeokhai AP. Global Warming and Climate Change: Realities, Uncertainties and Measures. International Journal of Physical Science. 2009;4(13):868-879.

[8] Jackson R. The Effects of Climate Change [Internet]. Climate Change: Vital Signs of the Planet. 2019 [cited 10 March 2020]. Available from: https://climate.nasa.gov/effects/

[9] Benjamin D, Por H, Budescu D. Climate Change Versus Global Warming: Who Is Susceptible to the Framing of Climate Change? Environment and Behaviour. 2016;49(7):745-770.

[10] Agarwal A, Narain S. Global Warming: in an Unequal World [Internet]. Indiaenvironmentportal.org.in. 2003 [cited 17 February 2020]. Available from: http://www.indiaenvironmentportal.org.in/files/GlobalWarming%20Book.pdf

[11] Climate change: Evidence, Impacts, and Choices. [Internet]. Ipclimatechange.trg-learning.com. 2013 [cited 4 March 2020]. Available from: http://ipclimatechange.trg-learning.com/wp-content/uploads/2013/11/NAS-Climate-Change-Evidence-Impacts-and-Choices.pdf

[12] Florides L. A., Christodoulides, P. and Messaritis, V., 2010. Global warming and CO2 Vs Sun. In: Harris, S. A. (Ed.). Global Warming, Chapter 3: 23-62. Croatia, IntechOpen. 249p. Doi: 10.5772/10283.

[13] Herring D. Couldn't the Sun be the cause of global warming? | NOAA Climate.gov [Internet]. Climate.gov. 2020 [cited 12 March 2021]. Available from: https://www.climate.gov/news-features/climate-qa/couldnt-sun-be-cause-global-warming

[14] Suonan J, Classen AT, Sanders NJ, He J. Plant phenological sensitivity to climate change on the Tibetan Plateau and relative to other areas of the world. Ecosphere. 2019;10(1).

[15] IPCC Special report "Global Warming of 1.5 degree Celsius-summary for teachers. [Internet]. Ipcc.ch. 2018 [cited 13 March 2020]. Available from: https://www.ipcc.ch/site/assets/uploads/sites/2/2018/12/ST1.5_OCE_LR.pdf

[16] Lineman M, Do Y, Kim JY, Joo GJ. Talking about Climate Change and Global Warming [Internet]. 2015 [cited 16 March 2020]. Available from: https://journals.plos.org/plosone/article?id=10.1371/journal.pone.0138996

[17] Weathers MR. Newspaper Coverage of Global Warming and Climate Change

(GWCC) as a Public Health Issue. Applied Environmental Education & Communication. 2013;12(1):19-28.

[18] Shapiro MA, Park HW. More than entertainment: YouTube and public responses to the science of global warming and climate change - Matthew A. Shapiro, Han Woo Park, 2015 [Internet]. SAGE Journals. 2014 [cited 11 March 2020]. Available from: https://journals.sagepub.com/doi/abs/10.1177/0539018414554730?journalCode=ssic

[19] Member Organizations [Internet]. US Climate Action Network. 2019 [cited 12 March 2020]. Available from: https://www.usclimatenetwork.org/member-organizations

[20] 100 Top Climate Change Sites [Internet]. World.org. 2013 [cited 23 March 2020]. Available from: http://www.world.org/weo/climate

[21] Liu X, Vedlitz A, Stoutenborough J, Robinson S. Scientists' views and positions on global warming and climate change: A content analysis of congressional testimonies [Internet]. 2015 [cited 17 March 2020]. Available from: https://link.springer.com/article/10.1007%2Fs10584-015-1390-6

[22] Allen DE, McAleer M. Fake news and indifference to scientific fact: President Trump's confused tweets on global warming, climate change and weather [Internet]. Eprints.ucm.es. 2018 [cited 12 January 2020]. Available from: http://eprints.ucm.es/47902/1/1817.pdf

[23] Global Climate Report - Annual 2018 | State of the Climate | National Centers for Environmental Information (NCEI) [Internet]. Ncdc.noaa.gov. 2018 [cited 17 March 2020]. Available from: https://www.ncdc.noaa.gov/sotc/global/201813

[24] Verheggen B. Global Average temperature increases. GISS, HadCRU and NCDC compared. [Internet]. My view on climate change. 2010 [cited 19 March 2020]. Available from: https://ourchangingclimate.wordpress.com/tag/ncdc/

[25] Sea Level | NASA Global Climate Change [Internet]. Climate Change: Vital Signs of the Planet. 2020 [cited 17 February 2020]. Available from: https://climate.nasa.gov/vital-signs/sea-level/

[26] Brawlower T, Bice D. Temperature: Ocean Warming | EARTH 103: Earth in the Future [Internet]. E-education.psu.edu. 2019 [cited 10 February 2020]. Available from: https://www.e-education.psu.edu/earth103/node/753

[27] Climate Change 2014 Synthesis Report Summary for Policymaker [Internet]. Ipcc.ch. 2014 [cited 1 March 2020]. Available from: https://www.ipcc.ch/site/assets/uploads/2018/02/AR5_SYR_FINAL_SPM.pdf

[28] Liu X, Hao F, Portney K, Liu Y. Examining Public Concern about Global Warming and Climate Change in China. The China Quarterly. 2019; 242:460-486.

Chapter 2

Causes and Mechanisms of Global Warming/Climate Change

Stuart A. Harris

Abstract

Comparison of the average mean surface air temperature around the world during 1951–1978 with that for 2010–2019 shows that the bulk of the warming is around the North Atlantic/Arctic region in contrast to the Antarctic ice sheet. Obviously, the temperature change is not global. Since there is a substantial difference between solar heat absorption between the equator and the poles, heat must be moving to the North Pole by surface ocean currents and tropical cyclones. The cold, dry Arctic air coming from Siberia picks up heat and moisture from the open oceans, making the sea water denser so that the warm water sinks slowly down to c. 2000 m. A deep-water thermohaline flow (THC) transports the excess hot (c. 18°C) water south to Antarctica. It is replaced by a cold (c. 2°C) surface water from that area. The latter quickly cool western Europe and Siberia, and glaciers start to advance in Greenland within about 10 years. The THC flow decreases in Interglacials, causing the increased build-up of heat in the Northern Hemisphere (c. 60% currently stored in the Atlantic Ocean), and the ice cover in the Arctic Ocean thaws. Several such cycles may take place during a single major cold event.

Keywords: global warming, major cold events, asymmetrical warming of the Earth, thermohaline currents, agents of transport of heat around the globe, minor cold and warm events

1. Introduction

Climate change has become a major consideration for the public and politicians alike because of its economic importance. Naugebauer [1] provides a good summary of the history of its study, but the conclusion reached is that current theories and models do not adequately explain what is taking place. There is a universal assumption that climatic changes are worldwide. During the last two decades, many parts of the land areas of the Northern Hemisphere have experienced increasing mean annual air temperatures together with more extreme weather. The public, media, and politicians have been assuming that the warming trends seen in many places on land are the result of increased concentrations of atmospheric carbon dioxide, following the lead of a panel of climatologists appointed by various European governments that produces periodic reports on the amount and causes of the presumed warming, *e.g.*, IPCC [2]. However, many scientists have disagreed (*e.g.*, [3–15]) citing too many causes that are closely correlated with recent major climate changes that are clearly not associated with carbon dioxide concentrations. The overall recent increase in mean average daily temperature on Earth is similar to that on the other planets in the solar system.

The purpose of this paper is to examine the evidence for a universal system of climatic change throughout the globe, together with some of the proposed causes and mechanics of the climatic changes that have taken place during the late Pliocene and Quaternary Periods. This should determine the most likely causes of the cold events and recent changes in weather around the globe.

1.1 Definitions

Global warming can be defined in two ways. Firstly, it can be interpreted literally as referring to increased surface temperatures throughout the whole world, but the IPCC defines it as the increase in mean annual air temperature of the Earth since the beginning of the Industrial Revolution, assumed to be between 1850 and 1900 A.D. There is some disagreement as to when this began, but global warming enthusiasts and modelers assume that it is occurring everywhere on the Earth's surface.

Recently, many scientists are using the term climate change in place of global warming since there is strong evidence that the changes vary from place to place as they have done in the more distant past. These changes are closely related to the nature of the local environment and the relationship to both the distribution of land and sea, the continuing plate tectonics altering the arrangement and dimensions of the land and sea together with the current climate.

1.2 The Earth within the solar system

The Earth is essentially a closed energy system, the heat is supplied from the Sun and the geothermal heat flows from the interior of the Earth. The latter is regarded as being fairly constant at 0.06 W/meter squared at 30 degrees C while the incoming solar radiation is potentially 6000 times greater in its effect ([16], p. 37). It is determined by the so-called solar constant which is 1.37 kW $\times$ m^{-2} [16]. The total heat flux through the Earth's surface due to energy generated in the mantle and the crust is approximately 0.0257% of the total Earth's solar irradiation. There is an 11-year cycle of variation in solar irradiance, but this has not increased since 1950. The incoming radiation is modified by clouds, surface albedo, snow cover, soil moisture, vegetation cover, latitude, thermal conductivity, and soil latent heat. The world total energy production is estimated to be about 0.0077% of the total solar irradiation reaching the Earth. Thus, solar radiation supplies more than 99.95% of total energy driving the world climate [17].

1.3 Current patterns of climate change

When the average mean surface air temperature around the world during 1951–1978 is compared with that for the period 2010–2019, it is obvious that although warming has taken place, it is asymmetric (**Figure 1**). The bulk of the warming is in the northern hemisphere while the temperature changes in the southern hemisphere decrease southward and become negative in and near Antarctica. The same result is obtained from space satellites although the area with negative temperature change in the higher latitudes of the southern hemisphere increases. There are also differences between the oceans and land masses as well as between continents. Obviously, the temperature change is not global, nor is it concentrated in the regions of the maximum industry but varies regionally. Accordingly, we must be dealing with climate change, not global warming.

The First Law of Thermodynamics states that energy and matter cannot be created or destroyed, although they can change their form. In a closed system, a decrease in energy in one area must result in a comparable increase somewhere else.

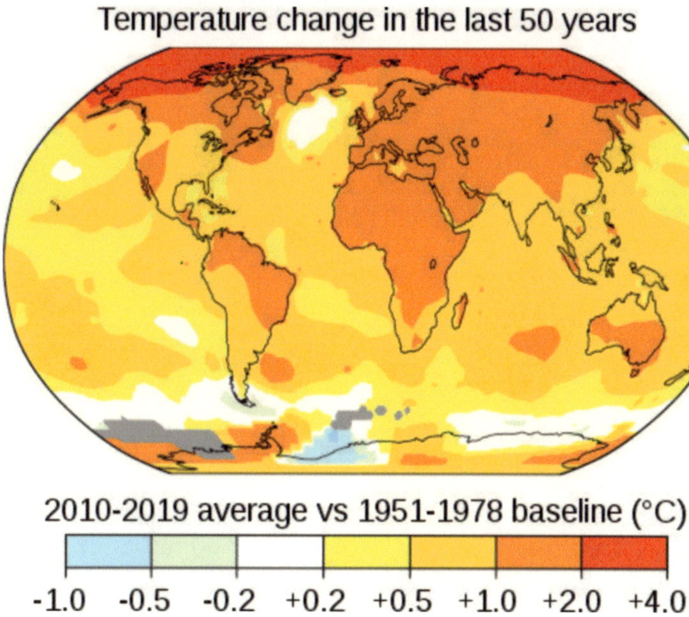

Figure 1.
The pattern of change in mean annual air temperature between the periods 1951–1978 and 2010–2019.

To get and maintain an asymmetric heat energy distribution as shown in **Figure 1**, there must be a set of forces preferentially moving heat energy towards the north polar regions as well as a mechanism to bring the energy back towards the South Pole. The rest of this paper will discuss the processes involved in the movement of heat energy around the globe in order to determine what may be happening. The conclusions must also explain the longer-term climatic history as we know it today.

2. Processes affecting the distribution of heat around the surface of the Earth

The Sun is far enough away that the solar radiation arrives at the surface of the Earth as a beam. Since the Earth is round, the bulk of the energy arrives within the Tropics, 23.5° north and south of the Equator. Under present-day conditions, the Sun does not rise for at least a day in winter at latitudes higher than 66.7° north and south due to the tilt of the axis of the Earth.

Figure 2 shows the typical fate of insolation in summer entering the atmosphere of the Earth at about 50°N latitude. In practice, the amount of insolation reaching the ground will also depend on the sum of the effects of numerous climatic cycles pulling in multiple directions [14, 18, 19]. There are a very large number of these interacting with one another, so they normally largely cancel one another out. However, they are also largely responsible for the continual noise in the climatic signal that makes it advisable to have a 30-year average of climatic data where possible.

There is one special group of cycles that are critical in determining the amount of solar radiation arriving at a given location on the surface of the Earth over long periods of time, *viz.*, the Milankovitch cycles first thought of by Adhemar (Croll [20, 21]). Milankovitch [22, 23] refined the calculations of the effect of the three types of Earth orbital movements that can alter the Sun's incoming radiation by up to 25% in the zone subtropical zones (30–60° north and south of the equator. They are the shape of the Earth's orbit (eccentricity, a 100,000 year cycle), the

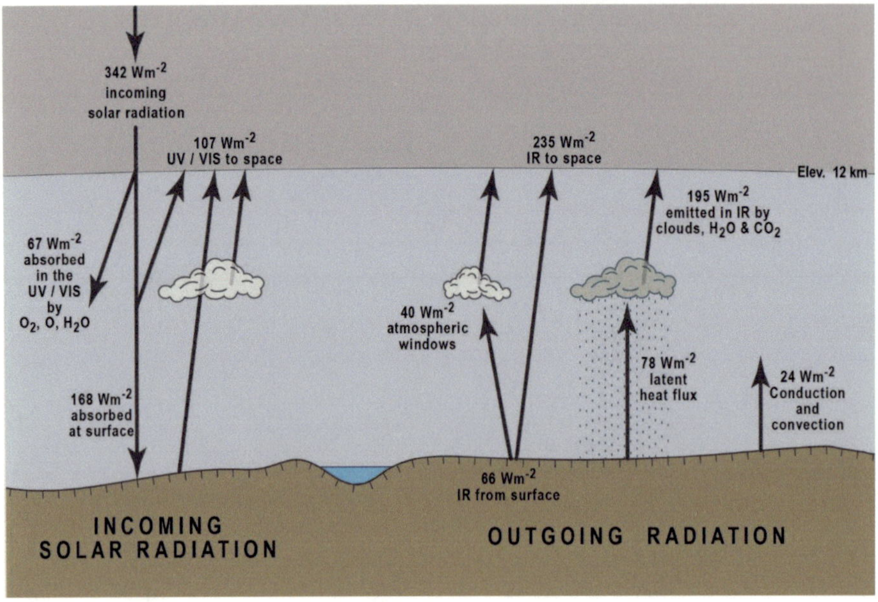

Figure 2.
Typical fate of the solar energy reaching the surface of the atmosphere and proceeding down towards the ground.

angle of tilt of the Earth's axis relative to the Earth's orbital plane (obliquity, from 21.1–24.5°in a 41,000 year cycle), and the direction of the Earth's axis of rotation is pointing (precession, in a 23,000 year cycle).

Milankovich calculated that cold events might occur approximately every 41,000 years and subsequent research confirms that they did occur at 41,000-year intervals between one and three million years ago. About 800,000 years ago, the cycle of the Ice Ages lengthened to 100,000 years, matching Earth's eccentricity cycle [24]. However, the cycles should not affect the data for the last two decades.

3. Movements of heat away from the Tropics

The fact that the bulk of the solar radiation arrives on the surface of the Earth along the zone between the Tropics of Capricorn and Cancer results in a tremendous imbalance of heat distribution between the Equator and the Poles. The amount of solar heating of the polar latitudes throughout the\year varies greatly, with the polar latitudes receiving considerably more solar energy in summer than in the winter when they receive no solar heat at all. As a result, in the winter hemisphere, the difference in solar heating between the equator and that pole is very large. This causes the large-scale circulation patterns observed in the atmosphere. The difference in solar heating between day and night also drives the strong diurnal cycle of surface temperature over land.

The seasonal imbalance results in about 30% of the heat absorbed in the Tropics currently moving towards the polar regions each year to partially make up the difference. As will be shown below, the effectiveness of the processes carrying out this transfer depends on the arrangement of the land masses and oceans as well as the connections between these two contrasting regions. Antarctica is situated over the South Pole and has a circular land mass with few indentations. The main exception is the Antarctic Peninsula that projects northwards into the path of the Antarctic Circumpolar Drift (**Figure 3**). This contrasts with the North Pole situated in a

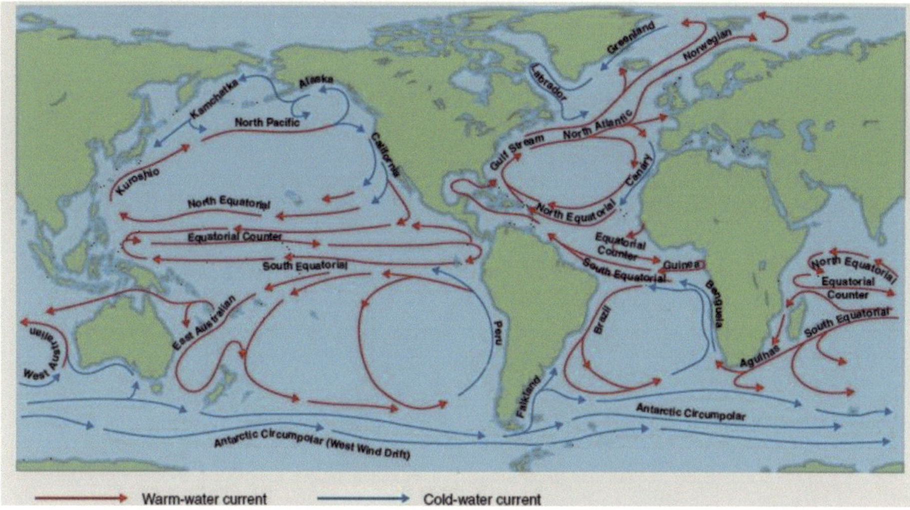

Figure 3.
Ocean currents, red being warm and blue being cold. The Gulf Stream/North Atlantic currents move the greatest amount of warm water, with the Kuroshio current moving the second-most amount. Both move north towards the north pole, but there are no comparable currents heating the shores of Antarctica.

generally frozen sea that is connected by north-south extending seas (gateways) to the tropical oceans ([14], **Figure 1**). In the northern hemisphere, there are large continents separated by oceans from one another. This distribution of land and sea causes a tremendous difference in solar warming of both land and sea as well as the transport of heat towards the poles. Without the ocean gateways, the northern polar areas would be as cold and inhospitable as Antarctica.

3.1 Agents of heat transfer around the Earth

There are two mediums for the transfer of large quantities of surface heat around the Earth, *viz.*, the water in the oceans and the air in the atmosphere. However, the thermal properties of the water make it far more effective in moving heat towards the higher latitudes. Dry air is not nearly as effective in moving heat, but it can transport heat over the surface of land masses. Hot, humid air is intermediate in transporting power since it contains up to 5% water vapor in extreme cases.

3.1.1 Transport of heat by water

Although water is confined to the lower levels of the globe except for lakes, its thermal properties make it a very important transporting medium for heat, *e.g.*, in bringing heat from central boilers to the houses in many cities in Russia and for cooling engines in automobiles, etc. Water has a very high heat capacity (4.187 mJ/m^3 K) so that it can store or transport large quantities of heat in a given volume of water [25]. In addition, it absorbs over five times as much heat as soil or rock since it is translucent. Currents, convection, and wave action mix the water whereas transmission into a rock or sediments must be by conduction. Thus, ocean currents transport an enormous amount of water polewards (**Figure 3**), primarily in the northern hemisphere where the Gulf Stream and North Atlantic currents transport heat to north Greenland. Note the five ocean gyres at tropical and subtropical latitudes in the Atlantic, Pacific, and Indian Oceans. These accumulate large quantities of heat in the upper layers of the seas, spawning the monsoons and tropical cyclones that move towards the poles.

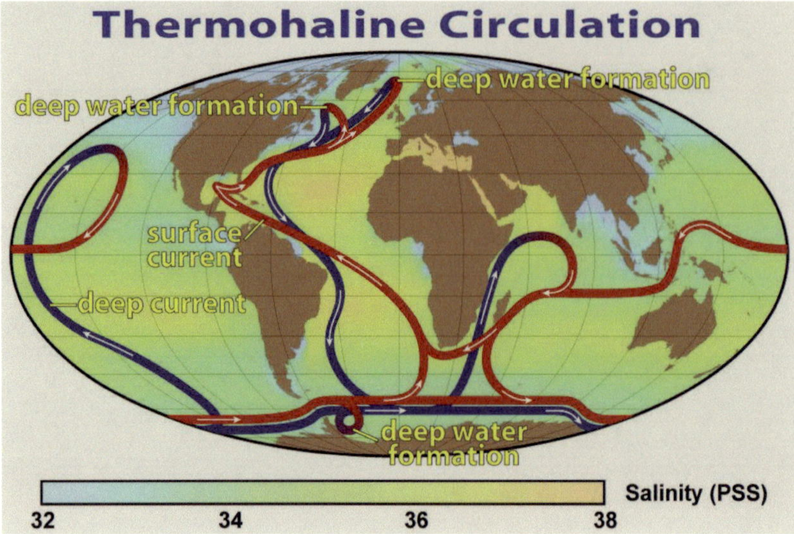

Figure 4.
The thermohaline circulation (THC) and the salinity of the surface waters (NASA).

Research during the past decades [26, 27] has shown that the North Atlantic Current is part of a vast system of fast-moving, deep thermohaline currents (THC) that moves heat down to the southern hemisphere and forms a global thermohaline circulation system (**Figure 4**). Periodically, the cold surface waters northward water off the coast of North America goes south to be replaced by cold water from Antarctica. Cold Antarctic surface waters must move north to replace this southerly current. This is believed to result in rapid cooling in Greenland and the Arctic regions whereas gradual warming take place in Antarctica [28–30]. This exchange was named the "bipolar see-saw" by Broecker [4].

The deep-sea thermohaline current goes south to the sea surrounding Antarctica via the Atlantic and its effects are discernible all the way to the North Pacific. This circulation pulls warm surface seawater north via the Atlantic because more water is needed to replace the dense, increasingly saline seawater that has sunk towards the ocean floor and subsequently participated in the conveyor belt. The total flow of the larger north-south exchange system is thought to be about 16–20 Sv, where Sv is a unit of ocean current flow, 1 Sv = 1 million m^3/s (million cubic meters per second).

During studies of the apparent warming on land in northeastern North America, it was found that about 60% of the energy increase was being stored in the adjacent seas. Levitus *et al.* [30] summarized the evidence showing that the warming was extending down to about 2000 m. Suddenly, the warm water disappeared [31], but was subsequently found having sunk and moved [32].

For the seawater to reach the great density required for it to become deep-sea water, the surface seawater must increase its salinity. This occurs when cold, dry Arctic air moves over the Arctic seas and sea ice, evaporating water, so concentrating the salts in the remaining waters. Deep-sea thermohaline water that forms in the northern oceans flows south over the ridge between Greenland and Iceland, Iceland and the Faroes, and between the Faroes and Scotland (**Figure 5**). Examination by a deep-sea array of sensors shows that the location of the sinking of the saline waters is no longer close to its source area near Newfoundland but has moved further east, perhaps due to an increase in movement of the cold Labrador current southwards [32]. The new path is shown in **Figure 5**. The second array of sensors at 25°N is showing a slowing of the Meridional flow at that location [33]. Galaasen *et al.* [34]

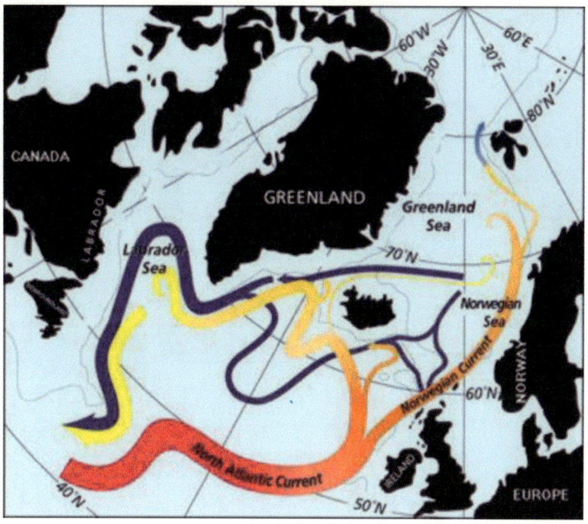

Figure 5.
New location of the saline North Atlantic water [32]. Blue is the return flow of deep warm water while the warm sinking waters are in orange.

reported that there was a rapid reduction in North Atlantic deep water (NADW) during the peak of the Last Interglacial Period (Eemian), and this may be what is happening again now.

Examination of the oxygen isotope variations in the skeletons of foraminifera accumulating on the sea floor shows an intricate pattern of change during the last 3.5 Ma B.P. (**Figure 6**). There are over 100 changes representing over 50 major cyclic fluctuations of ocean surface temperature. However, there are far fewer major cold events on land during that period [38]. This means that there must be at least two different temperature cycles operating simultaneously. The first is this oceanic heating and cooling cycle. This marine cycle produces glaciations with a periodicity of about 100 ka during the last 800 ka B.P. [39] although the exact height of the individual peaks varies somewhat. From c. 800 ka to 2.6 Ma, the cycles occurred every 41 ka but were of lower amplitude [24]. From 2.6 Ma until at least 3 Ma, this cycle was even smaller in amplitude and more frequent during the much warmer climates. These cycles may be related to the bipolar seesaw [4, 40]. Currently, we appear to be at or near the top of a warming cycle in the North Atlantic.

These cycles must be accompanied by rising and falling sea levels due to the expansion and contraction of the seawater due to their temperature changes and there would be the associated degassing of carbon dioxide into the air during the warming phases, however, the gas would re-enter the water during the cooling phase of these cycles. This would result in fluctuating contents of carbon dioxide in the air, which would change in tandem with the air temperature except for a minor delay. The exact cause of these marine temperature cycles and their fluctuations over time is nonproven but likely to be related to the export of warm (c. 18°C) North Atlantic deep water (NADW) in the high-speed hydrothermal bottom currents (THCs) to the South Atlantic (Antarctic Ocean) to be replaced by cold surface water from that area. The Antarctic cold water would cool Greenland and the Arctic regions in the north very quickly so that glaciers are thought to have developed in eastern Greenland within 12 years of the exchange. This contrasts with the warm water, which would warm the main Antarctic ice cap very slowly [28]. Heat would build up in the surface layers of the NW Atlantic Ocean until the next exchange, hence the name “bipolar see-saw” [4].

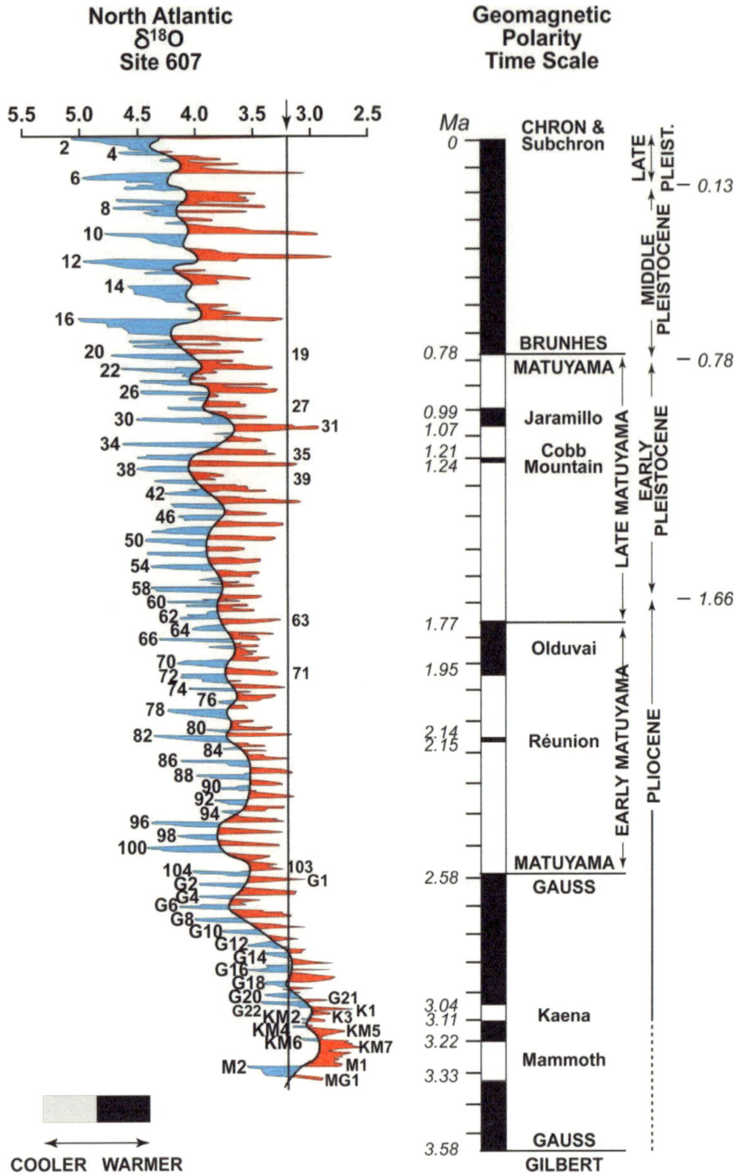

Figure 6.

Oxygen isotope palaeotemperature record [24, 35] and geomagnetic polarity timescale [36]. Black and white areas are normal and reversed polarity respectively. The arrow at the top indicates the mean Holocene oxygen isotope value. Numbers on the peaks and troughs are the isotope stages (modified from [37]).

The third set of cycles is seen in the ice cores from both the Greenland and Antarctic ice sheets. Seventeen of the smaller temperature cycles occurred in these between 65 ka and 5 ka B.P., spaced irregularly between 1.2 ka and 5 ka B.P. apart [41] during the first 14 peaks in **Figure 6**. They show an abrupt temperature change in Greenland cores but a gradual adjustment in the corresponding Antarctic ice cores, which also fits with the bipolar seesaw hypothesis. Thus it appears that the heating and cooling cycles experienced in the environs of the North Atlantic Ocean and eastwards through western Europe are part of the movement of part of the net solar energy from the Northern Hemisphere to the coastal regions of North America, the western European land mass and the Southern Hemisphere to partly warm those regions, preventing them from becoming more frigid than at present.

3.1.2 Transport of heat in the atmosphere

Dry air has a heat capacity of 0.00125 MJ/m^3 K and a thermal conductivity of 0.024 W/m K [25]. However, it contains variable amounts of water, *e.g.*, 2–3% on average at latitudes 50–60°N in the eastern Cordillera of the Rocky Mountains, under 1% in deserts, and up to 6% in Monsoons and tropical cyclones. It is the water content that determines the ability of the air to carry substantial quantities of heat, which it can subsequently unload as rain or snow. When an air mass moves over a water body such as a sea, it will increase its moisture content until the relative humidity reaches 100%, leaving the seawater more saline. Wet air changes temperature more slowly than dry air when rising over mountains since the heat given out by the condensing water vapor prevents the air from cooling as fast as unsaturated air. Dry, descending air warms more quickly as it descends producing the chinook/foehn effect, tending to dry and heat the ground over which it passes.

Unlike seas, the air masses can move in any direction over land or water bodies to places with lower air pressure. Wind speed depends mainly on the pressure gradient, which is also influenced by its temperature. Heavier gases such as carbon dioxide tend to collect in depressions. Lighter gases such as helium, hydrogen, and methane have low molecular weights and become lost to space with time. Those with the lowest molecular weights are lost most rapidly. The air becomes colder and decreases in pressure with altitude by expanding except where inversions occur. At higher elevations, there are fewer gas molecules to absorb incoming radiation.

The rotation of the Earth results in weather systems moving eastwards around the globe except at the Equator. It also causes moving air and water to slowly swing right in the Northern Hemisphere and left in the Southern Hemisphere due to the coriolis force.

There are several main climatic systems that transport large amounts of heat onto the nearby land areas, *viz.*, tropical and subtropical monsoons, hurricanes/typhoons, and air masses.

3.1.2.1 Tropical and subtropical monsoons

Monsoons are one of the dominant modes of heat transport in the Tropics [42]. The tropical monsoons originate over seas with surface temperatures above 27°C. They are the classic monsoons that bring enormous quantities of heat from the ocean gyres near the equator onto tropical landmasses during part of the year. The climate of these landmasses includes a marked dry season when the deciduous trees drop their leaves and the bare ground heats up enormously. These have a separate climatic regime to that moving heat to the Antarctic Seas. The main one is the Indian Monsoon that brings enormous amounts of rain to the Indian subcontinent in June–September. Large areas of western and central India receive more than 90% of their total annual precipitation during the period, while southern and northwestern India receive 50–75% of their total annual rainfall from the monsoons. Kathagat *et al.* [17] point out that the variations in the Indian Monsoon precipitation and air temperature are closely correlated with the success of the various Indian empires over the last 4600 years, with the higher precipitation correlating with periods of success. The main controlling factors are the degree of heating of the surface waters of the Gyre in the Indian Ocean and the presence or absence of either La Niña, which produces more precipitation, or El Niño, correlated with shorter periods of precipitation ending in drought [42]. Borah *et al.* [43] report that a cold anomaly in the North Atlantic can cause drought by a steep decline in late-season rainfall in India. The Indian monsoon cools the lower land but brings heat and moisture to the upper part of the Himalayan Mountains. The hot high-pressure cell over Tibet in summer aids the upper levels of the Monsoon to cross those mountains to deposit snow on the high peaks and ridges and then the air descends 4000 m like a chinook to the deserts of North China. Its main effect is to bring dry heat from just south of the equator northwards into the semi-deserts and deserts of south-central Asia.

An offshoot of the Indian monsoon is found in North Australia and represents the southward movement of Monsoon weather during the Australian summer. It is the main source of moisture for northern Australia and is related to the eastern section of the Indian Monsoon affecting the area west of Borneo and north to the Philippines. It represents a method of limited relocation of heat energy southwards over the adjacent seas rather than a substantial poleward movement of energy, unlike the situation in Tibet.

The second tropical monsoon is the West African Monsoon affecting primarily the west coast of Africa south of the Sahel deserts, starting at c. 10°N–18°S. The heat comes onshore from the seasonal shifts of the Intertropical Convergence Zone (ITCZ) and produces the great seasonal temperature and humidity differences between the Sahara and the equatorial Atlantic Ocean. The ITCZ migrates northward from the equatorial Atlantic in February, reaches western Africa on or near June 22, then moves back to the south by October. Various factors control the monsoon variability including the variability of ocean sea surface temperature, continental land surface conditions, and atmospheric circulation. It does not move energy polewards but offsets the variations in the extent of the Sahara.

The East Asian summer monsoon is subtropical and develops as the trade winds modified by the Coriolis Force pick up moisture from the China Sea and the western Pacific Ocean. It provides moisture to southern Japan and the eastern shores of China (**Figure 7**). It ceased during the last glacial maximum (30–19 ka, [37, 45]) due to the lowered sea level leaving much of the South China Sea as land [46]. It is affected by the Tibetan Plateau high-pressure cell in Spring, drawing the moist, hot air upslope onto the NE shoulders of the Qinghai-Tibet Plateau [47]. Currently, the temperature increase on the adjacent land is decreasing the extent

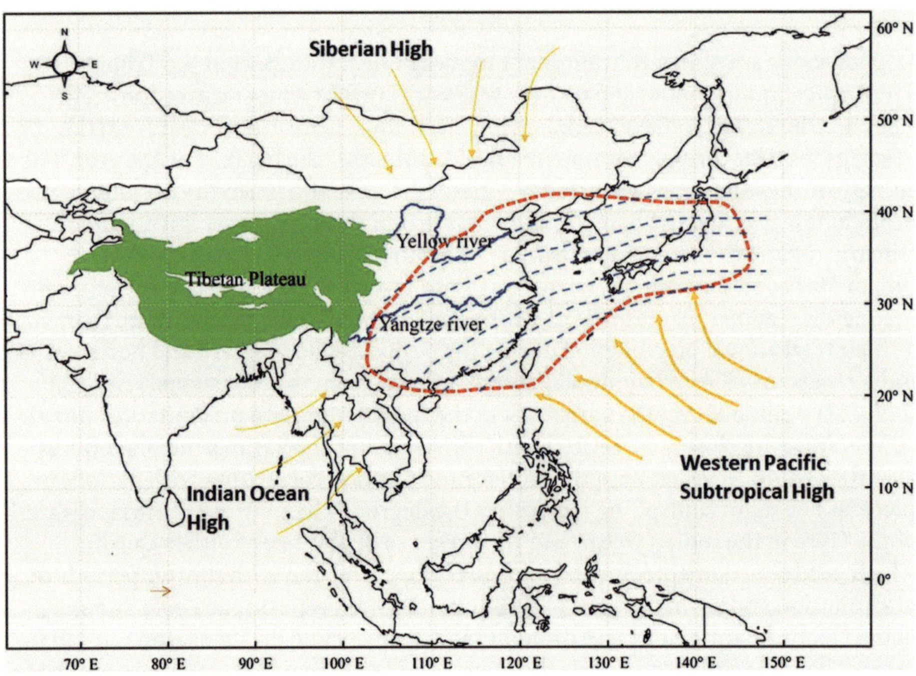

Figure 7.
Area affected by the East Asia monsoon (from [44]).

of this precipitation [48]. It brings moisture to the coastal areas and both heat and moisture to the higher land areas, which are otherwise semidesert.

The other subtropical monsoon is the North American Monsoon affecting the western and southern margins of the desert areas of Southwest USA and centered in northern Mexico [49]. It is a pattern of pronounced increase in thunderstorms and rainfall over large areas of the southern Cordillera, typically occurring between July and mid-September. During the monsoon, thunderstorms are fueled by daytime heating and build up during the late afternoon-early evening. Typically, these storms dissipate by late night, and the next day starts out fair, with the cycle repeating daily. The monsoon typically loses its energy by mid-September when much drier conditions are reestablished over the region. Lachnet *et al.* [50] found that it was much weaker during the last glacial maximum but strengthened after about 11 ka B.P. It is currently becoming more extreme but with fewer thunderstorms [51].

In summary, monsoons generally bring moisture onshore to areas that would otherwise be deserts or semideserts. Only on the north slope of the Tibetan Plateau does it bring heat energy northwards.

3.1.2.2 Tropical cyclones, hurricanes, and typhoons

Tropical cyclones are low-pressure centers that are widespread around the northern hemisphere (**Figure 8**) and include both hurricanes and typhoons. The difference is in the strength of cyclones, the stronger ones being referred to as hurricanes in North America and typhoons in Asia. There are divided into seven categories on the Saffir-Simpson scale on the basis of wind speed and strength.

They can vary in size from 100 to 2000 km in diameter. Unlike Monsoons, they move enormous quantities of heat polewards, primarily over the oceans in the Northern Hemisphere (**Figure 8**). Over land, they quickly decrease in strength since they no longer can replenish their heat and moisture from the ground.

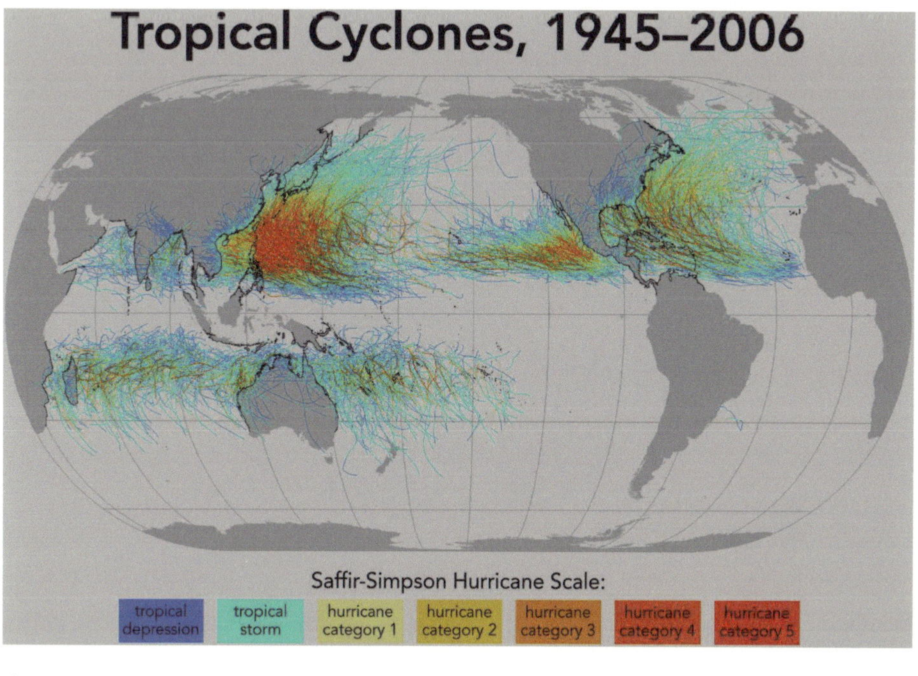

Figure 8.
Distribution of tropical cyclones between 1945 and 2006 around the world (NASA).

The strongest concentration of tracks of typhoons and hurricanes occurs along the east coast of Asia, where the warm precipitation is primarily dumped into the warm Kuroshio and North Pacific currents, thus augmenting their transport of heat to the North Pacific Ocean (see **Figure 3**). A combination of the Bering Land Bridge sealing off the connection between the Pacific and Arctic.

Oceans, when sea level drops more than 50 m and these warm currents, explain why the northern part of the Pacific Ocean did not freeze over during the last few glaciations. However, these two warm currents probably played a significant part in supplying the precipitation for Path 1 (**Figure 9**) that provided the relief valve for the Arctic air mass by producing the late (31–10 ka B.P.) glaciation in southern British Columbia. This greatly reduced the area covered by the Arctic air mass, allowing deglaciation of a substantial portion of the Northern Hemisphere ice sheets prior to 9 ka B.P. [52]. Currently, the typhoons affecting the Philippines are becoming more extreme [53]. A few typhoons also occur around the shores of India and near the Persian Gulf, but they are relatively infrequent.

South of the ITCZ, a band of Typhoons is found on the north coast of Australia west to Asia and Madagascar. They are mainly found over the Indian Ocean, but northern Australia and Queensland depend on the Monsoon rains from December to February to provide sufficient water for agriculture [54]. In El Niño years, there is increased precipitation whereas La Niña years result in shorter rainy seasons and drought. Similarly, Madagascar and the adjacent coast of southeast Africa depend on the tropical cyclones to provide enough water to sustain the seasonal growth of forest species. In all these areas, the monsoons cool down the land somewhat

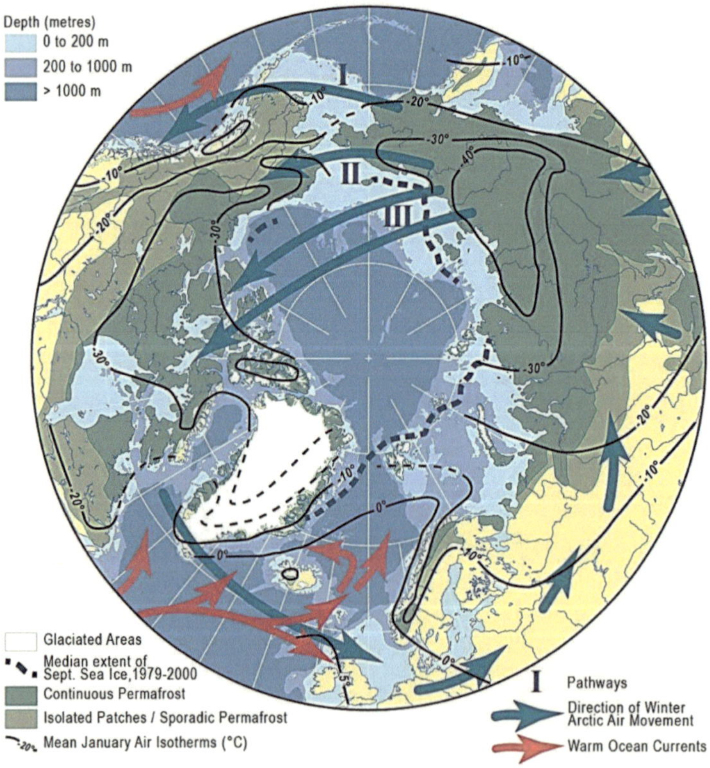

Figure 9.
Map showing the distribution of permafrost in the Arctic [37] together with the mean surface air January isotherms (°C) and the adjacent warm and cold ocean currents. Also shown are the three main paths (I–III) taken by the Arctic air as it moves from Asia to northern Canada, and the positions of the main warm ocean currents currently bringing heat from the Tropics.

along the margins of the adjacent deserts and/or semideserts. Over the ocean, these typhoons add heat to the underlying sea and may help supply heat energy to the warm surface thermohaline currents in the Indian Ocean.

The second-most concentration of hurricane tracks is found along the eastern side of North America, originating in the tropical Atlantic north of the ITCZ. These are becoming more numerous, of greater strength, and with high category numbers. They bring enormous amounts of heat and energy northwards with much of the precipitation falling into the Gulf Stream, thus adding to the heat energy bathing the east coast of the United States as well as the western shores of Europe. The warm currents enter the Arctic basin spreading westwards to eastern Greenland and eastwards towards Nova Zemla. The Arctic sea ice cover has been thawing since about 1920 (see Section 3.1.2.4 below) and Duk-Rodkin *et al.* [55] found evidence in the form of glacial and interglacial deposits on the eastern slopes of the Mackenzie Mountains that the Arctic Ocean had become ice-free prior to each of the last five glaciations. From then on, the Arctic air mass traversing the ocean will pick up moisture, increasing the salt content of the ocean in which case the water will tend to sink and later possibly take part in the deep-water thermohaline exchange.

3.1.2.3 General movement of air masses

The air masses can be divided into three main groups, *viz.*, the Equatorial air masses, the tropical air masses, and the Arctic air mass. As noted previously, the Equatorial air masses receive the most isolation and get sucked both north and south of the ITC to enter the dry, hot zone of the Horse Latitudes characterized by deserts on the land areas. The hot, dry desert surface heats the tropical air mass which forms an anticyclone over it, and the descending air then moves upwards and polewards or back along the surface to the Equator (**Figure 10**).

Each air mass operates as a distinct band around the globe, interacting with the others primarily along the boundaries between the air masses. Details of the operation of the Arctic air mass and its relationship with the tropical air mass will be found in [52]. It will be seen that the coldest air originates in northeast Siberia by reradiation of heat energy during the long, dark winter days. In cooling, it would

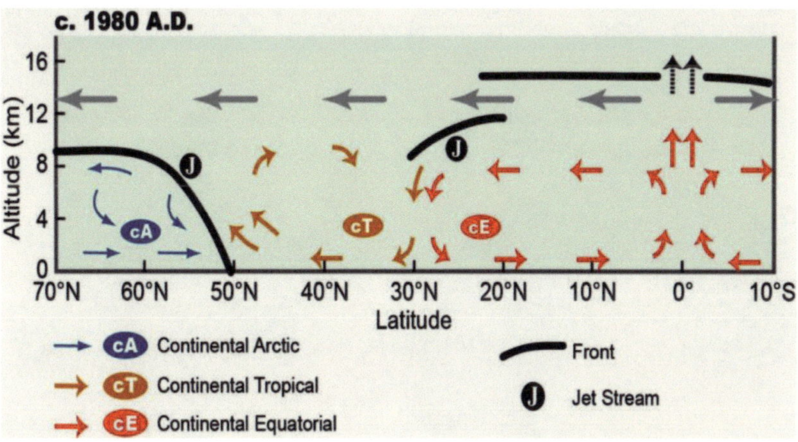

Figure 10.
North-south section through the lower atmosphere in western North America showing the circulation of the air masses under the climatic conditions occurring in 1980 A.D. [13]. Note that there are three basic air masses, viz., the Arctic, Tropical, and Equatorial, with jet streams along their margins. The more northerly black line marks the polar front, and the southern one is the subtropical front separating the tropical air (in orange) from the equatorial air mass (in red). The gray arrows indicate the air movement in the upper atmosphere.

deposit the excess water vapor as snow. The daily temperature can reach −66°C and stay there for several days before the air mass moves eastwards along the paths II and/or III in **Figure 9**. During the last major glaciation, paths II and III were followed to build up the Laurentide ice sheet. The cold air moved into northern Canada after picking up both heat and moisture crossing the Arctic Ocean. On arrival over the cold land, it cooled again by reradiation to become dense enough to push its way south, forming a lobe as it moved the tropical air out of the way, undercutting it and causing heavy snowfalls as the tropical air-cooled. This is the origin of the Rossby waves. **Figure 11** shows the evolution of the Rossby waves as they move eastwards.

When the Arctic air moves over the open warm water of the North Atlantic Ocean, it warms enormously while picking up large quantities of water thus aiding in the salinization of the surface seawater. Under the present climatic situation, it ensures that the northwest of Europe receives the warm, moist air at latitudes between southern Norway and central France with only relatively minor cold snowy spells in winter. It continues eastwards, cooling as it reaches the more continental parts of Central Europe, slowly modifying into dry, cold winter weather before it crosses the Urals and moves back to join the source area in NE Siberia.

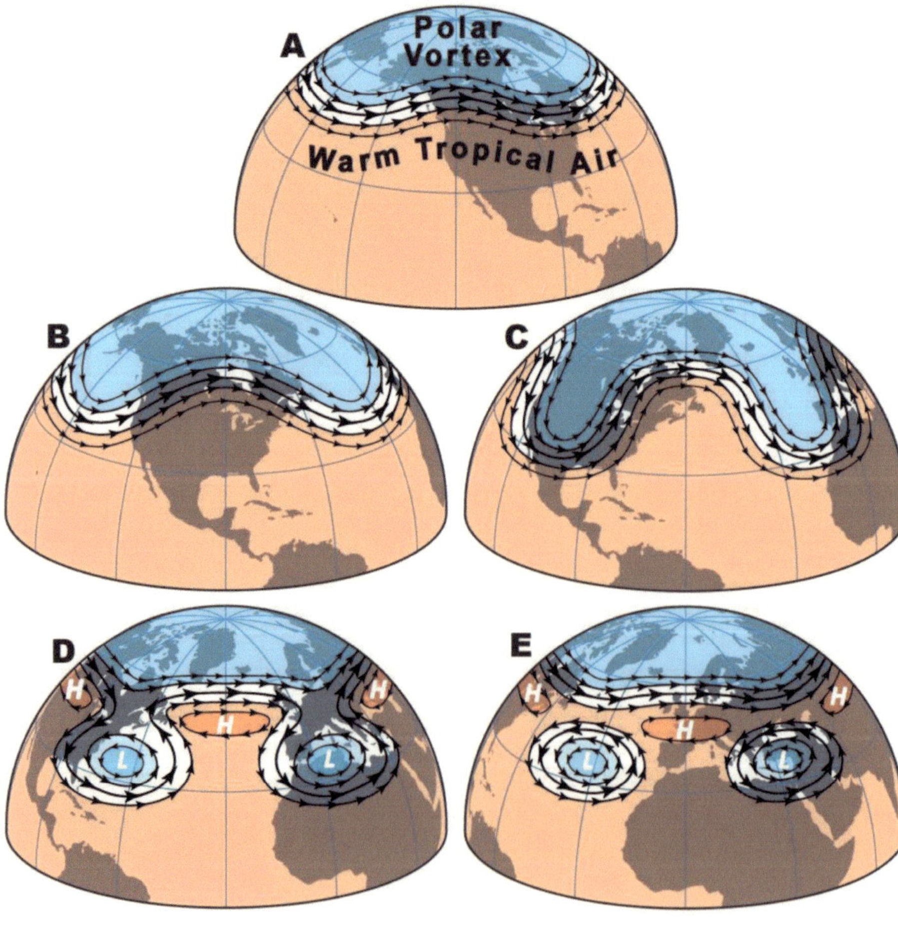

Figure 11.
Evolution of the southern edge of the Rossby waves as the cold air mass (in blue) moves eastwards [56]. H represents a high air pressure cell (an anticyclone) while L represents one of low pressure (a cyclone). Arrows show the direction of air movement while the Arctic Air mass is indicated in blue.

The tropical air of the semideserts and deserts on all the continents except Antarctica moves atmospheric heat polewards, *e.g.*, the Sahara, the Tibet-Plateau and Pamir Knot, the arid regions of the southwest United States and central Australia. If the supply of heat from the Equatorial regions increases, they expand into the surrounding areas. In Africa, the tropical air occupies today the vast Sahara desert and the semi-desert and savannah lands of the Sahel. The latter represents the transition zone between the desert in the north and the Equatorial and Monsoon lands of the west coast. The Sahara desert is tending to expand since the precipitation decreased in 1968, and the Sahel now has a more marked season [57, 58]. The Sahel-like conditions continue through Southern Sudan along with the interior of East Africa, but the precipitation there has decreased and become less reliable. In dry years, the anticyclone over the Sahara expands with the dry desert air extending over Italy, sometimes crossing the Alps to descend on to the north German Plain as a Foehn wind. This results in partial drying up of the Rhine River making it difficult for navigation. In East Africa, this expansion produces crop failure and famine, along with swarms of locusts.

3.1.2.4 Effect of sudden movements of large air masses

The coldest winter temperatures recorded in the northern hemisphere were obtained in the interior valleys of the mountains in northern Siberia at Verkhoyansk located at 67°33′N, 133°23′E. (−67.8°C on January 15, 1885), and equally cold temperatures were recorded at Oymyakon (located at 63°15′N, 143°09′E). These are sometimes referred to as the northern poles of cold. Similar cold temperatures (−65°C) have been recorded in a cold air drainage event in a mountain valley 100 km west of Fort Nelson, British Columbia [59]. These are still considerably warmer than the higher altitude stations inland in Antarctica.

In 2018, many stations in Antarctica experienced record colder temperatures, some 14°C colder than the previous record. The new current record low is −98°C at 81°S, 63.5°E. measured on the East Antarctic Plateau [60]. The bulk of Antarctica experienced a similar increase in cold, indicating that a considerable quantity of heat energy failed to go south during the Antarctic winter months. Since the deficit was near the South Pole, that missing energy must have gone north to the equatorial zone. This, in turn, would have upset the equilibrium between the Equatorial zone and the poles, and cause enhancement of the processes moving heat polewards from the Equator, primarily in the Northern Hemisphere.

This change appears to have triggered two particularly large, wide hurricanes that traveled along the East coast of North America dumping unusually large quantities of warm water into the Gulf Stream There was also a considerable increase in the number of Atlantic Hurricanes both in 2019 and 2020, and the hurricane season started earlier and lasted later than previously. The summer of 2019 saw unprecedented melting of the east side of the Greenland ice sheet which Tedesco and Fettweis [61] have tried to explain by changes in the atmospheric conditions over the Greenland ice sheet. This was also the time of the disappearance of the warm water from the Atlantic Ocean near the east coast of the United States, which has since been found near the west European coasts (**Figure 5**). The warm waters now go further north and bathe the northeast shores of Greenland and they may also be a cause of the exceptional melting of the ice sheet in 2019. The early arrival of the hurricanes in Quebec and New Brunswick meant that the warm rain fell on the remains of the isothermal winter snowpack resulting in major floods on the St John River and its tributaries. This resulted in exceptional flooding in many lowland areas.

In the case of the arid and semiarid areas, lack of precipitation caused by the expansion of the arid zone set up the conditions for large fires in Australia, Siberia, and the west coast areas of the United States. The wildlife was greatly affected in Australia, both by being caught in the fires and by the loss of suitable habitat. Expansion of the dry conditions in the Sahara and Sahel resulted in widespread famine in East Africa with the Victoria Falls dwindling to a trickle. Hordes of locusts compounded the problems in Somalia and Southern Sudan. In southern Europe, the drought severely affected the crops, *e.g.*, in Italy. In India, the Monsoon brought exceptionally heavy rains that resulted in\very difficult conditions for the inhabitants of the lowlands.

The most striking change that has occurred after the southern winter temperatures of 2019 is the enormous increase in the amplitude of the Rossby waves affecting the climate over the Northern Hemisphere, particularly North America. These waves have doubled in amplitude and resulted in remarkable changes in the autumn and winter weather. The western Prairie Provinces of Canada have also had exceptionally warm weather from mid November 2020 until late January 2021. The cold lobe of the Polar Vortex has missed this area but hit the Central and Southern United States and eastern Canada with heavy snowfalls and cold temperatures. Clearly, the lobes of extratropical heat have clashed with the Polar air mass-producing very high-pressure gradients extending far further south than usual.

From these results, it seems that any sudden movement of large quantities of energy to the equator has a very significant effect on the climate of the land areas particularly in the Northern Hemisphere. It can also bring about significant changes in the waters of the North Atlantic, probably speeding up the preparation of conditions suitable for a major seesaw exchange with the southern oceans.

3.2 Effects of mankind altering the processes of heat transfer

In general, the effects of the works of man in altering heat transfer between the hemispheres are both relatively puny and gradual. The processes discussed above are abrupt [62] and are not particularly susceptible to modification by humans. The main exception is the addition of heat artificially brought up from within the Earth during oil extraction from deep wells in Alaska [63]. The celebration commemorating the naming the Yakutsk Permafrost Research Station after Pavel Melnikov indicates that on completing studies at the top Russian School of Mines in the 1920s, he was posted to a head-up an Institute on Novo Sembla with the task of finding a way to thaw the sea ice along the Arctic coast of Russia. The heat from the oil and gas coming from within the ground needed to be largely removed prior to pumping through pipelines and could be used to thaw the ice. Within 8 years, he was appointed head of the fledgling Gas Prom which he continued to direct until his death. He also became President of the Russian Academy of Sciences. Under his guidance, many wells were situated near the Arctic coast and the pipelines were buried but not insulated in permafrost areas. They regularly float to the surface (see Figures 15.6 and 15.9 in [37]).

The end result for Russia was the establishment of a through route for shipping between European Russia the Orient after about 1990 A.D. This saved cargo vessels from having to go from Murmansk, around northwest Europe, and through the Suez Canal to get to the Orient. The effect of thawing the Arctic Ocean is that the open water will absorb more heat and so risk speeding the day when the warm saline waters of the North Atlantic Ocean take off south towards Antarctica and are replaced by cold Antarctic waters, thus precipitating a new cold event in the northern hemisphere. That would cause far greater problems for mankind than we face now.

A second problem is gas flaring. Around 2000 A.D., Russia flared off approximately 10 times the amount of natural gas as any other country. In 2009, the

Associated Petroleum Gas countries agreed to reduce the output of flared gas by 95% by January 2012. Unfortunately, Russia failed to meet these standards [64] although there had been a vast improvement. The effect of flaring is that it directly heats the passing Arctic air and combined with the geothermal heating of the air by the oil and gas in the pipelines, it causes a rise in the mean annual air temperature

Gas flaring volumes 2015–2019 (billion cubic meters)

	2015	2016	2017	2018	2019	Change 2019–2010
Russia	19.62	22.37	19.92	21.28	23.21	1.93
Iraq	16.21	17.73	17.84	17.82	17.91	0.09
United States	11.85	8.86	9.48	14.07	17.29	3.22
Iran	12.10	16.41	17.67	17.28	13.70	−3.50
Venezuela	9.33	9.35	7.00	8.22	9.54	1.32
Algeria	9.13	9.10	8.80	9.01	9.34	0.33
Nigeria	7.66	7.31	7.65	7.44	7.83	0.39
Libya	2.61	2.35	3.91	4.67	5.12	0.45
Mexico	5.00	4.78	3.79	3.89	4.48	0.59
Oman	2.43	2.82	2.60	2.54	2.63	0.09
Malaysia	3.72	3.16	2.83	2.25	2.37	0.12
Egypt	2.83	2.83	2.34	2.26	2.34	0.08
Angola	4.18	4.49	3.80	2.79	2.33	−0.46
Saudi Arabia	2.15	2.38	2.32	2.29	2.10	−0.19
China	2.08	1.96	1.56	1.82	2.02	0.20
Indonesia	2.90	2.77	2.33	2.06	2.00	−0.06
Republic of the Congo	1.18	1.14	1.14	1.58	1.67	0.09
Kazakhstan	3.69	2.67	2.42	2.05	1.57	−0.48
Gabon	1.56	1.56	1.50	1.38	1.46	0.08
Australia	1.14	0.73	0.66	0.86	1.39	0.53
Qatar	1.11	1.08	1.03	1.00	1.34	0.34
Turkmenistan	1.84	1.84	1.67	1.50	1.34	−0.16
India	2.20	2.06	1.50	1.34	1.31	−0.03
Brazil	1.33	1.44	1.10	1.00	1.14	0.14
United Kingdom	1.32	1.34	1.35	1.21	1.11	−0.10
Canada	1.81	1.30	1.34	1.33	1.05	−0.28
Cameroon	1.08	1.10	1.04	1.06	1.04	−0.02
Argentina	0.65	0.56	0.51	0.70	0.94	0.24
Syria	0.52	0.55	1.19	0.69	0.93	0.24
Ecuador	1.06	1.15	1.07	0.90	0.92	0.02
Rest of the world	11.30	10.45	9.22	8.72	8.49	−0.23
Total	**145.59**	**147.64**	**140.58**	**145.01**	**149.99**	**4.98**

As of July 21, 2020.
Source: World Bank Global Gas Flaring Tracker Report.

Table 1.
Gas flaring volumes for 2015–2019 by country in billion cubic metres [65].

Town or city	Deformation rate (%)
Chita	60
Pevek	50
Amderma	40
Dudinka	35
Dickson	35
Yakutsk	27
Tiksi	22
Anadyr	20
Salekhard	>10
Labytnangi	>10
Vorkuta	>8

Table 2.
Table showing the number of buildings showing structural damage in various Russian towns and cities located in areas of permafrost that will eventually result in the buildings being uninhabitable (from [66–69]).

in proportion to the oil and gas production in unit time. The latter cannot readily be predicted in advance, so cannot be suitably planned for in construction work.

Subsequently, Russia has successfully reduced its flaring by 2015 to lower levels (**Table 1**), but the unfortunate consequence in Russia was the damage to the infrastructure by the warming of the Arctic air mass. Most of the BAM railway across Siberia has had to be rebuilt due to thawing of the permafrost beneath the track, and similar problems have plagued the road network. **Table 2** shows the number of buildings that showed structural damage in various Russian towns and cities located in areas of permafrost that will eventually result in the buildings becoming uninhabitable. While some are near oil wells or the Arctic coast, others such as Yakutsk are far to the east of the main oil-producing regions.

4. Relationship to past cold events in North America

The first ocean water temperatures below freezing point were appearing in North America starting about 3.5 Ma [38]. Sea temperatures are currently estimated to be 7°C higher today than during the Last Glacial Maximum and the air temperature is 10°C higher. Drill cores from the Greenland ice cap and cores from sea bed sediment have been used to conclude that fluctuations between warm and cold periods during an ice age are caused by changes in the sea currents in the North Atlantic, particularly with relation to the process called the Thermohaline Circulation (THC). The current thinking by Oceanographers is that the changes in water temperature and salinity are probably the main triggers causing repeated glaciations in North America over the last 3.5 Ma [5, 70]. Krissek [71] noted late Cenozoic ice-rafting records from Leg 145 sites in the North Pacific Ocean, concluding that they started in late Miocene times and intensified during the late Pliocene times. Mudelsee and Raymo [72] suggested that the first glaciers had developed by 3.6 Ma B.P. based on patterns of globally distributed marine $\delta^{18}O$ records. Haug *et al*. [73] argued that a saline arctic halocline had developed in the North Pacific Ocean by 2.7 Ma B.P. with warm sea surfaces.

Since nowhere had a complete section showing the full sequence of glacial tills been found, Harris [38] summarized the dated evidence for glacial events and permafrost in North America to provide a guide as to when and where the known locations of the products of past cold events have been found. Methods of dating included tephrochronology, magnetostratigraphy, radiocarbon dating, and potassium-argon dating. The oldest dated deposits from North America included subglacial tuyas with glassy, fractured upper surfaces dated at 3.5 Ma B.P. from Wells Gray Park, British Columbia [74]. Cioppa *et al.* [75] reported that the lowest allow formation of the Kennedy Drift (Unit 1) in Southwest Alberta and Montana has Gauss normal polarization indicating that it is older than 2.6 Ma B.P. The first evidence of an ice cap consists of the Klondike gravels recently dated at 2.64 Ma B.P. with late Gauss magnetism [76]. This is believed to be the most extensive ice cap to have formed in Alaska while its magnetostratigraphy suggests that the increased amplitude of the 41 ka obliquity cycle had already developed [77]. This icecap is regarded as resulting from the warm sea surfaces at that time as well as being the source of the North Pacific ice-rated debris in the marine cores. This ice sheet also altered the course of the Yukon river [78]. About 2.8–2.4 Ma B.P., the Panama Gateway between North and South America closed up so the ENSO currents could not pass into the Atlantic Ocean to take part in warming the North Atlantic.

Altogether, 13 major cold events were recognized in the late Pliocene-Pleistocene record from North America, although some minor events were missed, and the dating of some of the evidence has been changed as a result of more recent work. The sequence started off with isolated events, followed by definite ice sheets separated by large interglacials that decrease in size until 1 Ma B.P. After 750 ka, the 100 ka eccentricity Milankovich cycle has controlled the frequency of the cold events, separated by short (11–20 ka) interglacials.

Barendregt and Irving [79] showed that each time there was a change in the direction of the Earth's magnetism, there was a significant change in the areas affected by glaciations. This implies that there were tectonic changes to the landscape at those times. In the case of the last change c. 1.0–0.8 ka B.P., the mountain range that had separated the Hudson Bay drainage from the Arctic drainage sank to form the present-day Arctic Islands. Meanwhile, the Western Cordillera tilted northwards so that the highest land lay in northern Mexico. This caused the incision of drainage causing the deep erosion of the Copper Canyon in northern Mexico and the Grand Canyon in the southwest United States.

Tephrochronology and magnetostratigraphy both have their limitations. The first is only helpful if there are enough tephras that can readily be recognized that are widespread and have known ages. Magnetostratigraphy sorts the deposits by normal and reversed categories but it cannot prove which deposit is which within each group. In spite of this, it can be useful when determining the ages of a pile of different sediments using the known age sequence corresponding to the Quaternary deposits. Barendregt and Duk-Rodkin [80] produced maps of the distributions of normal and reversed tills corresponding to the four main magnetic zones in the period back to 2.8 Ma B.P., but they do not exhibit the former distribution of each former ice sheet.

During the last seven glaciations, each glacial cycle has spanned about 4 different marine temperature cycles (see **Figure 4**) but varies somewhat independently of them. The best indicator of the volume of ice present on land is provided by the changes in sea level (**Figure 12**). On land, the Wisconsin Glaciation began with the development of extremely cold winter temperatures in Siberia, northeast China, and Greenland with the cold air generating heavy snow accumulations over parts of eastern and northern North America [52, 56] corresponding to a reduction in sea level of c. 30–50 m, starting after 120 ka B.P. The growth of the ice centers continued during the Early Wisconsin glaciation for about 30 ka before a warmer period, probably due to the bipolar seesaw, heralded the commencement of the

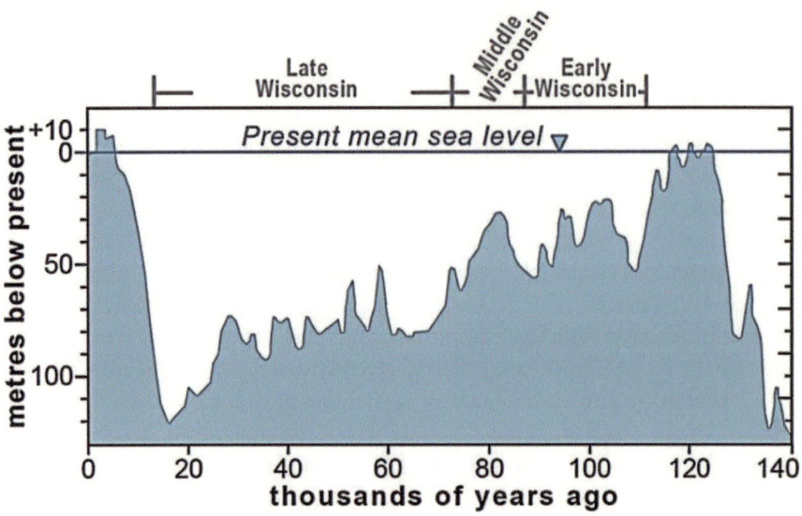

Figure 12.
Mean sea levels during the last 140 ka B.P. (http://rst.gsfc.nasa.gov/Sect16/Sect16_2html).

Middle Wisconsin Interglacial at about 75 ka B.P. (**Figure 12**). While the seawater warmed, the snow-covered ice caps reflected much of the incoming solar radiation back into space so that they only lost about half their ice mass. The next cooling of the seawater began around 70 ka B.P. with the commencement of the late Wisconsin Glaciation. Alternately cooling and warming seas resulted in fluctuating but alternating periods of glacial advances and retreats from the existing ice centers as the climate on land continued to cool. However, the southern Rocky Mountains south of the Peace River became ice-free by 59 ka B.P. and continued like that until 31 ka B.P. [81, 82]. It is thought that the climate there during that period was similar to today.

About 30 ka B.P., there was a major increase in the cold temperatures accompanied by a major drop in sea level of c.50 m that caused expansion of the ice sheets which advanced across the Canadian Prairies to the foothills of the Rocky Mountains. Concurrently in northeast China, there were extremely cold, dry conditions on the Tibetan Plateau, *e.g.*, [45, 83–86] and very cold air traveled eastwards to northern Canada along paths II and III in **Figure 9**, probably over open water.

At this time, the Arctic Front was well south of the USA border and the main ice sheets and displacement of biogeographic zones extended far to the south. In order to have deglaciation, a vast amount of Arctic air had to be turned into the tropical air. The tilt of the Earth was changing causing increasing insolation in the Northern Hemisphere resulting in expansion of the air masses, which was greatest over land areas in the south. This applied extra pressure on the Arctic air mass over the northern lands which were relieved by the activation of the path I in **Figure 9** because the air over the Pacific Ocean did not warm up much due to the increased heat being absorbed in the water [52, 53]. Since the cold Arctic air crossed the open Pacific Ocean, it picked up both heat and moisture, only to deposit the moisture as snow in the mountain cirques beginning in 31 ka B.P. The air then crossed the continental divide and continued east downslope as a dry Chinook wind, starting the retreat of the local glaciers. By 25 ka, the cirques on the west side of the Cordillera were full and the glaciers started to descend to the valley floors. After 20 ka B.P., tremendous mounts of snow were deposited covering the southern Cordillera [87, 88] and formed a dome covering the mountains at latitude 54°, so relieving the pressure on the dwindling Arctic air mass and causing the Arctic Front to migrate north to about its present position. From 15 ka to 10 ka, the flow from the Arctic air mass decreased so that the last active glaciers ceased expansion about 10 ka B.P. in the Fraser Valley and on the east coast of Haida Gwai. By then, the Laurentide Ice Sheet had retreated from the Prairies to form a mass occupying Hudson Bay. It is likely that this pressure release also permitted the Scandinavian-Russian ice cap and the bulk of the ice in the Swiss Alps to retreat at the same time [89–91].

5. Discussion and conclusions

There are marked differences in the thermal properties of water and air. Water has a very high heat capacity (4.187 mJ/m^3 K) so that it can store or transport large quantities of heat in a given volume and it absorbs over five times as much solar energy as soil or rock since it is translucent. Currents, convection, and wave action mix the water, whereas transmission of heat energy into a rock or sediments is primarily by conduction. Water also occupies about 70% of the surface of the Earth. Air has a relatively low heat capacity although this depends on the content of water vapor which can vary from under 1% over deserts to over 4% in monsoons and tropical cyclones. Soil and rock absorb only one-fifth of the solar energy absorbed by water, and they also do not reflect back into space nearly as much solar energy as snow. Thus, the distributions of these materials at the surface of the Earth determine what energy flows are needed to produce the thermal equilibrium between the equatorial regions that receive intense solar radiation and the polar regions that receive far less direct solar energy.

The pattern of temperature change in **Figure 1** is not that of global warming. Instead, it suggests that the Northern Hemisphere is receiving more heat than the Southern Hemisphere and this sets up a system of deep saline warm (c. 22°C) water flows to the Antarctic Ocean (THC's). These are replaced by a similar flow of cold (c. 2°C) Antarctic water with more normal salinity that then accumulates in the North Atlantic where it warms up. This is believed to produce a very rapid cooling of Greenland accompanied by the growth of the glaciers there. There would also be rapid cooling of the land to the east of the North Atlantic as occurred during the Younger Dryas event (>18°C cooling, according to Isarin [92]) whereas there would be only slow warming occurring on the Antarctic ice cap. This cold air flowing eastwards would result in greater winter cooling in Siberia, thus resulting in colder air moving into North America from the cold Siberian high-pressure center in winter.

The cause of this situation is the unequal heating of the land and sea coupled with the circular outline of the Antarctic shores compared with the north-south orientated oceans connecting through gateways into the Arctic Ocean. Warm ocean currents (Gulf Stream and the Kuroshio Current) carry large quantities of warm water northwards along the east coasts of continents in the Northern Hemisphere, aided periodically by deluges of warm water from Hurricanes and Typhoons. The North Atlantic stores a great amount of heat in its surface layers, but evaporation of water into the passing westerly winds increases their salinity. This causes the warm water to sink and accumulate in the deeper water channels.

There is evidence that there is a continual loss of THC water along the deep flow route south and a counter flow back to the North Atlantic. These have decreased in volume by 15% over the last 50 years, suggesting that the THC is weakest during the warmer interglacial times. The frequent rather random changes in water temperature indicated by the $\delta^{18}O$ data from marine cores indicate that periodically, abrupt outbursts of THC water occur relatively frequently but do not always indicate the developments of a cold event involving major glacial advances. However, they could well be involved in causing some or all of the relatively minor cold events whose origin is discussed by Warner *et al.* [93]. Deciding whether this is their origin or whether they are the result of Holocene outbursts of glacial meltwater into the northern oceans may be a problem.

These changes in ocean temperatures are aided by the warm Gulf Stream and Kuroshio currents that flow north along the east coasts of North America and East Asia. The shallow seas in the Bering Strait result in closing off the north Pacific Ocean from the Arctic Ocean when the sea level drops more than 50 m due to sequestering of water on land in the form of glaciers. The North Atlantic remained open to the Arctic Ocean during the last five glacial events, and the latter lost part or all of its ice cover. Cold, dry air flowing across open water picks up large amounts of moisture and heat, thus increasing the salinity and density of the sea surface. This denser water then sinks and becomes the water taking part in the THC.

The eccentricity of the Earth's orbit appears to be involved in causing the major cold events during the last 750 ka B.P., whereas obliquity caused a spacing of 41 ka between 750 ka B.P. and about 2 Ma B.P. There are numerous other cycles and one-time events that influence the day-to-day, week to week and other fluctuations in weather and climate that mask the underlying controls and movement of the plates making up the surface of the Earth slowly changed the overall pattern in terms of geological time.

There is still a lot to be found out about the thermohaline circulation in relation to climate change since the changes in sea temperature after at least 20 ka B.P. is the same for both the northern seas and the Antarctic ones [16]. It does, however, provide a viable mechanism for moving heat back from the Northern Hemisphere to the Southern Hemisphere when the difference in heat energy between the two hemispheres becomes too great.

These processes primarily affect the areas north of the Mediterranean Sea in Europe, the North Atlantic Basin, and Siberia [94]. They do not seem to affect the deserts nor the Sahel on the southern main landmasses [95, 96], while the climatic changes in Monsoon areas depend more on the climate over the source oceans.

There is no evidence to indicate that carbon dioxide is of any special importance in the processes so that the measures taken by governments to alleviate it as a problem are not needed. Society should not need to take special steps that damage or limit the economic health of economies unless it wishes to hurt itself. Data from the measuring devices in the arrays in the oceans and observations from the weather stations involved should make weather forecasting easier in the future.

Author details

Stuart A. Harris
Department of Geography, University of Calgary, Alberta, Canada

*Address all correspondence to: harriss@ucalgary.ca

References

[1] Naugebauer P. Climatic change: A review. 2019. Available from: https://www.researchgate.net/publication/333746714 [Accessed: September 18, 2021]

[2] IPCC. Special report on the impacts of global warming of 1.5 °C above pre-industrial levels. 2018. Available from: http://www.ipcc.ch/report/sr15/ [Accessed: December 02, 2019]

[3] American Chemical Society. Chemistry for Life. Available from: https://www.acs.org/content/acs/en/climate [Accessed: November 06, 2020]

[4] Broecker WS. The great ocean conveyor. Oceanography. 1991;**4**:89

[5] Broecker WS. Paleocean circulation during the last deglaciation. Paleoceanography. 1998;**13**:119-121

[6] Broecker WS. Was a change in thermohaline circulation responsible for the Little Ice Age? Proceedings of the National Academy of Sciences. 2000; **97**(4):1339-1342. DOI: 10.1073/pnas.97.4.1339

[7] Elrod MJ. Greenhouse warming potentials from the infrared spectroscopy of atmospheric gases. Journal of Chemical Education. 1999;**76**(12):1702-1705. DOI: 10.1021/ed076p1702

[8] Barrett J. Greenhouse molecules, their spectra and function in the atmosphere. Energy and Environment. 2005;**16**:1037-1046. DOI: 10.1260/095830505775221542

[9] Harris SA. Climatic change and permafrost stability in the Eastern Canadian Cordillera. In: Kane DL, Hinkel KM, editors. 9th International Conference on Permafrost. Extended Abstracts. Alaska: Fairbanks; 2008. pp. 93-94

[10] Harris SA. Climatic change and permafrost stability in the eastern Canadian Cordillera: The results of 33 years of measurement. Sciences in Cold and Arid Regions. 2009;**1**(5):381-403

[11] Harris SA, editor. Global Warming. Croatia: IntechOpen; 2010. p. 249. DOI: 10.5772/10283

[12] Harris SA. Greenhouse gases and their importance to life. In: Harris SA, editor. Global Warming. Croatia: IntechOpen; 2010. Chapter 3. 249 p. pp. 15-22. DOI: 10.5772/10283

[13] Harris SA. Climatic change in western North America during the last 15,000 years. The role of changes in the relative strengths of the air masses in producing changing climate. Sciences in Cold and Arid Regions. 2010;**2**:371-383

[14] Harris SA. Climatic change: Causal relationships over the last 240 Ma. Sciences in Cold and Arid Regions. 2013;**5**(3):259-274. DOI: 10.3724/SP.J1226.2013.00259

[15] Van Vleit-Lanoë B, Préat A. Arctique géologique 2/2 [in French]. 2020

[16] Judge AS. Deep temperature observations in the Canadian North. In: North American Contribution, Permafrost. 2nd International Conference; Yakutsk, USSR. Washington: National Academy of Sciences; 1973. pp. 35-40

[17] Kathagat G, Cheng H, Sinha A, Yi L, Li X, Zhang M, et al. The Indian Monsoon variability and civilization changes in the Indian Subcontinent. Science Advances. 2017;**3**(12):e.1701296. DOI: 10.1126/sciadv/1701246

[18] Anon. The sixty-year climate cycle. 2012. Available from: www.appinsys.com/GlobalWarming [Accessed: December 28, 2020]

[19] Avery D. Understanding climatic cycles: Here's how to\avoid climatic panics. Science News. August 08, 2017

[20] Croll J. On the change in the obliquity of the ecliptic, its influence on climate of the polar regions and the level of the sea. Philosophical Magazine. 1867;**33**:426-445

[21] Croll J. Climate and Time. New York: Appleton and Co; 1875. p. 388

[22] Milankovitch M. Théorie Mathématique des Phénomènes Produits par la Radiation Solaire [in French]. Paris: Gauthier-Villars; 1920

[23] Milankovitch M. Canon of Insolation and the Ice-Age Problem [in Serbian]. Belgrade, Yugoslavia: Royal Serbian Academy; 1941

[24] Ruddiman WF, Raymo M, McIntyre. Matayama 41,000-year cycles: North Atlantic and Northern Hemisphere ice sheets. Earth and Planetary Science Letters. 1986;**80**(1-2):117-129. DOI: 10/1016/0012-821x(86)90024-5

[25] Johnston GH, Ladani B, Morgenstern NR, Penner E. Engineering characteristics of frozen and thawing soil. In: Johnston GH, editor. Permafrost Engineering Design and Construction. Toronto: John Wiley and Sons; 1982. pp. 73-147

[26] Bryan K. Poleward heat transport by the oceans: Observations and models. Annual Revue of Earth and Planetary Science. 1982;**10**:38

[27] Bryan F. High-latitude salinity effects and interhemispheric thermohaline circulations. Nature. 1986;**323**:304

[28] Crowley T. North Atlantic deep water cools the Southern Hemisphere. Paleoceanography. 1992;7:489-497

[29] Blunier T, Brook GJ. Timing of millennial-scale climatic changes in Antarctica and Greenland during the last glacial period. Science. 2001; **291**(5501):109-112. DOI: 10.1126/science.291.5501.22109

[30] Levitus S, Antonius JI, Boyer TP, et al. World ocean heat content and thermosteric sea level change (0-2000 m). Geophysical Research Letters. 2012:1955-2010. DOI: 10.1029/2012GL051106

[31] Sullivan C. The North Atlantic Ocean's missing heat is found at depths. Eos. 2016:97. DOI: 10.1029/2016E00470009

[32] Lozier MS, Li F, Bacon S, Bahr F, Bower AS, et al. A sea level change in our view of overturning in the Subpolar North Atlantic. Science. 2019; **363**(6426):516-521. DOI: 10.1126/science.aau6592

[33] Bryden HL, Longworth HR, Cunningham SA. Slowing of the Atlantic meridional overturning circulation at 25°N. Nature. 2005;**438**:655-657. DOI: 10.1038/nature04385

[34] Galaasen EV, Ninnemann US, Irvali N, Kleiven KH, Rosenthal Y, Kissel C, et al. Rapid reduction in North Atlantic Deep Water during the peak of the Last Interglacial Period. Science. 2014:343. DOI: 101126/science.1248667

[35] Raymo ME. Global climate change: A three million year perspective. In: Kukla GJ, Went E, editors. Start of a Glacial. NATO ASI Series, Series 1. Vol. 3. NATO: Global Environmental Change; 1992. pp. 207-223

[36] Cande SC, Kent DW. Revised calibration of the geomagnetic polarity timescale for the Late Cretaceous and Cenozoic. Journal of Geophysical Research. 1995;**100**:6093-6095

[37] Harris SA, Brouchkov A, Cheng G. Geocryology. Bacon Raton: CRC Press; 2017. p. 765

[38] Harris SA. Chronostratigraphy of glaciations and permafrost episodes in the Cordillera of North America. Progress in Physical Geography. 1994;**18**:366-395. DOI: 10.1177/030913339401800305

[39] Lisiecki LE, Raymo ME. A Pliocene-Pleistocene stack of 57 globally distributed benthic $\delta^{18}O$ records. Paleoceanography. 2005;**20**. DOI: 10.1029/2004PA001071

[40] Oppo DW, Curry WB. Deep Atlantic circulation during the last glacial maximum and deglaciation. Nature Education and Knowledge. 2012;**3**(10):1

[41] Karlsson BD. Image of the week: The Bipolar Seesaw. EGU, Divisions, Cryospheric Sciences. 2016

[42] Indus Scrolls. Increased sea surface temperature affecting Indian monsoon: Study. Indus Scrolls Bureau. October 07, 2020

[43] Borah PJ, Venugopal V, Sukhatme J, Muddibihal P, Goswami BM. Indian monsoon derailed by a North Atlantic wavetrain. Science. 2020;**370**(6522): 1335-1338

[44] Yi G, Che Z, Piao S, Peng C, Ciais P, Wang QF, et al. High carbon dioxide uptake by subtropical forest ecosystems in the East Asian monsoon region. Proceedings of the National Academy of Sciences. 2014;**114**(13):4910-4915

[45] Harris SA, Jin HJ. Tessellons and "sand wedges" on the Qinghai-Tibet Plateau and their palaeoenvironmental implications. In: Hinckel K, editor. Proceedings of the 10th International Conference on Permafrost. Vol. I. Salekhard, Russia: Russian Academy of Sciences; 2012. pp. 147-153

[46] Voris HK. Maps of Pleistocene sea levels in Southeast Asia: Shorelines, river systems and time durations. Journal of Biogeography. 2000;**27**(5): 1153-1167. DOI: 10.1046/j.1365-2699.2000.00489.x

[47] Hsu H-H, Lin X. Relationship between the Tibetan Plateau heating and East Asian summer monsoon rainfall. Geophysical Research Letters. 2003;**30**(20). DOI: 10.1029/2003GL017909

[48] Loo YL, Bila L, Singh A. Effect of climatic change on seasonal monsoon in Asia and its impact on the variability of monsoon rainfall in Southeast Asia. Geophysical Frontiers. 2014. DOI: 10.1016/j.gsf.2014.02.009

[49] Adams D, Comrie A. The North American monsoon. Bulletin of the American Meteorological Society. 1997;**78**(10):2197-2213. DOI: 10.1175/1520-0477(1997)078<2197:TNAM>2.0.CO;2

[50] Lachnet MS, Asmeron Y, Bernal JP, Polyak VJ, Vazques-Selem L. Orbital pacing and ocean circulation-induced collapses of the Mesoamerican monsoon over the past 22,000 years. Proceedings of the National Academy of Sciences. 2013;**110**(23):9255-9260. DOI: 10.1073/pnas.1222804110

[51] Luong TM, Castro CL, Cheng M-I, Lahmary P, Adams DK, Ochra-Maya CA. The more extreme nature of North American monsoon precipitation in the southwestern United States as revealed by historical climatology of simulated severe weather events. Journal of Applied Meteorology and Climatology. 2017;**56**(9):2509-2529. DOI: 10.1175/JAMC-D-16-0358.1

[52] Harris SA. The relationship of sea level change to climate change in Northeast Asia and northern North America during the last 75 ka B.P. AIMS Environmental Science. 2019;**6**(1):14-40. DOI: 10.3934/environsci.2019.1.14

[53] Holden WH, Marshall S. Climatic change and typhoons in the Phllipines:

Extreme weather events in the Anthropogene. In: Samui F, Kim D, Ghosh C, editors. Integrating Disaster Science and Management: Global Case Studies in Mitigation and Recovery. 2018. Chapter 24. pp. 407-421

[54] Berry GJ, Reeder MJ. The dynamics of Australian monsoon bursts. Journal of Atmospheric Sciences. 2016;**73**(1): 55-69. DOI: 10.1175/JAS-D-15-0071.1

[55] Duk-Rodkin A, Barendreght R, Tarnocai C, Phillips F. Late Tertiary to Late Quaternary record in the Mackenzie Mountains, Northwest Territories: Stratigraphy, Paleosols, paleomagnetism and chlorine-36. Canadian Journal of Earth Sciences. 1996;**33**:875-895

[56] Harris SA. The subdivision, dynamics and history of the Late Wisconsin cold event in Northeast Asia and northern North America during the last 75 ka B.P. In: Bidadeau J-T, Nadeau DF, Fortier DF, Conciatori D, editors. Proceedings of the 18th International Conference on Cold Regions Engineering and the 8th Canadian Permafrost Conference; Quebec City, Canada; August 18-22nd 2019. Washington, D.C: American Society of Civil Engineers; 2019. pp. 570-578. DOI: 10.1061/9780784482599

[57] Nash DJ, de Cort C, Chase BM, Verschanan D, Nicholson SE, Shanahan TM, et al. African hydroclimate variability during the last 2,000 years. Quaternary Science Reviews. 2016;**54**:1-22

[58] Nicholson SE, Funk C, Fink AH. Rainfall over the African continent from the 19th century through the 21st century. Global and Planetary Change. 2018;**165**:114-127. DOI: 10.1016/j.gloplacha.2017.12.014

[59] Harris SA. Cold air drainage west of Fort Nelson, British Columbia. Arctic. 1982;**35**:537-541

[60] NSIDC. New study explains Antarctica's coldest temperatures. 2018. Available from: https://nsidc.org/news/newsroom/new-study-explains-antarctica-coldest-temperatures

[61] Tedesco M, Fettweis X. Unprecedented atmospheric conditions (1948-2019) drive the 2019 exceptional melting season over the Greenland ice sheet. The Cryosphere. 2020;**14**:1209-1223. DOI: 10.5m4/tc-14-1209-1223 http://www.science-climat-energie.be/2020/5/29/arctique-geologique-2-2/

[62] Jacobsson M, Ingólfsson Ó, Long AJ, Spielhagen RF. The dynamic Arctic. Quaternary Science Reviews. 2014;**92**:1-8

[63] Harris SA. Probable effects of heat advection on the adjacent environment during oil production at Prudhoe Bay, Alaska. Sciences in Cold and Arid Regions. 2016;**8**(6):00451-00460. DOI: 10.3724/SPJ.1226.2016.00451

[64] Loe JSP, Ladehaug O. Reducing gas flaring in Russia: Gloomy outlook in times of economic uncertainty. Energy Policy. 2012;**50**:507-517. DOI: 10.1016/j.enpol.2012.07.049

[65] Wheeler E. Global natural gas flaring growth in 2019. S. and P. Market Intelligence. December 21, 2020

[66] Velli YY. Stability of Buildings and Engineering Construction in the Arctic [in Russian]. Leningrad: Stroiizdat; 1973

[67] Velli YY. Studies, projecting and construction on frozen saline soils. In: Lisse et al. Fundamentals of Icy and Frozen Saline Soils [in Russian]. Leningrad; 1977. pp. 35-45

[68] Brouchkov AV. Frozen saline soils of the Arctic coast: Their distribution and engi- neering properties. In: Phillips M, Springman, Arenson, editors. Proceedings of the 8th International

Conference on Permafrost; Zurich, Switzerland. Lisse: Swerts and Zeitlinger; 2003. pp. 95-100

[69] Parmuzin CY. Land Use in the Cryolithozone [in Russian]. Moscow: Moscow University Press; 2008. p. 171

[70] Knutti R, Flückiger J, Stocker TF, Timmermann A. Strong hemispheric coupling of glacial climate through freshwater discharge and ocean circulation. Nature. 2004;**430**:851-856. DOI: 10.1038/nature02786

[71] Krissek L. Late Cenozoic ice-rafting records from Leg 145 sites of the North Pacific: Late Miocene onset, Pliocene intensification and Plio-Pleistocene events. Ocean Drilling Program. 1995:179-174

[72] Mudelsee M, Raymo ME. Slow dynamics of the Northern Hemisphere glaciation. Paleoceanography. 2005;**20**: PA4022. DOI. 10.1029/2005PA001153

[73] Haug CH, Canopolski A, Sigma DM, Rosell-Mele A, Swann GEA, Tiedemann R, et al. North Pacific seasonality and the glaciation of North America 2.7 million years ago. Nature. 2005;**433**:821-825

[74] Hickson CJ, Souther JG. Late Cenozoic volcanic rocks of the Clearwater Wells Gray area, British Columbia. Canadian Journal of Earth Sciences. 1984;**21**:267-277

[75] Cioppa MT, Karlestrom ET, Irving E, Barendregt RW. Paleomagnetism of tills and associated paleosols in southwestern Alberta and Northern Montana: Evidence of Late Pliocene-Early Pleistocene glaciations. Canadian Journal of Earth Sciences. 1995;**32**:555-564. DOI: 10.1139/e95-047

[76] Hidy HJ, Gosse J, Froese DG, Bond JR, Rook DH. A latest Pliocene age for the earliest and most extensive Cordilleran Ice Sheet in northwestern Canada. Quaternary Science Reviews. 2013;**61**:77-84. DOI: 10.1016/j.quascirev.201.11.09

[77] Berger A, Loutre MT. Insolation values for the climate for the last 10 million years. Quaternary Science Reviews. 1991;**10**:297-317. DOI: 10.1016/0277-3791(91)90033-Q

[78] Duk-Rodkin A. Tertiary-Quaternary drainage of the pre-glacial Mackenzie Basin. Quaternary International. 1994;**22**(23):221-241. DOI: 10.1016/1040-6182(94)90015-9

[79] Barendregt RW, Irving E. Changes in the extent of the North American ice sheets during the Late Cenozoic. Canadian Journal of Earth Sciences. 1998;**35**:504-509. DOI: 10.1139/cjes35-5-504

[80] Barendregt RW, Duk-Rodkin A. Chronology and extent of Late Cenozoic ice sheets in North America: Magnetostratigraphical assessment. Studies in Geophysical Geology. 2012;**56**:705-724. DOI: 1007/s11200-011-9019-3

[81] Fulton R. A conceptual model for growth and decay of the Cordilleran ice sheet. Géography physique et Quaternaire. 1991;**45**:281-286. DOI: 10.7202/013139ar

[82] Nichol C, Monahan P, Fulton R. Quaternary stratigraphy and evidence for multiple glacial episodes in the north Okanagan valley, British Columbia. Canadian Journal of Earth Sciences. 2015;**52**:338-356. DOI: 10.1139/cjes-2014-0182

[83] Harris SA, Jin HJ, He RX. Very large cryoturbation structures of Last Permafrost maximum age at the foot of the Qilian Mountains (NE Tibet Platea, China): A discussion. Permafrost and Periglacial Processes. 2017;**28**:757-762. DOI: 10.1002/ppp.1942

[84] Harris SA, Jin HJ, He RX, Yang SZ. Tessellons, topography and glaciations on the Qinghai-Tibet Plateau. Sciences in Cold and Arid Regions. 2018;**10**(3):187-206. DOI: 10.3724/SP.J.1226.2018.00187

[85] Jin HJ, Chang XL, Luo DL, He RX, Lü CZ, Yang SZ, et al. Evolution of permafrost in Northeast China since the Late Pleistocene. Sciences in Cold and Arid Regions. 2016;**8**(4):269-296

[86] Jin HJ, Jin X-Y, He RH, Lou DL, Cheng XL, Wang SL, et al. Evolution of permafrost in China during the last 20 ka. Science China, Series D, Earth Science. 2019;**62**:376-390. DOI: 10.1007/s11430-018-9272-0

[87] Patton H, Hubbard A, Andreasson K, Winsborrow M, Stockton AP. The build-up, configuration, and dynamical sensitivity of the Eurasian ice sheet complex to Late Weichselian climatic and ocean forcing. Quaternary Science Reviews. 2016;**153**:97-121. DOI: 10.1016/j.quascirev.2016.10.009

[88] Alley NF. Middle Wisconsin stratigraphy and climate reconstruction, South Vancouver Island, British Columbia. Quaternary Research. 1979;**11**:213-237. DOI: 10.1016/0033-5894(79)90005-X

[89] Alley NF. Late Pleistocene history and geomorphology, southwestern Vancouver Island, British Columbia. Canadian Journal of Earth Sciences. 1979;**16**(9):1645-1657. DOI: 10.1139/e79-154

[90] Patton H, Hubbard A, Andreassen K, et al. Deglaciation of the Eurasian ice sheet complex. Quaternary Science Reviews. 2017;**169**:148-172. DOI: 10.1016/j.quascirev.2017.05.019

[91] Monegato G, Ravizzi C. The Late Pleistocene multifold glaciation in the Alps: Update and open questions. Alpine and Mediterranean Quaternary. 2018;**31**:225-229

[92] Isarin RBF. Permafrost distribution and temperatures in Europe during the Younger Dryas. Permafrost and Periglacial Processes. 1997;**8**:313-333. DOI: 10.1002/(SICI)1099-1530(199709)8:3<313::AID-PPP255>3.0

[93] Warner H, Solomina ON, Gresjean M, Ritz SR, Jetel M. Structure and origin of Holocene cold events. Quaternary Science Reviews. 2011;**30**(21):3109-3123. DOI: 10.1016/j.quascirev.2011.07.010

[94] Inga R. The impact of global warming and climate change on the development of agriculture in the northern latitudes of the Eurasian continent. In: Harris SA, editor. Global Warming and Climate Change. Croatia: InterchopenOpen; 2021

[95] Epule T. Recent climatic change adaption strategies in the Sahal: A systematic review. In: Harris SA, editor. Global Warming and Climate Change. Croatia: InterchopenOpen; 2021

[96] Emmanuel O. Trends and persistence of meteorological drought in Sudano-Sahelian region of Nigeria under increasing global warming. In: Harris SA, editor. Global Warming and Climate Change. Croatia: InterchopenOpen; 2021

Section 2

Climate Change Education

Chapter 3

Social Representation of Climate Change among Young Spanish University Students

Amor Escoz-Roldán, Pablo Ángel Meira-Cartea and José Gutiérrez-Pérez

Abstract

Understanding the similarities and differences between what the population think and the available science on the causes, processes, consequences and solutions of anthropogenic alteration of climatic phenomena opens a window to demonstrate that the common culture surrounding Climate Change (CC) can have a greater influence than its scientific representation, since the weight of certain variables associated with the ways in which this common culture is constructed in societies generates differences in the degree of importance given to the phenomenon. Understanding the social representation of CC among university students is of great interest, since communication about it is usually restricted to the field of research and its threat potential has barely reached the general public, causing a discontinuous flow that does not connect significantly with climate science, resulting in different beliefs about climate change and a common culture around it that generates doubts and uncertainties.

Keywords: social representation, climate change, climate literacy, education, university

1. Introduction

We begin this chapter by taking the concept of cognition as a reference, as it is closely related to social representations as a useful study perspective for understanding how society appropriates the causes, consequences and solutions related to climate change. Cognition is the ability of living beings to process information and give it meaning, based on stimuli we receive from the environment through the senses. Thanks to this capacity, human beings assimilate and organise the information they receive in order to convert it into knowledge, through a series of mental processes such as perception, emotions, reasoning and language. These processes have a social dimension linked to communication, which makes it possible for us to interact with others to reach a consensus on ways of interpreting and explaining the world, its phenomena and events, and to articulate how to act accordingly.

These processes of interaction take on special relevance in a dynamic and changing world, constantly subject to social, environmental, economic and cultural transformations in which science and technology play an increasingly important role.

In order to internalise and interpret these changes, we rely unevenly on cultural constructions and scientific evidence that we incorporate into our cognitive baggage through frameworks of interpretation of facts, phenomena and events that give rise to shared representations in which the weight exerted by the influence of common culture is superimposed and often hinders an understanding of certain aspects of reality - for example, climate change - based on the appropriate appropriation of available scientific knowledge. Through processes of conceptual appropriation, linguistic meaning and cultural anchoring, we construct worldviews that translate into habits, behaviours, attitudes and lifestyles with which we confront the environmental problems of our surroundings on a daily basis, without realising that some of these cultural inertias are currently leading us to unprecedented environmental degradation that is causing, among other things, climate change. Clarifying the weight that these cultural constructions exert over scientific narratives in our way of understanding and explaining "objects" of a complex nature is an important challenge to promote changes that will allow us to reverse the impacts and alterations derived from anthropogenic climate change.

The research we present aims to understand to what extent the similarities and differences between what people think, from the individual to the collective, and the body of scientific knowledge available on climate phenomena and their anthropogenic alteration can be explained in relation to their causes, processes, consequences and solutions.

We justify the social relevance of this approach on the basis of the concept of a climate tipping point, i.e. the possibility of exceeding or having exceeded thresholds in climate change that can lead to abrupt or irreversible changes [1]. Several interconnected hotspots experiencing such changes have now been identified in the Amazon rainforest, the boreal forest, permafrost, Arctic sea ice and the Great Barrier Reef, among others [2].

The overshooting of these equilibrium limits is already causing millions of people to suffer the consequences of CC in the form of extreme poverty, migration, inequality, etc. Perhaps for this reason, shortly before COP25 in Spain, the European Parliament declared a climate emergency in the European Union, making it the first continent to do so.

Despite the alarming warnings provided by climate science and available scientific research, in general the governments of our countries have not incorporated anthropogenic CC as a priority issue in political agendas, so that communication of the emergency has been restricted to the spheres of scientific research and its threat potential has barely reached the general public. Nor has it been included as a priority in areas such as the economic, social, cultural or educational spheres, delaying the administrations from generating a flow of relevant information for society.

This flow has been discontinuous and, on many occasions, without meaningfully connecting society with climate science, resulting in different beliefs about climate change and a common culture around it that generates confusion, doubts and uncertainty.

And given that the social response to the threat posed by climate change depends more on how people and human collectives interpret and value it than on the representations that science constructs of it, it is therefore essential to understand how social representations of this phenomenon are being formed, as a basis for addressing the challenges involved in education and communication. Focusing this work on the study of representations of climate change in university students is justified because these are people who are at the most advanced levels of the educational system, which presupposes that they are people with a greater critical capacity to analyse the information they receive about CC and to discriminate its veracity and scientific rigour.

On the other hand, in a climate emergency scenario, universities must assume the responsibility of incorporating this situation into the training of their students in order to make them understand their responsibilities and their leadership role in the different disciplinary fields and social spheres in which they will have to work professionally.

Finally, this research is justified by the social diversity incorporated in the territorial and cultural contexts from which the samples of university students used are drawn. It is well known that the effects of climate change are cross-border and global, but they do not have the same impact in some areas of the planet as in others, so that the representation of the threat and the perception of risk may be different in each context.

In terms of the originality of the study, we can affirm that this research aims to explore other variants of what has already been studied so far in order to provide new data and broaden knowledge in this field.

On the basis that the study aims to deepen the understanding of the cognitive and socio-cultural processes involved in the representation of an "object" that is originally generated in the scientific field, the CC, this work aims to delve into the differences that may exist in the representation within the university community. More specifically, its novelty lies in checking how higher academic training can intervene and influence the social construction of SRs of CC, considering as a central hypothetical question the differences that may appear depending on the branch of knowledge, the degree and the academic year that students are studying and, therefore, their greater or lesser proximity to those scientific fields related to climate and its anthropic alteration.

On the other hand, we have innovated in the exploration of the territory/socio-cultural variable, making comparisons between university students who, being from different countries (i.e. Spain and Portugal), are considered at the same time within a common territory (which is the Iberian Peninsula) and yet experience different climatological characteristics on the north/south axis and belong to different socio-cultural and academic spheres.

Another novelty of this work is to use a measurement tool based on the best scientific literature on the causes, consequences, biophysical processes and solutions to climate change, relating it to some of the 17 Sustainable Development Goals proposed by the United Nations in Resolution 70/1 (2015), such as the SDGs associated with water and health.

Finally, another unique feature of this research is the multidisciplinary perspective it adopts. Climate change must be considered from an interdisciplinary perspective, as the causes that lead to it, the biophysical processes that shape it, its consequences and possible solutions are closely related, as it requires a complex approach that encompasses disciplines ranging from physics and biology to psychology and economics, among others. It is therefore essential to combine the contributions of the social and human sciences with those of the natural and technological sciences in order to tackle the climate crisis. In this way, and in the case in point, universities can be considered as tools for change, because education can change individual and social conceptions and enable the transformation of society. Furthermore, the perception of climate change is determined by many factors, but one of the most decisive in analysing it lies in education.

There is evidence based on research results that show that the common culture surrounding the phenomenon of climate change may have a greater influence than its scientific representation because the weight of certain variables associated with the ways in which this common culture is constructed in advanced societies generates differences in the degree of importance given to the phenomenon [3, 4]. Thus, we propose with this study to explore the social representation of climate change in university students through different variables, which allow us to clarify some relevant aspects of these processes linked to the context of university education:

1. To analyse whether university training conditioned by the branch of knowledge and the academic course being undertaken influences the social representation of climate change.

2. To assess knowledge about the causes, consequences, biophysical processes and solutions to climate change in university students in order to assess how the common culture that has been created around climate change and the culture stemming from academic training influences the construction of this knowledge.

3. To find out whether university students know how to identify, through the causes, consequences, biophysical processes and solutions to climate change, the risks involved in this problem and, specifically, those related to extreme atmospheric phenomena and to health.

4. To determine whether the cultural and/or territorial/climatological contexts of university students' places of residence have a significant influence on the generation of their social representation of climate change.

2. Background

Climate change and global warming are two different concepts. Global warming refers to an increase in the average temperature of the Earth's surface due to an increase in the level of greenhouse gases and climate change refers to a long-term change in the Earth's or a region's climate that includes global warming and everything that affects the increase in greenhouse gases. According to the IPCC [5] it is a statistically significant variation in the mean state of the climate or in its variability, persisting over a period of time, which may be due to natural internal processes or to changes in external forcing such as volcanic eruptions or changes in land use, for example. And according to the United Nations Framework Convention on Climate Change (UNFCCC), in its Article 1, it is the change in climate attributed directly or indirectly to human activity that alters the composition of the atmosphere and that is in addition to natural climate variability observed over comparable time periods.

The IPCC published a report in 2018 detailing the most important aspects of understanding the 1.5-degree temperature increase; the associated impacts and risks; the current trajectory of emissions; and possible transitions consistent with the projected warming [6]. On the other hand, there are the SDGs; a Plan of Action for all countries, poor, rich and middle-income, to promote prosperity and protect the planet. The Sustainable Development Goals [7]. Specifically, goal 13 is the one that refers to climate action, and specifically, goal 13.3 which is the one that is most adapted to this research, is based on improving education, awareness and human capacity in climate change mitigation, adaptation, impact reduction and early warning.

The main theoretical basis of this research is Serge Moscovici's Theory of Social Representations, which states that a representation is an organised set of knowledge and one of the psychic activities thanks to which people make physical and social reality intelligible, integrate themselves into a group or into a daily relationship of exchanges and release the powers of their imagination by acting in their immediate environment [8].

Denise Jodelet adds that the field of representation designates the knowledge of common sense, whose contents make manifest the operation of certain generative and functional processes with a social character in a way that alludes to a form of social thought [9].

Robert Farr assumes that social representations appear when individuals debate topics of interest or when there is an echo of events selected as significant or of interest by those who control the media [10].

Another interesting contribution is that of Dan Speber [11], who recognises that representation is the relationship between what is represented, what is represented and the user of the representation. He adds a fourth element if the representation is produced by something other than the user, in this case, a scientific object that is difficult to understand, such as climate change.

Basically, representation makes the strange familiar and the invisible perceptible. They are cognitive systems of values, ideas and practices with the function of establishing an order that enables individuals to orient themselves in their material and social world and to master it, and to enable communication between members of a community [10].

In this way we come to speak of social representations of CC, and for an object to be studied from this theory, the following conditions must be met [12] it must constitute a specific object with an entity, it must have social relevance and it must be strictly collective knowledge, arising from the social interactions of a group.

Thus, in order for society to give meaning to CC, it must do so by knowing the information they receive about it. And the socio-cultural contexts in which this information is distributed and recoded must be taken into account. Furthermore, the profile of the audiences, the media and the methodologies used for its dissemination must also be considered, as well as the interferences inherent to any communicative process that hinder its understanding when it becomes "an object of common culture" [13].

Some lines of contemporary research with respect to the social representations of climate change are along the following lines:

> *Environmental hypermetropia, i.e. climate change is difficult to perceive, we see it as something distant, as most of the time we see it through the media. But when we perceive it, it is because we are conditioned by our beliefs, i.e. subjectively, common sense interprets reality based on new information that we have shared, even if it is not linked to the scientific representation to which it is related. Heras [13] quotes Markova [14] to emphasise that "objectification" can turn a complex scientific fact into something less differentiated, into something conventional and similar to what is already known. Once simplified, thinking will become less reflexive or non-reflexive. It will be framed within the symbolic social environment and circulated and recycled through the activities of the subjects [14].*

The study of cognitive limitations is another line of research, as CC is a difficult phenomenon to understand and we find it more difficult to comprehend complex and evolving systems such as climate, which is why we are not aware of the distant consequences of our actions [15].

Another line is the value placed on risk, i.e. if we think that an activity is positive, we will think that the risks derived from it will be lower and vice versa [16]. On the other hand, there are the responses to climate change, i.e. intuitive decision-making processes will be favourable if the decision-maker has sufficient information to evaluate the available options [13].

There is also the information deficit which means that, although environmental problems are identified in science, information from scientists also needs to be incorporated into common sense so that they can be treated as real environmental problems [17].

And finally, there is the lack of public confidence in the scientific consensus and in the power that governments may have to put in place effective measures to mitigate the problem [18].

With this in mind, we now turn to some of the studies that have been carried out on the SRs of CC in university students. Some examples are those related to:

- Online courses or MOOCs on climate change [19].
- Project "The university in the face of climate change" [20].
- Social recognition of CC as a problem [21].
- Emotions towards climate change [22].
- Gender perspective focused on climate change [23].
- Geographical perception of CC [24].
- Confusions regarding biophysical processes of CC [25–27].
- Participation of university students in workshops or clubs related to CC [28].
- Procedural knowledge on biodiversity and climate change [29].

The study of social representations of climate change in groups of university students deserves special attention due to the fact that they are supposed to be educated people with sufficient capacity to consider, reason and, where appropriate, disagree with the information received in the media and, in some cases, in the degrees they are studying, and they are also attractive people to study the perception they have of this specific case since, from the results obtained, the future trend towards behaviour more or less involved in the intention to change behaviours that can reduce the phenomenon of climate change can be predicted. The role played by universities in this process deserves special attention as they are institutions of reference in the generation of scientific knowledge and in the training of both professionals and citizens who are sensitive to the threats of climate change and who are involved in the search for social, economic and technological alternatives.

The preamble of the 2015 Paris agreement recognises the importance of education, training and awareness-raising in access to information. But, in addition, article 12 of the agreement calls for cooperation in the adoption of measures to improve education, training, public awareness and participation and public access to information on climate change.

It is therefore hoped that, in the future, the inclusion of content on the climate crisis in school curricula will improve climate literacy, with climate literacy being understood as the understanding of the influence we have on the climate and vice versa. That is, a climate literate person will understand the essential principles of the Earth's climate system, know how to evaluate scientific information, be able to communicate this knowledge in a meaningful way, and be able to make informed decisions regarding actions that may affect the climate [30].

An international reference project in which this research is framed is the RESCLIMA Project which, in general terms, investigates how the social representation of climate change is internalised and expressed in different societies and groups, such as university students, paying special attention to the processes of scientific transposition, interpretation and negotiation of meanings that shape this representation. The work carried out in this research is above all an attempt to explore the mechanisms by which certain lay theories about CC spread and become shared by different societies and groups without taking scientific knowledge into account.

To conclude, we cannot fail to mention the social movements and collectives that have emerged since the declaration of the climate emergency. These include Fridays for Future, Mothers for Climate, Teachers for Future and Extinction Rebelion.

These are movements made up mainly of students who are calling on governments to act immediately to avoid exceeding 1.5°C. They reproach them for their inaction. They reproach them for their inaction and the long negotiations without agreements. They reclaim the value of politics and the public and call for useful policy solutions through non-violent civil disobedience in an attempt to stop mass extinction, global warming and minimise the risk of social collapse. Due to these kinds of movements, other movements that already existed such as "NOMO" (No Mather) movements based on not having children for ecological reasons or people opting for more sustainable diets are now being reinforced. This may be determined by a characteristic of culture, which lies in the idea that culture transforms itself through the progressive generation of subcultures, which are attempts to register a change in the environment or a new differentiation of the social.

3. Method

This research is described as Empirical-analytical as it focuses on the analysis of the acquisition of climate knowledge through experience. We have used the quantitative method in quantifiable values such as means or percentages to explain or predict the correlation of variables. It is exploratory in scope because it examines a research problem about which there are doubts and it is considered descriptive because quantifiable information has been collected to be used in the statistical analysis of the selected sample. We could also mention that the research is of an evaluative nature because in this case it is a tool that could be used for the continuous improvement of the quality of educational programmes and the individuals included in them. It should be pointed out that the aim of this research is not to evaluate the teaching programmes of the degree programmes used for the analysis of the data, but rather to have an impact on society with the results obtained, so that if, through the students' degree of knowledge of CC, they can clarify a little better the deficiencies that these programmes have in the subject in question, they could be used as a reference so that programme evaluation professionals take them into account when including this phenomenon in these programmes. Finally, the theoretical approach is the Theory of Social Representations.

3.1 Sample

We have used a sample of undergraduate students from the public universities of Granada (Andalusia. Spain), having collected 522 questionnaires there, Santiago de Compostela (Galicia. Spain) with 644 and Miño (Braga. Portugal) with 560, for a total of 1709 students.

The reason why these three cities have been chosen is because they belong to the project where this research is framed, the RESCLIMA project, but another reason is because we wanted to take into account the variable of territory, which is considered of interest to compare the social representations that are generated in students who, despite studying in universities located in the Iberian Peninsula, the weather conditions are different within the same seasons of the year.

We considered a probability sample by convenience and chose 16 different degrees: Biology, Environmental Sciences, Speech Therapy, Business Management and Administration, Civil Engineering, Translation and Interpreting, Statistics, Sociology, Law, Education, Economics, Forestry and Environmental Engineering, Agricultural and Food Engineering, Chemical Engineering, History and Geography.

For example, in a first analysis, only the part of the sample belonging to the University of Granada and all the branches of knowledge was used, or, for example, in another analysis, the part of the sample belonging to the University of Granada was used together with the part corresponding to the University of Santiago de Compostela and only including the variable of the "social sciences and humanities branch".

As for the instrument used to carry out this research, it is an ad hoc questionnaire, with 45 questions divided into two blocks.

The first block consists of 32 Likert-type questions that may or may not be related to climate change, so that the aim is to assess the student's climate literacy. The questions in this first block are divided into four different dimensions: causes, consequences, related biophysical processes and solutions to climate change. This first block covers all the dependent variables that have been used for the analyses that will be shown below.

The second block consists of the remaining 13 questions, where the questions are of a personal nature to find out to what extent the student feels responsible for climate change, whether he/she has participated in any specific activity on the subject in question or whether or not he/she believes that this phenomenon exists, for example. These questions are the ones we have used to carry out the last analysis as independent variables; however, in the rest of the analysis we have used the independent variables of degree, course, branch of knowledge and territorial context, which will be explained in more detail later on as to why they have been chosen.

The instrument used is a 2nd generation instrument with a reliability of 0.74 based on Cronbach's alpha coefficient. This reliability is not as high as would be

Analysis	Objetives	Sample	Dependent Variable	Independent Variable	Division of the questionnaire by dimensions	Statistics descriptive statistics of frequency	Post-Hoc test
1	1, 2, 3	University students from Granada of NST-SSH	1st block of questions	Degree Course Branch of knowledge	NO	% more answer chosen Media ($\bar{X}$)	Tukey
2	1, 2	University students of Granada and Santiago de Compostela de NST	1st block of questions	Degree Course	YES	% answer corrects and incorrects	NO
3	1, 2	University students of Granada and Santiago de Compostela NST-SSH	1st block of questions	Branch of knoweledge Course	YES	% answer corrects Media ($\bar{X}$) (of 32 to 128 points.) Standard deviation (σ)	Tukey
4	1, 2	University students of Granada and Santiago de Compostela SSH	1st block of questions	Degree	YES	% answer corrects Scores of 32 to 128 points. Media ($\bar{X}$) Standard deviation (σ)	NO
5	2, 3, 4	University stundents of Granada, Santiago de Compostela and Braga	15 questions concerning the water from the 1st block	7 questions of 2° block concerning to the perception of information/ training/ pro-environmental environmental	YES	% answers corrects of 1st block and % answers of 2nd answers block	Scheffé

Figure 1.
Specific methodology used for each of the analyses according to the objectives.

desirable, which is why it is currently being improved by means of a validation of the questionnaire by experts.

In the following image we can see the specific methodology used for each of the analyses according to the objectives we want to achieve (**Figure 1**).

3.2 Hyphotesis

Therefore, based on the above, 3 alternative hypotheses and a null hypothesis have been launched:

> *#H1: The knowledge and beliefs of university students in relation to the causes, consequences, biophysical processes and responses and/or solutions that have to do with climate change are influenced by the university education received. This knowledge will be more abundant and specialised in climate science when students come from natural science or engineering/technology backgrounds, and more so as they progress in their academic studies.*

> *#H2: Within climate change-related beliefs, climate change denial will be significantly associated with a representation that downplays the consequences of global warming, including health risks and other extreme weather events related to water, such that students from territorial contexts with high average rainfall and low average annual temperatures will tend to minimise the social representation of risks associated with climate change, and vice versa.*

> *#H3: The social representation of climate change around the causes, consequences, biophysical processes and responses and/or solutions of climate change and its link to water and health are created due to a significant relationship with the territorial and socio-cultural context, just as it is associated with the student's self-perceived information and pro-environmental attitude.*

Thus, if these hypotheses cannot be accepted, it is possible to think that the social representation of climate change in university students is not determined in a population group by the academic training received, the territorial and socio-cultural context so that, even if they are considered to belong to a scientifically literate environment, this is not significantly modifying the social representation of this phenomenon, at least in aspects that are related to the individual's cognition.

4. Results

We present the results through 5 different analyses:

4.1 Analysis 1

For the first analysis only students of the University of Granada from all branches of knowledge from first year to fourth year were surveyed. The results of this study are that 100% of the respondents recognise the existence of CC and its impact (global and individual) and that it is due to human causes and are highly confident about it. The students surveyed have very general knowledge about CC, but contradictions in climate science concepts. They give themselves medium individual and collective responsibility, however, the opinion regarding scientific consensus is 50%.

There are 3 questions in the questionnaire referring to health, 2 of which are answered incorrectly and only one correctly. It can be stated that the knowledge on CC causes/consequences/solutions is low and they add that the information received in the degree course/specific information on CC is also low.

Finally, significant differences were found between science and literature students, with the Environmental Science degree being the one with the highest knowledge of climate science and in questions referring to climate risks and health. Differences were also found between first-year and fourth-year students.

4.2 Analysis 2

The results of this analysis are the same as those of the previous analysis in terms of opinions on the existence of cc and the security in them, however, on this occasion we have compared it with a demoscopy of 2013 with the Spanish population in general and in terms of beliefs of the existence of cc increases by more than 8 percentage points and by 20 points in the security that it is happening.

In this second analysis, the questionnaire was divided into 4 dimensions. In the causes dimension it was observed that 83.5% believe that CC is mainly due to human causes. 77.3% think that there is scientific consensus on this, however, the percentage is divided between those who think that there is scientific consensus on the causes and those who do not.

On the other hand, 73.3% think that the greenhouse effect is of natural origin, which is correct, while 40.6% believe that the greenhouse effect is caused by human activity, which is not the case. We also found significant differences in the degree. We observe that in the dimension causes and origins of greenhouse gases (**Table 1**) practically all the questions are answered correctly, but in the dimension bio-physical processes (**Table 2**) we can observe that the percentage of correct questions is lower.

On the other hand, in the consequences dimension (**Table 3**), the percentage of correct answers is also quite high, as is the solutions dimension (**Table 4**), where all the questions are answered correctly.

4.3 Analysis 3

In this third analysis the questionnaire has also been divided by dimensions, but this time the sample includes university students from all branches of knowledge.

As can be seen in the following image, the average of the four dimensions is 2.77, with the answers dimension once again being the one that stands out from this average and once again it is the processes dimension that is below it (**Figure 2**).

	CORRECT	FALSE
Every time coal, oil or gas is used, we contribute to climate change (C)	96,0%	4,0%
Increased meat consumption contributes to climate change(C) (degree: χ^2 = 31,802; p <,05) / (course: χ^2 = 12,904; p <,05)	55,0%	45,0%
CO2 is the main gas responsible for climate change (C) (course: χ^2 = 10,20; p <,05)	81,9%	18,1%
CO2 is a natural component of the atmosphere (C)	95,3%	4,7%
Most of the greenhouse gases present in the atmosphere come from natural sources (C) (course: χ^2 = 10,185; p <,05)	31,9%	68,1%

Table 1.
Questions concerning the causes and origin of greenhouse gases.

	CORRECT	FALSE
The greenhouse effect occurs when gases retain part of the radiation reflected by the earth's surface (C)	93,1%	6,9%
If it were not for the greenhouse effect, life as we know it would not exist (C) (degree: χ^2 = 63,798; p <,05) / (course: χ^2 = 8,943; p <,05)	77,2%	22,8%
According to the Earth's climate history, there have been oscillations between colder and warmer periods (C)	96,8%	3,2%
Climate change is a consequence of the hole in the ozone layer (I) (degree: χ^2 = 35,049, p < .05)	55,2%	44,8%
Polar ozone hole causes melting of the poless (F)	87,2%	12,8%
CO2 causes ozone layer destruction (F) (degree: χ^2 = 22,220, p < .05)/ (course: χ^2 = 32,710, p < .05)	61,6%	38,4%
Acid rain is one of the causes of climate change (F)	61,2%	38,8%

Table 2.
Questions related to the bio-physical processes linked to CC.

	CORRECT	FALSE
All countries will suffer from climate change (C)	94,5%,	5,5%
A warmer planet will expand the area of incidence of tropical diseases (C)	89,8%	10,2%
Higher temperatures will favour the occurrence of extreme weather phenomena (cyclones, hurricanes, floods, etc.) (C)	96,1%	3,9%
Climate change will exacerbate desertification problems on the Iberian Peninsula (C)	93,8%	6,2%
(C)	95,5%	4,5%
Many islands and coastal areas will be submerged due to climate change (C) (degree: χ^2 = 23,701; p <,05) / (course: χ^2 = 13,995; p <,05	57,7%	42,3%
Climate change will decrease rainfall in my country (F)	64,2%	35,8%
The greenhouse effect puts life on Earth at risk (F) Climate change will increase the number of earthquakes and tsunamis (F)	63,8%	36,2%
Skin cancers will increase as a result of climate change (F)	83,2%	16,8%
Rising temperatures will affect all regions of the planet equally(F) (course: χ^2 = 15,512; p <,05)	77,9%	22,1%

Table 3.
Questions concerning the consequences of the CC.

	CORRECT	FALSE
If we stop emitting greenhouse gases we will be less vulnerable to climate change (C) (C)	80,2%	19,8%
Climate change would be reduced if we planted more trees (C)	75,7%	24,3%
Replacing private transport with public transport is one of the most effective ways of tackling climate change (C)	83,2%	16,8%
If we stop emitting greenhouse gases we will not be affected by climate change (F) (course: $\chi 2$ = 18.062; p <,05)	20,3%	79,7%

Table 4.
Questions concerning the solutions of the CC.

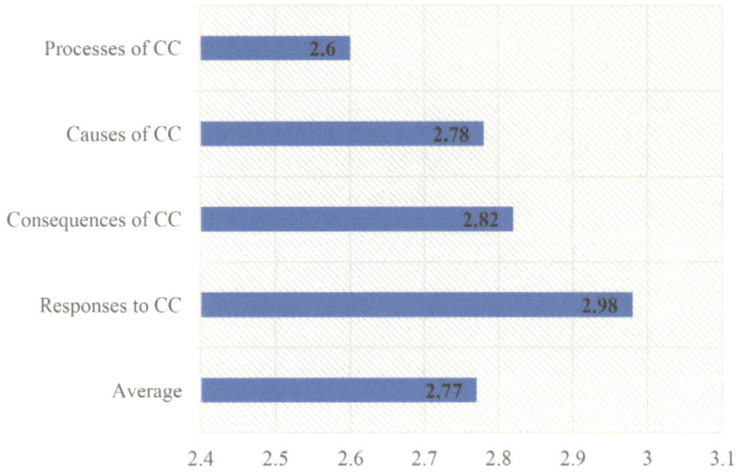

Figure 2.
Average score by dimension in the questionnaire.

	Academic Year				Academic Branch			
	Initial		Final		NST		SSH	
		SD		SD		SD		SD
Causes	2.76	0.31	**2.81**	0.33	**2.89**	0.33	2.70	0.29
Processes	2.57	0.38	2.65	0.41	2.71	0.40	2.53	0.38
Consequences	**2.79**	0.26	**2.86**	0.29	**2.89**	0.29	**2.77**	0.25
Responses	**2.99**	0.42	**2.98**	0.42	**3.02**	0.41	**2.96**	0.42
Total	2.75	0.22	**2.80**	0.25	**2.87**	0.24	2.71	0.21

Note: **bold** *text indicates above-average values, x = 2.77.*

Table 5.
Mean and SD by academic year and branch of knowledge.

If we compare this average by year and branch of knowledge (**Table 5**), once again, the answers dimension is the one that exceeds the average, but we can see that there are no differences as the year progresses and no major differences between those in science and those in the arts. When we look at the analysis of variance (**Table 6**) per item according to branch and year, we can see that there are no differences between science and arts, nor between first and last year in 28% of the questions in the questionnaire. On the other hand, there are no differences between science and arts, but there are differences in the year in 12% of the questionnaire, and on the other hand, there are differences in 43%. Finally, it is observed that there are differences in both the branch of knowledge and the academic year in 15% of the items.

So it can be concluded that:

Average number of correct answers to the total test is 65%, which means that the degree of climate literacy is medium.

There is a remarkable symmetry in the trends of the two independent variable handled: Course (64.78% at the beginning and 65.87% at the end) and Academic branch (68.1% for Natural Sciences and Technology (NST) and 61.68% for Social Science and Humanities (SSH).

The level of knowledge is not homogeneous.

A higher degree of Climate Literacy (CL) is observed in consequences of CC and responses to mitigate greenhouse gases (GHG) emissions or adapt to the

Statistical Significance	Items	No. items /total (and % of total)	Dimensions (n° items)
No differences by Branch or Year	2, 4,8, 10, 17, 21, 25, 28, 30	9/32 items (28.12%)	Causes (3)
			Consequences (5)
			Physical Processes (1)
Differences by Year but not by Branch	6, 22, 24, 32	4/32 items (12.5%)	Responses (2)
			Consequences (2)
Differences by Branch but not by Year	1, 2,7, 9, 12, 13, 15, 18, 19, 20, 23, 26, 29, 31	14/32 items (43.75%)	Responses (1)
			Causes (6)
			Consequences (2)
			Physical Processes (5)
Differences by Branch and Year	5, 11, 14, 16, 27	5/32 items (15,62%)	Responses (1)
			Causes (1)
			Consequences (1)
			Physical Processes (2)

Table 6.
Analysis of variance by item according to academic branch and year.

consequences of anthropogenic alteration of the Earth's climate; and a lower degree of CL in items referring to the causes and physical processes involved in CC.

Differences between NST vs. SSH are observed, reinforced when contrasting the results on the variable titration, with statistically significant differences in 60% of the items.

28% of the items show significant differences between 1st and last year and possible prevalence of common knowledge over scientific knowledge: there are items in which SSH students show a significantly higher competence than NST students.

4.4 Analysis 4

In this fourth analysis, the patterns are repeated in terms of questions referring to the existence of CC, in the certainty in it and that it is due to human causes. In this case, the social sciences and humanities students were used as the sample. The questionnaire was also divided into dimensions, but this time the results were shown as scores instead of averages. The maximum score that could be obtained was 128 and the minimum 32 (assigning 1 point to an incorrect answer and 4 to the correct one).

As can be seen in the figure (**Figure 3**), the average score is 86.71 points, and curiously, the degree with the highest score is History.

Once again, the consequences and solutions dimensions are the dimensions with the highest percentages of correct answers and therefore the highest levels of climate literacy.

Significant differences were also detected in all the categories analysed.

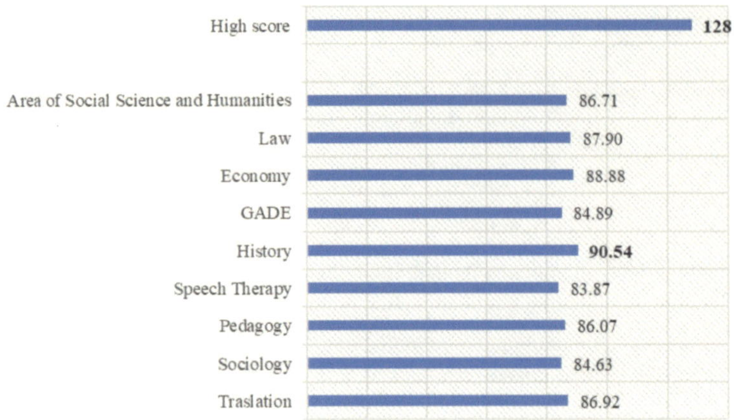

Figure 3.
Scores in the different analysed degrees.

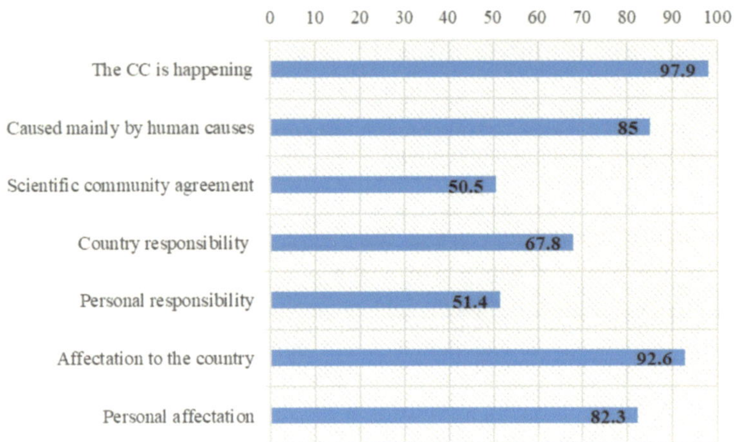

Figure 4.
General opinion questions on CC.

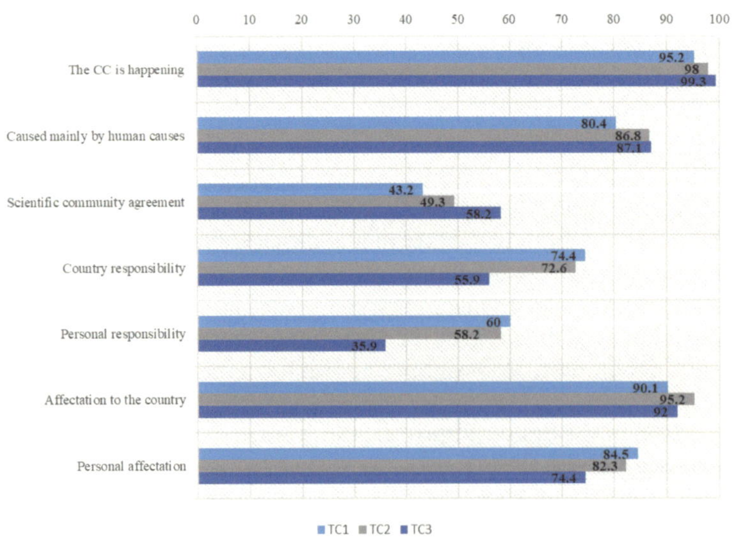

Figure 5.
General opinion questions on CC in the different context analysed.

4.5 Analysis 5

In this last analysis, branches of knowledge and academic year are ignored and importance is given to territorial context, self-perception of information and pro-environmental attitude. Based on this we make comparisons between these three different contexts which are students from Granada, Santiago de Compostela and Braga (in Portugal-which is considered the contrast context) through these variables according to those questions of the questionnaire that have to do with the water /CC relationship.

Figure 4 refers to the general opinion questions on CC, but analyses the results considering the three contexts together. As can be seen, the pattern of the existence of CC, caused by human causes, scientific consensus or responsibility, is repeated.

If we analyse these same questions, but now independently in each context (**Figure 5**), we can see that the results are practically the same, although in the

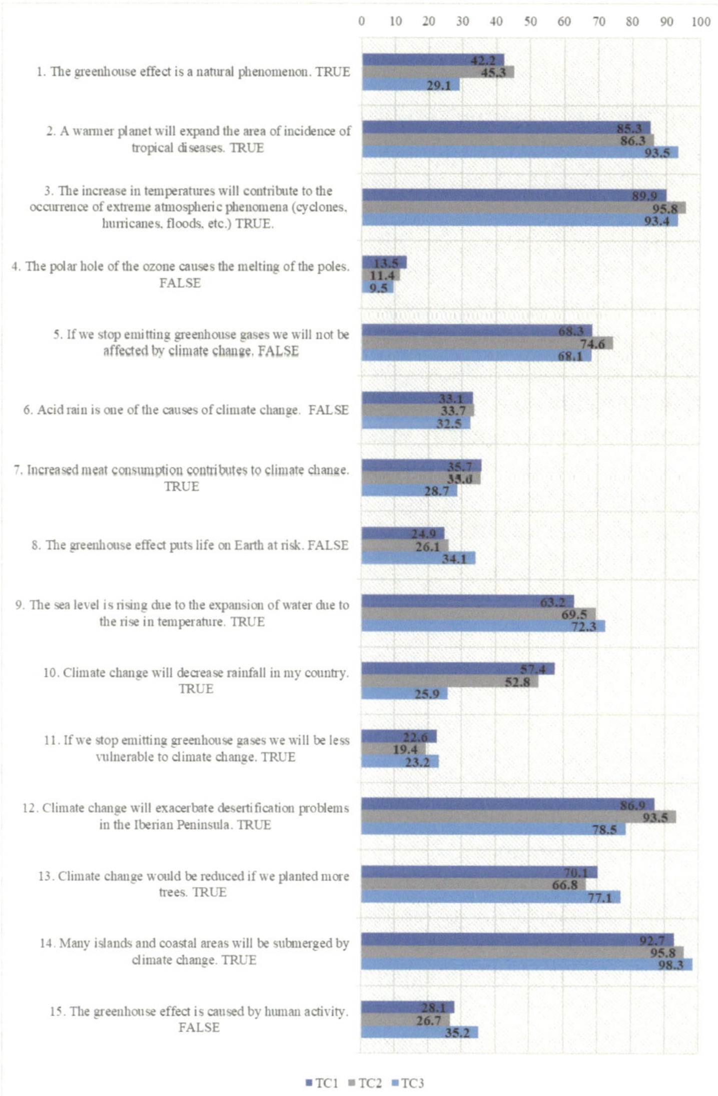

Figure 6.
Conceptual questions about CC.

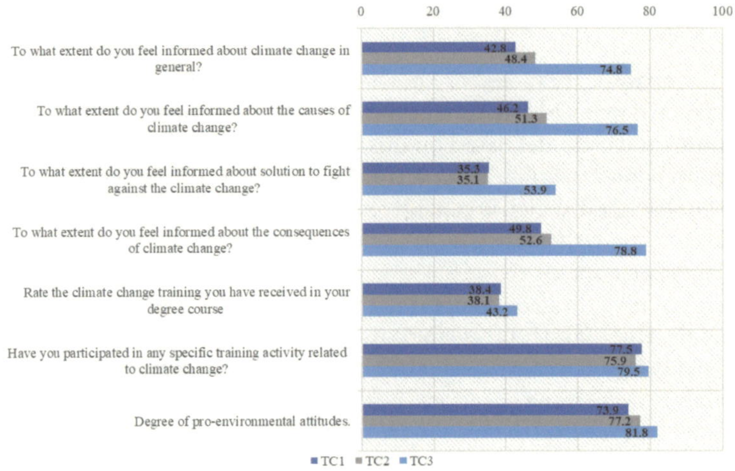

Figure 7.
Degree of information about CC and pro-environmental attitudes.

case of responsibility, it is the contrasting context that feels less responsible, both individually and collectively.

The next figure (**Figure 6**) shows the conceptual questions on CC, and as mentioned above, the questionnaire is now not divided by dimensions, but those questions that have to do with the relationship between water and CC have been chosen.

And although this is so, as can be seen, those questions that have not been answered correctly are also questions that, although related to water, also refer to biophysical processes that have to do with CC, so that we see again a reflection of the pattern of the previous results.

This is where the interpretation of the results of this analysis becomes more important. As can be seen (**Figure 7**) when questions are asked concerning the degree of self-perception about the information students have about the phenomenon, we see how the contrast context is the one that feels the most informed in all the questions asked, and the same happens when asked about the degree of pro-environmental attitude they have.

5. Discussion

This being so, the results are shown in terms of the three hypotheses that are put forward specifically for this analysis.

H1. Denying the existence of CC as a scientific phenomenon has a significant impact on the downward representation of the consequences of Global Warming and other extreme phenomena.

The results obtained are the same as those obtained previously and those consulted in other studies of the general population, which determines a denialism with a downward trend.

However, with regard to individual/collective responsibility, a greater "environmental hypermetropia" is denoted in territorial context 3.

H2. Territorial contexts with high average rainfall levels and low average annual temperatures exert a minimising influence on the social representation of the effects and perception of CC risk, whereas, on the contrary, territorial contexts with low average rainfall levels and high average annual temperatures will exert a maximising influence on the social representation of the effects and perception of CC risk.

In relation to the statements referring to CC effects and risk perception included in the consequences dimension (increase in temperature, rainfall and desertification in the Iberian Peninsula), it can be seen that territorial context 3 is the one with the highest self-perceived information; however, it is the sample that obtains the lowest values of correct answers and incorrectly assesses the item "Climate change will reduce rainfall in my country". Accordingly, the second hypothesis is also rejected.

H3. Different political and cultural contexts between two territories generate different pro-environmental attitudes and feelings of information and therefore different social representations about the causes, consequences and solutions of climate change and its relation to water.

For this hypothesis, three results are concluded:

1. The self-perception of the level of information is higher in territorial context 3, followed by territorial context 2 and 1 in the following variables:

To what extent do you feel informed about climate change in general, on the causes of climate change and on the consequences of climate change.

2. The self-perception of the level of information is higher in territorial context 3 but the same in territorial context 1 and 2 on the variables: To what extent do you feel informed about climate change in general, on the causes of climate change and on the consequences of climate change?

To what extent do you feel informed about measures to combat climate change? How do you rate the training you have received on climate change/Rates your degree of pro-environmental attitude?

3. Participation in specific training activities on CC in all variables is the same for all three territorial contexts.

It should be noted that territorial context 3 (contrast sample) is the territory with the same rainfall and temperature characteristics as territorial context 2, and yet it does not comply with the first two premises, and therefore the third hypothesis is not supported either.

Thus, we conclude that:

The relationship of information sensation in different political-cultural contexts between two territories, generate self-perceptions of the level of information but not different social representations about causes, consequences and solutions to the climate crisis and its relationship with water.

Repeated confusions are perceived in the three territorial contexts in relation to the items referring to greenhouse gases, greenhouse effect, acid rain, ozone layer, meat consumption. Thus, it can be concluded that the pattern is repeated.

6. Conclussion

Almost all the university students surveyed are aware of the phenomenon of climate change. Recognise what is happening and that the changes are mainly due to human causes. The level of climate literacy of the university students is medium-high.

All the academics involved in this research, regardless of where they come from, agree that the impacts of climate change will happen both globally and regionally (in their country). They identify that there will be episodes of extreme atmospheric. Phenomena and impacts on health, they are sure that the country where they live is more responsible for the causes of climate change than they are personally and they agree that they have received very little information on climate change in their training and little or no specific complementary training on the subject outside the university context.

There is a negative and erroneous belief about the fundamental biophysical processes for the development of life on Earth, as is the case of the greenhouse effect or errors in other ideas and para-scientific concepts.

The level of general knowledge about climate change stands out comparatively over other population groups less familiar with the scientific field in line with the available literature. The weight of the common knowledge of climate change is remarkable in the university students surveyed, probably due to the social interactions and informal communication that is generated around this phenomenon. There is a coexistence between the knowledge acquired in higher education and that acquired in non-academic contexts.

Relationship between the water crisis and climate change: the degree of knowledge of students about this relationship is much higher than when studying it by dimensions.

The self-perception of climate change information of university students in the context of contrast does not match the scientific correctness of the responses obtained in their vast majority.

When comparing the three territorial contexts: territory and climate has not an influence on the social representation of climate change.

Finally, we highlight the most relevant aspects of the results of this study:

1. Aspects to be highlighted regarding the Knowledge Branch:

   There are significant differences in general between students belonging to the branches of knowledge of natural sciences and engineering and students of social sciences and humanities, with the former having greater knowledge of the dimensions of climate change studied in this work.

2. Aspects to be highlighted in terms of the degree:

   Degrees that include in their educational offerings subjects that deal with climate sciences, in addition to social sciences, such as the Environmental Sciences degree, increase students' knowledge of climate change somewhat more than in the rest of the degrees.

3. Aspects to be highlighted regarding the course:

   The increase in the number of years of university education does not influence a better climate literacy of the student nor the competence to answer correctly the questions of the questionnaire.

4. Aspects to be highlighted in terms of territory:

   The weight of the common culture around the different aspects of climate change prevails over the influence of the territory.

Although in some aspects, such as the different perceptions of the relationship between the climate crisis and the water crisis, it can be noted in the three territorial contexts analysed, in general, it can be said that university students in these three contexts follow practically the same patterns in terms of the social representation of climate change.

To conclude, we would like to talk about the final considerations and limitations that have arisen in this work. The specific objectives have been achieved as it has been possible to explore the influence of common culture in relation to the variables studied. Important findings have been obtained in terms of understanding how social representations of climate change are formed in university students. In addition, future lines of research may be oriented towards studying how climate change is being dealt with in other university contexts in their curricula. Differences or discrepancies have been found between the theory studied and the data obtained, and it has been possible to determine that there is a relationship between the justification for this research and the results obtained.

With regard to the limitations of the study, it was initially decided to carry out this research, taking only Andalusian university students as a sample for the study, however, the lack of logistical and economic resources did not allow for this. This led to the limitation of the type of study due to the sample. Despite the fact that it was possible to survey a large number of university students, doubts sometimes arose as to whether the sample was representative or not. It should therefore be made clear that this study is, among other things, an exploratory study, which works with non-parametric descriptive statistics and, therefore, by quotas, i.e. it selects a sample from each region to allow comparisons. Another limitation of the study was not being able to make a more exhaustive comparison with respect to the territorial context of the student. That is to say, in this research we were able to work with two universities in the north of the Iberian Peninsula and only one in the south, so if we had had the opportunity to work with another university in the south of the Peninsula it would have been more correct, but the means did not allow it either. However, as this study is part of an international project, future lines of research can address this limitation in order to broaden the focus of the study and the project itself.

Acknowledgements

Article written as part of the Resclima Project, funded by the Spanish Ministry of Economics and Competitiveness (ref. EDU2015-63,572-P) and with European Regional Development Fund (ERDF). This research is also part of the Resclima-Edu2 Project, "Educacion para el cambio climatico en educación secundaria: investigación aplicada sobre representaciones y estrategias pedagogicas en la transición ecologica" funded by the Ministry of Economy and Competitiveness of the Government of Spain and the European Regional Development Fund (ERDF), Ref. RTI2018-094074-B-100; and the "Sustainability in Higher Education: Evaluation of the scope of the 2030 Agenda in curriculum innovation and teacher professional development in Andalusian Universities", Ref. B-SEJ-424-UGR18.

Conflict of interest

The authors declare no conflict of interest.

Author details

Amor Escoz-Roldán[1*], Pablo Ángel Meira-Cartea[2] and José Gutiérrez-Pérez[1]

1 University of Granada, Spain

2 University of Santiago de Compostela, Spain

*Address all correspondence to: amorescoz@correo.ugr.es

References

[1] IPCC. Climate Change 2001: Synthesis Report. Working Group Summaries and Working Group Summaries for Policymakers and Technical Summaries. Aviable from https://archive.ipcc.ch/ipccreports/tar/vol4/spanish/pdf/wg1sum.pdf

[2] Lenton TM, Rockström J, Gaffney O, Rahmstorf S, Richardson K, Steffen W, et al. Climate tipping points - too risky to bet against. Nature. 2019;575(7784):592-595. DOI: 10.1038/d41586-019-03595-0

[3] Meira-Cartea PA, Gutiérrez-Pérez J, Arto-Blanco M, Escoz-Roldán A. Influence of academic education vs. common culture on the climate literacy of university students / Formación académica frente a cultura común en la alfabetización climática de estudiantes universitarios. Psyecology. 2018;9(3): 301-40. DOI. 10.1080/21711976.2018.1483569

[4] Escoz Roldán, A. Gutiérrez Pérez, J, Meira Cartea, P. Water and Climate Change, Two Key Objectives in the Agenda 2030: Assessment of Climate Literacy Levels and Social Representations in Academics from Three Climate Contexts. Water. 2020;12:92. DOI: 0.3390/w12010092

[5] IPCC. Annex B. Glossary of terms of the synthesis report. Aviable from: https://archive.ipcc.ch/pdf/glossary/tar-ipcc-terms-sp.pdf

[6] IPCC. Global Warming of 1.5°C. An IPCC 2018 Special Report on the impacts of global warming of 1.5°C above pre-industrial levels and related global greenhouse gas emission pathways, in the context of strengthening the global response to the threat of climate change, sustainable development, and efforts to eradicate poverty.

[7] United Nations. Policy implications of the Sustainable Development Goals for population health. En: World Mortality Report 2015. UN; 2017. p. 53-55.

[8] Moscovici S. El psicoanálisis, su imagen y su público. Buenos Aires: Huemul; 1979.

[9] Mora M. La teoría de las representaciones sociales de Serge Moscovici. Athenea Digit Rev pensam investig soc. 2002;1(2). DOI: 10.5565/rev/athenea.55

[10] Farr R. Escuelas europeas de Psicología social, la investigación de representaciones sociales en Francia. Revista Mexicana de Sociología. 1983;45:641-657.

[11] Sperber D. Explicar la cultura: un enfoque naturalista. Ediciones Morata; 2005.

[12] Castorina JA, Barreiro A, En Heras Hernández F. Los usos de las representaciones sociales en la investigación educativa. Educación, Lenguaje y Sociedad. 2012;9(9):15-40.

[13] Heras Hernández F. Representaciones sociales del cambio climático en España: aportes para la comunicación. 2015.

[14] Markova I. En busca de las dimensiones epistemológicas de las representaciones sociales. En: Pez ED, Blanco A, editores. Madrid: Fundación Infancia y Aprendizaje; 1996. p. 163-182.

[15] Sterman E, Doran B, Aizley C, Russell S, Sciscione P, Solatorio R, et al. Effects of educational strategies on content retention and learner satisfaction when changing pain management practice. Pain Manag Nurs. 2011;12(2):e8. DOI: 10.1016/j.pmn.2010.10.024

[16] Slovic P, Peters E. Risk Perception and Affect. Curr Dir Psychol Sci. 2006;15(6):322-325. DOI: 10.1111/j.1467-8721.2006.00461.x

[17] Bulkeley H. ——Common knowledge? Public understanding of climate change in Newcastle, Australia||. Public Understanding of Science. 2000;9:313-333.

[18] González Gaudiano E, Meira Cartea. Educación, comunicación y cambio climático. Resistencias para la acción social responsable. Trayectorias. Aviable from: https://www.redalyc.org/articulo.oa?id=60712749003

[19] Ferrari E, Ballegeer A-M, Fuertes MA, Herrero P, Delgado L, Corrochano D, et al. Improvement on Social Representation of Climate Change through a knowledge-based MOOC in Spanish. Sustainability. 2019;11(22):6317. DOI: 10.3390/su11226317

[20] La universidad frente al cambio climático. Proyect. Aviable from: https://cambioclimatico.usal.es/

[21] González Gaudiano EJ, Cruz Sánchez GE, Medellín Milán P. Representaciones sociales del cambio climático en los estudiantes la Universidad Veracruzana. En: Casanova Cardiel H, editor. Mexico City, Mexico: Consejo Mexicano de Investigación Educativa; 2011.

[22] Flores RC, Amigón ET. Las emociones en las representaciones sociales del cambio climático. Educ em Rev. 2018;34(68):217-233. DOI: 10.1590/0104-4060.55684

[23] Flores R. Las representaciones sociales sobre el cambio climático de los estudiantes de pedagogía en México: un acercamiento desde la perspectiva de género. Educación. 2019;28(54):7-26.

[24] De Guttry C, Doring M, Ratter B. How Distant is Climate Change? Construal Level Theory Analysis of German and Taiwanese Students Statements. International Journal of Asian Social Science. 2017;7(5):434-447.

[25] Manolas EI, Tampakis SA, Karanikola PP. Climate change: the views of forestry students in a Greek university. Int J Environ Stud. 2010;67(4):599-609. DOI: 10.1080/00207233.2010.499208

[26] Wachholz S, Artz N, Chene D. Warming to the idea: university students' knowledge and attitudes about climate change. Int J Sustainability Higher Educ. 2014;15(2):128-141. DOI: 10.1108/IJSHE-03-2012-0025

[27] Botêlho LAV, Santos MF, Santos FK da S. A EDUCAÇÃO AMBIENTAL E A GEOGRAFIA ESCOLAR: DIMENSÕES CURRICULARES, POSSIBILIDADES E DESAFIOS CONTEMPORÂNEOS. Caminhos Geogr 2016;17(59). DOI: 10.14393/rcg175908

[28] Akrofi M, Antwi S, Gumbo J. Students in climate action: A study of some influential factors and implications of knowledge gaps in Africa. Environments. 2019;6(2):12. DOI: 10.3390/environments6020012

[29] Richter-Beuschel L, Bögeholz S. Student teachers' knowledge to enable problem-solving for sustainable development. Sustainability. 2019;12(1):79. DOI: 10.3390/su12010079

[30] GCRP, U. (2009). Climate literacy: the essential principles of climate science. US Global Change Research Program.

Section 3

Climate Change Effects on Agriculture

Chapter 4

The Relationship between Land Use and Climate Change: A Case Study of Nepal

Pawan Thapa

Abstract

Land Use and Climate change are interrelated to each other. This change influences one another at various temporal and spatial scales; however, improper land uses are the primary causal factor on climate change. It studies relevant literature and Nepal's case to assess the relationship between land use and climate change. Similarly focuses on how land-use impacts climate change and vice versa. In recent centuries land-use change significant effects on ecological variables and climate change. Likewise, understanding the research on both topics will help decision-makers and conservation planners manage land and climate.

Keywords: Land-use (LU), Climate Change, Review, Relationship

1. Introduction

During the few decades, land use and climate changes are two major global issues, and proper research highly demands finding their relationships and impacts for the future. However, the causes and consequences of human-induced climate change and land-use activities have rarely study independently earlier [1–3]. Researchers and stakeholders understand that these are closely linked to affect each other [4–6]. Begin comprehensive research on land use as one of the causal factors for climate change and vice versa. A concerted multidisciplinary effort needs to study this process, including socioeconomic and natural science [7, 8].

Land-use change is a process by which human activities transform the natural landscape. It refers to how the land used had transformed economic activities [9, 10]. Land-use change is responsible for increases in the human population, deforestation, food types, and demand for energy and fiber. Climate change involves global warming, precipitation, natural disasters like floods, storms, and droughts [11, 12]. Land-use impacts climate through deforestation and rapid population growth, whereas climate change impacts land use through unpredictable heavy rainfall and increasing temperature. The climate change effects on land use perceived by land use altered and land management strategies to mitigate climate change's adverse effects. For example, climate change affects crop production, which leads to land-use change. Both changes in driving forces vary in time and space. Land-use change (LUC) is fundamental to global adjustment that directly influences climate change. This review current practices in assessing land-use change related to climate change [13, 14].

Developing countries are the most affected by changes in climate and land use. They cannot prevent and respond to its impacts due to a lack of multiple factors

[7, 15, 16]. That increases devastating impacts and the rise of vulnerable groups on communities. Around 2030, climate change will affect the poorer significantly by rising death rate, malnutrition, tropical diseases, and urban heat stress [17–20]. Also, climate and land-use change contribute to natural disasters. Several studies justify this impact significantly by developing countries while limited resources, technology, and budget for combating and resilience.

2. Land-use effects on climate change

Humans have to change the land cover for centuries, but recent change rates are higher than ever [21–24]. Land-use change reflected in land cover change, the main component of Climate change, affects land-use decisions [8, 25–27]. It is pre-dominated by deforestation for cultivated land and then other land-use types like built-up industrial areas. Land use and land-cover change (LULCC) impact weather and climate at the synoptic scales and the mesoscale [28]. LULCC study provides essential data support for the research of human activities on environmental change [29–31].

Last three centuries, many developing countries are moving through transition economics, increasing demand for food and energy due to a growing population that has caused deforestation, cropland increased, and urbanization [32]. Several studies in China found that cropland area increased and forest decreased, a similar trend of cropland sharp raised and forest area declined in India, Nepal, and other South Asia [33, 34]. Every country from Asia, Africa, and Europe has faced land-use change due to economic and population growth. Since the past few decades, land-use change impacts on climate. The significant contribution is the precipitation cycle at a local and regional level [25, 35]. Climate change brings unpredictability of rainfall and extreme weather events, which will increase risk in the long term.

3. Climate change effects on land use

Climate change adversely impacted food security, terrestrial ecosystems, land degradation, and land use [36, 37]. Similarly, it is a significant driver for land degradation processes, which results in a land-use change. In many lower-latitude regions, yields of some crops (e.g., maize and wheat) have declined. In contrast, in many higher-latitude areas, yields of some plants (e.g., maize, corn, and sugar beets) have increased over recent decades [38]. Deforestation means land surface air temperature has increased from the growth of industries than the global land and ocean temperature [37].

Southeast Asia's wet and dry soil conditions affect climate and the result of alter in land use [39]. Similarly, increasing temperature and precipitation will increase warmer and shorter winters; this changes land-use patterns in many places. Climate change creates stresses on land, risks to livelihoods, biodiversity, human and ecosystem health, infrastructure, and food systems [40]. Increasing impacts on land projected under all future regions will face higher risks, while some areas will face risks previously not anticipated [41, 42]. In most cases, the climate response to land use and land-cover change may even exceed the contribution from increasing deforestation and land degradation.

4. Case studies on systems for climate change and land use in Nepal

Climate change and improper land use are significant in Nepal. Every year experience changes in temperature and precipitation faster than the global average.

Unplanned urban growth and terrain expose a range of climate risks and water-related disasters. Rapid snow and ice melt in the mountains and heavy rainfall in the foothills during the monsoon season. Around half of the greenhouse gas emits from agriculture, energy, land-use change, forest, and industries [43, 44]. Climate change and unplanned urban growth are severe issues as this placed fourth globally for vulnerability to climate change. Also, it ranked ninth most impacted countries by climate between 1999 and 2018 [45]. More than eighty percent of property loss and lives are related to climate hazards [33, 46–48]. A dependency on agriculture and tourism for income and livelihood made Nepalese people highly sensitive to climate change and land use. The land-use change key drivers of climatic factors are rainfall and temperature.

The map illustrates the Land Surface Temperature (LST) of Nepal of 2020 in **Figure 1**. The Land Surface Temperature (LST) data retrieves at 1 km pixels via a split-window algorithm and at 6 km grids by the day/night algorithm (NASA, Data Products: MODIS Land Surface Temperature (MOD 11), n.d.). The earth science data type for MODIS LST is MOD11A2 of an eight-day compositing period. It provides values at a per-pixel basis at 500 meters (m) spatial resolution [49]. It shows the highest with 34.1 for the majority party on the middle and lower part, i.e., Hilly and Terai regions of Nepal, whereas a lower value of −16.1 on the upper area, the Himalayan part. Low vegetation due to either snow cover or glaciers. Therefore, the slight variation in the surface temperature and driving factors are rapid increases in urbanization, the land cover by impervious land surfaces such as concretes. Land cover change can alter the thermal properties and surface radiation, which varies the urban area's humidity [50, 51]. Higher elevation is one factor in the spatial variation of temperature [52, 53]. The result shows the temperature decreases from the north direction (**Figure 1**).

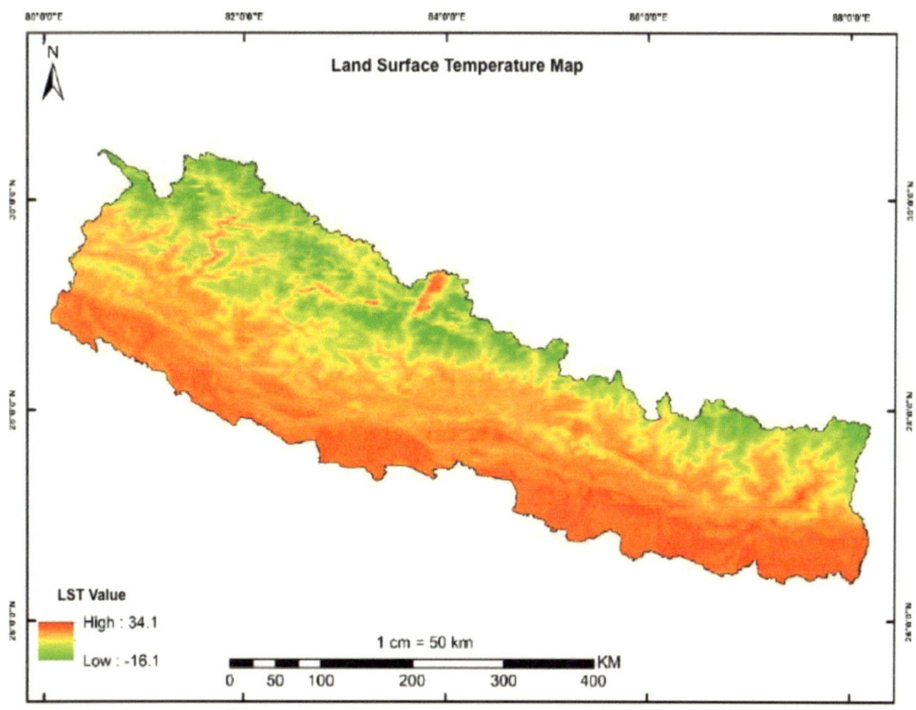

Figure 1.
Map of land surface temperature (LST) of Nepal.

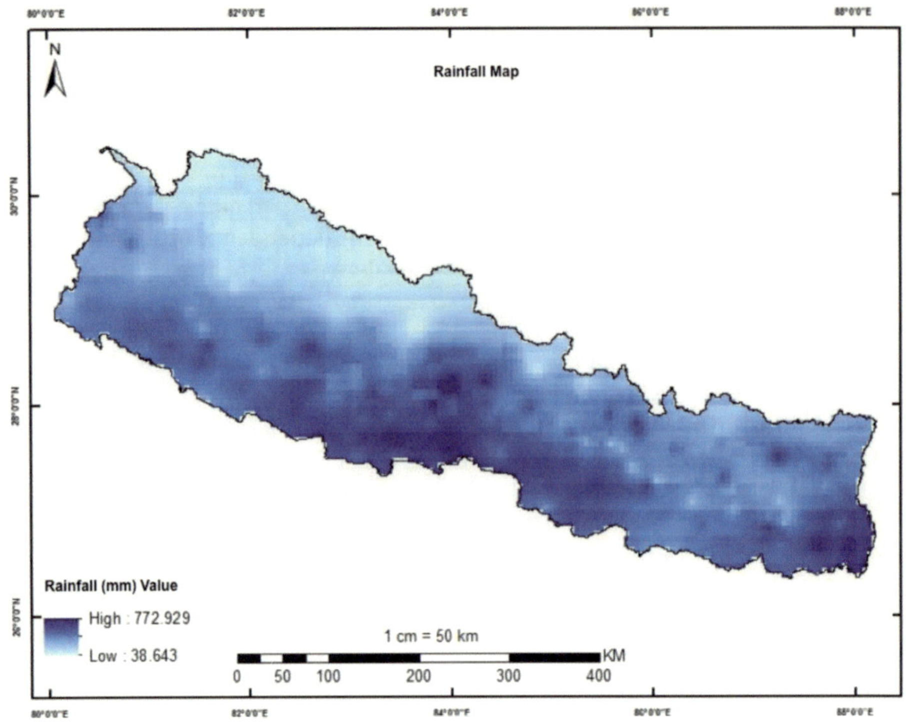

Figure 2.
Map of rainfall of Nepal.

Monthly Rainfall data apply from Climate Hazard's center InfraRed Precipitation with Station data (CHIRPS). The rainfall extremes are increasing, and the temperature rises in urban areas [54, 55]. There is a significant rise in temperature during summer and lower temperature in winter seasons (**Figure 2**).

The earth science data type for MODIS NDVI is MOD13A2, of a sixteen-day compositing period. It provides Vegetation Index (VI) values at a per-pixel basis at 1 kilometer (km) spatial resolution. The product of 2020 for the entire Nepal. NDVI values also changed significantly in the Kathmandu Valley's central core and outlying areas. The magnitude of change was high in the surrounding areas compared to the valley's core. It shows vegetation decreased significantly over the past two decades and newly developed urban areas (**Figure 3**).

The map global land cover at the 500-meter spatial resolution at annual time step for different land cover legends, from supervised classifications on that map of Spectro-temporal features derived from data from the MODIS. Final land use and land cover map of Nepal prepared (**Figure 4**).

In the year 1993, the growth rise at 2.77 percent and fall at −0.04 percent in 2014. From 2015, there is a gradual growth of population with 0.41 percent that plummeted around four times in 2020. The population growth rate touched 2 percent (**Figure 5**) and remains steady for approximately three decades between 1970 and 1998.

The pie chart shows Land Use Land Cover Change (LULCC) results for 2000 and 2019 (**Figure 6**). In 2000, forest area, agriculture, and built-up area presided land-use types, with 34%, 30%, and 10%, respectively. In contrast, in the year 2019, built-up area, barren area, and commercial area increases, with 16% followed by 9% and 5%, respectively. Landsat seven and eight satellite images acquired with no cloud option on one day: 1 January 2000 and 30 December 2020. LULC pie chart

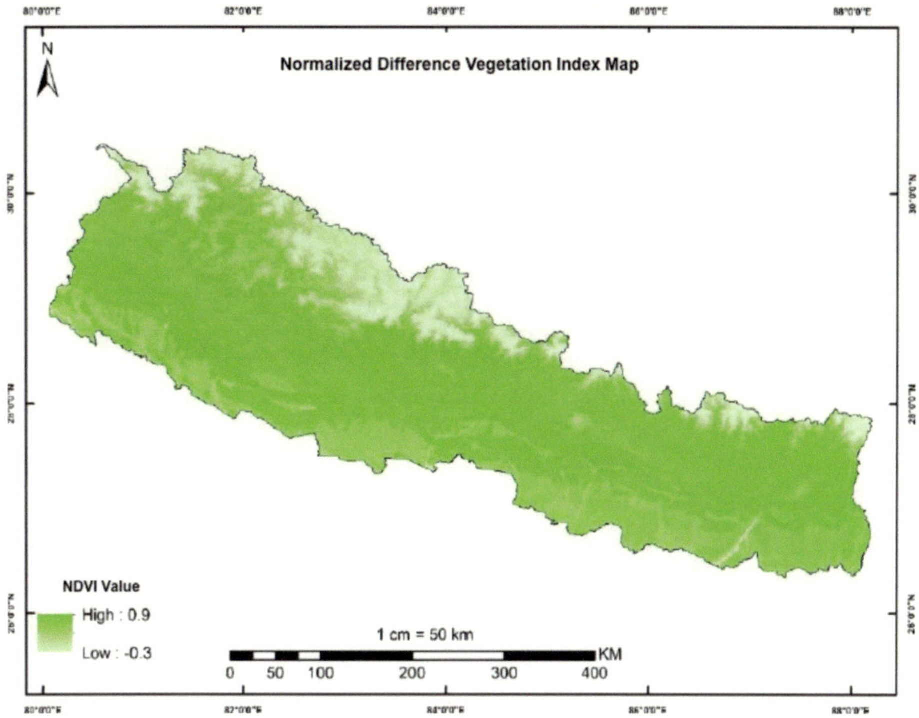

Figure 3.
Map of normalized difference vegetation index (NDVI) of Nepal.

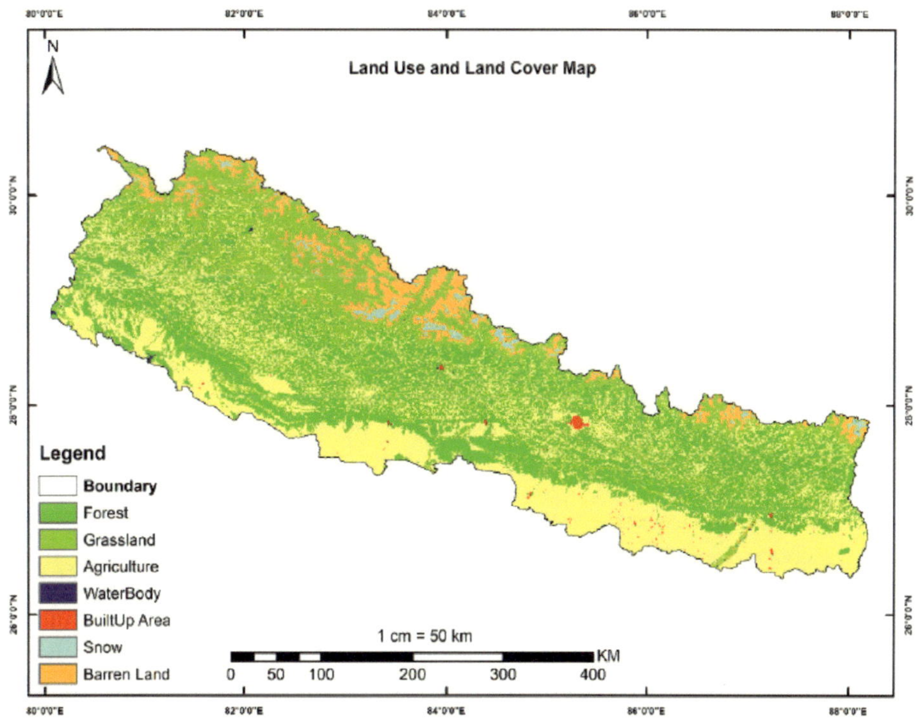

Figure 4.
Map of land use and land cover of Nepal.

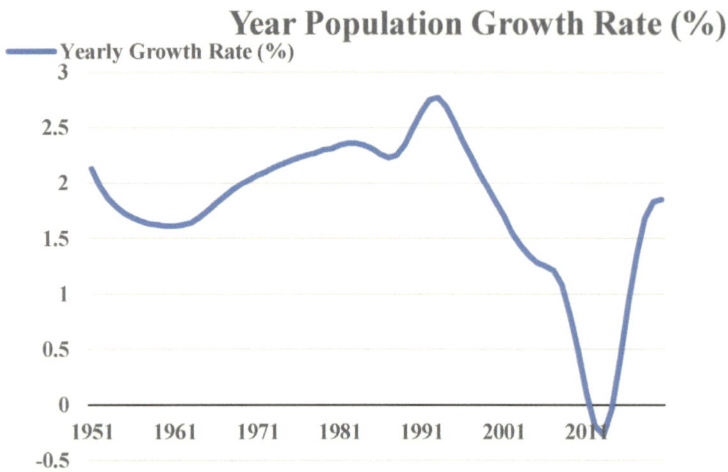

Figure 5.
Line graph of yearly population growth rate (source: Worldometers, 1950 and 2020).

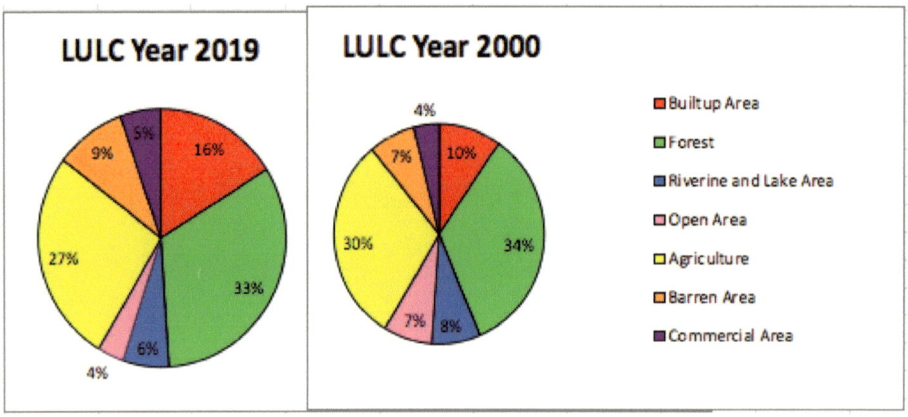

Figure 6.
Pie chart of land use land cover change of 2000 and 2019.

uses the data extracted from maximum likelihood supervised classification in ArcGIS 10.5. Land use and land cover seven types validate by ground truth samples from Google Earth. The overall accuracy for 2000 and 2019 around 70.93%, and the Kappa coefficients 0.7093.

5. Discussion

The relationship shows that both influence each other. It finds land-use change with an increase in significant built-up areas and a gradual decrease in a forest, agriculture, and open areas. It has an unprecedented rate of land use changes over three decades. LST estimates the surface temperature that increases with hot summer and cold winter [56]. MODIS is ideal for monitoring large-scale regional changes in land use and climate change. It can apply at the global or regional scale with a wide spectral range, high temporal resolution, low cost, and real-time [57]. Land Cover implies the physical or natural state of the Earths' surface. Land cover maps represent spatial information on different types (classes) of physical coverage of the Earth's surface, such as forests, grasslands, croplands [58]. It represents how much of a region is covered by mentioned land cover types. Climate change is a change in temperature and rainfall over three decades and focuses on human influences like land use. One factor of land use and climate change is the human population [1, 3]. Increasing population and urbanization had impacted rapid change in land and climate. According to the research results, Nepal is vulnerable to both changes due to rapid population growth, rainfall and temperature change, land covers, and land-use change [59, 60]. Both changes are comparatively more significant than in other world countries. Climate and land-use change impacts are similar to countries like India, Ethiopia, Bangladesh, Europe, and Africa [61]. Meanwhile, past and present comprehensive LUCC research in Nepal requires driving factors and future prediction.

6. Conclusion

The study shows that most studies examine global and regional impacts of land use on climate change. However, several studies take on every scale, such as global, regional, and local, to better understand this complex phenomenon. These studies suggested that proper research and its implementation are needed from every stakeholder to address this situation. The potential for land-related responses and the relative emphasis on adaptation and mitigation is context-specific, including the adaptive capacities of communities and regions. Changes in land conditions, either from land use or climate change, affect global and regional climate. Simple models based on climate change impacts on land use and vice versa will help give initial estimates of their relationship and effects on each other.

Author details

Pawan Thapa
Department of Geomatics Engineering, Kathmandu University, Nepal

*Address all correspondence to: pawan.thapa@ku.edu.np

References

[1] V. H. Dale, "The relationship between land-use change and climate change," Ecol. Appl., vol. 7, no. 3, pp. 753-769, 1997.

[2] B. L. Turner, "Local faces, global flows: the role of land use and land cover in global environmental change," Land Degrad. Dev., vol. 5, no. 2, pp. 71-78, 1994.

[3] B. L. Turner and W. B. Meyer, "Global land-use and land-cover change: an overview," *Chang. Land Use Land Cover Glob. Perspect.*, vol. 4, no. 3, 1994.

[4] R. E. Freeman, "Stakeholder theory," Wiley Encycl. Manag., pp. 1-6, 2015.

[5] L. Schmidt, T. Falk, M. Siegmund-Schultze, and J. H. Spangenberg, "The Objectives of Stakeholder Involvement in Transdisciplinary Research. A Conceptual Framework for a Reflective and Reflexive Practise," Ecol. Econ., vol. 176, p. 106751, 2020.

[6] P. Thapa and N. Dhulikhel, "Observed and perceived climate change analysis in the Terai Region, Nepal," GSJ, vol. 7, no. 12, pp. 35-43, 2019.

[7] P. A. Dirmeyer, D. Niyogi, N. de Noblet-Ducoudré, R. E. Dickinson, and P. K. Snyder, *Impacts of land use change on climate*. Wiley Online Library, 2010.

[8] K. Hibbard *et al.*, "Research priorities in land use and land-cover change for the Earth system and integrated assessment modelling," Int. J. Climatol., vol. 30, no. 13, pp. 2118-2128, 2010.

[9] J. Peng, Y. Liu, T. Li, and J. Wu, "Regional ecosystem health response to rural land use change: A case study in Lijiang City, China," Ecol. Indic., vol. 72, pp. 399-410, 2017.

[10] T. K. Rudel, "How do people transform landscapes? A sociological perspective on suburban sprawl and tropical deforestation," Am. J. Sociol., vol. 115, no. 1, pp. 129-154, 2009.

[11] S. Jamet and J. Corfee-Morlot, "Assessing the impacts of climate change: a literature review," 2009.

[12] N. Seddon, A. Chausson, P. Berry, C. A. Girardin, A. Smith, and B. Turner, "Understanding the value and limits of nature-based solutions to climate change and other global challenges," *Philos. Trans. R. Soc. B*, vol. 375, no. 1794, p. 20190120, 2020.

[13] W. E. Riebsame, W. B. Meyer, and B. L. Turner, "Modeling land use and cover as part of global environmental change," Clim. Change, vol. 28, no. 1, pp. 45-64, 1994.

[14] B. L. Turner and W. B. Meyer, "Global land-use and land-cover change: an overview," *Chang. Land Use Land Cover Glob. Perspect.*, vol. 4, no. 3, 1994.

[15] C. Bellard, C. Bertelsmeier, P. Leadley, W. Thuiller, and F. Courchamp, "Impacts of climate change on the future of biodiversity," Ecol. Lett., vol. 15, no. 4, pp. 365-377, 2012.

[16] P. Thapa, "Using Geospatial Technologies for Disaster Management in Developing Countries," 2021.

[17] J. L. Baker, *Climate change, disaster risk, and the urban poor: cities building resilience for a changing world*. The World Bank, 2012.

[18] S. L. Harlan and D. M. Ruddell, "Climate change and health in cities: impacts of heat and air pollution and potential co-benefits from mitigation and adaptation," Curr. Opin. Environ. Sustain., vol. 3, no. 3, pp. 126-134, 2011.

[19] A. J. McMichael, R. E. Woodruff, and S. Hales, "Climate change and human

health: present and future risks," The Lancet, vol. 367, no. 9513, pp. 859-869, 2006.

[20] Change, I. P. O. C. (2001). Climate change 2007: Impacts, adaptation and vulnerability. Genebra, Suíça.

[21] A. B. Estes, T. Kuemmerle, H. Kushnir, V. C. Radeloff, and H. H. Shugart, "Land-cover change and human population trends in the greater Serengeti ecosystem from 1984-2003," Biol. Conserv., vol. 147, no. 1, pp. 255-263, 2012.

[22] J. Hansen, M. Sato, and R. Ruedy, "Perception of climate change," Proc. Natl. Acad. Sci., vol. 109, no. 37, pp. E2415–E2423, 2012.

[23] R. A. Houghton *et al.*, "Carbon emissions from land use and land-cover change," Biogeosciences, vol. 9, no. 12, pp. 5125-5142, 2012.

[24] B. Yifang, P. Gong, and C. Gini, "Global land cover mapping using Earth observation satellite data: Recent progresses and challenges," ISPRS J. Photogramm. Remote Sens. Print, vol. 103, no. 1, pp. 1-6, 2015.

[25] E. F. Lambin, H. J. Geist, and E. Lepers, "Dynamics of land-use and land-cover change in tropical regions," Annu. Rev. Environ. Resour., vol. 28, no. 1, pp. 205-241, 2003.

[26] P. H. Verburg, "Simulating feedbacks in land use and land cover change models," Landsc. Ecol., vol. 21, no. 8, pp. 1171-1183, 2006.

[27] P. H. Verburg, K. Neumann, and L. Nol, "Challenges in using land use and land cover data for global change studies," Glob. Change Biol., vol. 17, no. 2, pp. 974-989, 2011.

[28] D. Niyogi, R. Mahmood, and J. O. Adegoke, "Land-use/land-cover change and its impacts on weather and climate," *Bound. Layer Meteorol.*, vol. 133, no. 3, p. 297, 2009.

[29] H. Chen, W. Zhang, H. Gao, and N. Nie, "Climate change and anthropogenic impacts on wetland and agriculture in the Songnen and Sanjiang Plain, Northeast China," *Remote Sens.*, vol. 10, no. 3, p. 356, 2018.

[30] R. Mahmood *et al.*, "Impacts of land use/land cover change on climate and future research priorities," Bull. Am. Meteorol. Soc., vol. 91, no. 1, pp. 37-46, 2010.

[31] M. Munsi, S. Malaviya, G. Oinam, and P. K. Joshi, "A landscape approach for quantifying land-use and land-cover change (1976-2006) in middle Himalaya," Reg. Environ. Change, vol. 10, no. 2, pp. 145-155, 2010.

[32] E. F. Lambin and P. Meyfroidt, "Global land use change, economic globalization, and the looming land scarcity," Proc. Natl. Acad. Sci., vol. 108, no. 9, pp. 3465-3472, 2011.

[33] P. Thapa, "Spatial estimation of soil erosion using RUSLE modeling: a case study of Dolakha district, Nepal," Environ. Syst. Res., vol. 9, no. 1, pp. 1-10, 2020.

[34] Y. Wang, N. Wu, C. Kunze, R. Long, and M. Perlik, "Drivers of change to mountain sustainability in the Hindu Kush Himalaya," in *The Hindu Kush Himalaya Assessment*, Springer, 2019, pp. 17-56.

[35] Q. Cao, Y. Liu, M. Georgescu, and J. Wu, "Impacts of landscape changes on local and regional climate: a systematic review," Landsc. Ecol., vol. 35, no. 6, pp. 1269-1290, 2020.

[36] P. Smith *et al.*, "Which practices co-deliver food security, climate change mitigation and adaptation, and combat land degradation and desertification?,"

Glob. Change Biol., vol. 26, no. 3, pp. 1532-1575, 2020.

[37] T. C. Sunderland and D. Rowland, "Forests, land use, and challenges to climate stability and food security," in *Sustainable Food and Agriculture*, Elsevier, 2019, pp. 95-116.

[38] B. Dawson and M. Spannagle, *The complete guide to climate change*. Routledge, 2008.

[39] R. A. Pielke Sr *et al.*, "Land use/land cover changes and climate: modeling analysis and observational evidence," Wiley Interdiscip. Rev. Clim. Change, vol. 2, no. 6, pp. 828-850, 2011.

[40] I. P. O. C. Change, "Ipcc," *Clim. Change*, 2014.

[41] H. T. L. Huong and A. Pathirana, "Urbanization and climate change impacts on future urban flooding in Can Tho city, Vietnam," Hydrol. Earth Syst. Sci., vol. 17, no. 1, pp. 379-394, 2013.

[42] J. A. Patz, D. Campbell-Lendrum, T. Holloway, and J. A. Foley, "Impact of regional climate change on human health," Nature, vol. 438, no. 7066, pp. 310-317, 2005.

[43] R. J. Plevin, A. D. Jones, M. S. Torn, and H. K. Gibbs, *Greenhouse gas emissions from biofuels' indirect land use change are uncertain but may be much greater than previously estimated*. ACS Publications, 2010.

[44] M. Stevanovic *et al.*, "Mitigation strategies for greenhouse gas emissions from agriculture and land-use change: consequences for food prices," Environ. Sci. Technol., vol. 51, no. 1, pp. 365-374, 2017.

[45] D. Eckstein, V. Künzel, L. Schäfer, and M. Winges, "Global climate risk index 2020," *Bonn Ger.*, 2019.

[46] R. P. Devkota, "Climate change: trends and people's perception in Nepal," *J. Environ. Prot.*, vol. 2014, 2014.

[47] G. Tuladhar, R. Yatabe, R. K. Dahal, and N. P. Bhandary, "Disaster risk reduction knowledge of local people in Nepal," Geoenvironmental Disasters, vol. 2, no. 1, pp. 1-12, 2015.

[48] E. K. Pradhan, K. P. West Jr, J. Katz, S. C. LeClerq, S. K. Khatry, and S. R. Shrestha, "Risk of flood-related mortality in Nepal," Disasters, vol. 31, no. 1, pp. 57-70, 2007.

[49] H. Dobesch, P. Dumolard, and I. Dyras, *Spatial interpolation for climate data: the use of GIS in climatology and meteorology*. John Wiley & Sons, 2013.

[50] Q. Meng, L. Zhang, Z. Sun, F. Meng, L. Wang, and Y. Sun, "Characterizing spatial and temporal trends of surface urban heat island effect in an urban main built-up area: A 12-year case study in Beijing, China," Remote Sens. Environ., vol. 204, pp. 826-837, 2018.

[51] I. R. Orimoloye, S. P. Mazinyo, W. Nel, and A. M. Kalumba, "Spatiotemporal monitoring of land surface temperature and estimated radiation using remote sensing: human health implications for East London, South Africa," Environ. Earth Sci., vol. 77, no. 3, pp. 1-10, 2018.

[52] S. Thakuri *et al.*, "Elevation-dependent warming of maximum air temperature in Nepal during 1976-2015," Atmospheric Res., vol. 228, pp. 261-269, 2019.

[53] T. Pawan, "Evaluating a Fit-For-Purpose Integrated Service Platform for Climate Change and Land Information of Rural Mountain Community," J Remote Sens GIS, vol. 9, p. 278, 2020.

[54] G. Naveendrakumar *et al.*, "South Asian perspective on temperature and rainfall extremes: A review," Atmospheric Res., vol. 225, pp. 110-120, 2019.

[55] A. Poudel, L. Cuo, J. Ding, and A. R. Gyawali, "Spatio-temporal variability of the annual and monthly extreme temperature indices in Nepal," Int. J. Climatol., vol. 40, no. 11, pp. 4956-4977, 2020.

[56] J. He, W. Zhao, A. Li, F. Wen, and D. Yu, "The impact of the terrain effect on land surface temperature variation based on Landsat-8 observations in mountainous areas," Int. J. Remote Sens., vol. 40, no. 5-6, pp. 1808-1827, 2019.

[57] R. S. Lunetta, J. F. Knight, J. Ediriwickrema, J. G. Lyon, and L. D. Worthy, "Land-cover change detection using multi-temporal MODIS NDVI data," Remote Sens. Environ., vol. 105, no. 2, pp. 142-154, 2006.

[58] M. Buchhorn *et al.*, "Copernicus Global Land Service: Land Cover 100m: Collection 3: Epoch 2019: Globe," *Version V3 01Data Set*, 2020.

[59] B. Paudel, Y. Zhang, J. Yan, R. Rai, and L. Li, "Farmers' perceptions of agricultural land use changes in Nepal and their major drivers," J. Environ. Manage., vol. 235, pp. 432-441, 2019.

[60] I. P. O. C. Change, "Climate change 2007: Impacts, adaptation and vulnerability," *Genebra Suíça*, 2001.

[61] K. K. Goldewijk, "Estimating global land use change over the past 300 years: the HYDE database," Glob. Biogeochem. Cycles, vol. 15, no. 2, pp. 417-433, 2001.

Chapter 5

Climate Change and Its Consequences in Agriculture

Golam Ali Mozaffari

Abstract

The process of global warming over the past two centuries has become a major and challenging topic among researchers and policymakers at the international level. The process of global warming has accelerated following the increase in greenhouse gas emissions due to excessive consumption of fossil fuels after the Industrial Revolution of the eighteenth century. The growing trend of the population due to scientific advances in the field of medical sciences, rising levels of education, and health among human societies have had a tremendous impact on reducing mortality and increasing the global population. Consumerism and diversity in consumption patterns among human societies have caused more pressure on the earth's natural resources and excessive use of fossil fuels for industrial production, which has exacerbated pollution and increased greenhouse gases, especially carbon dioxide has led. Global warming can have undesirable consequences in various fields, including agriculture, water resources, plant and animal life, and biodiversity. This chapter deals with the consequences and effects of global warming on agricultural climate indicators. Finally, some agricultural adaptation strategies with these changes are presented.

Keywords: climate change, agriculture, global warming, green house gases

1. Introduction

Today, the issue of climate change has been considered one of the most controversial problems in the world and has led to the reaction of governments and nations. Climate change is not a new issue, as in the past geological periods it has occurred frequently.

More than 35 types of greenhouse gases are produced by humans, which increases global warming. Perhaps the most important greenhouse gas is CO_2, much of which is due to fossil fuel consumption during the post-industrial revolution in the eighteenth century [1]. Recently, the increase in other greenhouse gases such as (N_2O), (CH_4), (C.F.C) have had the same effect as carbon dioxide effects on increasing the greenhouse effect. Overall, 60% of greenhouse effects have resulted from water vapor, 26% due to carbon dioxide and other gases accounted for 14% of this contribution [2]. **Figure 1** represents the contributions of the most important long-lived greenhouse gases to the increase in global radiative forcing from the pre-industrial era to 2019.

Scientists have found a new chemical threatening the atmosphere that appears to be a long-life greenhouse gas. This chemical - Perfluorotributylamine - is the most effective radiation chemical found to date and was more effective than other chemicals regarding its possible impact on the climate [3].

Asakere quotes Landsberg (1975), climate change occurs when reflected in atmospheric rotation patterns and global and at least regional meteorological

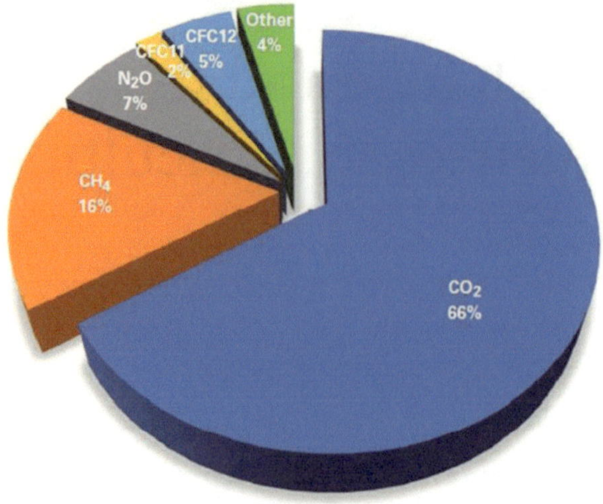

Figure 1.
Contributions of the most important long-lived greenhouse gases to the increase in global radiative forcing from the pre-industrial era to 2019 [3].

processes. He called the temporary deviation that ultimately leads to the return of the previous position or the opposite position as fluctuation. Many atmospheric scientists show little passion to use the term of climate change to explain such alterations. They mainly prefer to use expressions such as variation, long-term climatic variation, climatic anomalies (deviations or anomalies), or climatic fluctuation [4].

The climate in past geological eras: Climate surveys in past geological periods illustrate that important climate change has occurred in the past. Important events of climate change in the past are briefly [2]:

- Earth's climate is constantly changing. Evidence suggests that for most of Earth's history, the climate has been much warmer than today.
- The last glacial period (or ice age) began about 2.5 million years ago. Glacial periods are interrupted by warmer periods called interglacial periods. In North America, continental glaciers reached their maximum thickness and range from about 26,000 to 20,000 years and disappeared completely from North America about 6000 years ago.
- The glacial event called young dryas about 12,000 years ago caused the glaciers to return to north-east America and northern Europe.
- From 1880 to 2012, the earth's surface temperature increased by about 0.85°C. This trend of global warming accelerated in the 1980s and 1990s, then declined in the 2000s.

Figure 2 represents the relative air temperature variations (warmer and cooler periods) during the past 18,000 years. Some regions of the world experienced a cooling and other regions warming that either preceded or lagged behind the temperature variations shown in the diagram [2].

Figure 3 represents that the average temperature variations over the Northern Hemisphere for the last 1000 years relative to the 1961 to 1990 average (zero line). The blue line represents air temperatures constructed from tree rings, corals, ice cores, and pollen. Yearly temperature data measured by thermometers are in

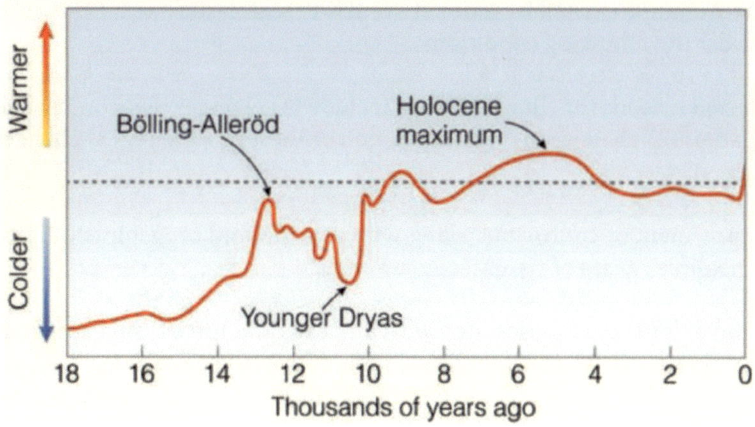

Figure 2.
Relative air temperature variations (warmer and cooler periods) during the past 18,000 years. These data, which represent temperature records compiled from a variety of sources, only give an approximation of temperature changes.

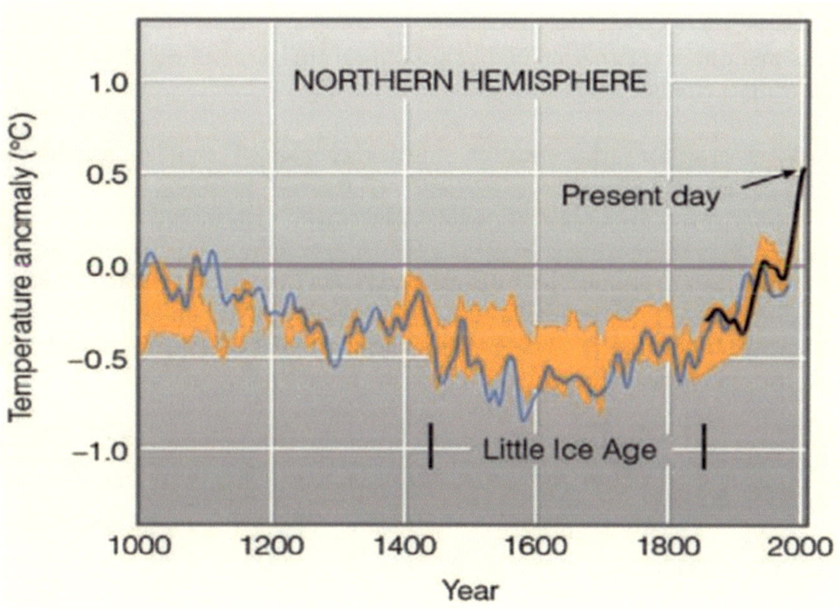

Figure 3.
The average temperature variations over the northern hemisphere for the last 1000 years relative to the 1961 to 1990 average (zero line). (reprinted by permission of the intergovernmental panel on climate change).

black. This reconstruction has been compared to other similar reconstructions. The area shaded orange represents where these reconstructions overlap the data by 50% or more. (Source: Adapted from Climate Change 2007: The Physical Science Basis. Contribution of Working Group I to the Fourth Assessment Report of the Intergovernmental Panel on Climate Change, 2007.

2. Possible causes of climate change

The main reasons for this change can be divided into two groups, including the changes caused by natural events and humans [2]:

Climate change caused by natural events: Research shows that climate change occurs under the following conditions.

- External reasons for climate change include (1) changes in incoming solar radiation (2) changes in the composition of the atmosphere, (3) changes in the earth's surface.
- Displacement of continents, along with volcanic and orogenic activities, are the possible causes of natural climate change.
- Milankovic's theory (associated with other natural forces) suggests that intermittent glacial and interglacial periods over the past 2.5 million years are the results of slight changes in the tilting of the Earth's axis and the geometry of the Earth's orbit around the Sun.
- Air bubbles trapped in the ice sheets of Greenland and Antarctica show that CO_2 levels and methane levels were lower in glacial periods and higher in warmer periods interglacial even when the level was higher, it was still much lower than today.
- The amount of fluctuation in the sun's output (brightness) may be for periods of climate change.
- Volcanic eruptions, rich in sulfur, may be responsible for colder periods in the geological past.

3. Climate change caused by human activities (anthropogenic)

Climate change caused by human activities, which has increased especially in the last two centuries, includes:

- Aerosols are injected into the bottom part of the atmosphere.
- Greenhouse gas
- Land Use Changes
- Nuclear War

4. Predicting climate trends based on climate models

Since the Earth's climate system is so large and does not allow controlled experiments, scientists have used mathematical models known as global circulation models (G.C.M.S) to evaluate known processes taking place and their possible interaction. At least ten global atmospheric circulation models have been developed by meteorological scientists in various research groups and used to predict the effects of rising greenhouse gases. The results of these simulations show an increase in average global warming in the range of 3 to 9 degrees Fahrenheit (1.5 to 4.5 degrees Celsius) by the end of the next century [5].

4.1. Weather models predict

- High latitudes and elevated areas experience greater heat continuity than the global average, especially in winter [5].
- Winter and night time temperatures will continue to rise disproportionately.
- The hydrological cycle is likely to be intensified, leading to more floods or droughts.
- Winter rainfall is mostly as rain to reduce snow, snow compaction; subsequently, runoff decreases in spring, and drought intensifies in spring and summer.
- Extreme weather events including very hot spell periods, flash precipitation, and droughts, under increasing conditions of the greenhouse effect, there will be changes in both the average amounts of weather parameters and the frequency of extreme weather events. Carbon dioxide concentrations in the atmosphere will reach twice the current level by 2050. Global temperatures are also expected to warm by 1.6°C. This increase in temperature will result in major changes in the natural and agronomic territory of the world, and some countries will benefit and others will be harmed.

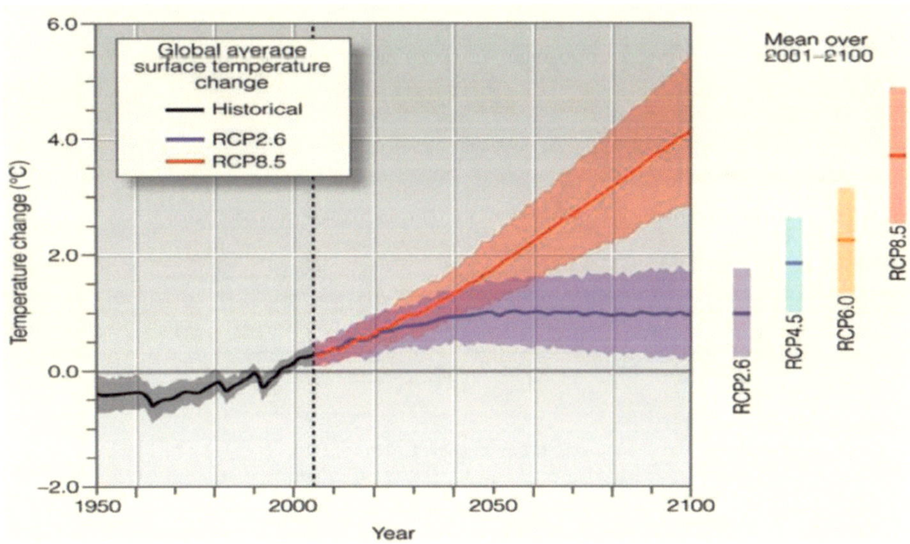

Figure 4.
Global average projected surface air temperature changes (°C) above the 1986–2005 average (dark purple zero line) for the years 2000 to 2100. Temperature changes inside the graph and to the right of the graph are based on dozens of climate models run with different scenarios, based on representative concentration pathways (RCPs). Each scenario describes how the average temperature will change based on different concentrations of greenhouse gases and various forcing agents. The black line shows global temperature change during the twentieth century. The shaded bars on the right side of the figure indicate the likely range of temperature change for each scenario. The thick solid bar within each shaded bar gives the best estimate for temperature change by the years 2081–2100 for each scenario. (see table 18.1 for additional information on the four RCPs.) (source: Adapted from the summary for policymakers, climate change 2013: The physical science basis. Contribution of working group I to the fifth assessment report of the intergovernmental panel on climate change, 2013. Reprinted by permission of the intergovernmental panel on climate change).

The reaction of diseases and pests to climate change can be directly and indirectly. The indirect reaction includes changes in crop type, crop soils, and agricultural operations as well as changes in plant and animal natural habitats that occur after climate change. Thus, changes will occur based on the resources required for pathogens and pests [6].

Under the 2015 Paris Agreement, countries agreed to reduce greenhouse gas emissions to reduce the rate of global average temperature rise to below 2 degrees, above pre-industrialization levels, and attempt to reduce this to 1.5 degrees above industrial levels.

While the overall intention to strengthen the global response to climate change is clear, the Paris Agreement has not specified exactly what the "average global temperature" means, or what period of history should be considered "pre-industrial." To answer the question of how close we are to 1.5 degrees of warming, we must first clarify how both expressions are defined in this special report [7]. The change in global temperature relative to 1850–1900 based on different scenarios (RCP) is shown in **Figure 4**.

Figure 5 represents that human-induced warming in 2017 has reached almost 1 degree Celsius above pre-industrial levels. At the current rate, global temperatures will reach 1.5 degrees Celsius around 2040. The 1.5 degrees Celsius path shown here

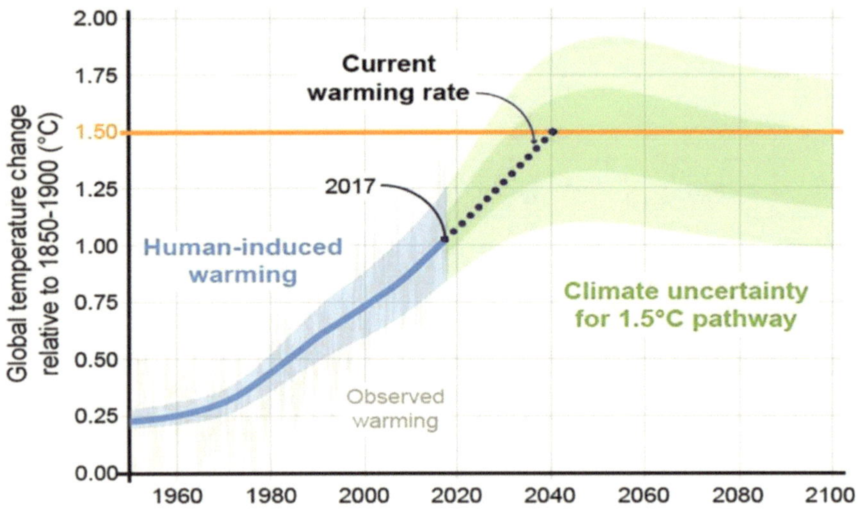

Figure 5.
Human-induced warming and predicting its future trend)IPCC, 2019, P82).

Name of pathway	Likely range of temperature range, °c	Mean estimated temperature change, °c
RCP2.6	0.3–1.7	1.0
RCP4.5	1.1–2.6	1.8
RCP6.0	1.4–3.1	2.2
RCP8.5	2.6–4.8	3.7

Temperature changes are relative to the average surface air temperature for the period 1986–2005. (Ahrens and Robert, 2014, p. 523).

Table 1.
Projected average surface air temperature increases: Ranges and best estimates for the period 2081–2100, using six representative concentration pathways (RCPs).*

includes emissions reductions that start immediately and reduce carbon dioxide emissions to zero by 2055 [8]. Each set of simulations uses a different representative concentration pathway (RCP), describing how the total radiant changes over this century (**Table 1**) [2].

5. Consequences and effects of climate change on agricultural climate indicators

Climate change is effective in geographical spread and plant growth in the world. Global warming will cause changes in temperature and humidity conditions in the world. Following that, some possible changes affecting the plant growth process will be mentioned.

6. Consequences and effects of climate change on the temperature characteristics affecting agriculture

The global warming process over the past two centuries, in addition to the effects on the amount of each atmospheric element, can also impact the time of each atmospheric component occurrence during the crop year. Investigation of possible changes in the time series of the onset and endset of 0 and 5°C temperatures and their trend indicates that these changes are global. It is estimated that the minimum temperature increase caused by climate change will increase wheat yields in Australia by 30–50% [9].

Changes in precipitation patterns and rising temperatures may alter land use for food production and consequently cause pathogens or new plant pests [9–11].

Research reveals that overnight temperatures have increased significantly in recent decades [12]. This has caused a decrease in diurnal variation. Naturally, with increasing temperatures during the year, seasonal fluctuations, i.e. the difference between winter and summer temperatures also decrease. Under these conditions, some crops such as sugar beet and sugarcane which react to diurnal variation [13] are affected and this increase in night temperature and decrease in the diurnal variation reduces sugar concentration in those plants.

Research shows that the temperature of winter has increased compared to the past and decreased the intensity of winter frosts [12, 14–17] and the start date of 0°C temperatures have also been postponed, which could have important implications on agricultural production. Roshan et al. [18] to predict the effects of global warming on wheat degree-day changes in Iran, used the general circulation model of INMCM-30 and the P50 scenario, which is the average SERS emission scenario. The results of this study showed that the degree-day values in most climatic regions of Iran have an increasing trend until 2100.

7. Consequences and effects of climate change over the growing season period

Plants, whether natural plants or crops and horticultural plants, to complete their growth and development, need to spend a period with optimum temperature and humidity conditions.

Various definitions and criteria have been proposed to determine the growing season, the first and last frost has been used in many studies [19–22].

The length of the growing season, the length of the days between the last hard frost and the first hard frost in spring are also defined. Hard frost refers to a minimum daily temperature that 50% of plants exposed to it are destroyed [23].

The length of the growing season, the length of time between the last killing frost in spring and the first killing frost in the fall are also defined (U.S. Army Agronomists, 1987). Killing frost is 28°F (−2.2°C) or colder [24].

According to the definition provided by the variation group of the Climatology Commission of the World Meteorological Organization, in the northern hemisphere, the interval between the first period after July 1st (10th of Tir), which is at least 6 consecutive days, the average daily temperature is more than 5°C and the first 6-day period with an average daily temperature of less than 5°C (in autumn) is considered as the growing period. In the southern hemisphere, this interval is considered from January 1st [25].

The length of plant growing season varies not only in terms of each plant species, but also there are differences in one species. Determining the growing season in each region has an effective role in selecting crops and cultivars and determining planting time and other crop decisions.

Many phenological, meteorological, and satellite researches report an increase in the length of the growing season due to rising temperatures in the northern regions during the twentieth century [26].

Chemielewki [27] investigated the relationship between annual and periodic changes of the growing season in Europe with air temperature changes, the results show that in Europe during 1989–1998 with an increase of 0.8°C the average temperature in the last decade, was occurred 8 days earlier at the beginning of the growing season. Research conducted in the United States [28] Australia, China [29], also shows a decrease in the number of freezing days at high and middle latitudes of the northern hemisphere and an increase in the length of the growing period compared to the twentieth century. Regarding the effects of climate change on growing period, researches such as temporal and spatial variation of phenological seasons in Germany from 1951 to 1996 [30], changes in the growing season in the last century [26] can be noted.

One of the important consequences of climate change is the change during the growing season, which follows the change in the beginning and termination of zero temperature and temperature of 5°C (thermal base of cold crops). Changes in the time and length of the growing season may not only have far-reaching consequences for plant and animal ecosystems, but a steady increase in the length of the growing season may lead to a long-term increase in carbon storage and changes in vegetation cover that may affect the climate system. The decrease in the length of the growing season leads to changes in the cultivation calendar and a decrease in yields of crops that have not yet fully reached their maturity and final growth, while increasing the length of the growing season may provide more opportunities for earlier cultivation, ensuring final growth and maturity, and even the possibility of further harvesting (if water is available) [30].

8. Implications and effects of climate change on the characteristics of effective rainfall in agriculture

Most agricultural activities such as land clearing, planting, and harvesting time need to be informed about the start and end dates of rainfall, rainfall amount, and the length of the growing season [1, 31, 32].

Understanding the variation of precipitation and its trend and the characteristics of the growing season is critical for planning and designing appropriate adaptation strategies at the basin level [33]. Increasing or decreasing rainfall and changes in rainfall distribution will affect water balance and will change the frequency of droughts and floods [34]. Research shows that precipitation and temperature changes are not uniform in all parts of the world and there are temporal and spatial fluctuations in different regions of the world) [35].

The amount of annual precipitation and it's seasonal distribution, as well as changes in the beginning and end date of precipitation in many regions of the world, have changed following the trend of rising temperatures on the planet. Many types of research have been conducted by researchers globally to understand the characteristics of precipitation, indicating that precipitation changes in some places increased and in some others have decreased, including [36–41] can be noted.

Dryland farming is defined as the production of crops without irrigation in semi-arid regions of the world, where annual rainfall is between 250 and 500 mm. The success rate of dryland farming in these regions depends on annual rainfall, appropriate distribution of precipitation during the growing period, start and end date of precipitation. If the date of rainfall delays in autumn, germination of crops, especially cereals, is delayed and this causes other growing stages to begin and end with delay. Under these conditions, the thermal needs of the plant are not provided in the autumn; Therefore, the plant, especially wheat, faces unfavorable conditions in the winter; and eventually, it may be destroyed due to freezing temperatures.

Van de Giessen et al. [42] believe that in West Africa, south of the coast, the start of the precipitation season has shifted forward. Rao et al. [43] believe that this evidence cannot always represent climate change. Other researches in this region did not illustrate a significant trend [44]. Future changes in the tropical rotation pattern may cause seasonal changes and lead to increased uncertainty at the start of the precipitation season [45].

9. Climate change and alters in agro-climatic zoning

According to the prediction of climate models that predict the continuation of more heat than the global average for high latitudes and elevated areas, especially in winter, as well as the increase in winter temperature and night temperature than in the past, certainly growing domains including natural and hand planting will be affected. Also, the changes that have taken place in agroclimatic indices in the world can be effective. New climatic conditions different from previous ones can alter the niche space (Ecological nest) of any living organism and disturb the favorable biological conditions for living organisms, including plants and animals. This change in environmental conditions, especially changing climatic conditions, may not be desirable for plant species. So that, by changing the optimum climatic conditions of the plant, new plant species to replace the previous species. Chamura et al. [46] used climate models and concluded that agro-climatic zones associated with major food crops of corn, sorghum, cassava, and peanuts in Ghana have changed. Research by Tranka et al. [47] indicates the deterioration of agricultural conditions in Ghana. Their results confirm that dryland farming in this country is facing serious risks and there is an increased risk of very undesirable years in many climatic zones of this country, leading to annual variation of crop yields. Ceglar et al. [48] studied the effect of climate change on the displacement of agro-climatic zones in Europe. The gradual warming of the European continent has led to the prolonged growing season, the cumulative increase in active temperatures combined with the events of the extreme hot events. This research, which has been conducted using climatic models and different emission scenarios, shows that much of Europe will be affected by the displacement of climate regions facing the north in the coming decades. In addition, the displacement of agro-climatic zones in Eastern Europe may reach twice the speed observed in the period 1975–2016. Some regions may lose the ability to grow specific crops for the benefit of northern European regions. This index-based assessment shows that the potential benefits of prolonging the thermal growing season in northern and eastern Europe are often unbalanced by the risk of late frost and heatwaves in early spring and summer.

Research reveals that due to the shortening of the growing period or due to lack of thermal requirements, agro-climatic zones have been displaced [49, 50–53]. The movement of vineyards from the main planting areas, i.e. the Mediterranean regions to the central and western regions of Europe, is evidence of this claim.

King et al. [54] indicated that suitable conditions for the degree-day of plant growth will experience a north-facing expansion of up to 1200 kilometers in northern regions by the end of the twenty-first century. The northerly expansion will provide favorable conditions for crop production, along with the relocation of agro-climatic zones to inland and other parts of Europe. Although ecological barriers (such as mountains and lakes) can temporarily stop movement, the speed of movement across the boundaries of climatic zones has a significant impact on the redistribution of ecological communities [55].

10. Climate change and increasing water-use requirement of plants

The physical process of converting liquid water into vapor and entering it into the air is called evaporation. This process is also carried out from the free surface of the water, soil surface, or wet surfaces of the plant. On the contrary, the process of transpiration is called water removal in the form of vapor from the leaves and inside the plant. Practically, the categorizing of transpiration and evaporation in two groups is difficult. Therefore, they are combined and used as evapotranspiration. The sum of evapotranspiration and the water needed for leaching soil is called the water-use requirement of the plant. Since the need for leaching is negligible compared to evapotranspiration, evapotranspiration is also defined as the water-use requirement of the plant [56]. The main factors affecting evapotranspiration depend on climatic elements such as temperature, precipitation, humidity, wind speed, and solar radiation. So, any changes in climatic elements will affect the amount of evapotranspiration and water required by the plant [57]. Due to the changes in climatic elements that have occurred following global warming, it can be expected that the amount of evapotranspiration in different geographical regions can also be changed.

Assessments conducted by various researchers indicate that the occurrence of climate change, evapotranspiration, and water-use requirements of plants will undergo serious changes [58]. Tao et al. [59] by investigating the effect of climate change on reference plant evapotranspiration in China's Xiangjiang basin, under RCP scenarios and using the SDSM model for downscale, concluded that in future periods, the amount of reference plant evapotranspiration will increase under all scenarios and the rate of this increase will vary depending on the region and scenario and the highest incremental rate will be observed under the RCP8.5 scenario..

Heidari Tasheh Kabood and Khoshkhoo [60] predicted future changes of reference evapotranspiration in western Iran based on RCP emission scenarios. They used the FAO-Penman-Monteith method to estimate evapotranspiration, canESM2 global circulation model to simulate climatic conditions, and for downscale of the data in this model, the SDMS method was used. The results showed that in all future periods and under all scenarios for all stations, the mean of reference evapotranspiration in annual scales and for autumn and winter will increase significantly at 0.01 level compared to the base period. In a similar study conducted using the CanESM2 model under RCP2.6, RCP4.5, and RCP8.5 scenarios in Mazandaran province in Iran, the results illustrated that the percentage of changes in evapotranspiration per different months varies from −16.1 to 25.7%. The highest and lowest percentage of reference evapotranspiration changes are in October and March respectively [61].

11. Climate change and its effects on plant pests and diseases

The importance of climate in the growth and development of plant diseases has been known for more than 2000 years. The ancient Greeks (286–370 BC) determined that cereals cultivated in the Highlands had less disease compared to cereals cultivated in low-lying areas [62]. Understanding the history of pests may shed light on possible future trends and strengthen the evidence base of national and international policies on plant conservation. Today, it is widely believed that increased international trade and travel will accelerate the pace of pest arrivals, which is caused by adverse globalization outcomes [63]. There is also the view that climate change may accelerate the establishment of new pests [63].

Increasing CO_2 levels associated with climate change may affect plant distribution, abundance, and yield, pests, and pathogens [64]. Research by Anderson et al. [63] on the origin of new plant, outbreaks showed that 56% of diseases were caused by new pathogens.

Lucke et al. [65] investigated pathogens in four major food crops of wheat, rice, soybeans, and potatoes. The limited data show that depending on the interaction between host and pathogen, the effect will be positive, negative, or neutral. Plant pathogens will have different responses to climate change while the life cycle of some pathogens is limited by rising temperatures, e.g., *Puccinia striiformis* f. sp. Tritici. In addition, other climatic factors such as increased CO_2 may provide more favorable conditions for pathogens such as *Fusarium pseudograminearum*.

Major factors of climate change that are likely to affect the severity and spread of plant diseases are increased CO_2 content, heavy and unseasonal rainfall, increased humidity, drought, hurricanes, and warmer winters [65, 66].

The reaction of diseases and pests to climate change can be directly and indirectly. The indirect reaction includes changes in crop type, crop soils, agricultural operations as well as changes in plant and animal natural habitats that occur after climate change. Thus, there will be changes in the resources needed for pathogens and pests. There are plenty of articles and writings on the interaction between pests, diseases, weeds, and climate change, but many of them are still disputed. Changes during the growing period have unfortunately affected the biological territory and life cycle of living organisms. Increasing the length of the growing season has caused changes in the pattern of bird migration, increased infection induced by insects, and changed the habitat of living organisms. The longer the growing season, the more insects such as locusts will be able to complete their reproductive cycle several times during spring, summer, and autumn. Higher winter temperatures may allow larvae to survive the whole winter in areas where the cold threatened them with death, therefore, this will lead to more infections in the next season [6].

Disproportionate warming of high latitudes and elevated areas in winter and at night can affect plant growth and development can change the geographical distribution pattern of production activities and will change the ecological balance between crops and pests. It is possible that, even without climate change, pest management will face major challenges over the coming decades [5].

Insects react to high temperatures by increasing the rate of growth and development and shortening the time between generations (very high temperatures reduce the lifespan of the insect). Warmer winters reduce insect mortality; consequently, the insect population will increase in the coming seasons. Warmer temperatures occur earlier in spring and cause the pest population to expand during sensitive stages of plant growth and development. Warmer winter temperatures will affect pests that cannot currently spend winter (do overwintering) in high latitudes, but these insects will spend winter in lower latitudes areas; then, migrate to agricultural fields in spring and summer [67].

As higher temperatures lead to longer growing seasons in temperate regions, this will provide an opportunity to increase insect damage. Prolonging the growing season will allow pest insects to create an extra generation and subsequently increase the pest population [67].

As temperatures rise, there will be a shift in the agroclimatic regions towards the pole, whereby many pathogens will spread into new geographic regions, where they will encounter potential new hosts [68].

Climate change can have various effects on natural enemies of pest species. Today, the effects of climate change on different aspects of pests and diseases activity on the plant growth process has been considered by researchers globally. Vector and pathogenic response to climate change, pathogen-host interaction response to climate change, the impact of climate change on host resistance, climate change and disease management, climate change effects on natural enemies agricultural pests, the distribution of plants, hosts and natural enemies, and the coincidence of enemy-hosts are among the topics that have attracted the attention of researchers [69].

12. The effects of climate change on biodiversity

Climate change poses new challenges to protect biodiversity. Species ranges and ecological dynamics are currently responding to recent climate change, and current reserves will not support all species designed to protect them. These problems are exacerbated by other global changes. While reviewing the past 22 years of biodiversity conservation research, Heller et al. [70] attempted to identify potential solutions, agreements, and goals to address climate change. In this study, 524 recommendations from 113 articles were published in 57 different journals and three books were identified and introduced.

Research illustrates that species respond to climate change challenges by moving their Niches space (ecological niche) along three axes of time (phenology), location (territory), and self (physiology). There is relatively little evidence of extinction caused by climate change. Studies prove that habitat destruction poses the greatest global threat to biodiversity over the coming decades [5].

At higher levels of biodiversity, climate can cause changes at the plant community level, predicting that this can affect the integrity of biomes enough. The 1000-year period predicts the displacement of 5–20% of the planet's ecosystems, especially the coniferous forests of temperate (regions, tundra, savannah, and northern forests) [71].

Recent assessments in tropical South America show that a large part of the Amazon rainforest is replaced by tropical savannas [72]. Coral coasts are expected to be threatened and destroyed by warmer and acidification of ocean water [73].

13. Agricultural adaptation to climate change

Climate change will have significant impacts on agricultural production and food security in the future. In the third assessment report, Intergovernmental Panel on Climate Change presented several scenarios and examines their implications for global regions. For Africa, it is predicted that many African countries will face a decline in crop yields, and due to droughts, floods, and other extreme events, there is more pressure on water resources, food security, and human health in these countries.

On the Asian continent, food security of the continent is also threatened in many countries located in arid, tropical, and temperate regions due to heat, water, rising sea levels, floods, droughts, and tropical storms. In Latin America, food security in countries, especially livelihood agriculture, is expected to be at risk. At higher latitudes, prolongation of the growing season and rising temperatures due to climate change will benefit agriculture.

Plants can cope with climatic conditions to some extent, some of them have natural adaptability, such as specific cultivars of rice that blossom in the morning to avoid the destructive effects of higher temperatures late in the day. Fewer studies have been conducted on the potential of plant resistance to high-temperature stress. The highest compatibility of products against environmental stresses has been made possible by humans. There are, fortunately, valuable experiences to deal with adverse weather events such as drought, floods, and salinity among farmer communities. For example, new irrigation methods, water stress and water salinity resistant species, and high-yielding plant species have been proposed.

The International Research Center (CGIAR) is one of the most important advisory groups that offers global experiences in agricultural adaptation to climate change conditions, and countries can share their research efforts with the center. Some compatibility methods are presented in **Table 2** [74].

Response strategy	Some adaptation options
• Use different crops	• Carry out research on new varieties
• Change land topography to improve water uptake and reduce wind erosion	• Subdivide large fields
	• Maintain grass waterways
	• Roughen the land surface
	• Build windbreaks
• Improve water use and availability and control erosion	• Line canals with plastic films
	• Where possible, use brackish water
	• Concentrate irrigation in periods of peak growth
	• Use drip irrigation
• Change farming practices to conserve soil moisture and nutrients, reduce run-off, and control soil erosion	• Mulch stubble and straw
	• Rotate crops
	• Avoid monocropping
	• Use lower planting densities
• Change the timing of farm operations	• Advance sowing dates to offset moisture stress during warm periods

Table 2.
Examples of adaptation options for agriculture.

14. Technologies for adaptation to climate change

The Intergovernmental Panel on Climate Change defines adaptation as compatibility in natural or human systems in response to actual or expected climatic adversity or their effects, which mitigate the damage or take advantage of useful opportunities [74, 75].

Different types of compatibility can be categorized into predicted, autonomous, and planned compatibility. There are several adaptation measures that the agricultural sector can take to tackle climate change in the future.

- Changes in planting date
- Cultivation of different species and plant varieties
- Develop and promote alternative crops
- Development of water-resistant and heat resistant species
- More use of intercropping use
- Use of sustainable fertilizers and tillage methods (improving soil drainage, non-plowing, etc.)
- Improved use of crop residues and weed management
- Further use of water harvesting techniques
- Improved control of pests and diseases for crops
- Implementing new or improved irrigation systems (reducing water leakage, maintaining soil moisture - mulching)
- Animal improvement management (providing shelter and shade, changes in heat-resistant breeds, grazing change, and rangeland rotation)
- Greater use of forestry crops practices
- Management of fire improvement in forests (changing layout status, landscape planning, cleanup underlying plants, insect control through planned burning)
- Development of early warning systems and protective measures for natural disasters (droughts, floods, tropical storms, so on).

15. Conclusion

Global warming has had undesirable effects on plant and animal life on Earth. The process of plant life on Earth is influenced by three natural factors: water, soil, and climate. Among these factors, the role of climatic elements is very important and any deviation from the favorable climatic conditions of the plant can pose serious risks to the plant and endanger plant life. Global Circulation Models (GCMS) predict average global warming in the range of 3 to 9 degrees Fahrenheit (1.5 to 4.5 degrees Celsius) by the end of this century. This increase in the average global temperature can have important consequences and effects on agricultural climate indicators. Research conducted by researchers around the world confirms this subject. In this chapter of the book, the consequences and effects of global warming on agricultural climate indicators are discussed, including the following, which are discussed in detail.

Consequences and effects of climate change on the temperature characteristics affecting agriculture.

- Consequences and effects of climate change over the growing season period.
- Implications and effects of climate change on the characteristics of effective rainfall in agriculture.
- Climate change and alters in agro-climatic zoning.
- Climate change and increasing water-use requirement of plants.
- Climate change and its effects on plant pests and diseases.
- The effects of climate change on biodiversity.

Author details

Golam Ali Mozaffari
Department of Geography, Yazd University, Iran

*Address all correspondence to: gmozafari@yazd.ac.ir

References

[1] Abaje IB, Sawa BA, Ati OF. Climate variability and change, impacts and adaptation strategies in dutsin-ma local government area of Katsina state Nigeria. Journal of Geography and Geology. 2014;**6**(2):103-112

[2] Ahrens CD, Hensens R. Meteorology Today. USA: CengageLearning. Available from: www.cengage.com/global 2014.

[3] W.M.O. The state of green house gases in the atmosphere based on glibal observations through 2019. Greenhouse Gas Bulletin. 2018;**14**:3

[4] Asakere H. Climate Change. Zanjan: Zanjan Uni Publications; 2007

[5] Rosenzweig C, Igles A, Yang XB, Epstein PR, Chivian E. Climate Change and U.S. Agriculture: Impact of Warming and Extreme Weather Events on Productivity, Plant Diseases and Pest. 2000. Available from: http://med.harvard.edu/chge/

[6] W.M.O. Climate Variability, Agriculture and Forestry. WMO Publications; 1994 NO. 802

[7] I.P.C.C. IPCC Special Report. 2021, Available from: ipcc.ch/sr15/download.

[8] I.P.C.C. Global Warming of 1.5° C. WMO Publications; 2019. p. 82

[9] Coakley S, Scherm H, Chakraborty S. Climate change and plant disease management. Annual Review of Phytopathology. 1999;**37**:399-426

[10] Cannon R. The implications of predicted climate change for insect pests in the UK, with emphasis on non-indigenous species. Global Change Biology. 1998;**4**:785-796

[11] Parker I, Gilbert G. The evolutionary ecology of novel plant–pathogen interactions. Annual Review of Ecology Evolution, and Systematics. 2004; **35**:675-700

[12] Insaf TZ, Lin S, Sheridan SC. Climate trends in indices for temperature and precipitation across New York state, 1948-2008. Air Quality Atmosphere and Health. 2012;**1**(1). DOI: 10.1007/s11869-011-0168-x

[13] Das UK. Agricultural meteorology in the U.S.A. Weather. 1950;**5**:383-388

[14] Fritch P, Alexander LV, Della-Marta P, Gleason B, Haylock M, Klein Tank AMG, et al. Observed coherent changes in climatic extremes during the second half of the twentieth century. Climate Research. 2002; **19**:193-212

[15] Scheifinger HA, Menzel EK, Peter C. Trends of spring time frost events and phonological dates in central Europe. Theoretical and Applied Climatology. 2003;**74**:41-51

[16] Karoly D. Observed trends in indices of daily temperature extremes in South America 1960-2000. Journal of Climatology. 2005;**18**:5011-5021

[17] Brown PJ, Raymond S, Bradley F, Keimig T. Changes in extreme climate indices for the Northeastern United States, 1870-2005. Journal of Climate. 2010;**23**:6555-6572

[18] Roshan GHR, Aoji R, Najafi MS, Shakohi E. Prospects for the effect of global warming on the degree-day changes required by wheat for different climates in Iran. Regional Planning Quarterly. 2011;**4**

[19] Wang JY. Agricultural Meteorology. Pacemaker Press; 1963. p. 693

[20] Cooter EJ, SK LD. Recent frost date trends in the North-Eastern U.S.A. International Journal of Climatology. 1995;**15**:65-75

[21] Brinkman WAR. Growing season length as an indicator of climatic variations. Climatic Change. 1979;**2**:127-138

[22] Kunkel KE, Easterling DR, Hubbard K, Redmond K. Temporal variations in frost free season in the united states: 1895-2000. Geophysical Research Letters. 2004;**31**:L03201. DOI: 10.1029/2003GL018624

[23] Baron WR, Gordon GA, Borns HW, Smith DC. Frost-free record reconstruction for Eastern Massachusetts, 1773-1980. Journal of Climate and Applied Meteorology. 1984;**23**:317-319

[24] Christiansen DE, Markstrom SL, Hay LE. Impacts of climate change on the growing season in the United State. Earth Interaction. 2011;**15**

[25] CCL/CLIVAR: Commission for Climatology\Climate Variability

[26] Linderholm HW. Growing season changes in the last century. Agriculture and Forest Meteorology. 2006;**137**:1-14

[27] Chemielewski FM, Rotzer T. Annual and spatial variability of beginning of growing season in Europe in relation to air temperature changes. Climate Research. 2002;**19**:257-264

[28] Deagaetano AT. Recent trends in maximum and minimum temperature threshold exceedences in Northern United States. Journal of Climate. 1996;**9**:1646-1657

[29] Zhai PM, Sun A, Ren F, Liu X, Gao B, Zhang Q. Changes of climate extremes in China. Climate Change. 1999;**42**:203-218

[30] Chen X, Tan Z, Schwartz MD, Xu C. Determining the growing season of land vegetation on the basis of plant phenology and satellite data in Northern China. International Journal of Biometeorology. 2000;**44**:97-101

[31] Dodd DE, Joliffe IT. Early detection of the start of the wet season in semi-arid tropical climates of Western Africa. International Journal of Climatology. 2001;**21**(10):1251-1262

[32] Audu HO, Bibi UM, Garba BG, Ibrahim AK, Hussaini A, Tarki SK, et al. Determination of onset, cessation of rains and hydrological growing season in Dadin Kowa and Gombe for agricultural planning. Nigerian Agricultural Journal. 2019;**50**(1):133-141

[33] Bekele D, Alamirew T, Kebede A, Zeleke G, Assefa M. Analysis of rainfall trend and variability for agricultural water management in Awash River Basin, Ethiopia. Journal of Water and Climate Change. 2017;**8**(1):127-141

[34] Kumar V, Jain SK. Trends in seasonal and annual rainfall and rainy days in Kashmir valley in the last century. Quaternary International. 2010; **212**:64-69

[35] Yue S, Hashino M. Long-term trends of annual and monthly precipitation in Japan. Journal of the American Water Resources Association. 2003; **39**:P587-P596

[36] Partal T, Kahya E. Trend analysis in turkish precipitation data. Hydrological Processes. 2006;**20**:2011-2026

[37] Taxak AK, Murumkar AR, Arya DS. Long term spatial and temporal rainfall trends and homogeneity analysis in Wainganga Basin, Central India. Weather and Climate Extremes. 2014;**4**:50-61

[38] Mengistu D, Bewketa W, Lalb R. Recent spatiotemporal temperature and rainfall variability and trends over the upper blue Nile river basin, Ethiopia. International Journal of Climatology. 2014;**34**:2278-2292

[39] Kassie BT, Rötter RP, Hengsdijk H, Asseng S, Van Ittersum MK,

Kahiluoto H, et al. Climate variability and change in the central rift valley of Ethiopia: Challenges for rainfed crop production. The Journal of Agricultural Science. 2013;**152**:58-74

[40] Wing HC, Senay GB, Singha A. Trends and spatial distribution of annual and seasonal rainfall in Ethiopia. International Journal of Climatology. 1734;**2008**:1723-1734

[41] Woldeamlak B, Conway D. A note on the temporal and spatial variability of rainfall in the drought-prone Amhara region of Ethiopia. International Journal of Climatology. 2007;**1477**:1467-1477

[42] Van de Giesen N, Liebe J, Jung J. Adapting to climate change in the Volta basin West Africa. Current Science. 2010;**98**(8):1033-1037

[43] Rao K, Ndegwa W, Kizito K, Oyoo A. Climate variability and change: Farmer perceptions and understanding of intra-seasonal variability in rainfall and associated risk in semi-arid Kenya. Experimental Agriculture. 2011;**47**(2): 267-291

[44] Sanogo S, Fink AH, Omotosho JA, Redl ABR, Ermert V. Spatio-temporal characteristics of the recent rainfall recovery in West Africa. International Journal of Climatology. 2015;**35**(15): 4589-4605. DOI: 10.1002/joc.4309

[45] Feng X, Porporato A, Rodriguez-Iturbe I. Changes in rainfall seasonality in the tropics. Nature Climate Change. 2013;**3**(9):811-815

[46] Chemura A, Schauberger B, Gornott C. Impacts of climate change on agro-climatic suitability of major food crops in Ghana. PLoS One. 2020. DOI: 10.1371/journal.pone.0229881

[47] Trnka M, Hlavinka P, Semenov MA. Adaptation options for wheat in europe will be limited by increased adverse weather events under climate change. Journal of the Royal Society Interface. 2015;**12**(112). DOI: 10.1098/rsif.2015. 0721

[48] Ceglar A, Zampieri M, Toreti A, Dentener F. Observed northward migration of agro-climate zones in Europe will further accelerate under climate change, AGU 100. Advancing Earth and Science. 2019. DOI: 10.1029/2019EF001178

[49] Hannah L, Roehrdanz PR, Ikegami M, Shepard AV, Shaw MR, Tabor G, et al. Climate change, wine, and conservation. Proceedings of the National Academy of Sciences. 2013;**110**(17):P6907-P6912. DOI: 10.1073/pnas.1210127110

[50] Marx A, Bastrup-Birk A, Louwagie G, Wugt-Larsen F, Biala K, Fussel HM, et al. Terrestrial ecosystems. Soil and Forests. 2017. DOI: 10.1007/978-3-540-88246-6

[51] Olesen JE, Trnka M, Kersebaum KC, Skjelvag AO, Seguin B, Peltonen-Sainio P, et al. Impacts and adaptation of European crop production systems to climate change. European Journal of Agronomy. 2011;**34**(2):96-112

[52] Olesen JE, Niemeyer S, Ceglar A, Roggero PP, Lehtonen H, Schonhart M, et al. Climate change impacts on environmental systems. Agriculture. 2017;**1**:223-243. DOI: 10.1007/978-3-540-88246-6

[53] Park T, Ganguly S, Timmervik H, Euskirchen ES, Higda KA, Karlsen SR, et al. Changes in growing season duration and productivity of northern vegetation inferred from long-term remote sensing data. Environmental Research Letters. 2016;**11**(8). DOI: 10.1088/1748-9326/11/8/084001

[54] King M, Altdorff D, Li P, Galagedara L, Holden J, Unc A. Northward shift of the agricultural climate zone under 21st-century global

climate change. Scientific Reports. 2018;**8**(1):1-10. DOI: 10.1038/s41598-018-26321-8

[55] Waldock C, Dornelas M, Bates AE. Temperature-driven biodiversity change: disentangling space and time. Bioscience. 2018;**68**(11):873-884. DOI: 10.1093/biosci/biy096

[56] Sabziparvar AK. Meteorology and Climatology in Agricultural Science. Hamedan: Bu Ali Sina University; 2019

[57] Houerou HN. Climatic changes and desertification. Secheresse. 1993;**4**(2): 95-111

[58] Goyal RK. sensitivity of evapotranspiration to global warming a case study. Agricultural Water Management. 2004;**69**:1-11

[59] Tao X, Chena H, Xua C, Houa Y, Jiea M. Analysis and prediction of reference evapotranspiration with climate change in Xiangjiang river basin China. Water Science and Engineering. 2015;**8**(4):273-281

[60] Haydarie T, Khoshkho Y. Predicting future changes in reference evapotranspiration in Western Iran based on Rcp release scenarios. Journal of Applied Research in Geographical Sciences. 2018;**19**(53):157-176

[61] Babolhkomi A, Gholami MA, Emadi A. The effect of climate change on reference evapotranspiration in Mazandaran Province. Iranian Journal of Soil Research. 2020;**51**(2):388-401

[62] Ghini R, Hamada E, Bettiol W. Climate change and plant diseases. Scientia Agricola. 2008;**65**:98-107

[63] Anderson P, Cunningham A, Patel N, Morales F, Epstein P, Daszak P. Emerging infectious diseases of plants. pathogen, pollution, climate change and agro technology drivers. Trends in Ecology and Evolution. 2004;**19**:535-544

[64] Chakraborty S, Luck J, Hollaway G, et al. Impacts of global change on diseases of agricultural crops and forest trees. CAB Reviews: Perspectives in Agriculture, Veterinary Science, Nutrition and Natural Resources. 2008;**3**:1-15

[65] Pimentel D, McNair S, Janecka J, et al. Economic and environmental threats of alien plant, animal, and microbe invasions. Agriculture, Ecosystems and Environment. 2001;**84**:1-20

[66] Rosenzweig C, Iglesias A, Yang X, Epstein P, Chivian E. Climate change and extreme weather events; implications for food production, plant diseases, and pests. Global Change and Human Health. 2001;**2**:90-104

[67] Stinner BR, Taylor RAJ, Hammond RB, Purrington FF, McCartney DA, Rodenhouse N, et al. Potential effects of climate change on plant-pest interactions. In: The Potential Effects of Global Climate Change in the United States. 1989 Washington: EPA-230-05-89-053, Appendix C Agriculture Vol. 2

[68] Baker RHA, Sansford CE, Jarvis CH, Cannon RJC, Mac Leod A, Walters KFA. The role of climatic mapping in predicting the potential geographical distribution of non-indigenous pests under current and future climates. Agriculture Ecosystems and Environment. 2000;**82**:57-71

[69] Linda J, Thomson LJ, Macfadyen S, Hoffmann AA. Predicting the effects of climate change on natural enemies of agricultural pests. Biological Control. 2010;**52**:296-306

[70] Nicole E, Heller NE, Zavaleta ES. Biodiversity management in the face of climate change: A review of 22 years of

recommendations. Biological Conservation. 2009;**142**:14-32

[71] Sala OE, Van Vuuren D, Pereira HM, Lodge D, Alder J, Cumming G, et al. Biodiversity across scenarios. In: Millennium Ecosystem Assessment. 2005. pp. 375-408, Chapter 10

[72] Lapola DM, Oyama MD, Nobre CA. Exploring the range of climate biome projections for tropical South America: The role of Co2 fertilization and seasonality. Global Biogeochemical Cycles. 2009;**23**:23

[73] Hoegh-Guldberg O, Mumby PJ, Hooten AJ, Steneck RS, Greenfield P, Gomez E, et al. Coral reefs under rapid climate change and ocean acidification. Science. 2007;**318**:1737-1742

[74] U.N.F.C.C.C. Technologies for Adaptation to Climate Change. U.N.F.C.C.C; 2006. pp. 22-23

[75] I.P.C.C. Climate change 2007. Synthesis Report. 2007

Chapter 6

Climate Smart Crops for Food Security

Saima Rashid, Muhammad Bin Mushtaq, Iqra Farooq and Zulqurnain Khan

Abstract

After ensuring the food security for over 50 years, the green revolution is eventually reaching its biological limits which are very much reflected by the ongoing stagnancy in yield increased over the past few decades. Meeting the increasing food demands due to increasing population is the greatest challenge for today's plant scientists. Changing climatic conditions are posing additional threats to crop growth, productivity and yield. After successfully deploying gene editing to modify simple traits, scientists are now embarked on more ambitious adventures in genomics to combat challenges of food security in the wake of increasing population and climate change adversaries. The chapter outlines use of new technologies in tailoring crops beyond simple traits aiming to harvest the desired diversity lost during domestication and manipulating complex traits, which evolved over evolutionary timescale with special emphasis on the development of climate smart crops.

Keywords: climate change, food security, climate smart crops, breeding, biotechnology

1. Introduction

Climate change including extreme weather and other associated events are representing challenges to agriculture of developing countries and global food security [1]. Crop production is very sensitive towards climate change. This is influenced by long-term trends in precipitation and average temperature, inter-annual climate variability, shocks in certain developmental stages and extreme weather events. Some plants are more tolerant towards certain types of stresses than others, and at each developmental stage, different types of stresses affect different plant species in different ways [2].

By 2050, it is expected that another 2.4 billion people be added to the population of developing countries of the world. Agriculture in developing countries is a key source of employment, but at present more than 20% of the population falls on an average, in the category of food-insecurity [3]. About 75% of the world's poor population is residing in the rural areas, and again agriculture is their ultimate source of earning [4]. Enhancing agricultural productivity and incomes in the small-scale production sector is very important to mitigate poverty and achieve food security, as a key component and driver of economic transformation and development, and within the wider perspective of urbanization and advances in the non-farm sector. It is estimated that globally by 2050, agriculture sector must have to expand by 60% to meet the increasing demand due to continuously increasing human population, and it can only be possible by increasing crop productivity under climate change [5].

2. Temperature shifts under climate change

Change in temperature can occur in different forms like fluctuation in overall average temperature, changes in the day and night temperatures, or changes in time, duration and intensity of extreme cold or hot weather. Generally, plants have been more vulnerable to the elevated temperature during the reproduction as well as grain filling or ripening stages. Response of plants to increasing temperature is species specific and facilitated by photosynthetic activity for the accumulation of plant biomass which control the plant growth, as well as managed by all changes in plant morphology and physiology that occur during all day. All kind of temperature stresses have their different impacts on harvesting time as well as on productivity of the crops. The impact of stress depends upon the sensitivity of every particular species to its developmental stage to the fluctuation in temperature. A different kind of response mechanism is needed to adapt for these effects. Increase in temperature during the growing season of plants caused a high respiration rate which means a low amount of energy left for support and growth of plant. Even an increase of 1 °C in average temperature can cause the reduction of 5-10% in major food crops [6].

3. Climate smart agriculture (CSA)

Climate Smart Agriculture (CSA) is an approach in which technological, strategic and investment conditions are developed to reach sustainable agricultural development for food security under climate change. The extent to which climate change is affecting agricultural systems necessitates ensuring comprehensive consolidation of these effects into national agricultural planning, investments and programs (**Figure 1**). CSA is transforming and reorienting sustainable agricultural systems to support food security under the new realities of climate change [8].

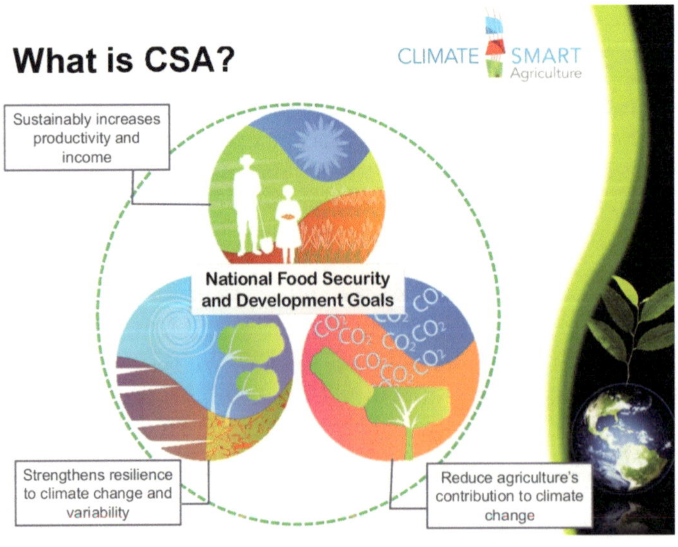

Figure 1.
Climate smart agriculture [7].

4. Objectives of CSA

CSA is striving to increase agricultural productivity in terms of climate smart crops, food security, and farmers' adaptive capacity and lowering greenhouse gas emissions as well [9]. The main objectives of CSA are given below:

- Sustainable increase of Food Security by agricultural productivity
- Building resilience and adapting to climate change
- Developing opportunities for reducing greenhouse gas emissions

4.1 Sustainable increase of food security by agricultural productivity

Agriculture is most important income source of around 75% of the world's poor living in rural areas. To improve the livelihood of this population, growth in the agricultural sector is highly effective that will increase food security in countries with a high percentage of the population dependent on agriculture [10]. Increasing productivity as well as reducing costs is important means of attaining agricultural growth which is possible through increased resource-use efficiency. "Yield gaps "is the difference between the yields that farmers obtain from their farms and the maximum yield potentials of that cultivar and such yield gaps are quite substantial for small farmers in developing countries [11]. Similar is the case with livestock productivity. Reducing these gaps by enhancing the productivity and efficiency of soil, fertilizer, water, livestock feed and other agricultural inputs, much higher returns can be obtained from this sector which in turn will reduce poverty and increase food availability. These same measures can often result in lower greenhouse gas emissions compared with past trends.

4.2 Building resilience and adapting to climate change

In the recent 5th assessment report by the Intergovernmental Panel on Climate Change (IPCC), it has been revealed that the impacts of climatic changes have been observed in different regions of the world. Results showed that drastic impacts of climate changes were more often as compare to the positive ones and underdeveloped countries have shown more vulnerability for the further negative effects of changing climate on agriculture [12]. In medium to long term, when average as well seasonal maximum temperature continuously increase, it led to a high average of rainfall, but these impacts are not distributed evenly as globally wet regions and seasons have higher rainfall as compare to the dry regions and seasons [13]. An increased frequency and intensity of extreme events like drought, high temperature, high rainfall and subsequent floods, have already been observed. Exposure of these increased climatic risks have already been observed in the different parts of the world, these risks put significant threat to potential for increased food security as well as reducing the poverty among the agriculture dependent populations having low-income. Formulation and implementation of effective adaptation strategies required to reduce and even to avoid these drastic impacts of climate change. According to the site-specific impacts of climate change, accompanied with a wide range of agro ecologic variations and farming, fishery system, and livestock, an effective adaption strategy will vary even with in country. For starting the development of an effective site-specific adaptation strategy, multiple potential measures have been identified already. These effect measures include enhancement of reliance of agro-ecosystems, through enhancing the ecosystem services by using landscape approaches as well as principles of agro ecology. Decreasing the risk exposure by building an input supply system, using the diversification of incomes or production, and by extension services for timely and efficient use of inputs, use of stress resistant or tolerant varieties, livestock breeds, use of forestry species and fishes, are some of the examples of that can be used to increase resilience.

4.3 Developing opportunities to reduce greenhouse gases emissions

Agriculture as well as has land use changes become a major source of greenhouse gas emissions, almost a quarter of overall anthropogenic GHG emissions has been produced from agriculture sector. Agriculture mainly contributes in GHS emissions through crop and livestock sectors, also it is the major factor of increasing deforestation as well as degradation of peat land. Under the business-as-usual growth of agriculture, non-CO2 emissions of agriculture sector are expected to increase. But there are multiple ways to reduce the emission of these gases from agricultural sector. Sustainable intensification is one of the main strategies of agricultural mitigation can reduce the emission intensity (e.g. the CO2 eq/unit product) of these gases. But this process includes the application of new techniques that can increase the efficacy of inputs used so that agricultural outputs increased more as compare to the emission increase [14]. Another significant pathway to reduce the emission is high carbon sequestration rate from agriculture sector. Plants as well as soils have ability to remove the carbon dioxide from the atmosphere and store into their biomass, this phenomenon is called carbon sequestration. Carbon sequestration can be performed through increased tree cover in livestock and in crop systems (e.g. Agroforestry) and by reducing the soil disturbance (e.g. reduced tillage). Still this kind of emission reduction might not be permanent as stored CO2 can be released if trees are cut or soil plowed. In spite of following challenges, high carbon sequestration has a significant potential of mitigation, particularly when the agricultural practices which generates the sequestration have also been important role in adaptation of food security. CSA pathways based on impact of climate change on agriculture are given in **Figure 2**.

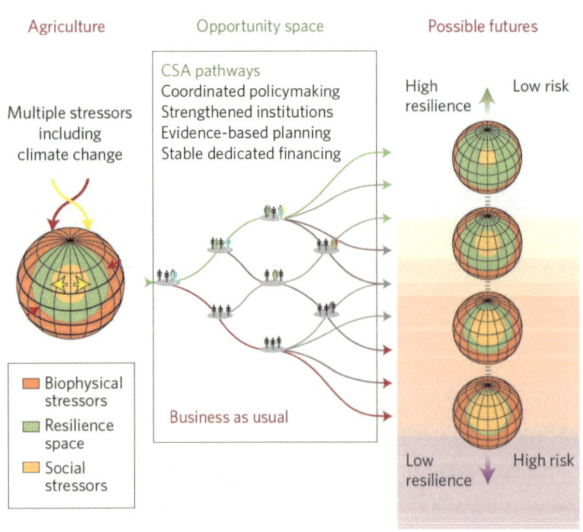

Figure 2.
CSA agriculture pathways [15].

5. Climate-smart food system and supply chain

Climate change effects on crop production have shown to have a strong and consistent global trend, which may have implications for food supply. Because of short-term supply fluctuations, the reliability of whole food systems could be jeopardized because of climate change. At regional scales, however, the potential effect is less obvious, but climate instability and transition are likely to intensify food insecurity in areas that are already vulnerable to hunger and malnutrition. Similarly, it is expected that food access and use will be influenced indirectly by collateral effects on household and individual incomes, and that food consumption will be hampered by a lack of access to drinking water and health problems. The evidence suggests that significant investment in adaptation and mitigation measures is needed to create a "climate-smart food system" that is more resilient to the effects of climate change on food security [16]. Food chain from pre-production to consumption has been elaborated in **Figure 3**.

6. Impacts of climate change on crop production

Due to climatic changes overall crop production system has affected, ultimately causing a challenge to global food security. But the more severe impact of these changes has been observed in underdeveloped countries. Over the next decade it is predicted that billions of people, particularly from underdeveloped countries may encountered with water as well as food scarcity, accompanied with a high risk to the life and health due to climate changes. Developing countries are more prone to the changing climatic conditions as these countries lack in social, financial as well as technological resources, which required facing the climate change [17].

Environmental conditions always cast an impact on either the succession or failure of crops, while the management of stresses caused due to these changes has been part of multidisciplinary studies. Global crop production system has shown continuous susceptibility to the risks of changing climatic conditions. Now farmers have been facing severe challenges than the normally experienced, due to changing climatic conditions. Global climatic conditions became extreme like, warmer temperatures, increased coastal waters, heavy precipitation, and geographical shifts in drought as well as storm patterns [18].

It is estimated that climatic changes may cause a considerable decrease in maize production in southern Africa. It may also cause up to 10% decrease in staple crops of south Asia, including rice while more than 10% decrease in millet and maize production [19]. With a slight increase of 1-3°C in the mean local temperature of some moderate- to high latitude areas, productivity may also be increased, depending on

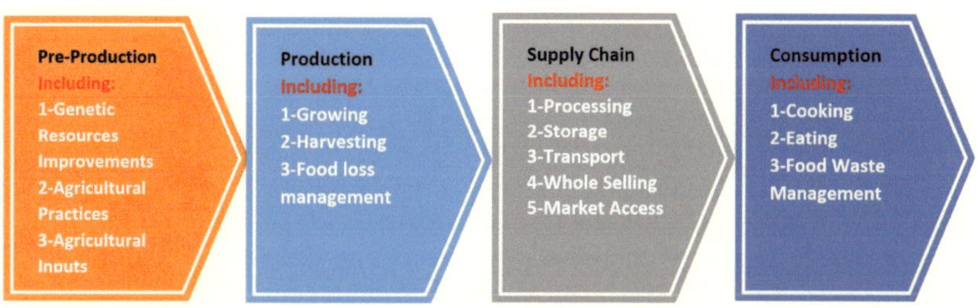

Figure 3.
Food chain from pre-production to consumption.

Event	Potential impact
Day and night temperature increased over most of the land areas as cold periods become shorter and warmer (virtually certain)	High yields in low temperature areas; while in high temperature areas yield reduces; increased outbreaks of different new insect pests as well as pathogens causing notable effect on crop production.
High frequency of precipitation over most areas (very likely)	Crop damages; soil erosion; waterlogged soils making land unable for cultivation
Increased drought affected areas (likely)	Soil erosion and degradation; reduced yields due to crop failure or damage; arable soil loss
Increased tropical cyclone frequency (likely)	Crop damage
Extremely increased level of sea water (excludes tsunami) (likely)	Saline irrigation water, fresh and estuaries water systems; arable land loss.

Table 1.
Impacts of climate change on crop production [22].

the crop. In contrasting, in areas of lower latitudes productivity of crop decreased with the even slightest change in relative temperature range [20].

Unpredictable seasonal as well annual, fluctuations have been observed in crop production system due to the abrupt outbreaks of disease and pest and other extreme events. This require an efficient adaptable management response towards these changing scenarios [21].

Agriculture crop production is facing a number of impacts due to climate change in the components of weather/climate such as temperature, precipitation, cyclones, sea level etc. (**Table 1**).

7. Modern tools of climate smart crop production

Climate change, which includes high temperatures and drought, is projected to have a detrimental effect on plant agronomic conditions as well as soil nutrients, diseases, and pests. As a result, climate-resilient varieties with broad spectrum and long-term tolerance to both biotic and abiotic stresses are required. The new genetic engineering method for crop enhancement is precise genome editing [23]. Climate change has put a pressure on researcher, farmers and scientists working in the field of agriculture to adopt new technologies to cope with the prevailing issues (**Figure 4**). For targeted genome editing in plants, several techniques have been developed, including zinc finger nucleases (ZFNs), TAL effector proteins (TALENs), RNA directed nucleases (RGENs), and CRISPR (clustered regularly interspaced short palindromic repeats)/Cas9 (CRISPR associated protein 9. Both of these approaches depend on the creation of double stranded breaks at particular loci and the activation of the DNA repair system [24].

7.1 Genome editing

Crops with higher yields and greater resistance to abiotic stress are needed to meet the demands of a growing global population and the effect of climate change on agriculture. Traditional crop improvement through genetic recombination or random mutagenesis, on the other hand, is a time-consuming process that cannot keep up with rising crop demand. Genome editing techniques including clustered regularly interspaced short palindromic repeat (CRISPR)/CRISPR-associated protein (CRISPR/Cas) allow for selective alteration of almost any crop genome

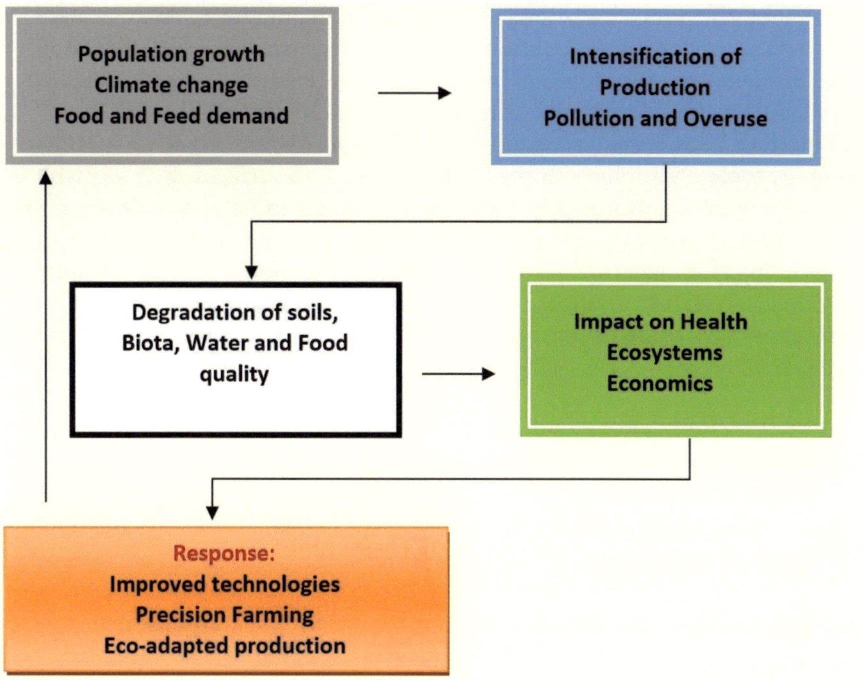

Figure 4.
Impact of climate change on agriculture.

sequence to generate novel variation and speed up breeding efforts. We anticipate a gradual transition away from conventional breeding and toward selective genome editing cycles in crop improvement. Crop enhancement by genome editing is not limited by existing variation or the need to pick alleles through several breeding generations. However, the lack of full reference genomes, a lack of awareness of possible modification goals, and the legal status of edited crops restrict current crop genome editing applications. We believe that overcoming the technological and social barriers to genome editing's implementation will allow this technology to produce a new generation of high-yielding, climate-ready crops [25]. At our lab, we are using different online platforms such as CHOPCHOP, CRISPR-P, MultiTargetor etc, and reagents provided by Addgene, Vectorbuilder, GeneCopoeia, Nootropics Frontline etc. for genome editing in crops.

7.2 Gene silencing

Gene silencing is a method of down regulating (or 'turning off') specific genes via the over expression of RNA sequences (RNAi), which prevents a gene's functional expression. Even though it has been available for many years, it is increasingly being used as a method for shutting off specific genes. Future food protection applications may involve shutting off pathogen attack receptors or stress response elements, which could be extremely useful in the face of climate change. Gene editing is a technique for making precise, targeted changes in genomes at a scale of one or a few nucleotides. Using clustered regularly interspaced short palindromic repeats (CRISPR) and the CAS9 nuclease, transcriptional activator-like effectors' nucleases (TALEN), two alternative systems currently provide state-of-the-art protocols for achieving these small-scale genomic adjustments. Precise genomic modification using CRISPR has been likened to a 'find and replace' function [26]. To precision edit genomes, TALENS employs a nuclease system based on the fusion of transcription activator-like effectors with target DNA binding domains and an endonuclease cleavage domain. Variable DNA binding domain sequences, like CRISPR/CAS9, enable different genomic targets to be addressed. In rice and wheat, the TALENS system has been effective in conferring powdery mildew resistance [27].

To monitor plant responses or stimulate pathogen resistance, RNA spraying technology topically applies complex synthetic RNA to surfaces, such as plant leaves. RNA spraying technology is known to be under investigation by a number of agricultural biotechnology firms. Since there is no alteration to the plant genome, RNA spraying eliminates the need for genetic modification in such applications. Instead, plant cells take up the sprayed synthetic RNA, temporarily silencing specific genes before the effect wears off, which can take anywhere from a few days to three months [28].

8. Use of genetic resources for climate smart crop

The sustainable use of plant genetic resources can help in adapting and mitigating the effects of climate change.

8.1 Genetic resources for climate change adaptation: a sustainable use

The sustainable utilization of plant genetic assets includes evaluation of genetic traits; identification of desirable traits; plant breeding, including epigenomics; variations in crop production; advancement and commercialization of hybrids; sustainable seed production and supply chain system; and establishment of new business sectors for the distribution of local varieties and related products. These exercises can play a key role to address the effects of climate change on sustainable crop production.

The sustainable utilization of plant genetic assets includes evaluation of genetic traits; identification of desirable traits; plant breeding, including epigenomics; variations in crop production; advancement and commercialization of hybrids; sustainable seed production and supply chain system; and establishment of new business sectors for the distribution of local varieties and related products. These exercises can play a key role to address the effects of climate change on sustainable crop production.

In their local production environments, farmer varieties and landraces are well adapted to current conditions and proved to be a successful source for adaptive genes in crop improvement [29]. However, they may lose this adaptation in the changing climatic conditions [30]. It may not be a practical solution to introduce more suitable crop varieties from elsewhere [31]. For this purpose, the only viable solution may be the breeding of new varieties. More genetic vulnerability renders crop potentially more susceptible to the impact of climate change. By incorporating novel traits into cultivars, this genetic vulnerability may be reduced. These novel traits are often found in wild relatives of the crops [32]. Pre-breeding is a source of introduction novel alleles from wild cultivars into crop varieties [33]. In this technique, intermediate materials are generated that are used as parents in plant breeding. Diversity and geographical locations of crop wild relatives and landraces can remotely be determined by using predictive characterization tools based on eco-geographic and climate data [34]. This method is known as the Focused Identification of Germplasm Strategy. In the changing climatic conditions, it would be a challenge for breeders and geneticists to increase the yields of major food crops or even maintain them that will definitely depend on their ability to improve local varieties by introducing adaptive traits through breeding [35]. It is also of much importance for the farmer community to actively participate in the varietal development process to increase the adoption rates of new varieties [36]. Use of a wide range of methodologies is required to develop crop varieties that are tolerant toward climate change induced stresses [37]. These methodologies include induced mutations, biotechnological applications, including cell and tissue biology, marker assisted selection and genetic engineering; and novel plant breeding techniques, including genome editing procedures. With the help of such techniques, Scuba Rice, a flood-tolerant variety of rice was developed for the flood prone areas, such as those found in Bangladesh, India and the Philippines. This is an excellent example of the successful breeding of a crop variety that supports climate-smart agriculture. In these areas where such extreme challenges are faced by crops, adoption of climate-ready varieties is expected to increase because of climate change. There are many neglected and underutilized edible plant species that are resilient and adapted to marginal areas [38]. For example, Moringa (Moringa oleifera), Yam bean (Pachyrhizuserosus) and Bambara groundnut (Vigna subterranea) etc. It would be strategically important to replace staple crops such as maize, with drought-resistant crops, such as cassava and millets in drought-prone regions of the world. However, this climate-smart agricultural adaptation strategy would only be possible if farmers are willing to adopt these new crops. Farmers can only get benefit from this strategy if the seed and planting materials of such crops are available in right quantity and quality and at an acceptable cost. The effectively availability of such resources is much important for these diverse crops and crop varieties to contribute to climate change adaptation and sustain rural livelihoods [39]. Variety approval and release procedures, seed production, quality control, and its marketing are important components of seed delivery systems. These systems usually fall under national and international policies and regulations, that involve diverse actors, such as government authorities, community-level cooperatives, private firms, input dealers, and contracted growers.

8.2 Genetic resources for climate change mitigation: a sustainable use

As compared to climate change mitigation, the contribution of plant genetic resources to climate adaptation process is more result oriented. However, to mitigate climate change, a number of strategies can improve the sequestration of greenhouse gases. More cultivation of C4 plants may be one of these strategies to maintain or increase carbon content in plants, such as maize, sugarcane, millets and sorghum [40]. It is found through various studies that increased carbon sequestration capacity through improved photosynthesis, is a heritable and it can be improved through conventional breeding [41].

Through breeding, improved varieties have been developed that are more productive and sequester more carbon. Legumes crops including pulses such as lentil, garden pea, chickpea, pigeon pea, groundnut etc., have diverse nitrogen-fixing capacity. Improved nitrogen fixation is also correlated with increased carbon sequestration, that's why cultivation of pulses and other legumes would provide additional support for the mitigation of climate change [42].

9. Climate smart crop production: practices and technologies

9.1 Sustainable crop production intensification

Crop production system has continuously been evolving from beginning of domestication of different crop species, almost 10,000 years ago. Crop production has improved and still improving through different means including, varietal selection, improved irrigation and crop planting methods, efficient use of cropping patterns and fertilizers, using the wild plants and wild relatives. In recent times crop production has relatively enhanced significantly which ultimately provides more food for a continuously increasing global population.

Green revolution has been a best and well documented example of improvement in crop production which revolutionized the crop production system almost in all developing countries during 1960s. Planting high-yielding varieties of crop and also using chemical and improved irrigation methods were the main components of green revolution. In the result of this, production of cereal food crops was increased over 2.2 billion tonnes from 800 million tonnes during the period from 1961 to 2000. It is estimated that almost one population of one billion was saved from famine due to this green revolution, but it costs a high price in the long run. Intensive cropping for several decades has caused a loss of fertility of agricultural soils, ground water depletion; induce resistance in pests, decreased biodiversity as well as air, soil and water pollution. But now this paradigm should be shifted to a new, as intensive cropping systems has not been sustainable and this is what Save and Grow – i.e. sustainable crop production intensification – is about [43].

It means a productive agricultural system not only conserve but also enhances the natural resources using an ecosystem approach which exploits the natural biological processes and inputs. This system not only reduces the negative impact on our environment but also enhances the flow of ecosystem services as well as natural capital. SCPI has also been contributing in the increasing of flexibility of system which is a critical factor, particularly under the aspect of climate change. SCPI can be achieved using better farming practices which are based on the improved efficiencies and well managed biological processes. It has been based on the agricultural production systems as well as management practices which include:

- maintaining the soil health to increase soil-related ecosystem services as well as crop nutrition.

- cultivate a diverse range of species and varieties with associations, rotations, and sequences.

- use of quality planting materials and seeds of high-yielding, and well adapted, varieties.

- integrated management of pest, diseases, and weeds; and

- efficient use of water.

SCPI has climate smart approaches and practices of crop production. Sustainable crop production systems have presupposed to address the vulnerabilities as well as risks caused by the climate changes. CSA has same purpose of achieving food security as the sustainable agriculture, using its own perspective of climate change. From crop production to preparation of land, crop planting and harvesting are basic parts of a farming system which ultimately form a broader agro-ecosystem and landscape. An actual crop is only one part of this agro-ecosystem. But crops may also be the integral part of other production systems like, agroforestry, rice-fish system and integrated crop livestock system. Other parts of agro-ecosystems include soil, biodiversity as well as ecosystem services.

9.2 Integrated pest management

Climate changes have been affecting the spread as well as the formation of different types of disease pests, and weeds. This phenomenon has a large consequence of change in the distribution as well as health of the naturally occurring plants, natural predators, hosts and adaptive variations in agricultural management. With an increase of globalization of trade as well as germplasm exchange, following changes present the pest control with new challenges. Integrated pest management (IPM), an ecosystem approach used for crop production as well as crop protection. This technique has based on the considerations of all possible pest control techniques. IPM considers the use of all possible and appropriate means, to prevent the development of pest population ultimately maintaining the levels of pesticide to economically justifiable limit. Thus, decreasing the risks to human health as well as to the environment through minimize agricultural ecosystem disturbance. Making comprehensive decisions at the field level have been essential for effective IPM [44].

9.3 Sustainable soil management through conservative agriculture

Conservative agriculture is technique which involves the covering of maintained land, reduced soil disturbance, and diversifying crop production. Even though conservation agriculture approach was developed to minimize soil erosion as well as to restore the degraded soils, but it also provides strategic initial base point for adaption against climate changes. Conservative agriculture focusses on the reproducing most stable soil ecosystem which can be attain in any agricultural ecosystem to minimize the dependence of producer on the external inputs to full fill the plant nutritional requirements and pest control (**Figure 5**). BY covering the soil, loss of soil moisture can be minimized; soil temperature can be stabilized, low erosion by water and wind, restoration of the soil carbon through plant debris breakdown and also provide the food material for beneficial soil organism. Using crop rotation and diversification technique disease and pest population will be minimized and soil nutritional value increase. Populations of different soil dwelling animals like earthworms, millipedes, and mites can be flourishing by avoiding mechanical tillage. These micro faunae will take over tillage and improve soil structure by building soil porosity. Conservation agriculture includes the surface organic matter. Soil aggregate improved through the excrement of these soil organisms, while worms create vertical channels which help in the removal of excess water. Soil micro fauna introduce their organic matter which helps in improvement of soil organic quality, structure and capacity to store water ultimately helping to survive under longer drought periods. Conservation agriculture system has ability to mitigate the climate change as untreated soil can work as carbon sink by storing and sequestering carbon. Untreated soil may also reduce the quantity of agriculture required to produce crops, ultimately reducing the fuel consumption [45].

9.4 Sustainable land management SLM

Both public as well as private benefits can be obtained through SLM innovations, making them a potential tool of finding the 'win-win' solutions for poverty, environmental issues and food scarcity. Farmers as a private beneficiary of SLM will get an increased productivity, lower costs, better production stability through growing as well as conserving natural capital (like water resources, soil organic matter, and different types of biodiversity). Through SLM practices soil fertility improved by using large quantity of biomass, reducing the soil disruption, conservation of water and soil, an increased activity as well as diversity of soil fauna, and supporting the

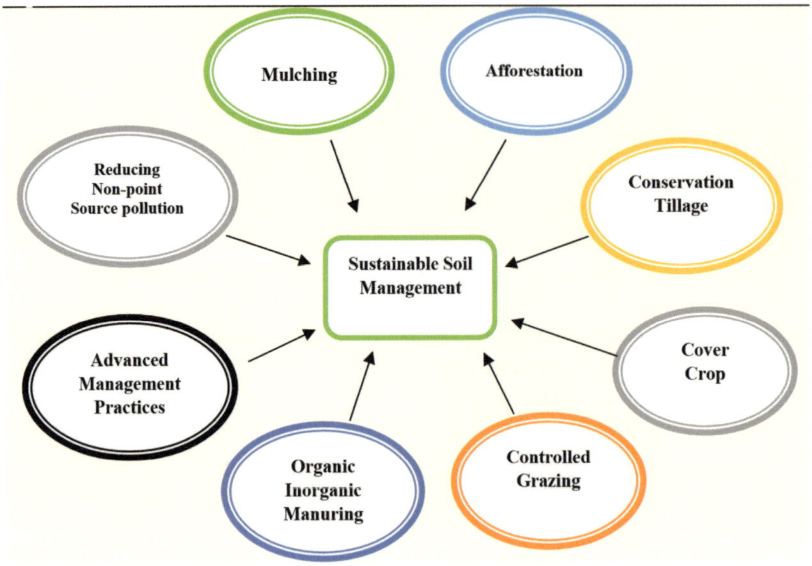

Figure 5.
Sustainable soil management.

elemental cycling mechanism. All this led to improved plant nutritional quality, high water retention ability, and improvement in soil structure contributing to increased yields as well as high resilience, ultimately resulting in improved food as well as livelihood [46].

9.5 Improved water management system

Loss of water can be countered and an improved water management can be achieved by the means of water and soil conservation; either by reduce irrigation which helped to maximize the yield per volume of water used; or through using more efficient irrigation technologies that can minimize the unproductive water loss through evaporation. Buy to attain a high irrigation efficiency and addition energy costs also required, because expansion of irrigation should have to be accompanied by the precise energy technologies (e.g. solar pumps). Strategy development and decision making for the water management and control should be accompanied with the water balance analysis, as for understanding of the impact of changes in water usage in agriculture on the water cycle, a precise assessment of water balance is required for both filed as well as catchment levels. But in upstream areas, introduction of rainwater harvesting technique on a large scale could adversely affects the downstream water users by affecting the groundwater recharge and flux.

9.6 Agro-ecosystem based cropping system approach

Changing climate, sustainable crop production and mitigation in agriculture are linked with each other. The management of ago-ecosystems for production of food, fodder and fuel as well as for management for adaption and mitigation to the changing climate have same fundamental principles and can also work together to attain the same goal: by ensuring the availability of enough, nutritious food for present as well as for future. A resilient ecosystem required for adaption and mitigation to changing climate as well as for crop production, this can be attain using practices and approaches basically based on the ecosystem services and sustainable management of biodiversity (**Figure 6**).

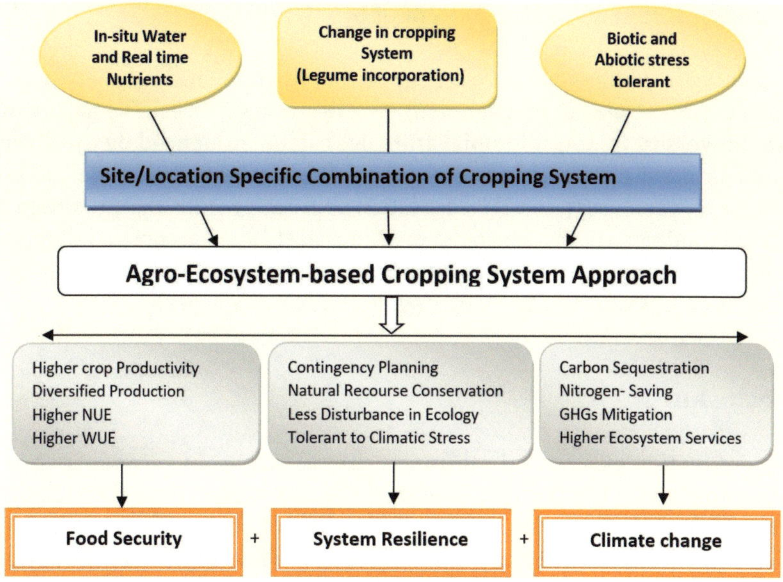

Figure 6.
Agro-ecosystem-based cropping system approach.

Climate smart crop production system is same as the sustainable crop production system as both concerned with climate change. Different opportunities for adaption to climate change and mitigation through contribution to the maintenance and delivery of different public goods like clean water, flood protection, carbon sequestration, ground water recharge and landscape amenity has been provided by sustainable agriculture system. Sustainable agriculture system has been less vulnerable to the stresses and shocks. Productive and sustainable agriculture systems make of the best crop varieties, livestock breeds as well as their biodiversity, agroecological and agronomic management [47].

The drastic impact of climatic changes on crop productivity has already been felt by agriculture sector. For example, in India, production of rice has been decreased 23% during the period of 2001-2002 due to water scarcity [48]. In Indonesia, about 1 344 million tonnes production of rice has been lost due to flooding [49]. While in Mississippi state of the USA, an estimated loss of up to US$ 8 billion were recorded due to flooding before the harvest season in 2008 [50].

For the security of future food production, crop production system needs to be adopted and mitigated the climate changes. To contrast the impacts of climate change, a better understanding of biological processes (below and above ground) which are involved in farm management practices, is needed. For this purpose, ecosystem management should integrate the different measures for building the resilience and mitigating risk in agriculture. All these elements have become critical under the changing climatic conditions. Biodiversity is essential to maintain the key functions of ecosystem (its structure and process) and to provide vital ecosystem services. It's a significant regulator of agro-ecosystem functions, not only due to its impact on production, but also for filling a variety of needs of the farmers as well as society at large. Biodiversity not only can increase the resilience of agro-ecosystems, but also act as means of risk reducing and adapting to the climate change. Agro-ecosystem managers, including the farmer, can enhance, build upon and manage the essential ecosystem services which have been provided by the biodiversity their efforts for a sustainable agricultural production system.

9.7 The conservation and enhancement of biodiversity

Sustainable farming practices have support both above and below ground cropping systems as well as management of ecosystem services. The nature of associated diversity (plant, microbial animals) can be influenced by the diversity and composition of planned biodiversity (e.g. selected crops) ultimately affecting delivery of ecosystem services. An ecosystem approach means that, to integrate the planned biodiversity that has been maintained through associated diversity e.g. more soil coverage and perennial cultivation, high on-farm plant diversity throughout the agro-ecosystem (e.g. resistance against noxious species).

10. Conclusion

Crop production plays a vital role under climate change by providing opportunities in adapting and mitigating the effects of climate change. Both the principles of sustainable crop production and the approaches for climate change adaptation and mitigation are in line. Climate-smart agriculture actually moves the agriculture from an unstable system towards a more efficient, resilient and sustainable system with the help of naturally auto-control mechanisms. Practices and approaches of climate smart crop production can be utilized by farmers, but the implementation of climate change adaptation and mitigation options not only rely on purely technical basis, but they also depend on social support from the population involved. It is very important to facilitate the farmers by giving the opportunities that are sustained by research institutions and policy. Well-built agricultural policies and research institutions at country level are crucial to counteract the effects of climate change in agricultural production systems and generating the income of the rural population, especially in developing countries. Strong government commitment is a dire need of the moment to formulate or adapt agricultural policies to overcome or minimize the impacts of climate change on crop production. Climate-smart systems are not only important in responding to changing climates especially to the increased unpredictability but can also contribute to mitigate any further change in the climate, thus making these systems more efficient, sustainable and productive. A production system can only be a climate smart, if it is productive and sustainable at the same time.

Modern technological tools such as gene editing, gene silencing and DNA sequencing have revolutionized crop improvement programs in terms of production. Information can be revealed that how plant responses towards stress by using genomics tools and this information can be translated to climate resilient crop. With the help of genomics, molecular markers that are linked to important agronomic traits can be identified; thus, helping to improve crop varieties in terms of quality production, stress tolerance and disease resistance. All these technologies will help to make the world more food secured.

An integrated approach is required to face the challenges of food security under climate change from global to local level as well as from research to policies and investment level. The whole agricultural sector can be shifted onto climate smart agriculture pathways with right policies, practices and investments. It will increase the food security by decreasing the impacts of climate change to global food security on long term.

Author details

Saima Rashid, Muhammad Bin Mushtaq, Iqra Farooq and Zulqurnain Khan*
Institute of Plant Breeding and Biotechnology, MNS University of Agriculture, Multan, Pakistan

*Address all correspondence to: zulqurnain.khan@mnsuam.edu.pk; ranazulqurnain@yahoo.com

References

[1] Neufeldt, H., Jahn, M., Campbell, B. M., Beddington, J. R., DeClerck, F., De Pinto, A., et al. (2013) Beyond climate-smart agriculture: toward safe operating spaces for global food systems, 2(1), 1-1.

[2] Rosenstock TS, Lamanna C, Chesterman S, Bell P, Arslan A, Richards M, Rioux J, Akinleye AO, Champalle C, Cheng Z, Corner-Dolloff C, Dohn J, English W, Eyrich AS, Girvetz EH, Kerr A, Lizarazo M, Madalinska A, McFatridge S, Morris KS, Namoi N, Poultouchidou N, Ravina da Silva M, Rayess S, Ström H, Tully KL, Zhou W. 2016. The scientific basis of climate-smart agriculture: A systematic review protocol. CCAFS Working Paper no. 138. Copenhagen, Denmark: CGIAR Research Program on Climate Change, Agriculture and Food Security (CCAFS).

[3] Wheeler, T. and von Braun, J. Climate change impacts on global food security. Science 341, 508-513 (2013).

[4] Rural Poverty Report 2011 (International Fund for Agricultural Development, 2011).

[5] Alexandratos, N. and Bruinsma, J. World Agriculture Towards 2030/2050: The 2012 Revision ESA Working paper No. 12-03 (FAO, 2012).

[6] Allara, M., Kugbei, S., Dusunceli, F. and Gbehounou, G. 2012. Coping with changes in cropping systems: plant pests and seeds. FAO/OECD Workshop on Building Resilience for Adaptation to Climate Change in the Agriculture Sector (23-24 April 2012).

[7] Presentation by Irina Papuso and Jimly Faraby, Seminar on Climate Change and Risk Management, May 6, 2013.

[8] FAO 2013. Climate Smart Agriculture-Sourcebook

[9] Campbell, B.M., Thornton, P., Zougmoré, R., van Asten, P., Lipper, L., 2014. Sustainable intensification: what is its role in climate smart agriculture? Curr. Opin. Environ. Sustain. 8, 39-43.

[10] (World Bank, World Development Report. 2008)

[11] (FAO, The State of Food and Agriculture. 2014)

[12] IPCC Summary for Policymakers. IPCC Climate Change 2014: Impacts, Adaptation, and Vulnerability. Part A: Global and Sectoral Aspects, eds Field, C. B. et al. Cambridge Univ. Press, 2014

[13] (Porter, J. R. et al. in Climate Change 2014: Impacts, Adaptation, and Vulnerability. Part A: Global and Sectoral Aspects, eds Field, C. B. et al. 485-533. IPCC, Cambridge Univ. Press, 2014

[14] Smith, P. et al. in Climate Change 2014: Mitigation of Climate Change Ch. 11. IPCC, Cambridge Univ. Press, 2014

[15] IPCC Summary for Policymakers Climate Change 2014: Impacts, Adaptation, and Vulnerability. Part A: Global and Sectoral Aspects (eds Field, C. B. et al.) (Cambridge Univ. Press, 2014).

[16] Wheeler, Tim, and Joachim Von Braun. 2013. "Climate Change Impacts on Global Food Security." Science 341 (6145): 508-513.

[17] United Nations Framework Convention on Climate Change (UNFCCC). 2007. Climate change: impacts, vulnerabilitiesand adaptation in developing countries (Available at http://unfccc.int/resource/docs/publications/impacts.pdf)

[18] IPCC. 2012. Summary for policy makers. In C.B. Field, V. Barros, T.F.

Stocker, D. Qin, D.J. Dokken, K.L. Ebi, M.D. Mastrandrea, K.J. Mach, G-K. Plattner, S.K. Allen, M. Tignor and P.M. Midgley, eds. Managing the risks of extreme events and disasters to advance climate change adaptation, pp. 1-19. A special report of Working Groups I and II of the IPCC. Cambridge, UK, Cambridge University Press.

[19] Lobell, D.B., Burke, M.B., Tebaldi, C., Mastrandrea, M.D., Falcon, W.P. and Naylor, R.L. 2008. Prioritizing climate change adaptation needs for food security in 2030. Science, 319 (5863): 607-610.

[20] Intergovernmental Panel on Climate Change (IPCC). 2007. Technical Summary. In Climate change 2007: mitigation. Contribution of Working Group III to the Fourth Assessment Report of the IPCC. Cambridge, UK, and New York, USA, Cambridge University Press.

[21] FAO-PAR (Platform for Agrobiodiversity). 2011. Biodiversity for food and agriculture: contributing to food security and sustainability in a changing world. Rome (Available at http://agrobiodiversityplatform.org/files/2011/04/PAR-FAO-book_lr.pdf)

[22] FAO. 2008a. Climate change adaptation and mitigation in the food and agriculture sector. High-level conference on food security-the challenges of climate change and bioenergy (Available at ftp://ftp.fao.org/docrep/fao/meeting/013/ai782e.pdf)

[23] Tripathi, Leena, Valentine Otang Ntui, and Jaindra Nath Tripathi. 2019. "Application of Genetic Modification and Genome Editing for Developing Climate-Smart Banana." Food and Energy Security 8 (4): e00168. doi:10.1002/fes3.168.

[24] Weinthal, Dan M., and Filiz Gürel. 2016. "Plant Genome Editing and Its Applications in Cereals." Genetic Engineering: An Insight into the Strategies and Applications. InTech, UK, 63-73.

[25] Scheben, Armin, Felix Wolter, Jacqueline Batley, Holger Puchta, and David Edwards. 2017. "Towards CRISPR/Cas Crops-Bringing Together Genomics and Genome Editing." New Phytologist 216 (3): 682-698.

[26] Ledford, Heidi. 2015. "CRISPR, the Disruptor." Nature News 522 (7554): 20.

[27] Zhang, Guojie, Cai Li, Qiye Li, Bo Li, Denis M. Larkin, Chul Lee, Jay F. Storz, Agostinho Antunes, Matthew J. Greenwold, and Robert W. Meredith. 2014. "Comparative Genomics Reveals Insights into Avian Genome Evolution and Adaptation." Science 346 (6215): 1311-1320.

[28] Robinson, Karl E., Elizabeth A. Worrall, and Neena Mitter. 2014. "Double Stranded RNA Expression and Its Topical Application for Non-Transgenic Resistance to Plant Viruses." Journal of Plant Biochemistry and Biotechnology 23 (3): 231-237.

[29] Mba, C., Guimaraes, E.P., & Ghosh, K. 2012. Re-orienting crop improvement for the changing climatic conditions of the 21st century. Agriculture and Food Security, 1:7

[30] Bellon, M. R., Hodson, D., and Hellin, J. 2011. Assessing the vulnerability of traditional maize seed systems in Mexico to climate change. Proceedings of the National Academy of Sciences of the United States of America, 108(33): 13432-13437.

[31] Bellon M.R. & van Etten, J. 2014. Climate change and on-farm conservation of crop landraces in centres of civersity In M. Jackson, B. Ford-Lloyd and M. Parry, eds. Plant genetic resources and climate change, pp. 137-150.

[32] Dwivedi, S.L., Stalker, H.T., Blair, M.W., Bertioli, D.J., Upadhyaya, H., Nielen, S. and Ortiz, R. 2008. Enhancing crop gene pools with beneficial traits using wild relatives. Plant Breeding Reviews, 30: 179-230.

[33] Nass, L.L. and Paterniani, E. 2000. Pre-breeding: a link between genetic resources and maize breeding. Scientia Agricola, 57(3): 581-558

[34] Redden, R. 2013. New approaches for crop genetic adaptation to the abiotic stresses predicted with climate change. Agronomy, 3(2): 419-432.

[35] Jarvis, A., Lane, A. & Hijmans, R. 2008. The effect of climate change on crop wild relatives. Agriculture, Ecosystems and Environment, 126 (1-2): 13-23.

[36] Ashby, J.A. 2009. The impact of participatory plant breeding. In S. Ceccarelli, E.P. Guimaraes, E. Weltzien, eds. Plant breeding and farmer participation, pp. 649-671. Rome, FAO.

[37] Shu, Q.Y. 2009. Induced Plant Mutations in the Genomics Era. Proceedings of an International Symposium on Induced Mutations in Plants, 11-15 August 2008. Plant Breeding and Genetics Section, Joint FAO-IAEA Division. Vienna, International Atomic Energy Agency. Rome, FAO.

[38] Ebert, A.W. 2014. Potential of underutilized traditional vegetables and legume crops to contribute to food and nutritional security, income and more sustainable production systems. Sustainability, 6(1): 319-335.

[39] Westengen, O.T. & Brysting, A.K. 2014. Crop adaptation to climate change in the semi-arid zone in Tanzania: the role of genetic resources and seed systems. Agriculture and Food Security, 3:3

[40] Lara, M.V. & Andreo, C.S. 2011. C4 plants adaptation to high levels of CO2 and to drought environments. In A. Shanker and B. Venkateswarlu, eds. Abiotic Stress in Plants-Mechanisms and Adaptations, pp. 415-428. Rijeka, Croatia, InTech.

[41] El-Sharkawy, M.A. 2016. Prospects of photosynthetic research for increasing agricultural productivity, with emphasis on the tropical C4 Amaranthus and the cassava C3-C4 crops. Photosynthetica, 54(2): 161-184.

[42] Jensen, E.S., Peoples, M.B., Boddey, R.M., Gresshoff, P.M., Hauggaard-Nielsen, H., Alves, B.J.R. and Morrison, M.J. 2012. Legumes for mitigation of climate change and the provision of feedstock for biofuels and biorefineries. A review. Agronomy for Sustainable Development, 32(2): 329-364.

[43] FAO. 2011. Save and grow: a policymaker's guide to the sustainable intensification of smallholder crop production. Rome.

[44] The new IPM paradigm for the modern ages and the growing world population. Journal of Integrated Pest Management, Volume 10, Issue 1, 2019, 12, doi:10.1093/jipm/pmz010

[45] Jarecki, M.K. and Lal, R. 2003. Crop management for soil carbon sequestration. Critical Reviews in Plant Sciences, 22: 471-502.

[46] Branca, Giacomo, Nancy McCarthy, Leslie Lipper, and Maria Christina Jolejole. 2011. "Climate-Smart Agriculture: A Synthesis of Empirical Evidence of Food Security and Mitigation Benefits from Improved Cropland Management." Mitigation of Climate Change in Agriculture Series 3: 1-42.

[47] Beddington, J., Asaduzzaman, M., Clark, M., Fernandez, A., Guillou, M., Jahn, M., Erda, L., Mamo, T., Van Bo, N.,

Nobre, C.A., Scholes, R., Sharma, R. and Wakhungu, J. 2012. Achieving food security in the face of climatechange. Final report from the Commission on Sustainable Agriculture and Climate Change. Copenhagen, CGIAR Research Program on Climate Change, Agriculture and Food Security (CCAFS) (Available at www.ccafs.cgiar.org/commission)

[48] FAOSTAT. 2012. FAO statistical yearbook (Available at http://www.fao.org/docrep/015/i2490e/i2490e00.htm)

[49] Redfern, S.K., Azzu, N. and Binamira, J.S. 2012. Rice in Southeast Asia: facing risks and vulnerabilities to respond to climate change. In Building resilience for adaptation to climate change in the agriculture sector. Proceedingsof a Joint FAO-OECD Workshop. Rome, FAO.

[50] United States Global Change Research Program (USGCRP). 2009. Global climate change impacts in the United States. Karl, T.R., J.M. Melillo, and T.C. Peterson, eds. United States Global Change Research Program. New York, USA, Cambridge University Press.

Chapter 7

Pulses Farming; An Adaptive Strategy to Climate Change in Arid and Semi-Arid Regions: A Case Study of Itigi District in Singida Region, Tanzania

Julius Jackson Nyaombo

Abstract

The study assessed pulses farming as an adaptation to climate change in semi-arid areas. FGD, Household Questionnaires, Key Informant Interview and Observation were used to collect primary data from four study villages; Itigi, Ipande, Damweru and Kitopeni. A total of 106 heads of households were involved in the study through questionnaires and quantitative and qualitative data were analyzed statistically and through content analysis technique respectively. Cereals among maize, millet and sorghum have lost their economic significance among farmers while pulses including chickpea, lentils and beans are potentially currently cultivated by more farmers. Climate change effects such as shifting and shortening of rainfall season and increasing drought have contributed to the decline of cereal crop yields. Specifically shifting of rainfall season and drought interaction vitally influence the decline of cereal crop yields, $F(1, 99) = 198.770$, $p = 0.000$ and the same causes have led to the adoption of chickpea and lentils farming in the study area as shifting of rainfall season was statically significant to the adoption of pulses farming, $X2 = 9.138$, $df = 2$, $p = 0.010$ and drought was, $X2 = 106.000$, $df = 3$, $p = 0.000$ as the pulses are found adoptive drought condition and shortened rainfall season. This has enhanced food and nutrition assurance to farmers, increased household income and farmers' livelihood.

Keywords: pulses farming, adaptive strategy, climate change, arid and semi-arid, Tanzania

1. Introduction

The world today is environmentally challenged by the emerging threats of climate change and variability [1]. The IPCC's Assessment Report 5 has shown the likely climate change impact to have affected greatly agricultural production, particularly in the tropics while enhancing food insecurity and malnutrition in several tropical countries [2, 3]. Tropics, semi-arid and arid regions are considered more vulnerable to climate change effects due to the population nature being exposed to drought and their livelihood source [2]. Furthermore, communities in these regions are largely depending in nature for livelihood, while they are geographically as well socially and economically at disadvantage [4]. Climate change has been affecting agriculture sectors in multiple ways including alteration of agro-ecological zones due to changes in temperature and rainfall, drought stress particularly in areas natured with semi-arid tropics including sub-tropics and flooding [5]. Adaptation strategies for climate change and its associated disasters are crucially important for sustaining the livelihood resilience for communities that are agriculture dependents [6]. Climate change extreme outcomes outlining increases heat and drought during the crops growing season enhances the decline of the top ground biomass and impactful loss of soil nutrients in which impacts largely agricultural production and soil healthy [2]. Adaptation and mitigation strategies differ depending on the local context specifying strategies that are effective to the area regarding the technical, social, natural, financial, physical capital and readiness of the people [6]. In alleviating the severity of climate change impacts adoption of adaptation strategies becomes a core value in agriculture and food production in general [7].

Most rural dwellers depend on agriculture as main the source of economic gains in which the sector is most vulnerable to climate change impacts [7, 8]. Poverty severity among rural dwellers poses difficulties to farmers in such areas in creating resilience to adaptation to climate change impacts [7]. Several adaptation strategies globally have been developed including the growing of new crops adaptive to climate stress, change of sowing time and others [7]. An adaptation to the changing climate impacts the cultivation of pulses globally has been growing and mostly the grown pulses include soybeans, dry pea, lentil and chickpea which are largely and globally grown in semi-arid areas [9]. These pulse crops have been much preferred due to their suitability in responding differently in the growing season's rainfall and temperature patterns as well pulses tend to increases cereal production when grown in rotation [9]. In several regions pulses have been grown in rotation with maize, wheat and rice while in some areas are intercropped with other crops such as marigold, sunflower, mustard, and coriander [10]. Pulses have been assuring global food security and enhancing the affordable attainment of protein among poor rural societies in which the level of protein contained in chickpea is estimated to be the same as the protein offered by meat [11]. It is globally within the context of climate change; pulses have shown a promising role in meeting world protein demand and food security [12]. Also, when intercropped or rotated, pulses have ensured the attainment of food varieties and increased soil productivity through nitrogen fixation [10]. Chickpea are mostly preferred recently due to their ability of the root trait variability in stress tolerance [11]. FAO has considered prioritizing the production of pulses as among strategies in meeting Sustainable Goal 2 through building resilience in agriculture [11]. Under climate change impacts, pulses particularly chickpea and lentils respond positively as an outcome of fertilization effects due to elevated CO_2 related to the lowering of leaf stomata conductance and a decreasing rate of transpiration leading to enhanced water use efficiency [13]. To respond to the impacts of climate change, the adaptation strategies have been developed in different areas including semi-arid areas of Tanzania where one among them is the cultivation of crops which are drought tolerant and those with short maturity period [14]. This study aimed at assessing the rate of pulses specifically chickpea farming adoption as adaptation measures to the impacts of climate change in semi-arid areas.

2. Materials and methods

This study was conducted in Itigi District (**Figure 1**) found in Singida region, Tanzania. Four villages were involved in the study in which chickpea farming has

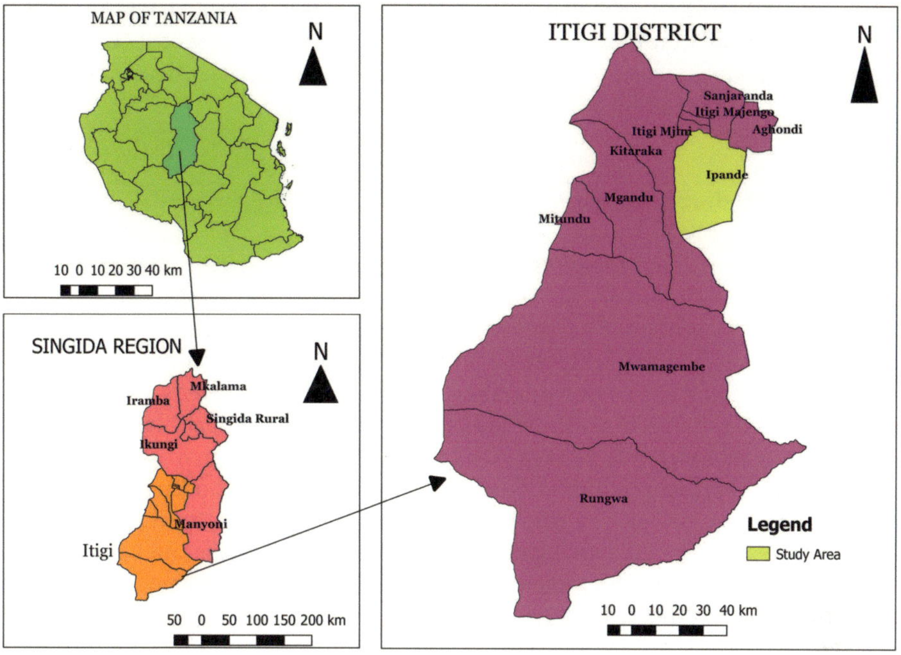

Figure 1.
Map showing the study villages. Source: AMUCTA Geography Unit.

been a dominant crop farming activity and the villages included Ipande, Damweru, Kitopeni and Itigi. Itigi district is found in coordinates 5°42′ S and 34°5′ E while topographically its altitude above the mean sea level is between 1244 m and 1300 m. Itigi is a semi-arid area experiencing low rainfall and shorty rain season which annually it receives rainfall ranging amount of 500–700 mm and normally experiences one drought season after every four years [15]. The minimum temperature is experienced during the night reaching 15°C while the maximum day temperature being 20°C during the rainy season while during the dry season being 30°C [16]. Sandy soils, grayish brown sands and black cracking soils characterize the soil dominance in the area [6]. Itigi is natured with seasonal streams which normally dry after the rainfall seasons. Based on the last census, Itigi district has a total population of 101,364 people males being 50,801 people and females being 50,563 people which when compared to the consecutive census (2002 and 2012) the population is rapidly growing threatening natural resources including land, water and forest resources found in an area [17, 18]. In terms of vegetation the study area is covered with of thickets famously named as Itigi thickets which are endemic to Itigi covering one-fifth of the total district area and miombo woodland with a variety of miombo species which are in danger due to excessive harvesting [19]. Gypsum deposits in the plains of Itigi offer employment to several dwellers particularly during the dry seasons [20].

The study used both quantitative and qualitative methods. Both primary and secondary data were used in the study. One hundred six households were engaged in the study where the sample was attained with the use of Israel sample calculation formulae [21] with the use of a set of prepared structured questionnaires. Two focus group discussions were conducted each having seven members and each included members from two close villages. Key respondents including agricultural officers, village local authorities were engaged through Key Informant Interview while collected information was ensured through transect observation. Data collected were coded in SPSS and analyzed through Chi Square technique to check the relationship between variables at $p > 0.05$ while the General Linear Model (Univariate Analysis of variance) was used to check the effect of two or more independent variables on one dependent variable.

3. Results and discussion

3.1 Socio-economic characteristics of respondents in the study villages

Assessed socio-economic parameters were sex status of respondents, education level, occupation and age of respondents (**Table 1**). The study involved both men and women whereby 63.2% were male and 36% were females which reflect the African cultural families that mostly are headed by men. The education level of respondents was as well reflective of African rural areas' situation in which most respondents had primary education (66%) followed by those who have not attended formal education (26.4%). The study involved more respondents aged above 36 years (74.5%) as are the ones matured enough to recall changes and variations of the climate in the study area and among them 24.5% were aged above 60 years who could easily assimilate the current climatic condition about agricultural activities compared to the past decades. 97% of respondents depend only on agriculture (crop farming and livestock) as a source of income and livelihood while only 3% are employed as local government authorities and other government sectors. During the dry season normally most of the people living in these villages are engaged in small-scale gypsum mining in which is normally becomes a main source of livelihood during the dry season.

3.2 Farming systems and crops grown in the study area

Several farming systems exist in the study villages where 61.3% of respondents are practicing crop farming while agro-pastoralism is practiced by 37.7% of respondents and agroforestry is practiced by 0.9% of respondents. Both cash crops and food crops including cereal and pulses (**Table 2**) are grown in the study area including cereals, root crops, pulses and oilseed crops. Cereals grown include maize and millet/sorghum while sunflower and groundnuts are the grown oil seeds and nuts. On the side of pulses include chickpea, beans, cowpeas and lentils.

Category		Frequency	Percentage (%)
Gender	Male	67	63.2
	Female	39	36.8
Age	18–35	27	25.5
	36–60	53	50
	60+	26	24.5
Education	Informal education	28	26.4
	Primary education	70	66
	Secondary education	7	6.9
	Higher learning	1	0.9

Table 1.
Socio-economic characteristics of respondents.

Category		Frequency	Percentage (%)
Crops are grown in the study villages	Cereal crops	59	55.7
	pulses	47	44.3
	Both cereal and pulses	59	55.7
Farming systems existing in the study village	Crop farming	65	61.3
	Agro pastoralism	40	37.7
	Agroforestry	1	0.9

Table 2.
Farming systems and crops are grown in the study area.

Maize is grown as the main source of food for farmers and the community around while sunflowers, chickpeas and lentils are grown as cash crops for commercial purposes.

In the four study villages (i.e., Itigi, Ipande, Damweru and Kitopeni), farmers have changed the type of crops particularly cereals they have been growing in the previous 10 years to grow cereal crops while some of them have adopted the farming of pulses in their respective farmland by reducing the farmland they have been using to grow cereal crops or they have expanded their farms (**Table 3**). Also, root crops particularly cassava, farmers have been declining their production due to their destruction imposed by wild animals particularly pigs and elephants which are suspected to be from the nearby game reserve.

3.3 Influence of climate change on the change of crops grown in the study area

The study identified impact of climate change has been a great driver to farmers in abandoning the growth of cereal crops and adoption of the growth of pulses particularly chickpea and lentils in which chickpea has gained more potentiality among the farmers due to its adaptive nature to drought. 100% of respondents agreed to have experienced impacts of climate change in their areas including change of rainfall reasons, increased drought and increased temperature as well as reduced raining season. 90.6% of respondents claimed the rainfall season to have changed while the remaining percentage (9.4%) were not aware of this shifting season in which the awareness on the changing of climate season was influenced by the age of respondents ($X^2 = 24.358$, df = 4, p = 0.000) whereby all the respondents claimed to have low awareness of the shifting rainfall season were aged between 18 and 30 years old. Formerly the rainfall season was starting in midst of November ending in the early of May but currently, the rainy season begins at the end of December ending at the end of March and in a few years early April.

Abandoned crops to adopt pulses farming			Adopted pulses		
Crops	Frequency	%	Crops	Frequency	%
Millet/sorghum and maize	48	45.3	Chickpea and lentils	106	100
Cassava	64	60			
Cotton	106	100			

Table 3.
Changed and adopted crops in the study villages.

The period of drought has extended as formerly it was from the late of May to early November while currently it starts from early of April to the midst of December. Also, the occurrence of insect pests was mostly noticed in which was more distractive to crops during the sunny days and when raining no insects were noticed. The trend of insect pest is increasing from year to year and the cereals particularly maize yield loss are more occurring due to these insects mostly army-worms and earwigs in which from 2015 to 2018 cereals production was rapidly declining due to farm crops attach caused by such pest insects. Increasing of temperature is currently experienced during the day and longer than the intensity experienced formerly. Few existing seasonal streams and dams are observed to dry earlier contrary to the formers decades in which they used to last to at least few days after the end of the rainfall season.

The impacts of climate change have significant contribution in the change and adoption of crops grown in the study area (see **Table 2**). 65.1% of respondents claimed to have changed/adopted the farming of pulses due to climate change impacts. Change of rainfall pattern has caused a loss of cereal crop yields for about 45.3% while increased drought is accounted for the loss of 47.2% of cereal crop yields. Shifting of rainfall season has a significant relationship with the declining of cereal crop production, $F (1, 99) = 27.710$, $p = 0.001$. Also, the drought caused a significant decline in the cereal production $F (1, 99) = 21.721$, $p = 0.041$. The interaction of shifting rainfall season and increased drought has significantly caused a loss of cereal crop yields, $F (1, 99) = 198.770$, $p = 0.000$.

The pulses that are mostly cultivated in the study area due to climate change are chickpea and lentils. Chickpea is grown by 100% of respondents. Statistically, the increase of drought has a significant relationship with the adoption of pulses farming in the study villages $X2 = 106.000$, $df = 3$, $p = 0.000$ and the shifting of rainfall season was found significantly to have caused the adoption of pulses farming in the study $X2 = 9.138$, $df = 2$, $p = 0.01$.

3.4 Impact of pulses farming in farmers' livelihood under climate change

Pulses in the study areas have become the main sources of income among community members in the study areas. Farmers have been selling chickpea and lentils to attain other basic needs. Income generated through chickpea has been used to buy enough food adequate to meet the demands of the household annually. Currently, most of the farmers are living in more improved houses (housed built with bricks) compared to formerly used huts. The income is as well used to send children to schools and in attaining their basic needs such as clothes and getting medical services in case of. Though few, some of the farmers are using income, generated through pulses farming to buy motorcycles, tractors for farming and cows for pulling plough.

3.5 Challenges to pulses production in the study areas

Although pulses seem to be alternative crops for semi-arid agriculture under climate change threats, still there are some challenges particularly in the study areas are posing difficulties to attain improvements in pulse crops production. Inaccessibility of improved seeds and lack of awareness of existing improved chickpea and lentil seeds among farmers has been a huge setback in which none of the farmers were found aware of the existence of improved varieties of seeds particularly of chickpea and lentils; therefore, all farmers in the villages are still using traditional seeds and seeds infected by bruchids.

Existing pests in the study area affecting chickpea including Rhizoctonia bataticola (dry root rot), Helicoverpa armigera (Gram pod borer) and collosobruchus chinensis (Bruchids) in which 95% of respondents identified such pests to cause a significant loss of their crop yields. These insect pests seem less responsive to numerous insecticides and pesticides used by farmers in these semi-arid areas. Poor storage facilities favour the harvest and seeds attach by bruchids (Collosobruchus chinensis) which lead to the great loss of harvests in the study villages. 93.6% of respondents claimed seeds loss due to bruchids attaches to their local storage facilities. Furthermore, seeds affected by bruchids are always observed to have been easily attached by other insects and diseases due to soil pathogens particularly when rotated with cereals consecutively in the same farming season.

The crops particularly chickpea and lentils are not included as among crops given first in national policy therefore farmers are not getting agricultural subsidies making difficulties in meeting financial demands for growing chickpea and lentils crops. Among others, minimal agricultural extension services, use of poor farming tools mostly hand hoes for cultivation due to lack of mechanization and a low level of agricultural awareness among farmers still pulls down the efforts of farmers in increasing pulses yields in their farmlands.

Poor farmers' perception of the fertility level of the soil where pulses are grown affects much pulses production. Pulses particularly chickpea is normally grown in vertisols (dark clay soil) in the all study villages which has a low level of phosphorous in which farmers perceive that the vertisols is always fertile as having a dark color making none of the farmers apply fertilizer in their farms but contrary to the demand of the crop which does well when the soil has a high level of phosphorous. Therefore, despite the ability of the pulses in nitrogen fixation, a low level of other soil nutrients among phosphorus and calcium limits the production of pulses in the study areas.

4. Conclusion

Climate change effects including drought, low rainfall, shifting of rainfall season and the increasing temperature still affects greatly agricultural production in terms of quality and quantity in several agro-ecological zones greatly in semi-arid and arid regions. These incidences have been leading to food insecurity and the dangers of malnutrition among rural communities in semi-arid areas. Several adaptation strategies have been developed including local ones and modernized ones such as climate and soil friendly agronomic activities, use of improved seeds and cultivation of short-season crops and drought adoptive crops. Farming of pulses and in particular chickpea is found effective to reduce the impacts of climate change effects particularly in semi-arid areas. This is due to the crop's ability to sustain the harsh climatic condition including drought and shortened rainfall season. Currently improving the adaptive behavior of the crop to resist the effects of increased insect pests as an outcome of climate change is very vital as will enhance crop productivity, ensure farmers availability of food security and nutrition, as well increase farmers household income in semi-arid regions in which its dwellers are more vulnerable to climate change impacts. More studies are needed particularly on insecticides and pesticides which are suitable for such localities. Improvement of other agricultural services including training farmers on better agronomic practices that are climate and soil friendly in attaining sustainable pulses production and soil productivity under climate change particularly in those arid areas and semi-arid areas.

Acknowledgements

The author wishes to thank all who were very cooperative in carrying this work. Community members in the study villages, agricultural officers and local authorities are highly appreciated for their due contribution in the completion of this work. Specially, appreciation and acknowledgement to District Agricultural officers and Professor Amos Enock Majule of Institute of Resources Assessment (University of Dar es Salaam) for their technical support in carrying this study.

Declaration of conflict of interest

The author wishes to declare no conflict of interest to report.

Author details

Julius Jackson Nyaombo
Archbishop Mihayo University College of Tabora (Constituent College of St. Augustine University of Tanzania), Tabora, Tanzania

*Address all correspondence to: juliusnyaombo@gmail.com; nyaombo@amucta.ac.tz

References

[1] Sarker M, Alam K, Gow J. Assessing the determinants of rice farmers' adaptation strategies to climate change in Bangladesh. International Journal of Climate Change Strategies and Management. 2013;**5**(4):382-403. DOI: 10.1108/IJCCSM-06-2012-0033

[2] Dhanya P, Ramachandran A. Farmers' perceptions of climate change and the proposed agriculture adaptation strategies in a semi-arid region of south India. Journal of Integrative Environmental Sciences. 2015. DOI: 10.1080/1943815x.2015.1062031

[3] URT, United Republic of Tanzania. National Adaptation Programme of Action (NAPA). Dar es Salaam: Vice President's Office—Division of Environment; 2007

[4] Islam M, Bray M. Adaptation to climate change in agriculture in Bangladesh: The role of formal institutions. Journal of Environmental Management. 2017;**200**:347-358. DOI: 10.1016/j.jenvmn.2017.05.092

[5] Shetty P, Ayyappan S, Swaminathan M. Climate Change and Sustainable Food Security. National Institute of Advanced Studies. 2013. ISBN: 978-81-87663-76-8

[6] Sarker M, Wu M, Alam G, Shouse R. Life in Riverrine Islands in Bangladesh: Local adaptation strategies of climate vulnerable riverine dwellers for livelihood resilience. Land Use Policy. 2020;**94**:104574. DOI: 10.1016/j.landusepol.2020.104574

[7] Alam G, Alam K, Mushtaq S. Climate change perceptions and local adaptation strategies of hazard-prone rural households in Bangladesh. Climate Risk Management. 2017:52-63. DOI: 10.1016/j.crm.2017.06

[8] Wall D, Plunkett M. Major and Micro Nutrients Advice for Productive Agricultural Crops. Johnstown Castles: Teagasc, Agricultural and Food Development Authority; 2020

[9] Cutforth H, McGinn S, McPhee E, Miller P. Adaptation of pulse crops to the changing climate of the northern Great Plains. Agronomy Journal. 2007;**99**:1684-1699. DOI: 10.2123/agronj2006.0310s

[10] Zafaranieh M. Effects of various combinations of chickpea/safflower intercropping on chickpea yield components. International Journal of Agricultural Innovations and Research. 2014

[11] Considine M, Siddique K, Foyer C. Nature's pulses power; Legumes, food security and climate change. Journal of Experimental Botany. 2017;**68**(8):1815-1818. DOI: 10.1093/jxb/erx099

[12] Bessada S, Barreira J, Oliveira M. Pulses and food security; Dietary protein, digestibility, bioactive and functional properties. Trends in Food and Technology. 2019;**93**:53-68. DOI: 10.1016/j.tifs.2019.08.022

[13] Zeinali E, Hajarpoor A, Soltani A, Faramarz S. Simulating climate change impacts on production of chickpea under water limited conditions. Journal of Agriculture Science Development. 2014;**3**(6):209-217

[14] Kangalawe R, Lyimo J. Climate change, adaptive strategies and rural livelihoods in semi-arid Tanzania. Natural Resources Scientific Research. 2013. DOI: 10.423/nr2013.43034

[15] Benedict E, Majule A. Climate change adaptation; role of local agriculture innovations in semi-arid Tanzania. Journal of Agriculture and Ecology Research International. 2015;**3**(4):147-159: Article no JAERI. 2015.043

[16] Makundi P, Lyimo J. influence of soil management practices on crop production in the changing climate, a voice from farmers; the case of Sanjaranda village in Manyoni District, Tanzania. Research Journal of Agriculture and Environmental Management. 2015;**4**(1):013-022

[17] URT, United Republic of Tanzania. National Sample Census of Agriculture 2002/2003—Regional Report, Singida Region. Dar es Salaam: Ministry of Finance and Economic Affairs, National Bureau of Statistics; 2007

[18] URT, United Republic of Tanzania. Population and Housing Census, Population Distribution by Administrative Areas. Dar es Salaam: Ministry of finance and economic affairs, National Bureau of Statistics; 2013

[19] Makero J, Malimbwi R, Eid T, Zahabu E. Models predicting above and below ground biomas of thicket and associate tree species in Itigi thicket vegetation of Tanzania. Journal of Agriculture, Forestry and Fisheries. 2016;**5**(4):115-125

[20] Makero J, Kashaigili J. Analysis of land cover changes and anthropogenic activities in Itigi thickets in Tanzania. Advances in Remote Sensing. 2016;**5**: 269-283

[21] Israel G. Determining Sample Size. Gainesville, Florida: Institute of Food and Agricultural Sciences, University of Florida; 2012. Available from: http:// edis.ifas.ufl.edu

Chapter 8

Carbon Sequestration in Agroforestry Technologies as a Strategy for Climate Change Mitigation

Lazaro Elibariki Nnko

Abstract

Worldwide agroforestry has been recognized as a potential greenhouse gases mitigation strategy under Kyoto protocol. And this is due to its potential in carbon sequestration. There are several agroforestry technologies with different rate in carbon sequestration. In that respect carbon sequestration can depend on type of technology, climate, time since land use change and previous land use. Our knowledge in this topic from the tropical countries such as Tanzania is how ever very limited. To address this challenge this study was undertaken in Kilombero District where the local community are practicing various agroforestry technologies. The objective of this study was to understand the carbon sequestration in different trees species in agroforestry technologies and also to understand which agroforestry technology provide the greatest benefit in term of carbon sequestration. Ecological survey was conducted and a total of 90 plot engaged in different agroforestry technologies were randomly selected from three villages of different altitudinal range. Pivot table was used in analysis and allometric equation was used for computing biomass and carbon. The result shows that *Mangifera indica* contributed highest carbon over all the tree species encountered during ecological survey with 189.88 Mg C ha^{-1}. Home garden, Mixed intercropping, Parkland and Boundary with 19 514.19 $MgCha^{-1}$, 648.44$MgCha^{-1}$,144.79 $MgCha^{-1}$ and 139.29 Mg C ha^{-1} respectively were the agroforestry technology practiced in Kilombero. From the results Home garden contributed more to carbon sequestration and this study results can be used to inform practitioners and policy makers on the most effective agroforestry technologies for carbon sequestration since agroforestry technologies are expected to play important role as climate change mitigation strategy.

Keywords: agroforestry technology, carbon sequestration, mitigation, climate change, carbon stock

1. Introduction

Agroforestry have been considered as a viable alternative to prevent and mitigate the climate change. Using trees as means of mitigating climate change have been achieved by maintaining the existing once on farm land and or by increasing the plantation of short rotation or increasing fast growing trees on the farm fields [1]. Mitigating climate change through increased carbon sequestration in the soil can particularly become useful especially when addressed in combination with other challenges that affect the people livelihood such as reverting land degradation and ensuring food security [2, 3].

Usually potential in carbon sequestration may occur in different land uses including Agricultural land use and forest land through improved land use management and convention to land use with higher carbon storage in harvested product [4]. On other hand IPCC recognized agroforestry having high potential for sequestering carbon under climate change mitigation strategies [1]. Within agroforestry carbon can be stored above and below ground biomass [5]. In that respect agroforestry practices accumulate more carbon than forest and pasture because they have both forestry and grassland sequestration and storage pattern active [6, 7] but, sequestration potential of agroforestry depend on plant characteristics, tree species, age, crop, biodiversity and tree density. Also depend on structural arrangement, management factor such as fertilization, residual, and harvesting regime. This factors together with agroecological condition as well as soil characteristics in the area where the agroforestry is implemented influence the above and below ground carbon sequestration [5, 8–10]. Jose and Bardhan [8] also pointed that if the agroforestry technologies is to be used for climate change mitigation through carbon sequestration, then better information is required about above and below ground biomass and carbon stock. The aim of this study was to determine the carbon stock in different agroforestry technologies and also to determine which technology has the best potential for long term carbon sequestration.

2. Material and methods

2.1 Description of the study area

2.1.1 Geographical location

The study was conducted in Kilombero District located in Morogoro Region between 08° 00′ 16" South and 36° 04′ 364″ East with elevation ranging from 262 m to 550 m above the sea level, See. Administratively, Kilombero District has five divisions, 19 wards and 46 villages. The district is bounded by Kilosa District in the North, South East by Ulanga district, South west by Iringa region as well as in the West and the East by Lindi Region [11].

2.1.2 Climate

The climate in the study area is marked by wet and dry seasons which are further categorized into four sub seasons, hot wet season from December to March, cool wet season April to June, cool dry season July to August and hot dry season September to November. The area receives between 1 200 and 1 800 mm of rainfall per year and temperatures ranging from 26 to 32°C [12].

2.1.3 Land use

Generally, the land use is categorized as village land, reserved land and general land as defined in the Village Land Act 1999 [13]. Meanwhile, Kilombero is considered as one of the fertile spots in Tanzania. The main economic activities in the area include cash crops, food crops, petty trading and fishing in Kilombero River [11]. Overall cereals of the Coast, such as rice, millet, and maize, are grown widely. Also, vegetables such as sweet potatoes, yams, ground-nuts, melons, pumpkins and cucumbers, and many other excellent articles of food. Tobacco is grown very abundantly, sugar-cane, the castor oil plant, cocoa and cotton, are also cultivated [14].

2.1.4 Population

According to the 2012 census the population of Kilombero was 407 880 with male 202 789 and female 205 091 [15]. This area is currently experiencing a doubling of the human population over the years. It has been demonstrated that within Tanzania population growth results in environmental degradation [16]. This increase of population has resulted from migration from various places for cultivation due to soil suitability for farming and livestock keeping migration.

2.2 Methods

2.2.1 Reconnaissance survey

Pre-visiting was conducted to as well as pre-testing of inventory equipment's. This was also conducted so as to familiarize with the study area and observe the nature of the agroforestry farmland.

2.2.2 Study design

The research design for this study was descriptive and cross-sectional. Descriptive design was involved as it gives thorough information concerning agroforestry technologies. Also, cross-sectional design was chosen as data were collected at once without repetitions.

2.2.3 Data collection

A total of 90 farm land were visited in this study. Data for above ground biomass were species, number of trees, Diameter at breast height (DBH), height and diameter at 0.3 m for cocoa trees [17, 18]. A systematic sampling design was expected to be used but during the field all the encountered farms were less than two hectare and there for all the trees within the farmland were all measured and all the farms were considered as a plot [19]. Height were measure using Suunto hypsometer and diameter using calipers.

2.2.4 Data analysis

2.2.4.1 Biomass and carbon stoking

Information obtained from the biophysical survey mainly inventory data was recorded in Microsoft excel for biomass calculation and carbon stock. Allometric equations were used to convert the field measurement attribute mainly heigh and diameter into stand biomass. Since in agroforestry there is diversification of wood perennial, then general allometric model for Cultivated land herbaceous, mixed tree, intercropping and grain crops (for all tree species) was used for the trees without specific equation. Most of these models have been developed for Tanzanian tree species and vegetation types [20]. Carbon stock was computed as the product of Total Biomass and factor of 0.5 [21].

Below are the species specific allometric equations and general allometric equation for computing the above and below Biomass [20, 22]

Tectona grandis

$$\mathrm{AGB} = 0.3356 \times D^{2.1651} \tag{1}$$

$$BGB = 0.0279 \times D^{1.7430} \times Ht^{0.7689} \tag{2}$$

Theobroma cacao

$$AGB = 0.1208 \times d^{1.98} \tag{3}$$

$$BGB = AGB \times 0.25 \tag{4}$$

Cocos nucifera

$$AGB = 3.7964 \times Ht^{1.8130} \tag{5}$$

$$BGB = 13.5961 \times Ht^{0.6635} \tag{6}$$

Cashew nuts

$$AGB = 0.3152 \times D^{1.7722} \times Ht^{0.5003} \tag{7}$$

$$BGB = AGB \times 0.25 \tag{8}$$

For other tree then general allometric equation for Cultivated land herbaceous, mixed tree, intercropping and grain crops (for all tree species) was used.

$$AGB = 0.051 \times \left(D^2 \times Ht\right)^{0.93} \tag{9}$$

$$BGB = AGB \times 0.25 \tag{10}$$

For carbon estimation then for each tree Total Biomass (TB) which is sum of AGB and BGB was computed by the equation below.

$$Carbon\ Stock = TB \times 0.47 \tag{11}$$

where
ABG = Above ground biomass
BGB = Below ground biomass
D = diameter at Breast height
d = diameter at 0.3 m above the ground
TB = total biomass

3. Result and discussion

3.1 Carbon stock in tree species

During biophysical survey total of 37 tree species from 16 families were found in the study area. There were equal species distribution in all villages. Carbon stock per individual tree were computed to identify the tree species which contribute more to the carbon stock in all the agroforestry technology. Over all the Tree species found during the field survey the *Mangifera indica* species were found to have high carbon stock 189.88 Mg C ha^{-1} followed by *Cocos nucifera* 98.44 Mg C ha^{-1}. *Theobroma cacao* and *Vertex doniana* had 0.0013 $MgCha^{-1}$ and 0.0008 $MgCha^{-1}$ respectively which is lesser in the list of all species studied **Table 1**.

Botanical name	Local name	Family	Biomass (Mg C/ha)	Carbon (MgC /ha)
Mangifera indica	Muembe	Anacardiaceae	379.7534795	189.8767397
Cocos nucifera	Mnazi	Aracaceae	196.8731288	98.43656442
Persea americana	Mparachichi	Lauraceae	146.1686097	73.08430484
Tectona grandis	Mtiki	Lamiaceae	62.30097903	31.15048952
Ficus stuhlmannii	Mkuyu	Moraceae	60.81134344	30.40567172
Citrus sinensis	Mchungwa	Rutaceae	19.48128997	9.740644984
Elaeis guineensis	Mchikichi	Aracaceae	17.87292025	8.936460123
Bauhinia thonningii	Msegese	Fabaceae	15.35726596	7.678632982
Canica papaya	Mpapai	Caricaceae	10.59125521	5.295627607
Milicia excelsa	Mvule	Moraceae	7.686083716	3.843041858
Psidium guajava	Mpera	Myrtaceae	5.621283726	2.810641863
Annona murcata	Mstafeli	Annonaceae	5.227879976	2.613939988
Cedrella odorata	Msedrela	Meliaceae	4.743090445	2.371545222
Senna siamea	Mjohoro pori	Fabaceae	4.73001597	2.365007985
Sorindeia obtusifolia	Mpilipili	Anacardiaceae	3.060067712	1.530033856
Senna Spectabilis	Mjohoro	Fabaceae	2.749675264	1.374837632
Khaya anthotheca	Mkangazi	Meliaceae	2.231995495	1.115997747
Azadirachta indica	Mwarobaini	Meliaceae	2.200888122	1.100444061
Anacardium occidentale	Mkoroshio	Anacardiaceae	1.965061796	0.982530898
Citrus lemon	Mlimao	Rutaceae	1.859403344	0.929701672
Sclerocarya birrea	Mng'ong'o	Anacardiaceae	1.170126345	0.585063172
Citrus reticulata	Mchenza	Rutaceae	1.138515669	0.569257834
Artocarpus heterophyllus	Mfenesi	Moraceae	1.086251846	0.543125923
Averrhoa bilimbi	Mbilimbi	Oxalidaceae	0.938353611	0.469176805
Delonix regia	Mkirismasi	Fabaceae	0.788099951	0.394049976
Olea europeana	Mzaituni	Oleaceae	0.601228169	0.300614084
Cinnamomum zeilanicum	Mdalasini	Lauraceae	0.41427903	0.207139515
Brachystegia boehmi	Myombo	Fabaceae	0.395745497	0.197872749
Citrus autatiifolia	Mndimu	Rutaceae	0.386040111	0.193020056
Syzygium cordatum	Mnyonyo	Myrtaceae	0.359024087	0.179512044
Tamarindus indica	Mkwaju	Fabaceae	0.330284545	0.165142272
Terminalia aemula	Mkulungu	Combretaceae	0.110472679	0.05523634
Syzygium cumini	Mzambarau	Myrtaceae	0.103151497	0.051575749
Saraca asoca	Mwashoki	Myrtaceae	0.102336024	0.051168012
Annona squamosa	Mtopetope	Annonaceae	0.059397123	0.029698561
Theobroma cacao	Mkokoa	Malvaceae	0.002515752	0.001257876
Vitex doniana	Mfuru	Verbenaceae	0.00166612	0.00083306

Table 1.
Biomass accumulated and carbon sequestered in different tree species.

Technology	Total biomass Mg/ha	Total carbon Mg/ha
Boundary	46.43 ± 7.85	23.22 ± 3.92
Home garden	813.09 ± 352.32	406.55 ± 176.16
Mixed intercropping	41.84 ± 10.67	20.92 ± 5.34
Park land	57.92 ± 14.75	28.96 ± 7.38

Table 2.
Average estimated biomass and carbon stock under different agroforestry system.

3.2 Carbon stock in different agroforestry technologies

Also, biomass and carbon stock were computed and presented based on agroforestry technology. Based on agroforestry technology Total biomass average were 46.43 ± 7.85 for Boundary, 813.09 ± 352.32 for Home garden, 41.84 ± 10.67 for mixed intercropping and 57.92 ± 14.75 for Parkland. Carbon stock for agroforestry technologies were also computed and the results shows that Boundary had 23.22 ± 3.92, Home garden 406.55 ± 176.16, Mixed intercropping 20.92 ± 5.34 and Parkland were 28.96 ± 7.38 (**Table 2**)

4. Discussion

4.1 Carbon stock in tree species

From the results all the 37 species obtained during biophysical survey *Mangifera indica* had the highest amount of carbon stock on its biomass 189.88 $MgCha^{-1}$ followed by *Cocos nucifera* with 98.44$MgCha^{-1}$. High amount of carbon in this species may be due to its dominance as a result of high demand of mango tree product as well as palm tree products (fruits and coconut juice) both domestic and local market demand [23]. This cannot be explained only by the total number of appearances of *Mangifera indica* which is 178 because there were other species which appeared mostly than *Mangifera indica* but by the superiority of the Dbh than the other species. Study conducted in Philippines on carbon sequestration revealed that *Mangifera indica* can sequester 100.71 $MgCha^{-1}$. This is lower than the amount obtained in this study. In fact, Brown [24] and Gibbs, [25] reported that Dbh is 95% of the total biomass, and in this study *Mangifera indica* present huge biomass and carbon stock which can be due to its high average diameter and height. Other species like *Theobroma cacao* and *Vertex doniana* had lesser carbon stock due to smaller average diameter and height. Age of the tree and number of occurrences of a tree in the plots also can be used to justify the amount of carbon stored in a particular species [26]. Usually, older tree undergone photosynthetic activities with much longer time compare to younger tree and consequently are absorbing and storing more carbon [26]. The top most tree with highest biomass were tree species used for food (*Cocos nucifera*), fruits (*Mangifera indica and Persea americana*), timber production, (*Tectona grandis*) and one mostly used for shade (*Ficus stuhlmannii*). Variety of species documented and observed during the field display the potential for agroforestry to enhance the resilience for farmer for present and future climate risk. For example, farmer in both villages maintain varieties of trees for timber, fruits, animal fodder to support livestock during drought. Similar study conducted in Kenya shows that majority of small holder farmers maintain trees not only for food support but also for the soil and water conservation [27, 28]. There was variable distribution of tree species on different land uses. High tree species diversity was found in home garden where multipurpose trees for various purpose such as shade, timber, and food are grown. For example, the tree with high frequency in home garden were *Mangifera indica*, *Cocos nucifera, Persea americana, Tectona grandis and Ficus stuhlmannii*. Moreover, Kindt *et al*., [29] pointed out that usually high economic values trees are widely spread in a farm land.

4.2 Carbon stock in agroforestry technologies

A number of studies have shown that agroforestry in tropics have high Carbon stock than any crop field or pasture [5, 30]. From the result home garden leads in the carbon sequestration with 19 514.2 Mg C ha^{-1}. This result is highly influenced by the mixture of component of agroforestry such as cattle's, high occurrence of trees and agricultural crops. In other way home garden have been observed as the potential technology for carbon sequestration due to the fact that it sequesters carbon in biomass as well as in soil, reduce fossils fuel burning by encouraging fuelwood production and reduce pressure on natural forest. More ever in home garden there is no complete removal of biomass [31, 32]. Similar study conducted in India shows that home garden of 12-17 years accumulate 55.8-162MgCha^{-1}[33]. Agroforestry technology study conducted in by Kumar [34] showed that based on species composition, soil and climate generally agroforestry can sequester carbon of 68–228 Mg C ha^{-1}. High amount of carbon stock in home garden maybe the progress of carbon sequestration which was estimated to be 391 000MgCha^{-1} by 2010 and 586 000MgCha^{-1} by 2040 [35]. Mixed intercropping which involves wood perennial and herbaceous crop were observed to store 648.44 Mg C ha^{-1}. High carbon within the mixed intercropping is higher than those of from the sole cropping system due to addition carbon pool in tree and increased carbon soil carbon pool as a result of carbon input from litter
fall and fine root turn over [36] mixed intercropping can store 121–125 Mg C ha^{-1} and this was explained by the higher growth and assimilation rates [37]. Parkland technology was observed to store 144 MgCha^{-1}. Parkland technology unlike mixed intercropping the trees are not arranged in accordance with crops but some little trees are left on the crop land. Study conducted in Guinea shows parkland carbon stock may also range from 22.22–70.8 Mg C ha^{-1} [38]. Parkland agroforestry are very stable (long standing) and high carbon storage [39]. On boundary agroforestry the tree is planted purposely for indicating the boundary or fencing. In this study boundary agroforestry technology stored 139.29
Mg C ha^{-1}. Study show that Boundary planting have positive effect on both soil character tics, crop production and carbon sequestration [30]. Hooda *et al*., [40] indicate that tree boundary and herbaceous crop can have carbon storage ranging from 18.53–116.29 Mg C ha^{-1}. Other study indicates that greater potential of carbon sequestration was found in the boundary plantation of *Populus deltoides* and Eucalyptus hybrid [41]. In this study the carbon stock was found to be higher compared to other studies conducted in various areas. The difference in the carbon stock can be explained by factors such as allometric equation which could be a limitation resulting in large variation in such estimate [42]. Low cutting of trees in the field could also be the source of high mount of carbon. In the study many trees observed were for various purposes such as food, fruits, shade, wind break and for boundary hence maintained for long time hence sequestering high amount of carbon.

5. Conclusion

Agroforestry technology can have important role in climate change mitigation. This study concludes that there are benefit in term of carbon sequestration form the implementation of agroforestry technologies and these are most relevant in the tropical climate. In this study we also found that carbon stock is determined by the number and average trees found in the farm land. The fruits trees were the most abundant trees suggesting multipurpose use and quick economic benefit. Therefore, understanding the drivers of tree selection can help to meet both local food, fuel and global climate regulation needs. Therefore, we recommend the identification of the benefit provided by different species in different agroforestry technologies and realize its economic benefit. This will establish co benefit of on farm carbon stock from economic value of benefit derived from trees.

Author details

Lazaro Elibariki Nnko
Department of Forest and Environmental Economics, Sokoine University of Agriculture, Morogoro, Tanzania

*Address all correspondence to: princekisa77@gmail.com

References

[1] Gupta, A., Dhyani, S. K., Ramnewaj, Handa, A. K., Prasad, R., Alam, B., Rizvi, R. H., Gupta, G., Pandey, K. K. and Jain, A. (2013). Modeling analysis of potential carbon sequestration under existing agroforestry systems in three districts of Indo-gangetic plains in India. Agroforestry system.87(3): 1129-1146

[2] Batjes, N.H. (2004). Estimation of soil carbon gains upon improved management within croplands and grasslands of Africa. Environment, Development and Sustainability. 6:133-143.

[3] Feliciano, D., Ledo, A., Hiller, J. and Nayale, D.R. (2018). Which agroforestry option gives the greatest soil and above ground carbon benefit in different world region. Agriculture Ecosystem and environment.254:117-129

[4] IPCC. (2000). Summary for Policymakers, Special Report on Land Use, Land Use Change and Forestry. Cambridge University Press, Cambridge, UK.

[5] Nair, P. K. R., Kumar, B. M. and Nair, V. D. (2009). Agroforestry as a strategy for carbon sequestration. Journal of Plant Nutrition and Soil Science. 172(1): 10-23.

[6] Udawatta, R.P. and Jose, S. (2011). Carbon sequestration potential of agroforestry practices in temperate North America. In: Carbon sequestration potential of agroforestry Systems: opportunities and challenges. Edited by Kumar, B.M, Nair, P.K.R. Springer, Dordrecht, pp. 17– 42.

[7] Sharrow, S.H. and Ismail, S. (2004). Carbon and nitrogen storage in agroforests, tree plantations, and pastures in western Oregon, USA. Agroforestry System. 60:123-130

[8] Jose, S. and Bardhan, S. (2012). Agroforestry for biomass production and carbon sequestration. Agroforestry. System. 86:105-111

[9] Baah-Acheamfour, M., Chang, S.X., Bork, E.W. and Carlyle, C.N. (2017). The potential of agroforestry to reduce atmospheric greenhouse gases in Canada Insight from pairwise comparisons with traditional agriculture, data gaps and future research. The Forestry Chronicle For. Chron. 93(2):180-189

[10] Marone, D., Poirier, V., Coyea, M., Olivier, A. and Munson, A.D., (2017). Carbon storage in agroforestry systems in the semi-arid zone of Niayes, Senegal. Agroforestry System. 91(5): 941-954

[11] United Republic of Tanzania. (2007). National sample census Security. Food Development, Livestock Office. Presidents Government, Local. Dar es salaam. Tanzania. 322pp.

[12] Balama, C., Augustino, S. and Makonda, F. B. S. (2016). Forestry adjacent household voices on their perceptions and adaptation strategies to climate change in Kilombero District Tanzania. Springer Plus 5(792): 1 – 21.

[13] United Republic of Tanzania (2017). Ramsar Advisory Mission Report Kilombero valley. Government print. Dar es Salaam. Tanzania.77pp.

[14] Bergius, M., Benjaminsen, T., Maganga, F. and Buhaug, H. (2020). Green economy, degradation narratives, and land-use conflicts in Tanzania. World Development.129: 104 – 850.

[15] United Republic of Tanzania (2013). 2012 Population and Housing census. National Bureau of Statistics Ministry of Finance, Dar es Salaam. 244pp.

[16] Madulu, N. F. (2004). Assessment of linkages between population dynamics and environmental change in

Tanzania. African Journal of Environmental Assessment and Management 9: 88 – 102.

[17] Mbobda, T., Bruno, R., Louis, Z., Valery, N. N., Boris, N., Glawdys, M. D. R., Roger, N. L. and Louis-paul, K. B. (2016). Plant Diversity and Carbon Storage Assessment in an African Protected Forest: A Case of the Eastern Part of the Dja Wildlife Reserve in Cameroon. Journal of Plant Sciences 4(5): 95.

[18] Nadège, M. T., Louis, Z., Cédric, C. D., Louis-Paul, K. B., Funwi, F. P., Ingrid, T. T., Clotex, T. V., Flore, N. Y. A., Bruno, T. M. R. and Julliete, N. (2019). Carbon storage potential of cacao agroforestry systems of different age and management intensity. Climate and Development.11(7): 543-554.

[19] Charles, R., Munishi, P. and Nzunda, E. (2014). Agroforestry as a Resilient Strategy in Mitigating Climate Change in Mwanga District, Kilimanjaro, Tanzania. Global Journal of Biology, Agriculture and Health Sciences 3(2):11-17.

[20] Malimbwi. R.E., Eid, t., Chamshama, S. A. O. (2019). Allometric tree Biomass and Volume Models in Tanzania. Department of Forest Mensuration and Management. Morogoro Tanzania.129pp

[21] Shirima, D.D. and Munishi P.K.T. (2013). Aboveground carbon stocks in Kilombero Nature Reserve. Journal of Tanzania Association of foresters.12:74-83.

[22] Malimbwi, R.E. and Mugasha, W.A.(Ed.) (2019). In Proceeding of the conclusive seminar of the infores Project; Morogoro; Tanzania;25-26 September 2019.12-29 pp.

[23] Mitra, A., Biswas, S., Pal, N., Pramanick, P., Datta, U., Biswas, P. and Mitra, A. (2018). Biomass and Stored Carbon in the Above Ground Structures of Coconut tree. International Journal of basic and applied research.8(2):60-65.

[24] Brown, S. (2002). Measuring carbon in forests: current status and future challenges. Environmental Pollution.116(3): 363-372.

[25] Gibbs, H. K., Brown, S., Niles, J.O., Foley, J.A. (2007). Monitoring and Estimating Tropical Forest Carbon Stocks: Making REDD a Reality. Environmental Research Letters 2(4): 1-13.

[26] Janiola, M.D.C and Marin, A.R. (2016). Carbon sequestration potential of fruit trees plantation in southern Philippines. Journal of biodiversity and environmental science.8(5):164-174.

[27] Reppin, S., Kuyah, S., Neergaard. A., Oelofse, M. and Rosenstock, T. S. (2020). Contribution of agroforestry to climate change mitigation and livelihoods in Western Kenya. Agroforest System 94: 203-220.

[28] Faye, M.D., Weber, J.C., Abasse, T.A. (2011). Farmers' preferences for tree functions and species in the West African Sahel. For Trees Livelihoods. 20:113-136.

[29] Kindt, R. and Coe, R. (2005). Tree diversity analysis. A manual and software for common statistical methods for ecological and biodiversity studies. World Agroforestry Centre (ICRAF), Nairobi

[30] Albrecht, A., Kandji, S.T. (2003). Carbon sequestration in tropical agroforestry systems. Agriculture Ecosystem and Environment 99:15-27.

[31] Kumar, B. M. (2006). Carbon sequestrations potential of tropical home garden. In Tropical home garden: A time tested example of sustainable

agroforestry. Edited by Kumar, B.M and Nair P.K.R. Netherlands.185-204.

[32] Gajaseni, J. and Gajaseni, N. (1999). Ecological rationalities of the traditional home garden system in the Chao Phraya Basin, Thailand. Agroforest System. 46: 3 – 23.

[33] Roshetko, M., Delaney, M., Hairiah, K. and Purnomosidhi, P. (2002). Carbon stocks in Indonesian homegarden systems: Can smallholder systems be targeted for increased carbon storage? American Journal of Alternative Agriculture. 17(2): 125 – 137.

[34] Kumar, A. and Sharma, M. P. (2015). Assessment of carbon stocks in forest and its implications on global climate changes. Journal of environmental science 6 (12) : 3548 - 3564

[35] Jose, S. (2009). Agroforestry for ecosystem services and environmental benefits. Agroforestry Systems.76: 1-10.

[36] Oelbermann, M. (2002). Linking carbon inputs to sustainable agriculture in Canadian and Costa Rican agroforestry systems. Ph.D. Thesis. Dept. Land Res. Sci., University of Guelph, Canada, 208 pp.

[37] Peichl, M., Thevathasan, N.V., Gordon, A.M., Huss, J. and Abohassan, R. A (2006). Carbon sequestration potentials in temperate tree-based intercropping systems. Agroforestry systems. 66:243-257.

[38] Luedeling, E. and Neufeldt, H. (2012). Carbon sequestration potential of parkland agroforestry in the Sahel. Climatic Change.115:443-461.

[39] Takimoto, A. (2007). Carbon sequestration potential of agroforestry systems in the West African Sahel: an assessment of biological and socioeconomic feasibility. Unpublished thesis submitted in fulfilments of the requirements for the Degree of Doctor of Philosophy at University of Florida.73-82.

[40] Hooda, N., Gera, M., Andrasko, K., Sathaye, J. and Gupta, M.K. (2007). Community and farm forestry climate mitigation projects: case studies from Uttaranchal, India. Mitigation and Adaptation Strategies for Global Change. 12: 1099-1130.

[41] Murthy, I. K., Gupta, M., Tomar, S., Munsi, M., Tiwari, R., Hegde, G.T. and Ravindranath, N.H. (2013). Carbon sequestration potential in Agroforestry system in India. Earth science and Climate Change.4 (131):1-7

[42] Swai, G., Ndagalasi, H.J., Munishi, P.K.T. and Shirima, D.D. (2014). Carbon stock of Hanang mountain forest reserve, Tanzania: An implication for climate change mitigation. In Proceeding of the second climate change impacts, mitigation and adaptation programme writer's workshop. 20th -21st February 2014, Morogoro. Tanzania.1-9pp.

Chapter 9

Macro Analysis of Climate Change and Agricultural Production in Myanmar

Thida Htoo

Abstract

This study attempts to examine the effect of climate change on agricultural production in Myanmar. The study is based on country level data of Myanmar cereal crops for the time span of 2009 to 2019 obtained from various sources such as FAO STAT and Department of Agriculture and Department of Meteorology. This study is mainly used by multiple regression model to find out the best fit in the model. The research found that a 1°C increase in temperature in the growing period may decrease with production of cereal by 3849347 ton. A 1mm increase in rainfall in the growing period may decrease production of cereal by 5762 ton. The research found that change in temperature is adversely affected on production of cereal crops in Myanmar. The policies aiming to enhance production of cereal crops should focus on adoption of climate change adaptation measures in Myanmar.

Keywords: Temperature, Rainfall, Cereal crops, Adaptation measures, agricultural production

1. Introduction

Agriculture plays a vital role in the Myanmar economy and contributes around (20.9%) of the total GDP and is a source of employment for rural population of the country. The performance of agricultural sector has had large impact on the tendencies of Myanmar's GDP. About 70% of the rural population employs in agricultural sector. Myanmar is a leading and second biggest producer of rice and pulses. Myanmar agriculture not only contributes food grains but also contributes mainly towards the exports. Myanmar has the largest area of agri-ecological land in Southeast Asia and is the largest exporter for beans and pulses in Asia and ranks second after Canada in the world [1]. The improvement in agriculture sector becomes thus important for the betterment of growth of the economy.

The fluctuation of Agricultural production mainly depends on climate change. The climate change is closely associated with food security and poverty of majority of population in the country which mainly depend on agricultural sector [2]. The climate change largely affects Myanmar agriculture and become a challenge for future agriculture production in Myanmar [3] due to the reason that Myanmar is one of the top ten nations prone to climate change and extreme events like drought [4] in central dry zone of Myanmar, consisting of three regions namely the lower Sagaing, Mandalay and Magyay Regions, occupying 13 percent of total land areas and 19 percent of total population of Myanmar [5]. The Central Dry Zone is a strategic area for the agriculture of the country as well as second leading rice producing region in which 22 percent of annual rice production provides for the requirement of the country. In the Central Dry Zone, Shwe Bo and Kyaut Se are the major rice producing areas where rice-based farming system is mainly adopted.

The Dry Zone is particularly at risk from floods and drought [6, 7] and has a serious water scarcity. As a result, Dry Zone becomes more vulnerable to the impact of climate change [8]. Coastal regions are also vulnerable to the impacts of rise in sea level and cyclones [6, 7]. Frequent extreme weather events have caused a decrease in productivity of agriculture and thereby leading to decrease in GDP and household income as well as a decrease in food security [9]. Agriculture and climate change are thus directly associated and have mutual effects. The climate change affects the agricultural production at both micro and macro scales [10].

This study attempts to examine the effect of climate change on agricultural production in Myanmar at macro scale. This study is divided into four parts: overview of Myanmar economy as part I, overview of climate change situation as part II, most common crops and production in Myanmar as part III, data source and empirical model as part IV and the effect of climate change on agricultural production in Myanmar as part V.

2. Overview of Myanmar economy

Myanmar's economic sector can be grouped into agricultural sector, industrial sector and service sector. Agriculture, livestock and fisheries, forestry sectors are considered as Agriculture sector; mining, processing and manufacturing, energy, electrical power and construction sectors are regarded as Industrial sector and finally transportation, communication, fiancé, social and administration, rents and services and trade sectors are grouped as Services sector.

Many efforts were being made by the government in attaining economic growth throughout the country. The government has been formulating and implementing national and regional level economic plans as well as sectoral development plans, aiming to accelerate growth, and to achieve equitable and balanced development all over the country through Framework for Economic and Social Reform (FESR) and Myanmar sustainable Development Plan (MSDP). Moreover, more investment on manufacturing sector were made by the government through providing technical assistance and loans to Small and Medium Enterprises (SMEs), and facilitating vocational training programs to labor force.

The economic growth was significant in 2010, reflecting the growth rate of GDP had been rapidly increasing till 2014–2015. The average growth rate of GDP had decreased to 9.5 percent due to the impact of Cyclone Komen in 2015. The economic growth had fluctuated and just above 5 percent since 2012, due to the reasons that the fluctuation in agricultural production associated with climate change and climate-related natural disasters and lack of capacity to adopt climate change adaptation strategies in response to climate change [11–13]. Over the same period, the share of the industry and service sector had increasing trend while the share of agricultural GDP had decreasing trend over the period between 2000 and 2019.

GDP of the country during the period between 2009 and 2010 and 2018–2019 can be seen in **Table 1**. During the period between 2009 and 2010 and 2014–2015, the economic growth rate was seen as upward trend, indicating an increase in growth rate due to government encouragement of manufacturing sector through providing technical assistance and loans to Small and Medium Enterprises (SMEs), and facilitating vocational training programs to labor force. Starting from 2015 to 2016, the

Year	Agriculture		Industry		Service		GDP (Kyats Million)	Growth rate of GDP
	Value (Kyat Million)	Percent	Value (Kyat Million)	Percent	Value (Kyat Million)	Percent	Value (Kyat Million)	Percent
2009–2010	15214305.5	30.2	19884328.3	39.5	15230298.2	30.3	50328932	5.7
2010–2011	15632203.6	30.3	20120658	39.0	15838569.3	30.7	51591431	5.9
2011–2012	16015275.5	29.1	22234265.6	40.2	16785769.9	30.5	55035311	6.7
2012–2013	16685355.3	27.9	24400089.5	40.8	18718696.1	31.3	59804141	8.7
2013–2014	17065701.6	26.1	26808190.3	41.0	21511938.1	32.9	65385830	9.3
2014–2015	18654719.7	25.3	30452170.9	41.3	24627179.4	33.4	73734070	12.8
2015–2016	19460590.9	24.1	33753224.1	41.8	27535524.9	34.1	80749340	9.5
2016–2017	20551566.4	23.2	37294006.3	42.1	30738765.3	34.7	88584338	9.7
2017–2018	20897983.7	22.1	40188430.1	42.5	33474598.2	35.4	94561012	6.7
2018–2019	21107523.6	20.9	43325969.4	42.9	36559442.8	36.2	100992936	6.8

Source: Ministry of Planning and Finance (2020).

Table 1.
Growth rate of gross domestic product and Sectoral contributions.

value of GDP had decreasing trend with 6.8 percent in 2018–2019 due to government encouragement on private sector in the service sectors as well as reduction in government budget on infrastructure for public sector projects. Though Myanmar economy still depends on agriculture sector, the share of agriculture in GDP is gradually decreasing with increasing share of industry and service in Gross Domestic Product.

3. Overview of climate change in Myanmar

Myanmar has a tropical climate with three seasons: summer, rainy and winter. Summer season starts from March and April, rainy season starts from May to October, and winter season starts from November to February. Seasonal temperatures vary largely in most of the regions/state in the country. The Central Dry Zone is highest in temperature and lowest in mean annual precipitation (500–1,000 mm/year). The maximum temperature has 40–43°C in the hot/dry season to minimum temperature has 10–15°C in the cool/relatively dry season and decrease from 0°C to −1°C in the hilly. The Coastal region, south of the country does not vary in seasonal temperature [14].

The climate trend over last six decades are found as a rise in mean temperature by around 0.08°C each decade; overall rise in rainfall throughout the country; rise in sea level and more extreme weather events such as drought and flood; and the monsoon period become shorter with more intense rainfall [15]. Over last three decade (1981–2010), temperature increased 0.14°C and 0.35°C per decade in coastal and inland regions. During the period between 1981 and 2010, total annual precipitation increased by 157 mm per decade in coastal areas and by 37 mm per decade inland [3].

Between year 2000 to 2010, the climate trend over last decades can be seen as large variation in highest temperature with 1.158°C and lowest temperature with 0.162°C. After year 2010, temperature had been an increasing trend from lowest temperature with 0.467°C to highest temperature with 1.552 [16].

Extreme weather events like droughts and flood are the main cause of natural disaster in Myanmar. The average annual rainfall of Myanmar is about 1630 mm, with half of the country lying within the 1500–2000 mm. Flooding and land slide occurred due to high intensity rainfall in dry zone areas [17].

The new climate change projections for Myanmar reveals a 0.8–2.7°C increase in minimum temperature and a 0.8–2.6°C increase in maximum temperature by the end of 2100 under representative concentration pathway (RCP) 4.5. Under RCP 8.5, minimum and maximum temperatures will increase by 0.9–4.6°C and 0.8–4.4°C respectively. Precipitation is projected to rise by 36 percent under RCP 4.5 and 40 percent under RCP 8.5.

Other key features of probable change at country level consists of increasing temperature with more extreme hot days and more extreme rainfall, resulting in more droughts and floods; increasing risk of flooding as a result of higher average rainfall intensity in monsoon events; more variable rainfall in the rainy season across the country (but particularly in the north) from March to November and a decrease between December and February [18]; more frequent and more intense extreme weather events, including cyclones/strong winds, flood/storm surge, intense rains, extreme high temperatures, drought and sea level rise [19].

4. Most common crops and production in Myanmar

Myanmar mainly produce the most common crops such as rice, pulses and beans, and maize and other crops in four different zones: Delta zone, Coastal zone,

Central dry zone, and hilly and mountainous zone. Rice is the main staple food crop, cultivating nearly 50% of Myanmar's agricultural land [15]. The Ayeyarwady Delta Region contributes the largest share of production of rice while central dry zone contributes the highest share of pulses and beans. Among the most common crops, rice is normally grown during the monsoon season and pulses and beans are largely grown during the dry season. The Delta zone including Yangon, Bago, Ayeyarwaddy, and Mon state largely grow major crops such as rice, and pulses; Coastal zone including Tanintharyi Region, Mon State, Kayin State mainly grow rice, rubber, oil palm and fruit tree; Central dry zone including Magway, Mandalay largely grow rice for subsistence, oil cops, pulses, sesame, groundnuts, vegetables and tea; hilly and mountainous zone including Shan, Chin typically grow rice, wheat, maize, sorghum, vegetables, sugarcane, and coffee.

The total arable land increased from 9909 hectare in 2000 to 11080.3 hectare in 2018. Major export crops are rice and pulses. The country has a huge potential for agricultural development due to abundant in natural resources but has been facing some constraints such as providing irrigation and provision of infrastructures, insecure land titles, limited financial resources, provision of farm inputs and extension services, limited use of farm machinery and low investment in the development of its agricultural sector. The annual variation of crop production and yields is caused by seasonal nature of farming, climatic variations and extreme events [17]. The precipitation, temperature and humidity influence the types, production and yield of crops depending on different spatial location, indicating that different climate in different location affect production of crop [20].

The crop yield has found significant annual variation over the past decade due to different factors such as holding with insufficient land appropriate for diversification covering with an average lot size of 2.27 ha [15] largely relying on rainwater for irrigation, frequent flooding and drought stress. The paddy production and yield had increasing trend over the past decade but decreasing trend after year 2010 due to the reason that paddy plantations were damaged in 2015 by flood, accounting for 79 percent of total damage areas and 89 percent of the destroyed crops [21]. More than a million acres of cultivated land were adversely affected by heavy monsoon-related flood in 2018 as well as about 250,000 acres were destroyed especially paddy, maize, sesame, various beans and pulses (Myanmar Time 2018). The production and yield had been increasing in 2017–2018 due to favorable weather and more provision of irrigated water to farmers [1, 16]. The 26.3 million tonnes of rice (paddy), 1.6 million tonnes of groundnut and 0.7 million tonnes of sesame seeds were produced under the crop land area of 110830.3 ha [16].

5. Data source and empirical model

This study is based on country level data of Myanmar cereal crops for the time span of 2009 to 2019 obtained from various sources such as FAO STAT and Department of Agriculture and Department of Meteorology. The data are classified into two parts: cereal crop production and meteorological variables (temperature and precipitation). The data on cereal crop production are obtained from FAO STAT and meteorological variables are obtained from Statistical Yearbook with various issues. The data on agricultural inputs such as use of pesticides, application of fertilizers was obtained from FAO STAT. Climate variables such as annual precipitation and maximum and minimum temperature have been taken into in the model. State and district-wise Annual rainfall data for the respective years (2009–2019) has been obtained from Central Statistical Yearbook from the Ministry of Planning, Finance and Industry in Myanmar [22].

6. Empirical model

The relationship between climate change and agricultural production for the years 2009–2019 is analyzed with an econometrics model using Panel Regression. The data were collected for the country as a whole. Cereal crop production in kg per hectare is taken into consideration as Dependent variable while climate variable (average annual maximum and minimum temperatures, and average annual precipitation), agricultural inputs (application of fertilizer consumption, and phosphate and potash) are regarded as Explanatory variables.

This study is mainly used by multiple regression model. This regression analysis has been conducted using "R" to find out the best fit in the model. The "R" was used to analyze the data collected, and results is presented in table. Linear regression inferential statistic was used to analyze the contributions of the independent variables to the dependent. The statistical modeling applies observed country data on production of cereal crops and historical weather records to fit Linear regression functions in order to predict the response of cereal crop production [23, 24]. As a statistical model, Linear regression equation is used to link variations in historical year-to-year cereal crop production to variation in particular climate variables.

The functional form of the equation may be written as [25].

$$(T\,Prod) = f\left(\text{TEMP, RAIN}, FC, \text{Phosphate, Potash}\right) \tag{1}$$

Where, TProd stands for total production for cereal crops. TEMP, RAIN, FC, Phosphate, Potash denotes annual temperature, rainfall, total fertilizer consumption, phosphate and potash respectively. Climate factors, fertilizers, phosphate and potash are assumed to be input factors for growth of production of cereal crops in multiple regression model.

The above equation can be written in the multiple regression form as:

$$\begin{aligned}\ln(T\,\mathrm{Pr}od) = \beta\,0 + \beta\,1\ln\left(\text{TEMP}\right) + \beta\,2\ln\left(\text{RAIN}\right) + \beta\,3\ln\left(FC\right) \\ + \beta 4\,\ln\left(\text{Phosphate}\right) + \beta\,5\ln\left(\text{Potash}\right) + \mu s\end{aligned} \tag{2}$$

Where, $\beta0$ is constant coefficient; $\beta1$, $\beta2$, $\beta3$, $\beta4$, and $\beta5$, are the coefficients for the respective variables and μs is the intercept term.

7. Effect of climate change on cereal production

Change in crop production are very sensitive to climate change particularly in changes in temperature and changes in rainfall [26, 27]. Climate change significantly affects the crop production due to the fact that crop production largely depends on temperature and water [28–30]. Climate change may lessen the period of crop maturation and increase in variation of crop yield, decrease suitable areas for crop production [31–33] decrease in crop yield, [34–36] and thereby leading to reduce crop production [37–39].

The result can be seen in **Table 2**. According to the result, the coefficient for temperature is found a negative and 1% significant in production of cereal crops in Myanmar. Temperature negatively affects the crop production due to the fact that temperatures in Dry Zone could reach 40–43°C during the hot dry season. This indicates that an increase in temperature reflects a decrease in cereal crop production in Myanmar. This means that a 1°C increase in temperature in the growing

	Estimates	Std. Error	T value	P(>t)
Intercept	144749537	37469082	3.863	0.00479*
Temperature	−3849347	1294063	−2.975	0.01775**
Rainfall	−5762	3095	−1.862	0.09966
fertilizer	−148386	122053	−1.216	0.25873
phosphate	−111175	449545	−0.247	0.81090
potash	63836	649849	0.098	0.92417
Multiple R-squared:				0.7575
Adjusted R-squared:				0.606
F-statistic:				4.999

**1% significant.*
***5% significant.*

Table 2.
Regression result of climate change on cereal production.

period may decrease with production of cereal by 3849347 tons. It is found that an increase in temperature strongly influences on crop production and the temperature-related extreme weather events have an association with the production of cereal crops. The result is the same with the finding of Lesk et al. 2016 that drought and extreme temperature adversely affected on production of agriculture across the world and also greater impact on production of cereal crops [40]; and finding of You L et al. 2009 who found that in term of cereal crops, the higher temperature can negatively effect on cereal production, reflecting a 1°C increase in temperature in the growing period may decrease production of wheat by about 3–10% [41]. The result is also consistent with the finding of [42, 43] in which higher temperatures may have negative impact on aggregate output. An increase in temperature-related extreme weather events have an association with reduction in production of crops in Myanmar. Due to the reason, irrigation is largely demanded in order to cope with reduction in cereal crops production. To cope with this condition, water storage need to be reinforced in order to capture rain to reduce water needs in growing season.

The result found that the coefficient for rainfall are also negative but not significant with production of cereal crop production. This indicates that an increase in rainfall reflects a decrease in cereal crop production due to the fact that heavy intense of rain may cause frequent floods in Central Dry Zone that greatly affect the production of crop. Mean annual rainfall is generally the lowest in the Central Dry Zone (500–1000 mm per year) that is prone to extreme heat events and drought. In term of cereal crops, the intense rain can negatively affect on production of cereal crops, reflecting a 1 mm increase in rainfall in the growing period may decrease production of cereal by 5762 tons. Changes in rainfall affects delaying planting date which keeps crops at risk under the condition of high temperature in the growing season [44].

The result also found that the coefficient for fertilizer and phosphate are negative but not significant with production of cereal crops, indicating that an increase in utilization of fertilizer and phosphate reflects a decrease in production of cereal crops. This means that improper use of fertilizer negatively affects the production of cereal crop production. One-kilogram increase in consumption of fertilizer per hectare decreases with 148386 tons of cereal crop production and One-kilogram increase in consumption of phosphate per hectare decreases with 111175 tons of cereal production respectively. In order to improve production of cereal crop production, the social capital become urgently needed to incorporate and coordinate with respective stakeholders in order to adopt climate change adaptation measures such as stress tolerance crop varieties selection, enhancing drought resilience, drip irrigation technique for enhancing water use efficiency, sharing on climate resilient farming method among farmers, soil and water conservation, conservation and cultivation with local adaptable crop varieties in response to climate change. The same experience is found that the adoption of new technologies, such as drought-tolerant seeds, and changing farm practices, as sowing dates, are moderating the impacts of climate variability and change on crop yields [45–47].

8. Conclusion and policy implication

The research found that an increase in temperature and rainfall negatively affects the production of cereal crops in Myanmar due to an increase in temperature that causes water stress in growing season due to reduction in moisture of soil and consequently reduction in cereal crop production [48, 49] and an increase in rainfall negatively affects the cereal crop production due to intense rainfall.

As a micro level, the farmers can adapt the effect of reduction in crop production by crop diversification as a farm-level response and/or non-farm income sources. So that farmers can have benefits with optimal utilization of their marginal land, and reduction in extreme climate risks and events and improving resilience in smallholder farming systems [50–53]. In order to stabilize and increase farm income of the smallholder farmers, essential requirement is to encourage crop diversification by the government that generates raising farm incomes of smallholder farmers [54–56]. The relevant stakeholders need to provide knowledge and awareness for systematic utilization of farm inputs such as fertilizer and pesticide.

At the macro level, it is imperative for Myanmar to diversity its economy. Government needs to paid attention on integrated farming system that can generate employment opportunities and increase total income for farmers in order to response to climate change risk. Government need to pursue economic diversification strategies including expanding in climate sensitive sectors such as agriculture, fishery, forestry and energy and tourism that increase resilient in response to climate change. These sectors can provide employment opportunities and generate total incomes for long-term adaptation strategies. Government need to build capacity building program for all stakeholders including extension agents and farmers in order to be expertise in their farming operation thereby leading to manage reduction in crops caused by climate change.

Thus, policies aiming to enhance production of cereal crops should focus on adoption of climate change adaptation measures in Myanmar. Social networking between farmers and extension agents should be enhanced for natural resource management and adoption of climate change adaptation measures. On the other hand, policies targeting to increase cereal crop production should also proceed by scaling up programs. Government need to promote climate change adaptation measures such as drought resistant high yield varieties seeds; provide fertilizer and pesticide and skilled extension agents and services. Irrigation need to be sufficiently provided in order to improve the production of cereal crops. This indicates that water resource management become urgent need in Myanmar in order to capture rain and to reduce water scarcity in summer. Seasonal water shortage should be coped with water resource management and thereby leading to more efficient utilization of water resource [57] and improvement in crop production can be achieved. In sum, diversification become play a vital role for increasing resilience within the sector for sustainable development and thereby providing country to employment opportunities and income which in turn leads to reduction in poverty in the long run.

Acknowledgements

Firstly, I would like to my deepest gratitude to Intech Open for giving me an opportunity to write and submit my paper. I would like to my great gratitude to all academic editors for chapter review for my paper. I also owe a great thanks to my parents.

Author details

Thida Htoo
Meiktila University of Economics, Myanmar

*Address all correspondence to: thidahtoo4@gmail.com

References

[1] European Chamber of Commerce in Myanmar (2019). Agriculture Guide 2020, Yangon, p.14

[2] Syed Ali Fazal, & Sazali Abdul Wahab (2013). Economic Impact of Climate Change on Agricultural Sector: A Review, CPS Journals Journal of Transformative Entrepreneurship Centre for Postgraduate Studies ISSN: 2289-3075 Universiti Malaysia Kelantan Vol. 1, Issue 1, pp: 39-49 (2013) p.40

[3] Horton R, De Mel M, Peters D, Lesk C, Bartlett R, Helsingen H, Bader D, Capizzi P, Martin S, Rosenzweig C (2017). Assessing Climate Risk in Myanmar. New York, USA: Center for Climate Systems Research at Columbia University, WWF-US and WWF-Myanmar.

[4] Kreft S, Eckstein D, Melchior I. Global climate risk index (2016). Who suffers most from extreme weather events? Weather-related loss events in 2015 and 1996 to 2015-2016

[5] Myanmar National Census Report (2014). The 2014 Myanmar Population and Housing Census: Mandalay Region, Sagaing Region and Magway Region. Nay Pyi Taw, Myanmar: Department of Population, Ministry of Immigration and Population.

[6] NAPA 2013; Myanmar's National Adaptation Programme of Action (NAPA) to Climate Change 2012.

[7] Porter JR, et al. (2014). Climate Change 2014: Impacts, Adaptation and Vulnerability: Contribution of Working Group II to the Fifth Assessment Report of the Intergovernmental Panel on Climate Change, eds Field CB, et al. (Cambridge Univ. Press, Cambridge, UK.

[8] National Commission for Environmental Affairs (NCEA). Myanmar's Initial National Communication Report. Nay Pyi Taw, Myanmar: Environmental Conservation Department, Ministry of Natural Resources and Environmental Conservation, Government of Myanmar; 2010a.

[9] Ministry of Agriculture & Irrigation [MoAI] (2015) Myanmar Climate-Smart Agricultural Strategy, available from: https://cgspace.cgiar.org/rest/bitstreams/ 63308/retrieve

[10] Quasem, 2011.

[11] Ministry of Natural Resources and Environmental Conservation (2017). Myanmar Climate Change Strategy and Action Plan (MCCSAP) 2016-2030.

[12] World Bank Group (2017). Capitalizing on Investment Opportunities.

[13] International Monetary Fund (IMF) (2017). 2016 Article IV Consultation – Press Release; Staff Report; and Statement by the Executive Director for Myanmar. Washington, D.C: International Monetary Fund (IMF).

[14] Egashira and Aye 2006 as cited by Myanmar Climate Change Strategy 2018-2030, pp.28-29).

[15] NAPA (2012). Myanmar's National Adaptation Programme of Action (NAPA) to Climate Change 2012, p.20).

[16] FAO STAT, 2018.

[17] Ramasamy Selvaraju, 2014. Managing climate risks and adapting to climate change in the agriculture sector in Nepal, FAO, Rome,p.18.

[18] INC Report 2012.

[19] MONREC 2012a. Ministry of Natural Resources and Environmental

Conservation (2017). Myanmar Climate Change Strategy and Action Plan (MCCSAP) 2016-2030.

[20] Molua E & Lambi C, (2006): "The economic impact of climate change on agriculture in Cameroon", CEEPA Discussion Paper No. 17, Centre for Environmental Economics and Policy in Africa, University of Pretoria, p.41

[21] Thinn Nay Chi Sun (2019): Why We Need to Talk About The Impact of Flood in Myanmar's Agriculture Sector, retrieved from http://www.changemag-diinsider.com/blog/why-we-need-to-talk-about-the-impact-of-flood-inmyanmar-s-agriculture-sector (02/12/2019)

[22] Japneet Kaur 2017, Impact of Climate Change on Agricultural Productivity and Food Security Resulting in Poverty in India, Master's Degree Thesis, Universita' Ca' Foscari, Venezia, p.23).

[23] Lobell DB, Field CB (2007) Global scale climate-crop yield relationships and the impacts of recent warming. Environ Res Lett 2:1-7.

[24] Lobell DB, Schlenker W, Costa-Roberts J (2011) Climate trends and global crop production since 1980. Science 333:616-620.

[25] Japneet Kaur 2017, Impact of Climate Change on Agricultural Productivity and Food Security Resulting in Poverty in India, Master's Degree Thesis, Universita' Ca' Foscari, Venezia, p.22

[26] Rosenzweig C, et al. (2014) Assessing agricultural risks of climate change in the 21st century in a global gridded crop model inter comparison. Proc Natl Acad Sci USA 111: 3268-3273.

[27] Wheeler T, von Braun J (2013) Climate change impacts on global food security. Science 341:508-513.

[28] Aggarwal PK, Mall RK, (2002)'Climate Change and Rice Yields in Diverse Agro-Environments of India. II. Effects of Uncertainties in Scenarios and Crop Models on Impact Assessment', Clim. Change *52*, 331-343.

[29] Naylor RL, Battisti DS, Vimont DJ, Falcon WP, Burke MB. Assessing risks of climate variability and climate change for Indonesian rice agriculture. Proceedings of the National Academy of Sciences. 2007;104(19):7752-7757

[30] Deutsch CA et al (2018). Increase in crop losses to insect pests in a warming climate. Science. 2018;361(6405):916-919

[31] Saseendran SA, Singh KK, Rathore LS, Singh SV, Sinha SK. Effects of climate change on rice production in the tropical humid climate of Kerala, India. Climatic Change. 2000;44(4):495-514

[32] Deryng D, Conway D, Ramankutty N, Price J, Warren R. Global crop yield response to extreme heat stress under multiple climate change futures. Environmental Research Letters. 2014;9(3):34011

[33] Schlenker, W. & Roberts, M.J. (2009). Nonlinear temperature effects indicate severe damages to US crop yields under climate change. Proceedings of the National Academy of sciences, 106(37), 15594-15598, p.4

[34] Masutomi Y, Takahashi K, Harasawa H, Matsuoka Y. Impact assessment of climate change on rice production in Asia in comprehensive consideration of process/parameter uncertainty in general circulation models. Agriculture, Ecosystems and Environment. 2009;131(3-4):281-291

[35] Jones MR, Singels A, Ruane AC. Simulated impacts of climate change on water use and yield of irrigated sugarcane in South Africa. Agricultural Systems. 2015;139:260-270

[36] Challinor AJ, Watson J, Lobell DB, Howden SM, Smith DR, Chhetri N. A meta-analysis of crop yield under climate change and adaptation. Nature Climate Change. 2014;4(4):287

[37] Kumar A, Sharma P. Climate change and sugarcane productivity in India: An econometric analysis. Journal of Social and Development Sciences. 2014;5(2):111

[38] de Carvalho AL et al. Impact of climate changes on potential sugarcane yield in Pernambuco, northeastern region of Brazil. Renewable Energy. 2015;78:26-34

[39] Zhao D, Li Y-R. Climate change and sugarcane production: Potential impact and mitigation strategies. International Journal of Agronomy. 2015;2015

[40] Lesk, C., P. Rowhani and N. Ramankutty (2016), "Influence of Extreme Weather Disasters on Global Crop Production", Nature, 529, p.110

[41] You L, Rosegrant MW, Wood S, Sun D. Impact of growing season temperature on wheat productivity in China. Agricultural and Forest Meteorology. 2009;149(6-7):1009-1014

[42] Dell, M., B.F. Jones and B.A. Olken (2012), "Temperature Shocks and Economic Growth: Evidence from the Last Half Century", American Economic Journal: Macroeconomics, 4 (3), 66-95.

[43] Acevedo, S., Mrkaic, M., Novta, N., Pugacheva, E., & Topalova, P. (2018), "The Effects of Weather Shocks on Economic Activity: What are the Channels of Impact?", IMF Working Paper, WP/18/144s

[44] Bowling et al., 2018 as cited in Donald Wuebbles, An Assessment of the Impacts of Climate Change on the Great Lakes, Environmental law and policy center, p.23.

[45] Suri, T. Selection and Comparative Advantage in Technology Adoption. Econometrica 2011, 79, 159-209.

[46] Di Falco, S. Adaptation to climate change in Sub-Saharan agriculture: Assessing the evidence and rethinking the drivers. Eur. Rev. Agric. Econ. 2014, 41, 405-430.

[47] Alvi, S.; Jamil, F. Impact of Climate Change and Technology Adoption on Cereal Yields in South Asian Countries. Eur. J. Sustain. Dev. 2018, 7, 237-246. [CrossRef] }as sited in Roberto Roson and Martina Sartori 2020, Do Farmers Adapt to Climate Change? A Macro Perspective, 9 June 2020, p.7

[48] Asseng S, Foster IAN, Turner NC (2011) The impact of temperature variability on wheat yield. Glob Change Biol 17:997-1012.

[49] Zhao D, Li Y-R. Climate change and sugarcane production: Potential impact and mitigation strategies. International Journal of Agronomy. 2015;2015

[50] Bradshaw B, Dolan H, Smit B. 2004. Farm-level adaptation to climatic variability and change: Crop diversification in the Canadian prairies. Climatic Change, 67, 119-141.Chavas J-P, Holt T (1990) Acreage decisions under risk: the case of corn and soybeans. Amer J Agr Econ 72(2):529-538

[51] Huang, J.K., Jiang, J., Wang, J.X. and Hou, L.L., 2014. Crop diversification in coping with extreme weather events in China. Journal of Integrative Agriculture, 13(4), pp.677-686

[52] Mijatović, D., Van Oudenhoven, F., Eyzaguirre, P., & Hodgkin, T. 2011. The role of agricultural biodiversity in strengthening resilience to climate change: towards an analytical framework. International journal of agricultural sustainability, 11(2), 95-107

[53] Smit, B., Skinner M W. 2002. Adaptation options in agriculture to climate change: A typology. Mitigation and Adaptation Strategies for Global Change, 7, 85-114

[54] FAO 2012. Making Agriculture work for Nutrition: Synthesis of Guiding Principles. Food and Agriculture Organization of the United Nations, Rome, Italy

[55] Osmani, S. R., Ahmed, T., Hossain, N., & Huq, S. 2016. Strategic review of food security and nutrition in Bangladesh. World Food Programme.

[56] Islam, A.H.M.S., von Braun, J., Thorne-Lyman, A.L. and Ahmed, A.U., 2018. Farm diversification and food and nutrition security in Bangladesh: empirical evidence from nationally representative household panel data. Food Security, pp.1-20.

[57] Robert I et al. 2017 Urban growth, climate change, and freshwater availability, PNAS, April 12, 2011, vol. 108, no. 1 p.6314.

Section 4

Climate Effects on Hydrology

Chapter 10

Rainfall Trends in Humid Temperate Climate in South America: Possible Effects in Ecosystems of Espinal Ecoregion

Julian Alberto Sabattini and Rafael Alberto Sabattini

Abstract

In central Argentina, the annual rainfall regime shows increasing since the 2nd half of the 20th century. The aim of this work was to evaluate the long-term changes in the intensity of rainfall in the central-north region of Entre Ríos between 1945 and 2019, based only on daily precipitation records aggregated at yearly, monthly and seasonal levels. We used monthly rainfall data for the period 1945–2019 from 6 localities in Province of Entre Rios, Argentina. The change detection analysis has been conceded using Pettitt's test, von Neumann ratio test, Buishand's range test and standard normal homogeneity (SNH) test, while non-parametric tests including linear regression, Mann-Kendall and Spearman rho tests have been applied for trend analysis. Like the regional results, this study observed a sustained increase in monthly rainfall to the breaking point in the 1970s, but then the annual rate of increase was even higher. The average annual rainfall in the region prior to that date was 946 mm, while after the same 1150 mm, equivalent to 21.5% higher than the 1945–1977 average and 8.5% higher according to the historical average 1945–2019.

Keywords: nonparametric trend tests, climate change, natural ecosystems, biodiversity

1. Introduction

There was a general increase in air temperature worldwide during the twentieth century, albeit with some differences between the hemispheres, corresponding to global warming. Global warming affects the hydrological cycle over land, resulting in observed changes to precipitation frequency, intensity, duration and amount [1, 2]. Although significant attention is paid to how changes in seasonal and annual precipitation sums affect ecosystems, relatively less is known about the ecological impacts of heavy rainfall events [3]. The evaluation of past trends of meteorological parameters at various spatial and temporal scales plays a crucial role in understanding climate change and its impact on food security, energy security, natural resource management, and sustainable development [4, 5]. Detailed analysis of rainfall trend is useful to rainfall forecasting, planning water resources development and management, designing water storage structures, irrigation practices and crop

IntechOpen

choices, drinking water supply, industrial development, and disaster management for current and future climatic conditions [6, 7].

The analysis of different global rainfall databases shows a change in an anomaly that was positive between 1950 and 1980 and became negative later [8]. While some studies show increasing rainfall, in other regions the evaluations show the opposite results. For instance, in Europe, the rain series show an increase in annual precipitation between 1940 and 1990 [9]. The climate of Italy, in turn, seems to be warmer and drier at the moment with a decrease in rainfall attributed to a reduction in the number of days of rain, as rainfall intensity shows a positive trend [10]. In different regions of South America somethings similar happens and has been studied by several authors. More recently [11] studied summer precipitation variability over Southeastern South America in a global warming scenario.

In central Argentina, the annual rainfall regime shows increasing rates from approximately the 1940s until the end of the century [12–14] with statistical and spectral analysis show that there is significant evidence that rainfall has increased in central Argentina since the 2nd half of the 20th century [15] analyzed breakpoints in annual rainfall trends in Córdoba, Argentina in the period 1930–2006, they observed from negative to positive in the 1950s in the north area of the region, while in the other areas the opposite change occurs in the 1970s. From the mid-1970s, a sharp increase in rainfall regime provided most of the area with a supply of moisture higher than previously reported [16–19]. Recently results in changes annual rainfall in five sub-regions of the Argentine Pampa Region indicate that the Western Pampas are more vulnerable to abrupt changes than the Eastern Pampas [20]. While different indicators in central Argentina reflect a change for precipitation at some sites, the intensity and variability of rainfall show significant long-term trends [21]. The rainfall cycle hypothesis has been supported by recent studies showing an abrupt negative change in the water regime of Pampas Region in recent years [17, 18] as well as by studies linking changes in rainfall with regular or recurring oceanic indices [19–21].

A strong increase in agricultural activity in central of Argentina [22] is a possible cause that would explain the climate change. The central-north region of Entre Ríos (Argentina) had a strong fragmentation of the landscape due to deforestation [23]. These changes are environmentally and economically important, as they have a direct impact on hydrological and soil resources, as well as on the agricultural potential of the region. The central-north of Entre Ríos has a humid temperature climate, Cf in the Koppen-Geiger classification, as revised by [24]. In this way, the Pampa Region (where the province of Entre Rios is located) receives sea winds throughout the year, with a moisture gradient decreasing from east to west [20].

The statistical trend detection in climatic variables and precipitation time series is one of the interesting research areas in climatology and hydrology as it impacts spatial and temporal distribution of water availability across the globe [25]. The parametric or non-parametric method under statistical approach is used to detect if either a data of a given set follows a distribution or has a trend on a fixed level of significance. Various non-parametric tests, including Mann-Kendall test and Pettit's test, are widely used to detect trend and change point in historical series of climatic and hydrological variables [26–28]. To understand the magnitude of trends many techniques have been proposed in the past, including t-tests [29, 30], Mann–Whitney and Pettitt's tests [31] and standard normal homogeneity test [32, 33].

The aim of this work was to evaluate the long-term changes in the intensity of rainfall in the central-north region of Entre Ríos between 1945 and 2019, based only on daily precipitation records aggregated at yearly, monthly and seasonal levels. In more specific terms, the quality of the rainfall series is first analyzed in terms of its homogeneity to assess the reliability of the meteorological information used.

Secondly, the existence of a trend in the indicators of the intensity and variability of rainfall is evaluated during a period showing a generalized increase in atmospheric temperature. Finally, the occurrence of a breakpoint that expresses a long-term trend change in the annual rainfall series in the region is assessed.

2. Material and methods

2.1 Region of study and rainfall data

We used monthly rainfall data for the period 1945–2019 from 6 localities (**Figure 1** and **Table 1**) in the southern of department La Paz (Province of Entre Rios, Argentina): Hasenkamp (HAS), Las Garzas (LGA), Alcaraz Norte (ALN), Bovril (BOV), Hernandarias (HER), El Solar (ELS). This data were collected with

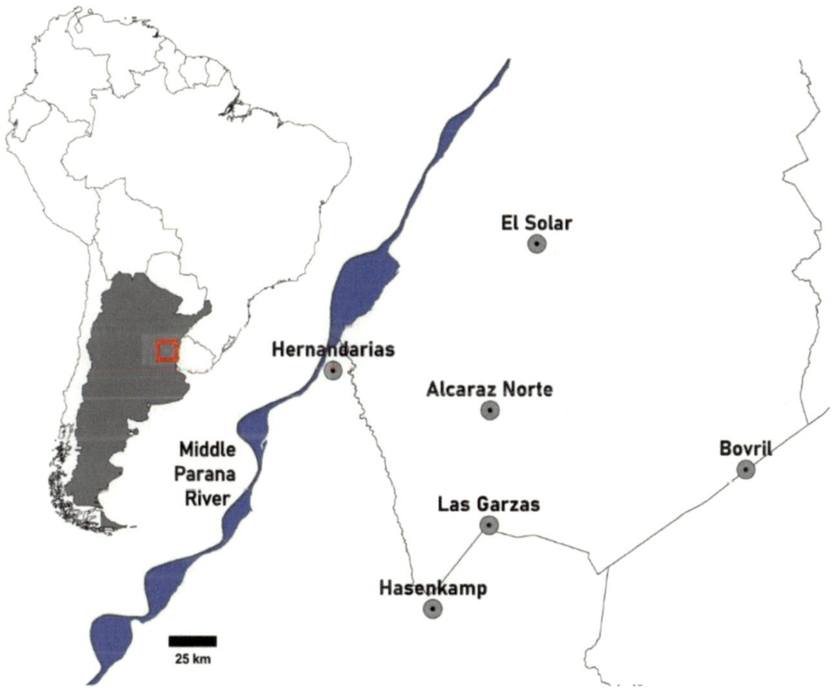

Figure 1.
Location map of southern of department La Paz (province of Entre Rios, Argentina) with localities analyzed.

Meteorological station		Latitude	Longitude	Altitude (m a.s.l)	Period and Entirety (%)
Hasenkamp	HAS	31°30′32.94"S	59°50′9.37"W	88	1945–2019 (93.8%)
Las Garzas	LGA	31°25′43.54"S	59°44′36.09"W	82	1945–2019 (100%)
Alcaraz Norte	ALN	31°19′37.49"S	59°45′15.88"W	68	1945–2019 (98.3%)
Bovril	BOV	31°20′26.89"S	59°26′30.97"W	79	1945–2019 (94.5%)
Hernandarias	HER	31°13′51.34"S	59°59′10.35"W	52	1945–2019 (96.5%)
El Solar	ELS	31°10′32.96"S	59°43′56.73"W	50	1945–2019 (99.2%)

Table 1.
Meteorological station used and the period analyzed.

conventional rain gauges, from the official records of Hydraulic Directorate (Direccion Hidraulica de Entre Rios, in spanish) and Cereal Bag (Bolsa de Cereales de Entre Rios, in spanish) of the Province of Entre Rios.

The data from the 6 locations was subjected to a process of quality control for possible errors. All data above the third quartile plus three times the interquartile range and located more than five standard deviations from the mean was treated as outliers. These outliers were then contrasted climatographically with readings from nearby stations. If the same reading was labeled as out of range for more than two seasons, the value was correct. Months classed as outliers and those without data were treated as gaps. Both types of gaps were filled but no missing data was completed if there were more than three gaps in one year.

Stations with missing data techniques linear regression were used. The filling of missing data by the linear regression technique consisted in using data from neighboring stations that presented coefficients of significant linear correlations with the station to be used in the study [34, 35],

$$P_x = a_o + \sum_{i=1}^{n} a_i P_i \tag{1}$$

where a_o and a_i are the coeffcients of adjustment of the linear model, obtained in the processing of correlation. In this case, stations that presented an R^2 greater than 0.90 were included. The two techniques are widely used to fill gaps in historical series and present low average deviations suitable for climatic studies on monthly and seasonal scales [36].

After the treatment of the time series, the monthly values of all rainfall stations were grouped into scales, according to the following definitions: a) autumn (March, April, and May), b) winter (June, July, and August), c) spring (September, October, and November), and d) summer (December, January, and February). For selecting the change point for a particular parameter, the method presented below has been used [37]: a) no change point or homogeneous (HG), series may be considered as homogeneous, if no or one test out of four tests rejects the null hypothesis at 5% significant level; b) doubtful series (DF), series may be considered as inhomogeneous and critically evaluated before further analysis if two out of four tests reject the null hypothesis at 5% significant level; and c) change point or inhomogeneous (CP) when series may has change point or be inhomogeneous in nature, if more than two tests reject the null hypothesis at 5% significant level.

2.2 Homogeneity tests for change point detection

Homogeneity testing is very crucial in climatological studies to represent the real variations in weather and climate. Inhomogeneity occurs in climate data due to several reasons including instrumentation error, changes in the adjacent areas of the instrument, and mishandling of the human. If the homogeneity is not tested prior to trend analysis, the results will indicate erroneous trends. In this study, the absolute homogeneity tests were performed on individual station records and calculating the ratio of observed series to the reference series. Four widely used statistical tests mentioned below were applied to the data to test for homogeneity. All the following four tests used in this study assume the null hypothesis of data being homogeneous. The change point detection is an important aspect to assess the period from where significant change has occurred in a time series. Pettitt's test, von Neumann ratio test, Buishand range test and standard normal homogeneity tests have been applied for change point detection in climatic series. The details of various change point tests applied in the study are presented here.

2.2.1 Pettitt's test

The Pettitt's test for change detection, developed by [38], is a non-parametric test, which is useful for evaluating the occurrence of abrupt changes in climatic records [39, 40] because its sensitivity. According to Pettitt's test, if $x_1, x_2, x_3, \ldots x_n$ is a series of observed data which has a change point at t in such a way that $x_1, x_2 \ldots, x_t$ has a distribution function $F_1(x)$ which is different from the distribution function $F_2(x)$ of the second part of the series $x_{t+1}, x_{t+2}, x_{t+3} \ldots, x_n$. The non-parametric test statistics U_t for this test may be described as follows:

$$U_t = \sum_{i=1}^{t} \sum_{j=t+1}^{n} sign\left(x_t - x_j\right) \tag{2}$$

$$sign\left(x_t - x_j\right) = \begin{bmatrix} 1, & if\left(x_i - x_j\right) > 0 \\ 0, & if\left(x_i - x_j\right) = 0 \\ -1, & if\left(x_i - x_j\right) < 0 \end{bmatrix} \tag{3}$$

The test statistic K and the associated confidence level (ρ) for the sample length (n) may be described as:

$$K = Max|U_t| \tag{4}$$

$$\rho = \exp\left(\frac{-K}{n^2 + n^3}\right) \tag{5}$$

When ρ is smaller than the specific confidence level, the null hypothesis is rejected. The approximate significance probability (p) for a change-point is defined as given below:

$$p = 1 - \rho \tag{6}$$

The test statistic K can also be compared with standard values at different confidence level for detection of change point in a series. The critical values of K at 1 and 5% confidence levels for different tests used in the analysis has been presented in **Table 2** [37].

2.2.2 von Neumann ratio test

The von Neumann ratio test has been described by [41, 42] and others. The test statistics for change point detection in a series of observations $x_1, x_2, x_3 \ldots x_n$ can be described as:

Number of observation	Critical values for test statistic at different significance level							
	Pettit Test		SNHT		Buishand Range test		Von Neumann Ratio Test	
	1%	5%	1%	5%	1%	5%	1%	5%
50	293	235	11.38	8.45	1.78	1.55	1.36	1.54
70	488	393	11.89	8.80	1.81	1.59	1.45	1.61
100	841	677	12.32	9.15	1.86	1.62	1.54	1.67

Table 2.
Critical values of test statistics for different change point detections tests.

$$N = \frac{\sum_{i=1}^{n-1}(x_i - x_{i-1})^2}{\sum_{i=1}^{n-1}(x_i - \overline{x})^2} \tag{7}$$

According to this test, if the sample or series is homogeneous, then the expected value $E(N)$ = 2 under the null hypothesis with constant mean. When the sample has a break, then the value of N must be lower than 2, otherwise we can imply that the sample has rapid variation in the mean. The critical values of N at 1 and 5% confidence levels given in **Table 2** can be used for identification of non-homogeneous series with change point.

2.2.3 Buishand's range test

The adjusted partial sum (S_k), that is the cumulative deviation from mean for kth observation of a series $x_1, x_2, x_3. \ldots x_k. \ldots x_n$ with mean ($\overline{x}$) can be computed using following equation:

$$S_k = \sum_{i=1}^{k}(x_i - \overline{x}) \tag{8}$$

A series may be homogeneous without any change point if $S_{k\sim}$ 0, because in random series, the deviation from mean will be distributed on both sides of the mean of the series. The significance of shift can be evaluated by computing rescales adjusted range (R) using the following equation:

$$R = \frac{Max(S_k) - Min(S_k)}{\overline{x}} \tag{9}$$

The computed value of $R/\sqrt{n}$ is compared with critical values given by [37, 41] and has been used for detection of possible change (**Table 2**).

2.2.4 Standard normal homogeneity (SNH) test

The test statistic (T_k) is used to compare the mean of first n observations with the mean of the remaining (n-k) observations with n data points [32].

$$T_k = kZ_1^2 + (n-k)Z_2^2 \tag{10}$$

$Z_1$and Z_2 can be computed as:

$$Z_1 = \frac{1}{k}\sum_{i=1}^{k}\frac{(x_i - \overline{x})}{\sigma x} \tag{11}$$

$$Z_2 = \frac{1}{n-k}\sum_{i=k+1}^{k}\frac{(x_i - \overline{x})}{\sigma x} \tag{12}$$

where, $\overline{x}$ and σx are the mean and standard deviation of the series. The year k can be considered as change point and consist a break where the value of T_k attains the maximum value. To reject the null hypothesis, the test statistic should be greater than the critical value, which depends on the sample size (n) given in **Table 2**.

2.3 Test for trend analysis

All the trend tests in this section assume the null hypothesis of no trend and the alternative hypothesis of monotonic increasing or decreasing trend existence. When the time series are serially independent, the Mann–Kendall test [43, 44] and Spearman's Rho test [45, 46] were applied to test for trends. The magnitude of the trend was estimated using Sen's slope method [47]. Always suggested to apply various statistical tests to analyze the trends in serially correlated data.

2.3.1 Mann-Kendall test

The Mann–Kendall test is a nonparametric test for monotonic trend detection. It does not assume the data to be normally distributed and is flexible to outliers in the data. The test assumes a null hypothesis, H_0, of no trend and alternate hypothesis, H_a, of increasing or decreasing monotonic trend. For a time series $X_i = x_1, x_2, \dots, x_n$, the Mann–Kendall test statistic S is calculated as

$$S = \sum_{i=1}^{n-1} \sum_{j=i+1}^{n} sign(x_j - x_i) \tag{13}$$

where n is the number of data points, x_i and x_j are the data values in timeseries i and j $(j > i)$, respectively, and $sign(x_j - x_i)$ is the sign function as

$$sign(x_i - x_j) = \begin{bmatrix} 1, & if\,(x_j - x_i) > 0 \\ 0, & if\,(x_j - x_i) = 0 \\ -1, & if\,(x_j - x_i) < 0 \end{bmatrix} \tag{14}$$

Statistics S is normally distributed with parameters $E(S)$ and variance $V(S)$ as given below:

$$E(S) = 0 \tag{15}$$

$$V(S) = \frac{n(n-1)(2n+5) - \sum_{k=1}^{m} t_k(k)(k-1)(2k+5)}{18} \tag{16}$$

where n is the number of data points, m is the number of tied groups, and t_k denotes the number of ties of extent k. Standardized test statistic Z is calculated using the formula below.

$$Z = \begin{cases} \dfrac{S-1}{\sqrt{var(S)}} & if\ S > 0 \\ 0 & if\ S = 0 \\ \dfrac{S+1}{\sqrt{var(S)}} & if\ S < 0 \end{cases} \tag{17}$$

To test for a monotonic trend at an α significance level, the alternate hypothesis of trend is accepted if the absolute value of standardized test statistic Z is greater than the $Z_{1-\alpha/2}$ value obtained from the standard normal cumulative distribution Tables. A positive sign of the test statistic indicates an increasing trend and a negative sign indicates a decreasing trend.

2.3.2 Spearman's rho test

The Spearman's rho test is a non-parametric widely used for studying populations that take on a ranked order. If there is no trend and all observations are independent, then all rank orderings are equally likely. In this test, the difference between order and rank (d_i) for all observations $x_1, x_2, x_3, \ldots x_n$ can be used to compute and Spearman's ρ, variance $Var(\rho)$ and test statistic (Z) using following equations. The null hypothesis is tested in this test considering the statistic is normally distributed.

$$\rho = 1 - \frac{6\sum d_i^2}{n(n-1)} \tag{18}$$

$$Var(\rho) = \frac{1}{(n-1)} \tag{19}$$

$$Z = \frac{\rho}{\sqrt{Var(\rho)}} \tag{20}$$

3. Results and discussion

Tables 3–8 show the results of the statistical analyzes carried out to know the point of change in monthly, seasonal and annual rainfall in each locality. A marked variability was observed in the months that changed significantly between the localities, fundamentally from January to May, even though the proximity between them does not exceed 40 km. This means, a priori and in subjective terms, that the climatic changes reported worldwide have a direct influence on a microspatial scale, as well as on the temporal window. However, in the region there was no heterogeneity in the breaking point between the localities evaluated for the months of November and December during the study period analyzed.

In relation to the statistical tests used, it is possible to conclude that the Von Neumman's test is more robust when establishing the heterogeneity of the time series, while the Standard Normal Homogeneity test a priori would require less demand from the variability of the time series. to set a breaking point. Based on the results of the SNH Test, it is observed that the month of May presents marked heterogeneity in all localities, but the year that defines the point of change differs significantly. When comparing and analyzing all the tests for each period of time, only Las Garzas and Hernandarias present a significant, but doubtful point of change in the year that followed.

In seasonal analysis, summer is the season of the year that presented marked heterogeneity in the time series in all localities. The year of break point was different by location. However, El Solar and Hernandarias presented significant modifications in the heterogeneity of the time series with breaking points during the 1970s and 1980s, respectively. Both locations are adjacent to the Middle Paraná River, a situation that could be influenced by local atmospheric conditions [48]. There is even greater concern today about the future of rivers worldwide due to a multitude of stressors that impact running waters including climate change [49]. We draw on the growing literature related to climate change to illustrate potential impacts rivers may experience and management options for protecting riverine ecosystems and the goods and services they provide. Regional patterns in precipitation and temperature are predicted to change and these changes have the potential to alter natural flow regimes. One of the key ways in which climate change or other stressors affect river ecosystems is by causing changes in river flow. Rivers vary geographically

Period	Standard Normal Homogeneity Test				Pettitt's test				Buishand Range test				von Neumann's test		
	T	*p*	*sig*	k	U*	*p*	*sig*	k	R/sqrt(n)	P	*sig*	k	Statistic	*p*	*sig*
JAN	13.94	0.003	**	2018	241	0.885	ns	1958	1.047	0.591	ns	1958	1.374	0.170	ns
FEB	7.55	0.104	ns	1976	504	0.057	**	1976	1.513	0.082	***	1976	−0.373	0.795	ns
MAR	4.44	0.427	ns	1949	218	1.026	ns	1990	0.713	0.969	ns	2007	0.704	0.482	ns
APR	5.23	0.309	ns	1978	500	0.060	**	1978	1.233	0.319	ns	1978	−0.457	0.648	ns
MAY	15.32	0.002	**	2009	448	0.120	ns	1980	1.565	0.061	***	1980	0.311	0.756	ns
JUN	2.42	0.843	ns	2006	316	0.493	ns	1986	0.844	0.871	ns	1974	−1.098	0.272	ns
JUL	1.59	0.968	ns	1968	238	0.903	ns	1987	0.876	0.833	ns	1968	−0.589	0.556	ns
AUG	11.01	0.018	**	2014	198	1.154	ns	2014	0.873	0.845	ns	2014	1.208	0.227	ns
SEP	1.77	0.949	ns	1988	257	0.792	ns	1988	1.036	0.608	ns	1985	−0.039	0.969	ns
OCT	3.49	0.613	ns	1955	526	0.041	**	1982	0.953	0.737	ns	1983	−1.414	0.157	ns
NOV	13.17	0.005	**	1976	740	0.001	**	1976	1.964	0.003	**	1976	−1.993	0.046	**
DEC	8.73	0.056	*	1988	494	0.065	**	1986	1.455	0.116	ns	1988	−0.116	0.908	ns
Summer	10.78	0.020	*	1976	564	0.023	**	1976	1.624	0.044	*	1976	−0.592	0.554	ns
Autumn	6.37	0.186	ns	1997	465	0.096	ns	1974	1.348	0.196	ns	1974	1.519	0.129	ns
Winter	2.74	0.775	ns	2016	172	1.320	ns	1986	0.896	0.814	ns	1986	−0.219	0.826	ns
Spring	4.40	0.438	ns	1977	520	0.045	**	1977	1.106	0.498	ns	1977	−0.404	0.686	ns
Annual	12.92	0.006	**	1977	668	0.004	**	1976	1.784	0.016	*	1977	−0.959	0.338	ns

*References: k: year to shift, sig: * 0.05%, ** 0.01%, *** 0,1%, ns: no signification.*

Table 3.
Results of change point analysis with all test used in Las Garzas location.

Period	Standard Normal Homogeneity Test				Pettitt's test				Buishand Range test				von Neumann's test		
	T	*p*	*sig*	k	U*	*p*	*sig*	k	R/sqrt(n)	p	*sig*	k	Statistic	*p*	*sig*
JAN	16.42	0.000	**	2018	192	1.192	ns	1959	1.009	0.646	ns	1994	1.121	0.262	ns
FEB	5.42	0.282	ns	2006	416	0.176	ns	2002	1.208	0.349	ns	1980	0.709	0.478	ns
MAR	4.11	0.487	ns	2019	221	1.008	ns	1993	0.633	0.993	ns	2007	2.081	0.037	*
APR	5.07	0.333	ns	2015	524	0.042	**	1978	0.965	0.713	ns	1978	−0.526	0.599	ns
MAY	10.37	0.002	**	2017	460	0.103	ns	1973	1.452	0.119	ns	1982	0.849	0.396	ns
JUN	3.11	0.695	ns	2006	320	0.475	ns	1986	1.059	0.572	ns	1975	−1.320	0.187	ns
JUL	1.74	0.949	ns	1948	357	0.334	ns	1988	0.903	0.805	ns	1987	−0.568	0.570	ns
AUG	10.54	0.020	**	2014	172	1.320	ns	2014	0.868	0.847	ns	2014	0.682	0.495	ns
SEP	2.08	0.901	ns	1988	267	0.730	ns	2006	1.036	0.610	ns	1988	0.365	0.715	ns
OCT	2.85	0.754	ns	2010	331	0.430	ns	1999	1.273	0.269	ns	2000	−0.135	0.892	ns
NOV	14.39	0.003	**	1976	718	0.001	**	1976	2.098	0.001	**	1976	−1.983	0.047	*
DEC	7.65	0.101	ns	1976	453	0.112	ns	1976	1.368	0.179	ns	1976	−1.802	0.072	***
Summer	8.85	0.532	*	2004	518	0.046	**	1976	1.464	0.110	ns	1976	−0.357	0.721	ns
Autumn	4.73	0.386	ns	1979	391	0.234	ns	1979	1.143	0.442	ns	1979	1.833	0.067	***
Winter	2.92	0.732	ns	1948	200	1.141	ns	1955	0.880	0.831	ns	1955	−1.339	0.181	ns
Spring	6.93	0.143	ns	2010	400	0.212	ns	1977	1.049	0.591	ns	1999	−0.273	0.785	ns
Annual	10.05	0.003	**	1977	582	0.017	**	1977	1.574	0.057	*	1977	−0.373	0.709	ns

*References: k: year to shift, sig: * 0.05%, ** 0.01%, *** 0,1%, ns: no signification.*

Table 4.
Results of change point analysis with all test used in Alcaraz Norte location.

Period	Standard Normal Homogeneity Test				Pettitt's test				Buishand Range test				von Neumann's test		
	T	*p*	*sig*	k	U*	*p*	*sig*	k	R/sqrt(n)	p	*sig*	k	Statistic	*p*	*sig*
JAN	11.30	0.015	**	2016	199	1.147	ns	1958	0.958	0.721	ns	1958	2.039	0.041	**
FEB	7.17	0.125	ns	2009	477	0.082	***	1999	1.291	0.248	ns	1980	−0.577	0.564	ns
MAR	3.80	0.547	ns	1949	199	1.147	ns	1980	0.721	0.968	ns	1949	2.246	0.025	*
APR	3.56	0.595	ns	1998	442	0.128	**	1978	1.138	0.447	ns	1978	−1.101	0.271	ns
MAY	10.85	0.019	**	2017	392	0.231	ns	1979	1.195	0.366	ns	1979	0.061	0.952	ns
JUN	2.66	0.792	ns	2006	253	0.815	ns	2006	0.959	0.723	ns	1973	−1.321	0.186	ns
JUL	4.19	0.468	ns	1978	400	0.212	ns	1987	1.265	0.283	ns	1978	−0.339	0.735	ns
AUG	12.68	0.006	**	2014	248	0.844	ns	1982	0.927	0.765	ns	2013	1.756	0.079	***
SEP	2.27	0.868	ns	1956	358	0.331	ns	1985	1.192	0.366	ns	1985	−0.926	0.354	ns
OCT	3.19	0.678	ns	1989	447	0.121	ns	1983	1.253	0.291	ns	1989	−1.270	0.204	ns
NOV	9.08	0.051	**	1992	586	0.015	**	1992	1.592	0.051	**	1977	−1.745	0.081	***
DEC	8.09	0.080	***	1989	496	0.063	***	1989	1.394	0.155	ns	1989	−0.765	0.444	ns
Summer	7.58	0.103	*	2008	516	0.048	**	1995	1.236	0.319	ns	1989	−0.248	0.808	ns
Autumn	4.04	0.510	ns	1997	452	0.114	ns	1989	1.040	0.596	ns	1989	0.275	0.784	ns
Winter	3.08	0.711	ns	1968	234	0.927	ns	1968	1.250	0.302	ns	1970	−0.259	0.796	ns
Spring	5.61	0.258	ns	1999	480	0.079	***	1992	1.305	0.232	ns	1922	−1.277	0.202	ns
Annual	10.24	0.026	**	1999	598	0.013	**	1997	1.453	0.118	ns	1997	−0.766	0.443	ns

*References: k: year to shift, sig: * 0.05%, ** 0.01%, *** 0,1%, ns: no signification.*

Table 5.
Results of change point analysis with all test used in Bovril location.

Period	Standard Normal Homogeneity Test				Pettitt's test				Buishand Range test				von Neumann's test		
	T	*p*	*sig*	k	U*	*p*	*sig*	k	R/sqrt(n)	p	*sig*	k	Statistic	*p*	*sig*
JAN	11.12	0.015	**	2018	254	0.809	ns	1961	1.085	0.532	ns	1958	1.501	0.133	ns
FEB	6.49	0.176	ns	1976	466	0.095	ns	1976	1.422	0.138	ns	1976	−0.291	0.771	ns
MAR	3.77	0.559	ns	2014	273	0.703	ns	1980	0.846	0.875	ns	2007	1.047	0.295	ns
APR	5.16	0.319	ns	1980	481	0.078	***	1978	1.302	0.237	ns	1980	−0.112	0.911	ns
MAY	14.09	0.003	**	2012	318	0.484	ns	1973	1.154	0.421	ns	2009	0.180	0.858	ns
JUN	2.55	0.815	ns	2006	416	0.176	ns	1982	0.939	0.750	ns	1982	−0.378	0.705	ns
JUL	1.34	0.985	ns	1958	195	1.173	ns	1988	0.984	0.692	ns	1968	0.160	0.873	ns
AUG	9.86	0.037	*	2014	177	1.288	ns	2014	0.841	0.877	ns	2014	1.410	0.159	ns
SEP	1.68	0.955	ns	1985	220	1.014	ns	1988	0.945	0.746	ns	1985	0.190	0.849	ns
OCT	3.90	0.537	ns	1988	504	0.057	***	1988	1.023	0.623	ns	1988	−1.550	0.121	ns
NOV	8.98	0.049	*	1975	581	0.018	**	1975	1.710	0.024	*	1976	−1.842	0.065	*
DEC	7.58	0.102	ns	2001	403	0.205	ns	1986	1.353	0.197	ns	1988	0.281	0.779	ns
Summer	7.74	0.098	***	2004	492	0.067	*	1995	1.252	0.295	ns	1980	0.302	0.763	ns
Autumn	5.24	0.309	ns	1969	435	0.141	ns	1969	1.162	0.412	ns	1969	0.943	0.346	ns
Winter	2.21	0.880	ns	2016	177	1.289	ns	1986	0.901	0.808	ns	1986	−0.193	0.847	ns
Spring	4.18	0.480	ns	1955	468	0.092	***	1988	1.001	0.664	ns	1976	−0.874	0.382	ns
Annual	12.83	0.006	**	1997	652	0.005	**	1997	1.635	0.041	*	1997	−1.063	0.288	ns

*References: k: year to shift, sig: * 0.05%, ** 0.01%, *** 0,1%, ns: no signification.*

Table 6.
Results of change point analysis with all test used in Hasenkamp location.

Period	Standard Normal Homogeneity Test				Pettitt's test				Buishand Range test				von Neumann's test		
	T	*p*	*sig*	k	U*	*p*	*sig*	k	R/sqrt(n)	p	*sig*	k	Statistic	*p*	*sig*
JAN	19.16	0.000	**	2018	214	1.052	ns	2006	1.148	0.440	ns	2006	0.643	0.520	ns
FEB	9.30	0.044	*	2009	401	0.209	ns	2004	1.200	0.368	ns	2004	−0.187	0.852	ns
MAR	3.85	0.539	ns	1949	190	1.205	ns	1949	0.851	0.866	ns	2007	0.688	0.491	ns
APR	8.29	0.071	***	1980	610	0.011	**	1978	1.501	0.087	***	1980	−0.832	0.405	ns
MAY	12.21	0.010	**	2017	418	0.172	ns	1973	1.279	0.264	ns	1973	0.895	0.371	ns
JUN	2.49	0.827	ns	2006	255	0.803	ns	2006	1.036	0.612	ns	2006	−1.626	0.104	ns
JUL	2.89	0.738	ns	2002	291	0.609	ns	1987	1.045	0.596	ns	2002	−0.234	0.815	ns
AUG	13.44	0.004	**	2014	234	0.927	ns	1984	1.096	0.509	ns	1999	0.561	0.575	ns
SEP	2.36	0.853	ns	1986	295	0.590	ns	1988	1.097	0.509	ns	1986	−1.232	0.218	ns
OCT	3.98	0.516	ns	2000	422	0.164	ns	1982	1.179	0.387	ns	2000	−0.719	0.472	ns
NOV	8.75	0.059	*	1985	560	0.025	*	1985	1.657	0.034	*	1985	−1.415	0.157	ns
DEC	15.81	0.001	**	1989	689	0.003	**	1989	1.948	0.004	**	1989	−1.264	0.206	ns
Summer	15.53	0.001	**	2004	528	0.040	*	1995	1.636	0.042	*	2004	−0.684	0.494	ns
Autumn	8.14	0.081	***	1979	568	0.022	*	1979	1.453	0.114	ns	1979	0.138	0.890	ns
Winter	2.65	0.794	ns	1997	249	0.838	ns	1997	0.977	0.699	ns	1997	−0.841	0.400	ns
Spring	5.89	0.232	ns	1999	481	0.078	ns	1982	1.198	0.356	ns	1983	0.204	0.839	ns
Annual	18.45	0.000	**	1999	767	0.001	**	1982	1.976	0.002	**	1982	−2.087	0.037	*

*References: k: year to shift, sig: * 0.05%, ** 0.01%, *** 0,1%, ns: no signification.*

Table 7.
Results of change point analysis with all test used in El solar location.

Period	Standard Normal Homogeneity Test				Pettitt's test				Buishand Range test				von Neumann's test		
	T	*p*	*sig*	k	U*	*p*	*sig*	k	R/sqrt(n)	p	*sig*	k	Statistic	*p*	*sig*
JAN	23.28	0.001	**	2018	186	1.231	ns	1985	1.036	0.604	ns	2018	0.577	0.564	ns
FEB	7.11	0.129	ns	2009	326	0.450	ns	2004	1.101	0.506	ns	2006	−0.253	0.800	ns
MAR	4.03	0.505	ns	1949	198	1.154	ns	1949	0.943	0.739	ns	1949	1.060	0.289	ns
APR	8.36	0.071	***	1978	614	0.010	**	1978	1.474	0.103	ns	1978	−1.744	0.081	***
MAY	13.67	0.004	**	2017	488	0.071	***	1973	1.421	0.136	ns	1980	−0.274	0.784	ns
JUN	2.75	0.773	ns	2006	341	0.391	ns	1986	0.914	0.776	ns	1974	−1.571	0.116	ns
JUL	2.18	0.086	***	1948	331	0.430	ns	1988	1.024	0.624	ns	1988	−0.425	0.671	ns
AUG	9.98	0.028	*	2014	241	0.885	ns	1966	0.900	0.804	ns	1966	0.447	0.655	ns
SEP	1.24	0.989	ns	1988	307	0.533	ns	1999	0.982	0.683	ns	1988	−1.191	0.234	ns
OCT	4.58	0.407	ns	1983	527	0.041	*	1982	1.343	0.201	ns	1983	−1.700	0.089	***
NOV	13.09	0.004	**	1977	712	0.002	**	1977	2.001	0.002	**	1977	−2.878	0.004	**
DEC	10.00	0.030	*	1996	525	0.042	*	1989	1.542	0.068	***	1989	−0.550	0.583	ns
Summer	10.85	0.018	*	2018	408	0.193	ns	1995	1.193	0.377	ns	1975	0.212	0.832	ns
Autumn	9.16	0.048	*	1985	629	0.008	**	1985	1.535	0.073	***	1985	−0.357	0.721	ns
Winter	2.61	0.803	ns	1948	219	1.020	ns	1992	0.870	0.842	ns	1992	0.009	0.993	ns
Spring	7.36	0.117	ns	1999	562	0.024	*	1977	1.384	0.165	ns	1977	−1.459	0.145	ns
Annual	13.77	0.004	**	1999	636	0.007	**	1989	1.699	0.024	*	1977	−1.476	0.140	ns

*References: k: year to shift, sig: * 0.05%, ** 0.01%, *** 0,1%, ns: no signification.*

Table 8.
Results of change point analysis with all test used in Hernandarias location.

with respect to their natural flow regime and this variation is critical to the ecological integrity and health of streams and rivers and thus a great deal has been written on the topic [50, 51]. The ecological consequences and the required management responses for any given river will depend not only on the direct impacts of increased temperature. Otherwise how extensively the magnitude, frequency, timing, and duration of runoff events change relative to the historical and recent flow regime for that river, and how adaptable the aquatic and riparian species are to different degrees of alteration.

The results resume depicting the homogeneity state of different series have been presented in **Table 9** (See Supplementary Appendix with results of Test's trend). The change point analysis on long-term series in all localities has indicated that a significant change point in the annual rainfall. The breaking point occurred in 1977 for the LGA, ALC and HER locations; year 1997 for BOV and HAS; and 1982 for the ELS locality. **Figure 2** shows the average annual precipitation of all the localities evaluated in each year for the region, as well as the historical annual during the period. On the other hand, since the breaking point occurred in 1977 for most of the localities, it was established that the average annual rainfall in the region prior to that date was 946 mm, while after the same 1150 mm, equivalent to 21.5% higher than the 1945–1977 average and 8.5% higher according to the historical average 1945–2019. In addition, an important piece of information results from the linear model that made it possible to establish that the region's average rainfall increased 4.9 mm per year from 1945 to 2019.

These results are consistent with those obtained in the north of the country where the rainfall change was concentrated in a step change during the 1970s [52]. In this region, half or more of the annual rainfall trend occurred in the months of El Niño phase, with less contribution from La Niña and the neutral phases. However, in the rest of subtropical Argentina and especially south of 30°S, increased precipitation occurred mostly during months of the neutral phase of El Niño/Southern Oscillation (ENSO), with only small trends during months of El Niño and La Niña phases [53]. Accordingly, most of the annual precipitation trends since 1960 in subtropical Argentina can be accounted for by two modes. The first mode, which is positively correlated with precipitation in northern Argentina and with ENSO indices, had a steep increase in precipitation at the end of the 1970s. The second mode, which has a maximum positive correlation with annual precipitation between 30 and 40°S, had a regular positive trend starting in the early 1960s and it is correlated with the southward displacement of the South Atlantic high [53, 54]. In addition, several researchers analyzed the changes in the isohyets, showing that the rainfall regime in Argentina is subject to a positive fluctuation in the 1950s and that it reached maximum values in the 1970s [55], data that coincide with this manuscript.

Average rainfall increased, favoring the expansion of agriculture [16, 22]. This conclusion is obtained primarily because the studies of the time have been hampered by the low significance shown by statistical tests when applied to climatic data, especially precipitation. In the study region mention that one of the factors of change in precipitation is agrarian transformation and claim that the technological innovation of the sector was accompanied by a process of change in the water regime [16]. Furthermore, confirm that the expansion of agricultural structure of Entre Rios, is favored by increased precipitation, generating crops of the marginal territory.

The behavior of historical series of monthly rainfall confirm that November and December, as and summer season, have significant change point in all localities. The annual rainfall in all localities showed a significant increase such as summer season (**Table 9**). November and December showed and significant rise in contrast to the rest of months.

Period	LGA			ALN			BOV			HAS			ELS			HER		
	a	b	c	a	b	c	a	b	c	a	b	c	a	b	c	a	b	c
JAN	HG	—	~	HG	—	~	DF	2016	~	HG	—	~	HG	—	~	HG	—	~
FEB	DF	1976	↑	HG	—	~	HG	—	↑	HG	—	↑	HG	—	~	HG	—	~
MAR	HG	—	~	HG	—	~	HG	—	~	HG	—	~	HG	—	~	HG	—	~
APR	HG	—	¿?	HG	—	~	HG	—	~	HG	—	~	CP	1978 1980	↑	CP	1978	↑
MAY	DF	1980 2009	↑	HG	—	↑	HG	—	~	HG	—	~	HG	—	↑	DF	1973 2017	↑
JUN	HG	—	~	HG	—	~	HG	—	~	HG	—	~	HG	—	~	HG	—	~
JUL	HG	—	~	HG	—	~	HG	—	~	HG	—	~	HG	—	~	HG	—	~
AUG	HG	—	~	HG	—	~	DF	2014	~	HG	—	~	HG	—	~	HG	—	~
SEP	HG	—	~	HG	—	~	HG	—	~	HG	—	~	HG	—	~	HG	—	~
OCT	HG	—	↑	HG	—	~	HG	—	~	HG	—	↑	HG	—	~	DF	1982	↑
NOV	CP	1976	↑	CP	1976	↑	CP	1992 1997	↑	CP	1975 1976	↑	CP	1985	↑	CP	1977	↑
DEC	DF	1986 1988	↑	HG	—	↑	DF	1989	↑	HG	—	↑	CP	1989	↑	CP	1989 1996	↑
Summer	CP	1976	↑	DF	2004 1976	↑	DF	1995 2008	↑	DF	1995 2004	↑	CP	1995 2004	↑	HG	—	↑
Autumn	HG	—	↑	HG	—	~	HG	—	¿?	HG	—	↑	DF	1979	↑	CP	1985	↑
Winter	HG	—	~	HG	—	~	HG	—	~	HG	—	~	HG	—	~	HG	—	~
Spring	HG	—	↑	HG	—	~	HG	—	¿?	HG	—	↑	HG	—	↑	HG	—	↑
Annual	CP	1977	↑	CP	1977	↑	DF	1997 1999	↑	CP	1997	↑	CP	1982	↑	CP	1977 1989 1999	↑

Reference: homogeneous series (HG), change point (CP), doubtful point (DF). Trends: ~ none, ↑ increase, ↓ decrease, ¿? Doubtful.
Reference: a- Nature Serie, b- Year shift, c- Trend, LGA- Las Garzas, ALN- Alcaraz Norte, BOV- Bovril, HAS- Hasenkamp ELS- El Solar, HER- Hernandarias.

Table 9.
Results of change point detection analysis and trends of rainfall for all localities.

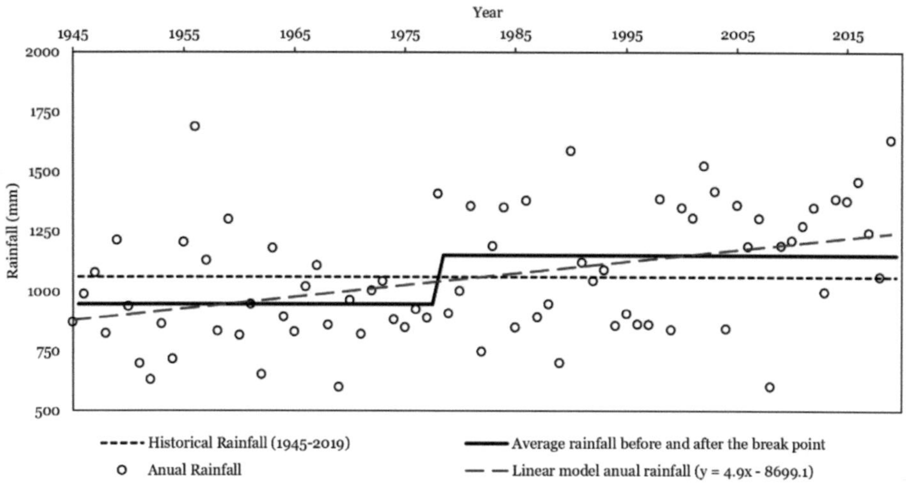

Figure 2.
Variation in the average annual rainfall of all the localities of the analyzed region.

In the last decade, a substantial change in the average climate conditions was observed in many regions of Argentina, particularly in the southern region of Mesopotamian Pampa that showed two abrupt shifts [20]. The first of these was positive, with annual average rainfall increasing from 1062.9 mm during the 1941–1999 sub-period to 1568.9 mm during a short sub-period between 2000 and 2003. The second abrupt change, which began in 2004, was negative, with average annual rainfall dropping to 1108.0 mm, only slightly higher than what it had been in the initial 1941–1999 sub-period (**Figure 3**).

Like the regional results, this study observed a sustained increase in monthly rainfall to the breaking point in the 1970s, but then the annual rate of increase was even higher. In South America [56], observed increasing trends in total annual precipitation values in Ecuador, Paraguay, Uruguay, northern Peru, southern Brazil, and northern and central Argentina. Qualitatively there was a change that indicated a significant increase in summer precipitation, and a decrease in the number of annual frosts, concentrating the winter season (July and August), assuming a “tropicalization of the region”. Rainfall tropicalization can be understood as local and regional processes and impacts of climate change, which can be observed mainly by changes in the precipitation regime and the intensification of tropical climatic characteristics [57]. This process is not exclusive of Espinal Ecorregion. It has been observed in other contexts and scales in tropical and subtropical regions that show an important increase in precipitation during the rainy season in tropical regions [58, 59].

Climate change can also indirectly affect organisms by altering biotic interactions, which can have profound consequences for populations, community composition and ecosystem functions [60]. Other aspects of biodiversity management will be affected by global change and will need adapting, including wildlife exploitation, e.g. forestry [61], pest and invasive species control [62] or human and wildlife disease management [63]. Indirect effects may occur: (i) via generation of new biotic interactions, as range-shifted species appear for the first time in naive communities [64]; (ii) by removing existing interactions when species shift out of their existing range [65]; or (iii) by modulating key behavioral, physiological or other traits that mediate species interactions [66]. When climate-driven changes in biotic interactions involve keystone or foundation species, impacts can cascade through

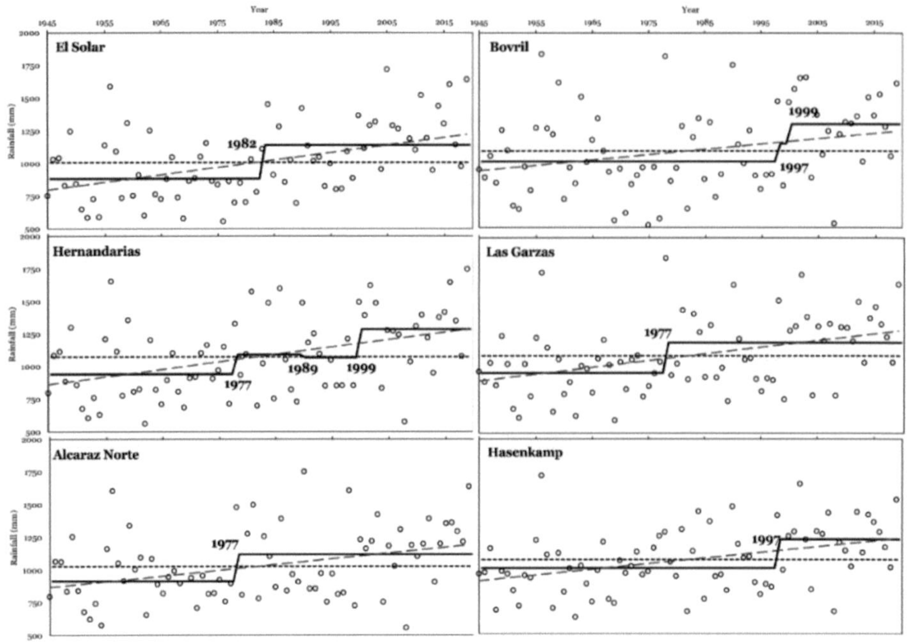

Figure 3.
Variation in the average annual rainfall in each locality of the analyzed region. Reference: Black dash line (−−) show historical rainfall (1945–2019), black solid line (—) the average rainfall before and after the break point and gray dash line (−−) show a linear model annual rainfall.

the associated community [61]. In this region, studies that have not yet been published for the province of Entre Ríos are showing indications of changes in the productivity of natural grasslands in native forests. Recently reports show that change the growth cycle has change in this ecosystem [67, 68], and mainly attributed to changes in precipitation regimes. These observations are like yields changes of the main crops, were the frequency of extreme weather events constitutes a growing risk.

Acknowledgements

This study was carried out in the framework of research and development projects UNER-PID No. 2196 "Ecological succession of a native forest intervened in the Spinal Ecorregion" and UNER-PID No. 2238 "Evaluation of the current and potential state of the native forests of Entre Rios in its productive and conservation aspect".

Conflict of interest

The authors declare no conflict of interest.

Supplementary appendix

Period	Average rainfall	Spearman's Rank Rho Test				Mann-Kendall Test			
		S	rho	p	z-value	Sen's slope	S	p	tau
JAN	121	70352	−0.001	0.995	0.041	0.008	1.000	0.967	0.004
FEB	131	50419	0.283	0.014*	2.470	1.000	541.000	0.014*	0.195
MAR	142	76056	−0.081	0.485	−0.686	−0.245	−0.015	0.493	−0.054
APR	110	54132	0.230	0.047*	1.880	0.761	0.041	0.060	0.148
MAY	68	51235	0.271	0.019*	2.360	0.576	0.057	0.018*	0.186
JUN	49	76397	−0.087	0.459	−0.778	−0.117	−0.017	0.436	−0.062
JUL	35	72477	−0.031	0.792	−0.357	−0.024	−0.790	0.721	−0.028
AUG	42	66369	0.056	0.634	0.357	0.050	0.790	0.721	0.028
SEP	69	73918	−0.051	0.661	−0.439	−0.108	−0.970	0.661	−0.349
OCT	108	52579	0.252	0.029*	2.140	0.800	0.047	0.033*	0.169
NOV	109	46105	0.344	0.002*	2.946	0.800	0.065	0.032*	0.232
DEC	106	50439	0.283	0.014*	2.402	0.891	0.526	0.016*	0.190
Summer	358	48743	0.307	0.007*	2.657	2.285	0.058	0.008*	0.209
Autumn	320	51593	0.266	0.021*	2.260	1.730	0.050	0.024*	0.178
Winter	126	69514	0.011	0.924	0.069	0.038	1.600	0.915	0.006
Spring	286	50059	0.288	0.012*	2.452	1.355	0.054	0.014*	0.193
Annual	1091	40638	0.422	0.000*	3.449	5.133	0.076	0.001*	0.272

References: (*) test with significant differences of 0.05%.

Table 1A.
Result of trend analysis rainfall at Las Garzas locality.

Period	Average rainfall	Spearman's Rank Rho Test				Mann-Kendall Test			
		S	rho	p	z-value	Sen's slope	S	p	tau
JAN	112	68775	0.022	0.853	0.297	0.154	0.660	0.766	0.024
FEB	119	56223	0.200	0.085	1.729	0.684	0.038	0.084	0.137
MAR	139	77544	−0.103	0.379	−0.883	−0.391	−0.019	0.377	−0.699
APR	102	54987	0.218	0.060	1.866	0.650	409.000	0.062	0.148
MAY	61	50832	0.277	0.016*	2.384	0.557	522.000	0.017*	0.189
JUN	41	78357	−0.115	0.328	−0.915	−0.120	−0.020	0.360	−0.073
JUL	36	76383	−0.087	0.460	−0.679	−0.056	−0.015	0.497	−0.054
AUG	45	65950	0.014	0.598	0.545	0.082	0.012	0.586	0.044
SEP	63	69320	0.103	0.906	0.087	0.007	0.200	0.931	0.007

Period	Average rainfall	Spearman's Rank Rho Test				Mann-Kendall Test			
		S	rho	p	z-value	Sen's slope	S	p	tau
OCT	101	63028	0.103	0.377	0.883	0.313	0.019	0.377	0.071
NOV	108	48056	0.316	0.006*	2.713	0.931	0.059	0.007*	0.214
DEC	99	52013	0.260	0.024*	2.093	0.833	0.046	0.036*	0.166
Summer	330	51074	0.273	0.018*	2.223	2.008	0.049	0.026*	0.175
Autumn	302	56165	0.201	0.084	1.670	1.216	0.037	0.095	0.132
Winter	123	69288	0.014	0.902	−0.091	−0.040	−0.210	0.927	−0.008
Spring	273	55183	0.215	0.064	1.715	1.107	0.038	0.086	0.136
Annual	1029	45357	0.355	0.002*	2.887	4.322	632.000	0.004*	0.228

Table 2A.
Result of trend analysis rainfall at Alcaraz Norte locality.

Period	Average rainfall	Spearman's Rank Rho Test				Mann-Kendall Test			
		S	rho	p	z-value	Sen's slope	S	p	tau
JAN	113	69650	0.009	0.937	0.128	0.043	0.290	0.898	0.010
FEB	119	51962	0.261	0.024*	2.347	0.920	0.051	0.019*	0.186
MAR	117	75068	−0.068	0.563	−0.677	−0.284	−0.015	0.498	−0.054
APR	136	58350	0.170	0.145	1.533	0.729	0.034	0.125	0.121
MAY	61	58123	0.173	0.137	1.491	0.350	0.033	0.136	0.118
JUN	43	76302	−0.085	0.466	−0.750	−0.118	−0.017	0.453	−0.060
JUL	44	82282	−0.170	0.144	−1.611	−0.263	−353	0.107	−0.128
AUG	50	61718	0.122	0.297	1.131	0.184	0.025	0.258	0.090
SEP	70	77711	−0.105	0.368	−0.860	−0.226	−0.019	0.390	−0.068
OCT	114	56116	0.202	0.083	1.752	0.655	0.038	0.080	0.139
NOV	105	49986	0.289	0.012*	2.265	0.804	0.050	0.024*	0.179
DEC	107	53477	0.239	0.039*	2.091	0.923	0.046	0.037*	0.165
Summer	339	51398	0.269	0.020*	2.306	2.054	0.051	0.021*	0.182
Autumn	314	54902	0.219	0.059*	1.876	1.431	0.041	0.061	0.141
Winter	148	74374	−0.058	0.621	−0.572	−0.204	−0.013	0.568	−0.045
Spring	289	55044	0.217	0.061*	1.715	1.215	0.038	0.086	0.136
Annual	1090	49195	0.300	0.009*	2.438	4.451	0.053	0.015*	0.192

Table 3A.
Result of trend analysis rainfall at Bovril locality.

Period	Average rainfall	Spearman's Rank Rho Test			Mann-Kendall Test				
		S	rho	p	z-value	Sen's slope	S	p	tau
JAN	114	71802	−0.021	0.856	−0.156	−0.010	−0.350	0.876	−0.013
FEB	129	51202	0.272	0.018*	2.424	1.029	0.053	0.015*	0.191
MAR	141	77861	−0.108	0.258	−0.910	−0.380	−0.020	0.363	−0.072
APR	114	55361	0.212	0.067	1.784	0.794	0.039	0.074	0.141
MAY	72	58926	0.162	0.166	1.409	0.431	0.031	0.159	0.116
JUN	43	82218	−0.169	0.146	−1.281	−0.184	−281	0.200	−0.102
JUL	33	71809	−0.021	0.856	−0.188	−0.013	−0.420	0.851	−1.523
AUG	41	67234	0.044	0.710	0.334	0.048	0.740	0.738	0.027
SEP	65	73176	−0.041	0.727	−0.371	−0.098	−0.820	0.711	−0.030
OCT	113	52731	0.250	0.031*	2.100	0.822	0.046	0.036*	0.166
NOV	110	50320	0.284	0.013*	2.447	0.840	0.054	0.014*	0.192
DEC	97	54742	0.221	0.056*	1.875	0.724	0.041	0.061*	0.148
Summer	340	50482	0.282	0.014*	2.502	2.032	0.058	0.012*	0.198
Autumn	327	54775	0.221	0.059*	1.844	1.279	0.040	0.063*	0.146
Winter	117	71460	−0.017	0.883	−0.238	−0.075	−0.530	0.812	−0.019
Spring	287	51115	0.273	0.018*	2.250	1.272	0.049	0.024*	0.178
Annual	1071	41184	0.414	0.000*	3,531	4,625	0.077	0.000*	0.279

Table 4A.
Result of trend analysis rainfall at Hasenkamp locality.

Period	Average rainfall	Spearman's Rank Rho Test			Mann-Kendall Test				
		S	rho	p	z-value	Sen's slope	S	p	tau
JAN	119	66804	0.050	0.672	0.435	0.207	0.960	0.664	0.035
FEB	118	58492	0.168	0.150	1.565	0.548	0.034	0.118	0.124
MAR	128	72232	−0.024	0.815	−0.165	−0.083	−0.370	0.869	0.013
APR	107	52121	0.259	0.025*	2.342	0.929	0.051	0.019*	0.185
MAY	56	53917	0.233	0.044*	1.982	0.462	0.043	0.048*	0.158
JUN	52	74096	−0.054	0.645	−0.224	−0.030	−1	0.823	−0.018
JUL	35	77632	−0.104	0.373	−0.865	−0.103	−0.019	0.387	−0.069
AUG	39	63108	0.102	0.382	0.948	0.146	0.021	0.343	0.076
SEP	58	71122	−0.012	0.921	−0.160	−0.026	−0.360	0.873	−0.013
OCT	105	59335	0.156	0.182	1.322	0.460	0.029	0.186	0.105
NOV	98	52646	0.251	0.030*	2.319	0.760	0.051	0.020*	0.184
DEC	95	43704	0.378	0.001*	3.208	1.105	0.070	0.001*	0.253
Summer	332	50288	0.284	0.013*	2.575	2.344	0.056	0.010*	0.203

Period	Average rainfall	Spearman's Rank Rho Test			Mann-Kendall Test				
		S	rho	p	z-value	Sen's slope	S	p	tau
Autumn	291	50852	0.277	0.017*	2.278	1.541	0.050	0.023*	0.180
Winter	126	66250	0.058	0.624	0.526	0.250	0.012	0.599	0.042
Spring	260	54160	0.230	0.048*	1.972	1.195	0.043	0.049*	0.156
Annual	1009	38960	0.446	0.000*	3.925	6.126	0.086	0.000*	0.309

Table 5A.
Result of trend analysis rainfall at El solar locality.

Period	Average rainfall	Spearman's Rank Rho Test			Mann-Kendall Test				
		S	rho	p	z-value	Sen's slope	S	p	tau
JAN	125	60702	0.009	0.942	0.102	0.050	0.240	0.916	0.009
FEB	130	61884	0.120	0.306	1.240	0.547	0.027	0.215	0.098
MAR	135	72915	−0.037	0.751	−0.421	−0.179	−93.000	0.674	−0.034
APR	112	50504	0.282	0.014*	2.314	0.957	0.051	0.031*	0.183
MAY	60	49393	0.297	0.010*	2.571	0.593	563.000	0.010*	0.203
JUN	47	79615	−0.133	0.257	−1.075	−0.125	−0.024	0.282	−0.085
JUL	34	73537	−0.046	0.695	−0.412	−0.041	−0.910	0.680	−0.033
AUG	42	65897	0.063	0.594	0.490	0.066	0.011	0.624	0.039
SEP	62	65597	0.067	0.569	0.590	0.143	0.013	0.555	0.047
OCT	104	56327	0.199	0.087*	1.766	0.608	0.039	0.077*	0.140
NOV	114	45954	0.346	0.002*	2.978	1.067	0.065	0.003*	0.235
DEC	108	50090	0.287	0.012*	2.280	0.939	0.050	0.023*	0.181
Summer	363	54028	0.231	0.046*	1.985	1.888	0.044	0.047*	0.157
Autumn	307	47158	0.329	0.004*	2.763	1.805	0.061	0.006*	0.219
Winter	123	70554	−0.004	0.976	0.009	0.000	3.000	0.993	0.001
Spring	280	48297	0.313	0.006*	2.722	1.700	596.000	0.006*	0.215
Annual	1073	40694	0.421	0.000*	3.778	6.420	827.000	0.000*	0.298

Table 6A.
Result of trend analysis rainfall at Hernandarias locality.

Author details

Julian Alberto Sabattini[1*] and Rafael Alberto Sabattini[2]

1 National Scientific and Technical Research Council, Department of Ecology, School of Agriculture, National University of Entre Rios, Oro Verde, Argentina

2 Department of Ecology, School of Agriculture, National University of Entre Rios, Oro Verde, Argentina

*Address all correspondence to: julian.sabattini@fca.uner.edu.ar

References

[1] Sherwood S, Fu Q. A drier future? Science 2014; 343: 737–739. DOI: 10.1126/science.1247620

[2] Stocker T. Climate change 2013: the physical science basis: Working Group I contribution to the Fifth assessment report of the Intergovernmental Panel on Climate Change. Cambridge, UK: Cambridge University Press. 2014.

[3] Li L, Zheng Z, Biederman JA, Xu C, Xu Z, Che R, Wang Y, Cui X, Hao Y. Ecological responses to heavy rainfall depend on seasonal timing and multi-year recurrence. New Phytol, 2019; 223: 647-660. DOI: 10.1111/nph.15832

[4] Sen Z, Zekâi S. Innovative Trend Analysis Methodology. J. Hydrol. Eng. 2011; 17:1042–1046. DOI: 10.1061/(ASCE)HE.1943-5584.0000556

[5] Milly PCD, Betancourt J, Falkenmark M, Hirsch RM, Kundzewicz ZW, Lettenmaier DP, Stouffer RJ. Stationarity Is Dead: Whither Water Management? Science. 2005; 319:573–574. DOI: 10.1126/science.1151915.

[6] Hoekema DJ, Sridhar V. Relating climatic attributes and water resources allocation: A study using surface water supply and soil moisture indices in the Snake River basin, Idaho. Water Resour. Res. 2011; 47:1–17. DOI: 10.1029/2010WR009697.

[7] Abbot J, Marohasy J. Skilful rainfall forecasts from artificial neural networks with long duration series and single-month optimization. Atmos. Res. 2017; 197:289–299. DOI: 10.1016/j.atmosres.2017.07.015.

[8] Trenberth KE. Changes in Precipitation with Climate Change. Climate Research. 2011; 47(1-2):123–138. DOI: 10.3354/cr00953.

[9] Kiely G, Albertson JD, Parlange MB. Recent trends in diurnal variation of precipitation at Valentia on the west coast of Ireland. Journal of Hydrology. 1998; 207(3–4): 270-279. DOI: 10.1016/S0022-1694(98)00143-7.

[10] Cislaghi M, De Michele C, Ghezzi A, Rosso R. Statistical assessment of trends and oscillations in rainfall dynamics: Analysis of long daily Italian series. Atmospheric Research. 2005; 77:188–202. DOI: 10.1016/j.atmosres.2004.12.014.

[11] Junquas C, Vera C, Li L, Le Treut H. Summer precipitation variability over Southeastern South America in a global warming scenario. 2012; 38(9-10):1867–1883. DOI: 10.1007/s00382-011-1141-y.

[12] Lucero OA, Rozas D. Characteristics of aggregation of daily rainfall in a middle-latitudes region during a climate variability in annual rainfall amount. Atmospheric Research. 2002; 61:35–48. DOI: 10.1016/S0169-8095(01)00101-6.

[13] Ravelo AC, Irastorza RO, Zanvettor RE. Zonificación de la región Pampeana Argentina según tendencias de series pluviométricas (1931–2000). Rev. Argent. Agrometeorol. 2003; 3–4: 85–90.

[14] Pasquini AI, Lecomte KL, Piovano EL, Depetris PJ. Recent rainfall and runoff variability in central Argentina. Quat Int 2006; 158:127–139. DOI: 10.1016/j.quaint.2006.05.021.

[15] de la Casa AC, Nasello OB. Breakpoints in annual rainfall trends over Córdoba, Argentina. Atmos. Res. 2010; 95:419–427. DOI: 10.1016/j.atmosres.2009.11.005.

[16] Sierra EM, Hurtado R, Spescha L. Corrimiento de las isoyetas anuales medias decenales en la Región Pampeana 1941–1990. Rev. Fac. Agron. 1994; 14:139–144.

[17] Pérez S, Sierra EM, Casagrande G, Vergara G. Incremento de la precipitaciones (1921/1998) en el Centro-Este de la provincia de La Pampa (Argentina). Rev. Fac. Agron. 1999; 19 (2):193-196.

[18] Forte Lay JA, Scarpati OE, Capriolo AD. Precipitation variability and soil water content in pampean flatlands (Argentina). Geofísica Internacional. 2008; 47(4):341-354.

[19] Pérez S, Sierra E, López E, Nizzero G, Momo F, Massobrio M. Abrupt changes in rainfall in the Eastern area of La Pampa Province, Argentina. Theor. Appl. Climatol. 2011; 103: 159–165. DOI: 10.1007/s00704-010-0290-y.

[20] Pérez S, Sierra E, Momo F, Massobrio M. Changes in Average Annual Precipitation in Argentina's Pampa Region and Their Possible Causes. Climate. 2015; 3:150–167. DOI: 10.3390/cli3010150.

[21] De la Casa AC, Ovando GG, Nasello OB. Changes in the Intensity and Variability of Precipitation in the Central Region of Argentina between 1960 and 2012. Climate. 2018; 6:66. DOI: 10.3390/cli6030066.

[22] Viglizzo EF, Roberto ZE, Filippin MC, Pordomingo AJ. Climate variability and agroecological change in the Central Pampas of Argentina. Agric. Ecosyst. Environ. 1995; 55:7–16. DOI: 10.1016/0167-8809(95)00608-U.

[23] Sabattini JA, Sabattini RA, Ledesma S. Caracterización del bosque nativo del centro norte de Entre Ríos (Argentina). Agrociencia. 2015; 19: 8-16.

[24] Kottek, M.; Grieser J, Beck C, Rudolf B, Rubel F. World Map of the Köppen-Geiger climate classification updated. Meteorologische Zeitschrift. 2006; 15(3):259-263.

[25] Jaiswal RK, Lohani AK, Tiwari HL. Statistical Analysis for Change Detection and Trend Assessment in Climatological Parameters. Environ. Process. 2015;2:729–749. DOI: 10.1007/s40710-015-0105-3.

[26] Jaiswal AK, Prakasa Rao GS, De US. Spatial and temporal characteristics of evaporation trends over India during 1971–2000. Mausam. 2008; 59(2): 149–158.

[27] Eslamian S, Hassanzadeh H, Khordadi MH. Detecting and evaluting climate change effect on frequency analysis of wind speed. Int J Global Energy. 2009; 32(3):295-304. DOI: 10.1504/IJGEI.2009.030657.

[28] Salarijazi M, Ali Mohammad AA, Adib A, Daneshkhan A. Trend and change-point detection for the annual stream-flow series of the Karun River at the Ahvaz hydrometric station. Afr J Agric Reserv. 2012; 7(32): 4540–4552. DOI: 10.5897/AJAR12.650.

[29] Staudt M, Esteban-Parra, Castro-Diez Y. Homogenization of long-term monthly Spanish temperature data. Int J Climatol 2007; 27:1809–1823. DOI: 10.1002/joc.1493.

[30] Marengo JA, Camargo CC. Surface air temperature trends in southern Brazil for 1960–2002. Int J Climatol 2008, 28: 893–904. DOI: 10.1002/joc.1584.

[31] Yue S, Wang C. The Mann-Kendall Test Modified by Effective Sample Size to Detect Trend in Serially Correlated Hydrological Series. Water Resour Manag 2004; 18:201–218. DOI: 10.1023/B:WARM.0000043140.61082.60.

[32] Stepanek P, Zahradnek P, Skalak P. Data quality control and homogenization of air temperature and precipitation series in the area of Czech Republic in the period of 1961-2007. Int J Global Energy. 2009; 3:23-26. DOI: 10.5194/asr-3-23-2009.

[33] Bates BC, Chandler RE, Charles SP, Campbell EP. Assessment of apparent non-stationarity in time series of annual inflow, daily precipitation and atmospheric circulation indices: A case study from southwest Australia. Water Resour Res. 2010; 46: W00H02. DOI: 10.1029/2010WR009509.

[34] Tabony RC. The estimation of missing climatological data. Journal of Climatology 1983; 3:297-314. DOI: 10.1002/joc.3370030308.

[35] Xia Y, Fabian P, Stohl A, Winterhalter M. Forest climatology: Estimation of missing values for Bavaria, Germany. Agricultural and Forest Meteorology. 1999; 96: 131-144. DOI: 10.1016/S0168-1923(99)00056-8.

[36] Paulhus JL, Kohler MA. Interpolation of missing precipitation records. Monthly Weather Review. 1952; 80:129-133. DOI: 10.1175/1520-0493(1952)080<0129:IOMPR>2.0.CO;2.

[37] Winingaard JB, Kleink Tank AMG, Konnen GP. Homogeneity of 20th Century European Daily Temperature and Precipitation Series. Int J Climatol 2003; 23:679–692. DOI: 10.1002/joc.906.

[38] Pettitt AN. A non-parametric approach to the change point problem. J Appl Stat 1979; 28(2):126–135. DOI: 10.2307/2346729.

[39] Zhang S, Lu XX. Hydrological responses to precipitation variation and diverse human activities in a mountainous tributary of the lower Xijiang, China. Catena. 2009; 77:130–142. DOI: 10.1016/j.catena.2008.09.001.

[40] Gao P, Mu XM, Wang F, Li R. Changes in streamflow and sediment discharge and the response to human activities in the middle reaches of the Yellow River. Hydrol Earth Syst Sci. 2011; 15:1–10. DOI: 10.5194/hess-15-1-2011.

[41] Buishand TA. Some methods for testing the homogeneity of rainfall records. J Hydrol. 1982; 58(1–2):11–27. DOI: 10.1016/0022-1694(82)90066-X.

[42] Kang HF, Yusof F. Homogenity test on daily rainfall series in Peninsular Malasiya. Int J Contemp Math Sci. 2012; 7(1):9–22.

[43] Mann, H.B. Nonparametric Tests Against Trend. Econometrica. 1945; 13: 245–259. https://www.jstor.org/stable/1907187

[44] Kendall, M.G. Rank Correlation Methods; Charles Griffin: London, UK. 1955

[45] Spearman, C. The proof and measurement of association between two things. Am. J. Psychol. 1904; 15:72–101. DOI: 10.2307/1422689.

[46] Lehman, E.L. Nonparametric Statistical Methods Based on Ranks; Holden-Day: San Francisco, CA, USA. 1975.

[47] Sen, P.K. Estimates of the Regression Coefficient Based on Kendall's Tau. J. Am. Stat. Assoc. 1968; 63:1379.

[48] Iriondo MH, Paggi JC, Parma MJ. The Middle Paraná River. Limnology of a Subtropical Wetland. Springer-Verlag Berlin Heidelberg. 2007, 382 pp.

[49] Palmer M, Lettenmaier DP, LeRoy Poff N, Postel SL, Richter B, Warner R. Climate Change and River Ecosystems: Protection and Adaptation Options. Environmental Management. 2009; 44: 1053–1068. DOI: 10.1007/s00267-009-9329-1.

[50] Poff NL, Allan JD, Bain MB, Karr JR, Prestegaard KL, Richter BD, Sparks R, Stromberg J. The natural flow regime: a new paradigm for riverine conservation and restoration. BioScience. 1997; 47: 769–784. DOI: 10.2307/1313099.

[51] Postel S, Richter B. Rivers for life: managing water for people and nature. Island Press, Washington, DC, 2003; p 240.

[52] Barros VR, Boninsegna JA, Camilloni IA, Chidiak M, Magrín GO, Rusticucci M. Climate change in Argentina: trends, projections, impacts and adaptation. WIREs Climate Change. 2015; 6:151-169. DOI: 10.1002/wcc.316.

[53] Dai A, Fung IY, del Genio AD. Surface observed global land precipitation variations during 1900–88. J. Clim. 1997; 10:2943–2962. DOI: 10.1175/1520-0442(1997)010<2943: SOGLPV>2.0.CO;2.

[54] de la Casa AC, Nasello OB. Análisis periódico de las condiciones hidrológicas en la Provincia de Córdoba, Argentina. Anales AFA. 2015; 26:170–176. DOI: 10.31527/analesafa.2015.26.4.170.

[55] Hoffmann JAJ, Nuñez S, Gomez A. Fluctuaciones de la precipitación en la Argentina, en lo que va del siglo. II Congreso Interamericano de Meteorología. V Congreso Argentino de Meteorología. Anales 1987; 12-1.5.

[56] Haylock MR, Peterson TC, Alves LM, Ambirizzi T, Anunciação YMT, Báez J, Barros VR, Berlato MA, Bidegain M, Coronel G, Corradi V, Garcia VJ, Grimm AM, Karoly D, Marengo JA, Marino MB, Moncunill DF, Neche TD, Quintana J, Rebello E, Rusticucci M, Santos JL, Trebejo I, Vincent LA. Trends in total and extreme South American rainfall in 1960-2000 and links with sea surface temperature. Journal of climate. 2006; 19: 1490-1512. DOI: 10.1175/JCLI3695.1

[57] Nascimento Jr, Rodrigues Silvestre M, Santanna Neto JL. Trends and rainfall tropicalization in Paraná State, south of Brazil. Atmósfera. 2020; 33(1):1-18. DOI: 10.20937/ATM.52441.

[58] Hasan Z, Akhter S, Kabir A. Analysis of rainfall trends in the South-east Bangladesh. Journal of Environment. 2014; 3:51-56. https://www.jstor.org/stable/24870334.

[59] Addsisu S, Selassie YG, Fissha G, Gedif B. Time series trend analysis of temperature and rainfall in lake Tana sub-basin. Ethiopia. Environmental Systems Research. 2015; 4:25. DOI: 10.1186/s40068-015-0051-0.

[60] Zarnetske PL, Skelly DK, Urban MC. Biotic multipliers of climate change. Science 2021; 336:1516–1518. DOI: 10.1126/science.1222732.

[61] Dale VH, Tharp ML, Lannom KO, Hoges DG. Modeling transient response of forests to climate change. Sci. Total Environ. 2010; 408:1888–1901. DOI: 10.1016/j.scitotenv.2009.11.050.

[62] Ziska LH, Blumenthal DM, Runion GB, Hunt ER, Diaz-Soltero H. Invasive species and climate change: an agronomic perspective. Clim. Change. 2011; 105:13–42. DOI: 10.1007/s10584-010-9879-5.

[63] Harvell CD, Mitchell CE, Ward JR, Altizer S, Dobson AP, Ostfeld RS, et al. Ecology – Climate warming and disease risks for terrestrial and marine biota. Science. 2002; 296: 2158–2162. DOI: 10.1126/science.1063699.

[64] Ling SD. Range expansion of a habitat-modifying species leads to loss of taxonomic diversity: a new and impoverished reef state. Oecologia. 2008; 156:883–894. DOI: 10.1007/s00442-008-1043-9.

[65] Smale DA, Wernberg T. Extreme climatic event drives range contraction of a habitat-forming species. Proc. R. Soc. B. 2013; 280:20122829. DOI: 10.1098/rspb.2012.2829.

[66] Bellard C, Bertelsmeier C, Leadley P, Thuiller W, Courchamp F. Impacts of climate change on the future of biodiversity. Ecol. Lett. 2012; 15:365–

377. DOI: 10.1111/j.1461-0248.2011.01736.x.

[67] Sabattini JA, Sabattini RA, Urteaga Omar F, Bacigalupo M, Cian JC, Sabattini IA, Dopazzo VM. Recuperación del pastizal natural en un bosque nativo degradado del Espinal argentino: persistencia del control químico de arbustivas. Investigación Agraria. 2019; 21(2):93-107.

[68] Sabattini RA, Sabattini JA, Cian JC, Lindt MA. Mechanical recovery of a native forest with shrubs of the Espinal Ecoregion (Argentina). Journal Forestist. Istanbul University. 2018; 68 (2):78-86. DOI: 10.26650/forestist.2018.392323.

Chapter 11

Delineation and Dimension of Deforestation

Moses Z. Sithole and Isaac A. Agholor

Abstract

Deforestation is induced by human activities, cascading into associated cost and economic benefits. The concepts, dimensions and, deforestation caused by deliberate human activities were extensively examined. The chapter also highlighted the rationale for deforestation, environmental dimension to deforestation and contributions of forestry and forest by products to livelihoods. The segmented cases and experiences to create awareness on the need to discourage deforestation were explored. While recognizing that forest provides useful support for majority of rural dwellers, the justification for forest protection is advocated. Some communities depend on forests for their main livelihood, and use medicinal therapies derived from indigenous plants found in the forest. Forest remains an important pool of biodiversity and deforestation must be avoided. The chapter conclude that awareness of the impact of deforestation by the citizenry should be encouraged and supported by policies.

Keywords: deforestation, biodiversity, water cycle, soil erosion, air quality, flooding and natural habitats, climate change and human health. Rationale for deforestation

1. Introduction

Trees play a very pivotal role in the lives of both human beings and animals. Particularly, to human beings, trees plays a huge role. Trees takes up carbon dioxide and make use of it to produce the oxygen needed by human beings [1]. Not only that, trees are a source of food, medicine, shade and supports biodiversity. Moreover, trees provide the wood we need to make papers, furniture and shelter. Trees also helps control temperatures, reduces soil erosion as well as soil salinity [2]. Significantly, trees remain the source of and provides food. Shade, medicine, and enhancement of biodiversity is realized from trees [1]. Therefore, it is beyond any shadow of doubt that trees are important for human life to be sustained. However, there is an ever-growing challenge across the world. The challenge attributed to, is the clearing of our forests for some reasons. Human settlement, construction of roads, and medicinal extraction as well as agricultural purposes are some reasons, among others, that lead to clearing of our forests.

A forest is an enormous area that is covered with trees and vegetation, which is also known as undergrowth [3]. It is also regarded as a very multifaceted ecosystem dominated with trees that safeguards the earth and also support a myriad of life forms [4]. Trees are very much important as far as environmental components is concerned. This is mainly because of the very benefits of trees on the myriad of life forms it supports. In as much as a forest is of great importance to life, there are two processes which determines its existence. That is the processes of afforestation and deforestation. Afforestation is defined in simple terms as a process of planting trees in a particular land [5]. This promotes vegetation and canopy for soil protection and leads to the balance of the ecosystem. Afforestation is a method sought, so incredible, by humankind, in the quest to fighting the effects of climate change, dealing with issues of soil degradation and maintenance of soil organic carbon levels [6]. This is on technique of humankind to prevent desertification.

On the other hand, deforestation as defined by [1], is the clearing of trees in forests. Significantly, deforestation can be traced to manmade and natural processes. The natural process of deforestation includes: plant diseases, damage of trees by other plants, damage of trees by animals feeding or habiting on the trees, storms and wind, as well as other climatic conditions such as flood and veld fires [4, 7, 8]. While manmade processes involves: construction developments, agricultural engagements, and clearing of trees for medicinal purposes and usages [9]. It is furthermore, of paramount importance to note that deforestation is the major and lead factor to climate change. Climate change refers to the alterations of climatic patterns regionally as well as worldwide [10]. Deforestation comes with a number of consequences. During the clearing the forests, good and developmental cost can be tabled. However, the cost and consequences of deforestation are punitive to both human kind and animals (both aquatic and wildlife).

The rationale for deforestation are highlighted and it includes: soil degradation, increased chances and levels of soil erosion, increase in the reflected radiation of solar, increase in the migration of animal species in search for habitat and greener pastures, and increased economic instability [1–2, 8]. Moreover, it is evident from research that the increase in the rate at which population grows as well as the changes in the ways and technologies of cultivation are the primary reasons of deforestation [11]. This research facts points to us the danger that continues to loom around the myriad of life forms. Therefore, the task of preaching the importance and practice of sustainable development remain pivotal. Hence, this chapter, draws on the meaning of deforestation, shades light on the rationale and consequences of deforestation.

2. Delineation and dimensions of deforestation

2.1 Delineation of deforestation

The conversion of forested areas into permanent non-forest areas and lands is known as deforestation [6]. Deforestation is defined also as a process of shrinking the areas of tropical forests for agricultural, grazing and construction purposes [4]. Other authors have presented some more definitions of deforestation. Tariq and Aziz [1] opined that deforestation is the degradation of the environment that threatens the quality of the existing forests. Notably, deforestation can be intentional and non-intentional clearing of the existing forests, through human activities [3, 12]. These land use changes come with enormous economic benefits for improved livelihoods across the world [13]. However, these benefits, could only be sustainable with good forest management practices [1, 6]. Therefore, it is important to create sustainable use of forest resources across the world, for sustainable development.

Deforestation involves the process of reducing vegetal cover in the forest's lands. This reduction of vegetation can either be man-made or be naturally enforced [14]. Deforestation occurs because of natural phenomenon such as wind and storm, veld fire and plant diseases. Moreover, deforestation comes in two groups, that is: 1) the change in land use and 2) the change in the land cover [3, 8]. These two groups can then be described as the categories of deforestation. Therefore, the **Table 1**, below, shows the examples of the two categories of deforestation:

Pareta and Pareta [12] argued that deforestation is the wasting of existing forests vegetation and resources. Vegetation refers to all characteristics of plant cover in an area. Therefore, deforestation is the process of engaging in removing and reducing vegetation in the existing forest. The resources involve: stable soils, wildlife, balanced biodiversity, wildlife habitats, air and clean water. Significantly, in the process of clearing forests, humankind deprive themselves the wonderful God-given resource that is the beauty of the environment [14]. Of which, if it is taken care of, it offers great opportunities for recreation and tourism [6]. Recreation and tourism can be of economic importance as far as livelihood is concerned. Therefore, it is certain that the use of forest resources must be accompanied with sustainable use of readily available resources, for the betterment of the livelihoods of the current generation, without compromising the future generation.

2.1.1 Deforestation and human activities

In the quest for the betterment of the current generation's livelihood, mankind exploits the readily available resources. In as much as powerful and convincing reasons can be tabulated, it remains a fact that most of the activities of mankind are in a way leading to more environmental problems. Hence, some authors came up with some definitions of deforestation in relation to human activities: As the alteration of afforested lands to non-forested lands which can be intentional and non-intentional [15]. Some of the activities of humankind that leads to deforestation are reflected in **Table 1** above. Significantly, the clearing of trees in the existing forests for commercial usage is entails deforestation. Commercial usages of the tree materials collected from the forest includes: firewood harvesting, production of charcoal, mining of timber, logging, hunting, tourism and agricultural activities [1–3, 6].

On the other hand, [9] consider deforestation as the clearing of a significant number of trees in a forestland. Particularly, in the context of human activities as compared to natural causes. Deforestation is the process of degrading trees in the forest by human activities. Notably, the collection of all unsustainable forestry practices, such as use of firewood and timber, and these practices neglect the principle of enduring development which involves the sustainable use of resources while ensuring that the needs of the future generations are not compromised [16]. Human, also cause reduction in forest resources through clearing all vegetation from the existing forestlands without recognizing the need of environmental conservation [17].

Furthermore, human activities transcends the cutting down of trees to mining, logging, agriculture, human habitation, and industrialization [7–8, 18]. Therefore, deforestation is anti-livelihoods support and development. Afforestation is the

Categories of deforestation	
Change in land use (man-made)	**Change in land cover (man-made)**
Alteration of forest land for agricultural purposes (ranching and farming activities).	Clearing of trees for firewood and charcoal production
Alteration of forest lands for mining purposes (mining of underground minerals, timber mining and stone mining)	Clearing of trees for logging
Alteration of forest lands for areas of settlement (formal and informal settlement)	Clearing of trees for medicinal purposes (both authorized and non-authorized persons)
Alteration of forest lands for road construction and industrialization	Loss of trees, plants and grasses due to veld fires

Table 1.
The categories of deforestation.

only way out of the devastating consequences of deforestation, backed with implementation of policies on afforestation. As a matter of fact, if trees are extinct, then human and animal life itself is endangered and destroyed.

Human and animal life is encored and supported by plant life and the shortage of trees, will result to inadequate supply of woods for shelter, furniture, and the paper products used today [2–4, 19]. Furthermore, the availability and flow of oxygen is dependent on vegetation used to sustain human life. Therefore, understanding what deforestation is, becomes of paramount importance. Hence, [6–9, 16] examined deforestation as the clearing of trees from the existing forest lands due to bad agricultural practices, lumbering activities, change in climatic patterns, urbanization, air pollution, tourism, and overgrazing. On the other hand, [2–4] submits that the destruction of vegetation cover in the existing forestlands is referred to as deforestation. While [6–8] delineates deforestation to be the quest for economic productivity through agricultural endeavors, mining, and urbanization leading to land degradation. Therefore, it is clearer and undisputable, that the standing delineation of deforestation in the context of manmade processes is that deforestation is the cutting down of a huge number of trees leading to clearing of the existing forests for the purpose of livelihood betterment while compromising the future generations, by avoiding replenishing the cleared forests in ignorance or intentionally.

2.2 Rationale for deforestation

Numerous studies, [1–4, 9, 13] indicated the flux inherent in environmental deforestation. Nevertheless, increases in environmental awareness and activism across the world, on deforestation have also been reported to be on the increase as well. Therefore, there is a need to create continuous awareness campaign of deforestation. Significantly, deforestation is traceable to the following reasons: Change in ways of farming, intensified ranching and overgrazing, tourism, urbanization, wars and military affairs, mining, population growth (overpopulation), medicinal purposes, undervaluing forestry, logging, firewood harvesting and charcoal production, policy failure, poor land-use planning, damage of trees and plants by mammals, damage of trees by other plants, death and decay of trees, wind and storms, and veld fires [1–9, 13, 20].

2.2.1 The deforestation caused by deliberate human activities

The rationale for deforestation as shown on **Figure 1** below, occur in two broad categories: the intentional and the non-intentional. Intentional deforestation involves the human activities resulting to deforestation. It encompasses all the processes and activities that humankind engaged in without careful and sustainably approach to protect the environment. For example, clearing forest lands for agricultural purposes, mining, and industrialization. This is an intentional approach to deforestation, since, there are no careful considerations towards afforestation.

Agriculture is one of the major drivers of deforestation [12]. On the other hand, the agricultural sector is one of the major contributors towards economic growth and it employs most of the rural people [1]. As for employment and wealth creation, 60% of the world's population depend on the agricultural sector [8]. However, agricultural activities contributes 60% of deforestation worldwide [13]. Gorte and Sheikh [20] argued that agricultural activities leading to massive deforestation is mostly associated with cultivation shifts. This happens when a forest is cleared by means of burning trees and cutting down trees so to avail space for cultivation. Agricultural activities varies from nation to nation, but it occurs at three levels,

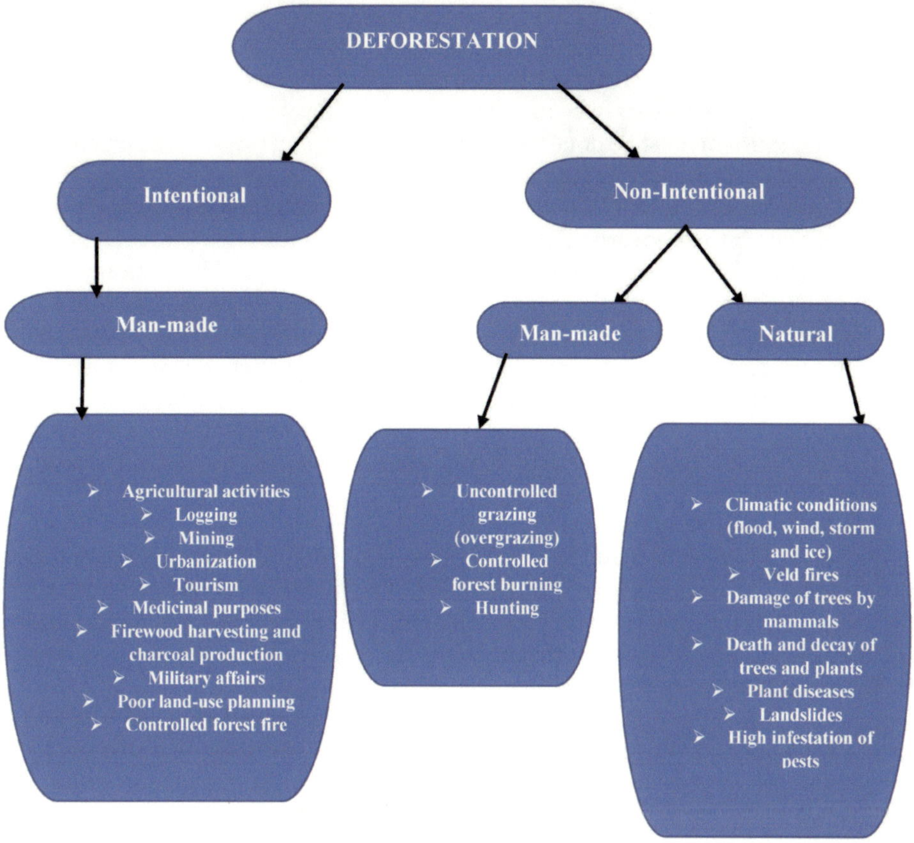

Figure 1.
Schematic illustration of deforestation.

namely: commercial, small-scale and subsistence farming levels [1–2]. Moreover, agriculture is known to contribute 12% of towards the greenhouse gas emissions worldwide [6]. This mainly happens when the forest is burnt down for agricultural purposes as well as when farm implements are in use [4].

On the other hand, logging and mining are reported to be other reasons of deforestation. Logging is the commercial clearing of forest by cutting trees down for the purpose of making paper [13]. While mining involves all activities whereby minerals are extracted either from the ground or forest. Mining in forest clearing happens when people are mining timber for construction and paper making. However, mining also happens in terms of underground minerals mining, such includes mining of gold, iron ore, coal, diamond and oil [1–3]. Moreover, mining of stone and sand does happen from the surface of the earth. Particularly, trees are cleared to give space for sand and stone mining. Logging is one of the economic growth contributors, globally. However, unsustainable commercial logging, results in deforestation. Pareta and Pareta [12] argued that logging is not the cause of deforestation. On the contrary, [3] argued that logging practiced in an unsustainable way is a direct cause of deforestation. Therefore, sustainable extraction of wood form the forests is to be practiced.

Population growth increases urbanization and rural settlement, and there are activities that creates ever-growing demands for space in the environment [1]. In the quest for human settlement, industrialization, employment, and wealth creation, causes deforestation to occur [20]. Urbanization is a necessity for development. However, it must follow a planned, managed and controlled manner [18]. On the other hand, tourism sites are preferred for recreation and wealth creation [2]. However, if unsustainable practices are employed, that will result in massive vegetation cover loss [1–2, 20]. Therefore, it is an undisputable fact that urbanization and tourism are key role players in economic growth and development. Hence, urbanization and tourism must be done with careful planning, management and control.

Firewood and charcoal production is reported by [1] to be among the main drivers of deforestation, globally. Firewood harvesting is one of the reasons of deforestation that needs to be governed. As a matter of fact, [7] reported that their study found that 90% of the respondents in their study, uses firewood for cooking. This follows findings by [1], that there is a cumulative demand of firewood and timber for domestic use as the populations of nations continues to grow. This further leaves the tress at mercy of the ever-growing population. Kissinger et al. [15], asserted that timber accounts for almost 70% of deforestation, globally. The timber collected is then used for both firewood and charcoal production. However, the major drivers of deforestation includes: firewood harvesting, charcoal production and overgrazing of livestock [2].

Timber is, significantly, needed and utilized almost by everyone in construction of shelter. While firewood and charcoal is used for cooking and warming in winter season [15, 16]. Considering the human demand on forest and forestry products, the enactment of policies to govern the harvesting of timber becomes critical. More so, to create an awareness about the impact of deforestation and the apparent need for forest conservation.

Another reason for deforestation is the uncontrolled cutting down of trees and shrubs for medicinal purposes [1]. In this case, trees and plants with medicinal benefits are destroyed. Notably, this action does not only result in deforestation, but it also endangers the environment for human habitation. In most instances, trees are used for medicinal purposes by villagers and people surrounding forestlands. While others sell herbs obtained from the forest for income generation.

2.2.2 Deforestation: non-intentional activities and processes

As illustrated in **Figure 1**, above, deforestation also is as a result of non-intentional activities and processes. These processes and activities involves both natural and man-made activities and processes. Natural processes leading to non-intended deforestation involves: death and decay of trees and plants, plant diseases, landslides, damage of trees by other plants, high infestation of pests as well as climatic conditions such as veld fires, floods, and winds and storms [2–4]. On the other hand, man-made processes and activities involves: uncontrolled grazing which leads to overgrazing, hunting as well as controlled forest burning [4–5, 9]. Therefore, this section examines the natural and manmade processes and activities leading to deforestation at a non-intended approach.

Climatic conditions and landslides causes serious environmental damage to lands used for forestry and agricultural activities. The massive environmental harm comes in the form of deforestation and land degradation [13]. Climate change is one of the environmental issues across the whole world. Significantly, climatic conditions and climate change are both regarded as the drivers and the effect of deforestation and degradation of forests. As drivers of deforestation, climatic conditions are recognized in the form of tropical storms and winds, floods, ice, droughts, heat waves and veld fires [4, 7, 8]. On the other hand, landslides refers to mass wasting. Mass wasting is an occurrence of the movements of mass rocks, debris and earth downslope. This occur in five forms, namely: falls, topples, slides, spreads and flows. As a result of this phenomenon, destruction of forests occurs as well. Hence, landslides and climatic conditions are both drivers of deforestation across the world.

Deforestation occur as a result of unavoidable death and decay of trees and plants, plant diseases, high infestation of pests, damage of trees by other plants in the forestlands [8, 13–16]. With robust attention and huge budgets given to plant protection in the agricultural sector [7, 20]. Therefore, the lack of forest management policies, inadequate and ineffective management of the existing forests puts our existing forests at a risk of deforestation.

These are human activities resulting in unintended deforestation. These include uncontrolled grazing resulting in overgrazing, uncontrolled forest burning and hunting [20]. Hunting is one of the means of livelihood, particularly, for rural communities. In as much as hunting, poaching and snaring provides rural people with food harvested from wildlife, in terms of bush meat, it remains a factor causing deforestation [16, 20]. Hunters cut down trees in the forests to setup traps and this traps and snares are sometimes made on living plants and trees [21]. Furthermore, the traps set on living trees, then causes damage which can result in death and decay of the affected tress in the forest.

2.3 Dimensions of deforestation

Human activities tend to justify deforestation as good actions because it support livelihoods. However, as depicted on **Figure 2**, below, deforestation comes with its attendant environmental and socio-economic consequences. The rationale for deforestation includes: shortages of food, shortages of wood, mass destruction of wildlife and high wildlife migration rate, alteration of vegetation cover and structure, damage and loss of flora and fauna essential for recreation and tourism, desertification, disturbed water cycle, soil erosion, flooding, landslides, and land degradation, reduced atmospheric moisture, destruction of biodiversity, hardening of iron-rich soils to form laterite, changes and imbalances in nutrient cycle and energy, declined pollination, increased poverty levels, decline in soil fertility, decline of medicinal herbs production, decline in income generation, water loss as well as loss of natural flood and erosion control [2–7, 22]. All these activities contribute to deforestation.

2.3.1 Environmental dimension of deforestation

As shown in the **Figure 2**, below, deforestation have a number of environmental impact. These are: loss of biodiversity, mass destruction of wildlife and high rate of wildlife migration, alteration in the composition and structure of vegetation, damage of flora and fauna, desertification, soil erosion, flooding and the hardening of iron-rich soils to form laterite.

Biodiversity refers to all the desirable and important variety of animal and plant life in a habitat. The variety of animal and plant life is also referred to as the fauna and flora. Biodiversity is also known as the foundation of the ecosystems linked to humankind [9, 20]. Furthermore, within an ecosystem, there are wide range of animals and plants that supports production of biomass, water and nutrient cycle and energy, as well as soil formation and maintenance [7, 16]. However, deforestation and the degradation of forests destroys the biodiversity. When the biodiversity is destroyed, animals may migrate while some animal species may be extinct.

Significantly, the loss of biodiversity, because of deforestation have harsh impact particularly, on human health. This is due to the fact that biodiversity supports and benefits human life in many different ways [16]. Conservation of nature, is therefore, threatened by the destruction of biodiversity [8]. Which also leads to human-animal conflict. Human-animal conflict happens due to the declines in the

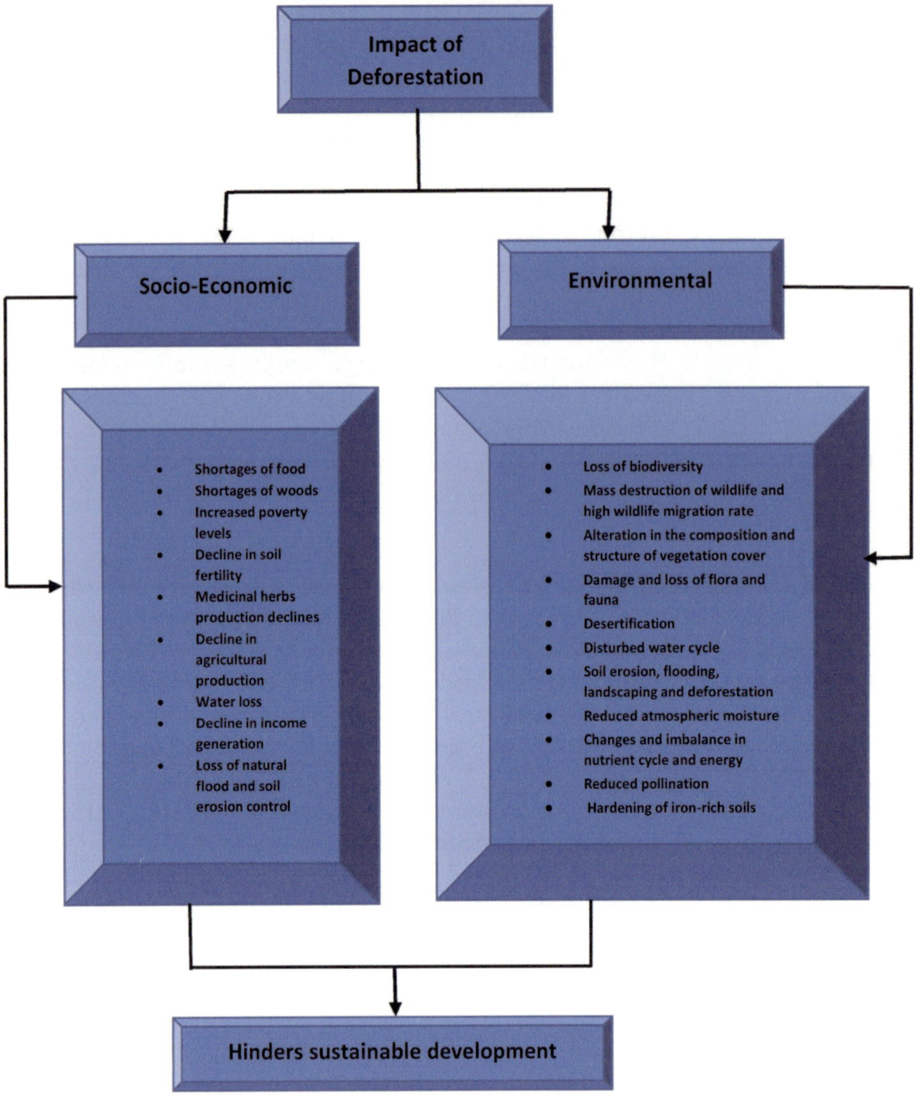

Figure 2.
Impact of deforestation.

agricultural productivity. These decline in agricultural productivity results from the effects of climate change which resulted from deforestation [23]. Thus, hunting, poaching and snaring becomes the way out of poverty for many rural people. Therefore, conservation approaches must be considered so to conserve the biodiversity that supports our health system and human existence.

Forests are not only the source of wood, but are important for human life support [15]. The trees growing in the forests make use of the carbon dioxide to produce the oxygen we need to sustain life [1]. In as much as trees are beneficial to human life, there is an ongoing threat to forest and plant life. This is because of all the natural and man-made processes resulting in deforestation. Deforestation has the ability to disrupt the normal weather conditions and create extremely hot and drier weather conditions [17]. Hence, it is believed that deforestation leads to the change in climate. Climate change manifest in increased temperatures, increased drought and prolonged drought seasons, delayed rain seasons, floods and displacement of vegetation cover, very heavy winds, and tropical storms [2–4, 16]. All these changes in climatic conditions have serious implications on human and animal life. The imbalances of the atmospheric nutrient cycle, reduction of atmospheric moisture, increase in the high concentrations of Chloro-Floro Carbons (CFSs) in the atmosphere [4]. All these are results of deforestation and have serious implications on human health and existence. Therefore, deforestation is not only a theoretical concept, but a reality with detrimental impact to human health and life. Hence, measures to replenish our existing forests are critical to be implemented with careful consideration for sustainable development.

2.3.2 Socio-economic impact of deforestation

As shown in the **Figure 2**, above, deforestation have a number of socio-economic consequences. These are: shortages of food, shortages of wood, increased levels of poverty, decline in soil fertility, loss of medicinal therapies, loss of valuable agricultural land, water loss and, inadequate income. These consequences are intertwined with sustainable development. In most communities, forest is valued to an extent that, its destruction amounts to the destruction of their values, beliefs, norms and cultures [4]. This further shows that deforestation is not only an environmental issue but is a social and economic issue. Deforestation result in shortages of woods [21]. Woods from the forests are used for different purposes which includes making and production of paper, making of furniture, construction as well as firewood and charcoal production.

However, sustainable management of our existing forests will ensure that we do not run short of wood. Significantly, [1] presents that deforestation results in decline of soil fertility. However, land degradation because of deforestation render the soil infertile. Furthermore, decline in soil fertility leads to failure in agricultural productivity. And poor agricultural productivity means that we are unable to produce adequate and quality food [24]. This then result in food shortages. Food shortages at the extreme leads to food insecurity which is linked with many heath issues. Therefore, conservation of natural resources is necessary for economic and social needs of communities.

Daramola et al. [24] defines sustainable development as a process of economic development that does not compromise the future generations but conserves the natural resources for them. Murphy [22] suggests that the development that satisfies the needs of the current generation, no depriving the future generations the ability to satisfy their needs is sustainable development. Sustainable development is mostly known for its three (3) pillars, namely: social, environmental and economic sustainability [2, 22–24]. This means that any development happening, must be economic equitability, it must be socially bearable, and environmental viability (**Figure 3**).

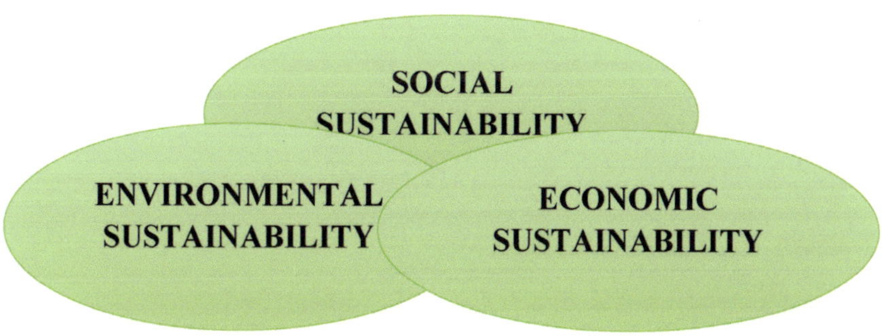

Figure 3.
Pillars of sustainable development (SD).

Therefore, sustainable development aims to promoting conservation of nature and all natural resources for the use by future generations. However, **Figure 2**, reveals that deforestation, as one of the environmental issues, globally, and hinders the achievement of sustainable development objectives.

3. Conclusion

Deforestation result from human activities in the environment. It may be unexpected, and sometimes deliberate action supported by livelihood needs. The concepts and dimensions of deforestation, human activities in relation to deforestation, the rationale for deforestation, environmental dimension to deforestation and contributions of forest and forest by products to livelihoods were examined in this chapter. The authors segmented cases and experiences and, highlighted efforts to create awareness on the need to discourage deforestation. Nevertheless, forest remain an important pool of biodiversity and deforestation must be avoided. The chapter conclude that awareness of the impact of deforestation should be supported by policies. Therefore, it is the authors believe that policy discourse will assist to provide enduring forest conservation and management.

Conflict of interest

The authors declare no interest.

Author details

Moses Z. Sithole* and Isaac A. Agholor
Faculty of Agricultural and Natural Sciences, University of Mpumalanga, Mbombela, South Africa

*Address all correspondence to: apostle12352@gmail.com

References

[1] Tariq, M., and Aziz, R. 2015, "An Overview of Deforestation Causes and Its Environmental Hazzards in Khyber Pukhtunkwa", Journal of Natural Sciences Research, Vol. 5 (1), 52-58.

[2] Kassa, H., Dondeyne, S., Poesen, J., Frank, A., and Nyssen, J., 2017, "Impact of Deforestation on Soil Fertility, Soil Carbon, and Nitrogen Stocks: the Case of the Gacheb Catchment in the White Nile Basin, Ethiopia", Journal of Agriculture, Ecosystems and Environment: http://dx.doi.org/10.1016/j.agee.2017.06.034.

[3] Tejaswi, G., 2007, "Manual on Deforestation, Degradation, and Fragmentation Using Remote Sensing and GIS", Monitoring, Assessment and Reporting on Sustainable Forest Management (MAR-SFM) Working Paper 5/2007.

[4] Okia, C.Z., 2012, "Global Perspectives on Sustainable Forest Management". Intech Publishers, Rijeka, Croatia.

[5] Bennett, L., 2017, "Deforestation and Climate Change", the Climate Institute, Washington, DC.

[6] FAO (Food and Agriculture Organization of the United Nations), 2010, "Global Forest Resources Assessment", Rome, Italy.

[7] Tariq, M., Rashid, M., and Rashid, W., 2014, Causes of Deforestation and Climatic Changes in Dir Kohistan", Journal of Pharmacy and Alternative Medicine, Vol. 3 (2), 28-37.

[8] Gebru, T.D., 2016, "Deforestation in Ethiopia: Causes, Impacts and Remedy", International Journal of Engineering Development, Vol. 4 (2), 204-209.

[9] Hosonuma, N., Herold, M., Sy, V., Fries, R.S., Brockhausg, M., Verchot, L., Angeslsen, A., and Romijn, A., 2012, "An Assessment of Deforestation and Forest Degradation Drives in Developing Countries", Environmental Research Letters, Open Access: doi:10.1088/1748-932/7/4/044009.

[10] Bindoff, N.L., Stott, P.A., AchutaRao, K.M., Allen, M.R., Gillett, N., and Gutzler, D., 2013, 'Detection and Attribution of Climate Change: From Global to Regional'. In: Stocker TF et al. (eds.) Climate Change 2013: The Physical Science Basis', Contribution of Working Group I to the Fifth Assessment Report of the Intergovernmental Panel on Climate Change, Cambridge University Press, Cambridge, United Kingdom and New York, NY, USA.

[11] Abere, S.A., and Opara, J. A., 2012, "Deforestation and Sustainable Development in the Tropics: Causes and Effects", Journal of Educational and Social Research, Vol. 2 (4), 105-109.

[12] Pareta, K., and Pareta, U., 2011, "Forest Carbon Management using Satellite Remote Sensing Techniques: A case Study of Sagar District", International Scientific Research Journal, Vol 3 (4), 335-348.

[13] Goll, N.B., Li, J., McKay, J., and John, S., 2014, "Analysis on the Causes of Deforestation and Forest Degradation in Liberia: Application of DPSIR Framework", Research Journal of Agriculture and Forestry Science, Vol. 2 (3), 20-30.

[14] Phua, M., Tsuyuki, S., Furuya, N., and Lee, S., J., 2008, "Detecting Deforestation with a Spectral Change Detection Approach using Multitemporal Landsat Data: A Case Study of Kinabalu Park, Sabah, Malaysia". Journal of Environmental Management, Vol. 88, 784-795.

[15] Kissinger, G., Herold, M., and Sy, D.V., 2012, "Drivers of Deforestation and Forest Degradation", A Synthesis Report for REDD and Policymakers, Lexeme Consulting, Vancouver, Canada.

[16] Moutinho, P., 2012, "Deforestation around the World", Intech publishers, Rijeka, Croatia.

[17] Ekiti, A., 2017, Effects of Deforestation on Land Degradation", ResearchGate: https://www.researchgate.net/publication/318921682.

[18] Mujuri, E. K., 2007, "Deforestation and Afforestation, A World perspective". A Class Paper, International Resource Management.

[19] Ahmad, S.S., Abbasi, Q, Jabeen. R., and Shah, M. T., 2012, "Decline of Conifer Forest Cover in Pakistan: A GIS Approach", Pakistan Journal of Botany, Vol. 44 (2), 511-514.

[20] Gorte, R.W., and Sheikh, P.A., 2010, "Deforestation and Climate Change", Congregational Research Service (CRS) Report for Congress, March 2010.

[21] Boateng, R., Heeks, R., Molla, A. and Hinson, R., 2008, "E-commerce and Socio-Economic Development: Conceptualizing the Link", Journal of Internet Research, Vol. 18 (5), 562-592.

[22] Murphy, K., 2012, "The Social Pillar of Sustainable Development: A Literature Review and Framework for Policy Analysis", Journal of Sustainability, Science, Practice and Policy, Vol. 8 (1), 15-29.

[23] Porter-Bolland, L., Ellis, E.A., Guariguata, M.R., Ruiz-Mallen, I., Negrete-Yankelevich, S., and Reyes-Garcia, V., 2011, "Community Managed Forests Protected Areas: An Assessment of their Conservation Effectiveness Across the Tropics", Forest, Ecology and Management Journal: doi:10.1016/jforeco.2011.05.034.

[24] Daramola, A., Adeoye, A., Akinola, R., Alagbe, W., and Ajayi, S., 2010, "Sustainable Development and Environmental Protection: Strategies and Procedures for Developing Nations", Institute for Environment Research and Development, Benja Ota, Nigeria.

Chapter 12

Meteorological Drought and Temperature in Sudano-Sahelian Region of Nigeria under Increasing Global Warming

Ojonigu F. Ati, Kayode Aremu, Adewale F. Olatunde, Iliya B. Abaje and Emmanuel O. Oladipo

Abstract

This study examinesmeteorological drought and temperature in the Sudano-Sahelian region of Nigeria under increasing global warming. Rainfall data of 60 years (1961 to 2020) and temperature of 40 years (1981–2020) for nine stations were used to quantify drought occurrences as well as the pattern of temperature. The Bhalme and Mooley Drought Index (BMDI) and Student's *t*-test were used to indicate if changes have occurred in the amount of rainfall and temperature (increase or decrease) over space and time in the study region. The BMDI was used because it is simple and able to depict historically documented droughts better than other indices. The student's *t*-test was used to determine if the sub-period means of the series have altered significantly over time. Results show that the spatial and temporal occurrence and duration varied from one sub-area to another. Also, its intensity was mainly from mild to moderate. Results also show a consistent alternation between drought and wet years over the study period. The trend of temperature shows a tendency towards decreasing temperature during the period of study although it was not significant. The study concludes that the study region experiences climate variability that may well be part of global climate change.

Keywords: trends, drought persistence, global warming, Sudano-Sahelian region, meteorological drought, climate change

1. Introduction

Global warming and climate change have dominated conferences, seminars, and discourses for several decades in almost all parts of the earth. This is because several destructive atmospheric, weather, and climatic conditions such as inundation by water, absence or inadequate rainfall, and rising temperatures are being attributed to them [1]. As a result, several studies are being provided to critically look at how these extreme events relate to the duo. Some of these studies on the continent of Africa that have tried to understand the rainfall cycle and pattern include [2–19] amongst others. Those on the trend of temperature include but are not limited to [20–22]. The findings of these studies have generally not been definite and have resulted in diverse opinions and views on the relationship between climate variability and change on one hand and drought and temperature on the other. One of such opinions is that drought that has occurred and may likely occur in the Sahel Region of West Africa is an indicator of climate change. To the proponents of this thought, its long history of occurrence and the probability of reoccurrences in the future are enough evidence [23–25]. In contrast to these arguments are those that argue that though the occurrences of drought and fluctuations in temperature seem unusual in some years and decades past, they generally have not deviated from the long-term probabilities and climate of the area. Olatunde [26, 27] argued that the droughts that have occurred and persisted in the Sahelian part of Africa should be taken as part of the region's climate. This argument is especially cogent when the definition of climate change and/or discontinuity as identifiable changes in the average temperature, precipitation, and wind pattern that persists for a long period typically decades or longer [28] or an abrupt and permanent change in the mean values of these variables during the period of record [29, 30] taken into consideration and the fact that wet periods consistently alternated with the drought periods in most sub-areas of the region. These arguments as well as the issue of limited years of available data suggest the constant need for studies of this nature especially in the Sudano-Sahelian region of Nigeria. This study intends to ascertain if the lack or inadequate rainfall and change in average temperature imply a change in the climate of the area under study.

2. Study area

The study area is made up of the Sudan and Sahel Savanna zones in Nigeria which covers about one-third of its entire land area. It occupies from latitude 10°N to latitude 14°N of the Equator and from longitude 4°E to longitude 14°E of the Prime Meridian (**Figure 1**). The region lies south of the Sahara Desert, extending from the Sokoto plains on the west through the northern parts of the plains of Hausaland to the Chad Basin on the east [32].

The climate of the study area is the tropical wet/dry type, also referred to as Aw. The annual mean rainfall for the zone varies from less than 500 mm in the extreme northeastern part to approximately 1323 mm in the southern part [33, 34]. The wet

Figure 1.
The Sudano-Sahelian Zone of Nigeria (Adapted from [31]).

season occurs from April to October with the highest rainfall amount in August, while the dry season is from November to April [35, 36]. The intensity of rainfall is high in July and August, from about 60 mm hour^{-1} to 99 mm hour^{-1} [37]. The rainfall pattern is variable over space and time with inter-annual variability of between 15% and 20% [37, 38].

The climate of the study area is influenced by three air-masses, namely: the tropical maritime (mT); the tropical continental (cT), and the equatorial easterly. The mT and cT meet along a sloping surface known as the Intertropical Discontinuity (ITD). The equatorial easterly is an unstable cool air mass from the east in the upper troposphere along the ITD [32, 34]. The equatorial easterly sometimes undercut the mT or cT air masses to result in line squalls or dust devils respectively [39, 40]. The location of the ITD depends on the season in association with short-period fluctuations. Overall, it is located to the north of the Sudan-Sahel region in July and August, thereby making the area to be completely under the effects of the mT air mass. The reverse is the case from October to May when the ITD is located south of the region causing the cT air mass to prevail over the region [32].

The movement of the ITD is not regular, it varies according to the season. Its retreat towards the south is rapid than its northward progress. Its northward movement is at about 160 km per month while that of the southward retreat is at about 320 km per month [41]. This explains the gentle onset of the rainy season in the study area and northern Nigeria as a whole and its sudden end [42]. The mean highest air temperature occurs during the hot season from March to May while the mean lowest air temperature occurs during the cold season from December to February.

Temporal occurrence of temperature is steadier than that of rainfall, however, the highest diurnal range of temperature is in the dry season. The highest air temperature occurs either in April or May and the lowest between December and February [43]. The mean atmospheric relative humidity is between 70% and 90% during the rainy season and between 25–30% during the dry season [44]. Evapotranspiration is high year-round, the highest amount occurring during the dry season [45]. Sunshine hours depend on latitude and season. The mean annual number of sunshine hours in the study area is high in contrast to the southern part of the country, with very high values in the extreme northeast part. For example, Nguru has a mean annual total of 3177 hours [46]. Overall, values are high during the dry season in the area as a result of low cloudiness compared to the rainy season [47].

The entire study area is covered by Savanna vegetation and classified into Sudan and Sahel with the density of trees and other plants decreasing towards the north. The Sudan and Sahel zones are occasionally together referred to as the Sudano-Sahelian Ecologically Zone (SSEZ).

3. Materials and methods

3.1 Data collection

Sixty years (1961–2020) of rainfall data and forty years (1981–2020) of temperature from nine (9) stations were analysed. These data are secondary and from the archives of NiMet, Abuja, and NASA Data Access Viewer. The criteria used in choosing the stations are; a long period of available rainfall and temperature data; those stations without large missing data for the time under study; those stations that have not been moved since they were established and also the data were subjected to normality distribution test (**Table 1**).

Station	Station No.	Latitude ('N)	Longitude ('E)	Altitude (m)
Yelwa	1004.54	10°53	04°45	224.00
Potiskum	1111.40	11°43′N	11°07′E	487.68 m
Maiduguri	1113.50	11°51′N	13°05′E	348.00 m
Kano	1208.03	12°03′N	08°32′E	475.80 m
Gusau	1206.14	12°10′N	06°42′E	468.00 m
Sokoto	1205.51[A]	12°55′N	05°12′E	309.00 m
Nguru	1210.52[E]	12°58′N	10°28′E	341.00 m
Katsina	1307.04	13°0i′N	07°41′E	516.63 m
Kaduna	1007.32	10O41′N	07O19XE	634.00 m

Table 1.
Meteorological stations used.

3.2 Data analysis

The rainfall amounts for months with effective rainfall for plant growth (April to October) for the nine (9) chosen stations were analysed with the Bhalme and Mooley Drought Index (BMDI) to determine drought intensity [48]. For farming activities, the growing season from April to October is taken to be very crucial in drought research and assessments. This is so, as these months are when about 95% of the entire annual rainfall is received in the study area [49].

Generally, the BMDI for a particular month K is determined with the formula:

$$I_k = (Mk/d) + (1 + C)\ I_{k-1} \tag{1}$$

Where:
C is taken as a constant.
d is taken as aconstant.
I_K = drought severity for the Kth month.
I_{k-1} = drought severity for the (K-1) month.
M = moisture index is determined by.

$$M = 100\ (x–\bar{x})/S \tag{2}$$

In Eq. (2),
X is the monthly rainfall amount;
$\bar{x}$ is the long period average monthly rainfall; and.
S is the standard deviation for the initial month being considered (K-1).
Eq. (1) is therefore given as:

$$I = M/d \tag{3}$$

The values of C and d in Eq. (3) for the Northern area in Nigeria are 0.43 and 38.84, respectively. These constants were derived by [48]. They were used in Eqs. (1) and (3) to determinethe monthly values of BMDI of the stations used. From the values, the averages or seasonal drought index (SDI) series were derived for each year in the study period. The seasonal indices were used to classify a year into any of the following wetness/dryness categories, using [48] classification chart (**Table 2**).

BMDI	The character of anomalous moisture conditions (CAMC)
4.00 or more	Extremely wet
3.00 to 3.99	Very wet
2.00 to 2.99	Moderately wet
1.00 to 1.99	Slightly wet
0.99 to −0.99	Near normal
−1.00 to −1.99	Mild drought
−2.00 to −2.99	Moderate drought
−3.00 to −3.99	Severe drought
−4.00 or less	Extreme drought

Source: Shuiabu and Oladipo [31].

Table 2.
BMDI classification chart.

The rainfall data were divided into non-overlapping decades; 1961–1970, 1911–1980, and so on to the present decadeof 2011–2020 and temperature decade of 1981–1990, 1991–2000 to the present decade of 2011–2020. Cramer's test as given by [50], was utilized in the compiling of the means of the decades with the mean of the entire study period. In the application of Cramer's test, the mean ($\overline{x}$), and the standard deviation (δ), were determined for the stations and the total number of years under study, *N*. As mentioned earlier, this was to find out the difference (in terms of a moving *t*-statistic) between the mean ($\overline{x}k$), for each successive n-year period and the mean ($\overline{x}$) for the entire study period. The *t*-statistic was taken as:

$$t_k = \left(\frac{n(N-2)}{N - n\left(1+\tau_k^2\right)}\right)^{1/2} \tau_k \tag{4}$$

with τ_k as a standardized measure of the difference between means given as:

$$\tau_k = \frac{\overline{x}k - \overline{x}}{\delta} \tag{5}$$

with $\overline{x}k$ as the mean of the sub-period of n-years. $\overline{x}$ and δ as the mean and standard deviation of the entire series respectively and τ_k as the value of the student *t*-distribution with *N*-2 degrees of freedom. It was then compared against the "students" *t*-distribution table, at a 95% confidence level as regards the two-tailed form test. When t_k was outside the bounds of the two-tailed probability of the Gaussian distribution (equal to 1.96 at 95% confidence level), a significant shift from the mean was assumed.

The temperature series were divided into 30-year overlapping sub-periods of 1981–2010 and 1991–2020, while the rainfall series were divided into 30-year non-overlapping sub-periods of 1961–1990 and 1991–2020 as recommended by the World Meteorological Organisation. The student's *t*-test, t_d, was later used to ascertain if the sub-period means have deviated significantly in the occurrence of wet and dry years through time. The statistics, t_d, was calculated using:

$$t_d = \frac{\left(\overline{X}_2 - \overline{X}_1\right) - (\mu_2 - \mu_1)}{\left[\frac{N_2 S_2^2 + N_1 S_1^2}{N_2 + N_1 - 2} \cdot \frac{1}{N_2} + \frac{1}{N_1}\right]^{½}} \tag{6}$$

with $(\overline{X}_1 - \overline{X}_2)$ being the difference in group means, $(\mu_2 - \mu_1)$, was the expected differences (set equal to 0), N_2 and N_1 being the number of cases in each sub-sample, and S_2 and S_1 were the respective standard deviations. When t_d was outside the bounds of the two-tailed probability of the Gaussian distribution, equal to 1.96 at 95% confidence level, a significant deviation and shift from the mean were assumed.

4. Results and discussion

4.1 Frequency distribution of drought series

The results of the analysis of the occurrence of drought, in the study area, using BMDI are depicted in **Figures 2–10**. They show that the occurrences and intensity of drought in sub-areas of the study area vary over time and space. This is exemplified by some sub-areas with intermittent years of drought and wetness. Yelwa sub-area experienced such a situation, even though droughts of mild and moderate intensities prevailed between 1965 and 1991. The later years of study, 2010–2020 also experienced a drought of mild intensity. Potiskum sub-area values also indicated years of drought and wetness throughout the study period with intensity varying between mild and moderate and a reduction in the drought intensity towards the end of the study period. This same situation also occurred in Gusau and Kaduna sub-areas. For Kaduna, the years between 1961 and 2011 were those of alternating wet and drought with intensity being mild.

Some sub-areas experienced very clear and distinct periods of drought. Generally, Maiduguri drought for example can be categorized into before the 1980s with

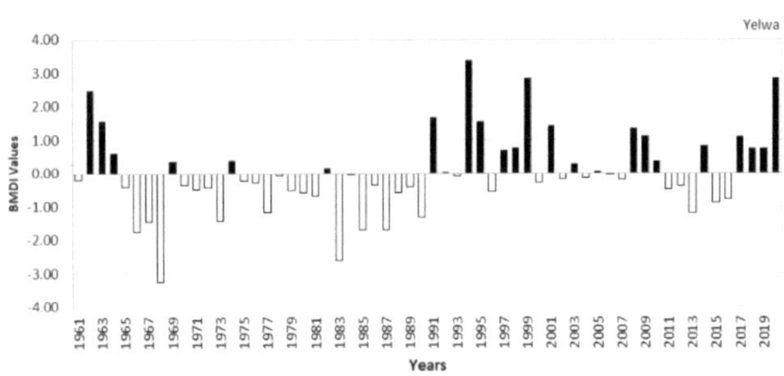

Figure 2.
Annual fluctuations in Bhalme and Mooley drought index for Yelwa.

Figure 3.
Annual fluctuations in Bhalme and Mooley drought index for Potiskum.

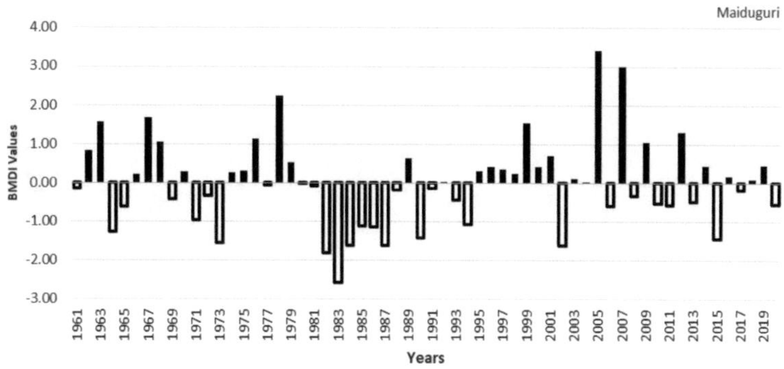

Figure 4.
Annual fluctuations in Bhalme and Mooley drought index for Maiduguri.

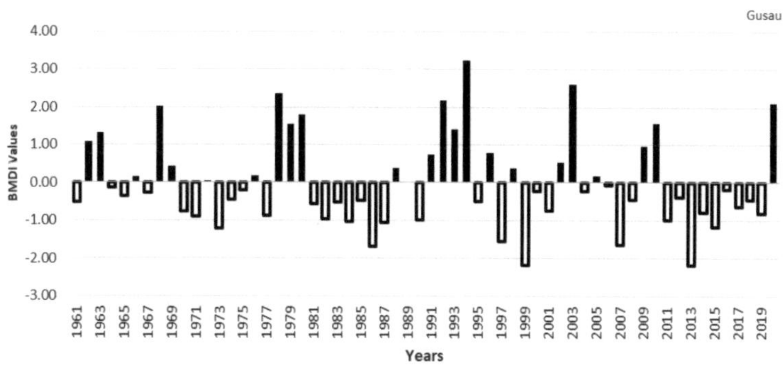

Figure 5.
Annual fluctuations in Bhalme and Mooley drought index for Gusau.

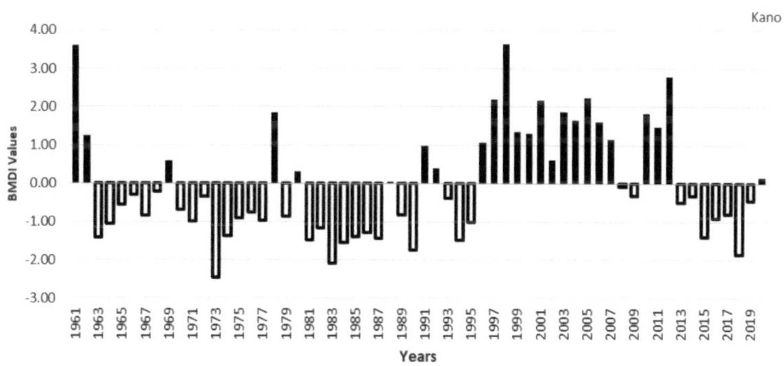

Figure 6.
Annual fluctuations in Bhalme and Mooley drought index for Kano.

drought years and intensity that were less than moderate. This was followed by the drought of the 1980s with an intensity of between mild and moderate and the post-1980s with intermittent drought and wet years and reduced drought intensity of less than moderate. In Kano, drought dominated from the beginning of the study period to 1995, while for Sokoto and Katsina the drought years were 1967 to 1995 and 1967 to 2000 respectively with the intensity ranging from mild to moderate. Nguru also experienced similar droughts situation between 1969 and 2015.

The occurrences of drought as discussed above and wet years in the study area as exemplified in Kano between 1996 and 2010, in Sokoto and Katsina after 1995 and

Figure 7.
Annual fluctuations in Bhalme and Mooley drought index for Sokoto.

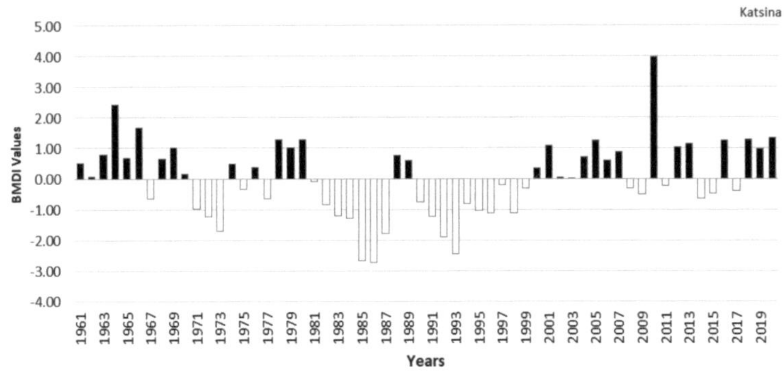

Figure 8.
Annual fluctuations in Bhalme and Mooley drought index for Katsina.

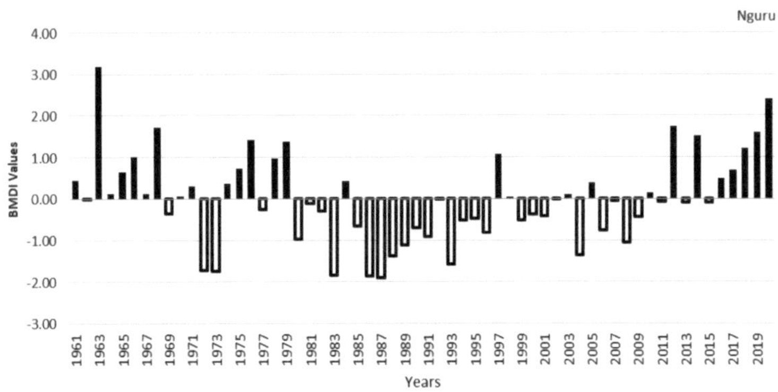

Figure 9.
Annual fluctuations in Bhalme and Mooley drought index for Nguru.

2000 respectively (with few drought years and reduce intensities), before 1969 and after 2015 in Nguru and between 2011 and 2020 in Kaduna indicate a constant shift in the climate of the study area. This shift from findings seems more temporary than permanent and therefore, indicates climate variability than climate change. This variability in climate in more recent years has been exemplified in the Kaduna sub-area that before 2013 had indicated wet years in earlier studies [19, 50] had by 2013 to 2020 turned to drought. This however is in agreement with other studies that indicated the return of drought in the sub-area [14, 17, 27]. This notion of

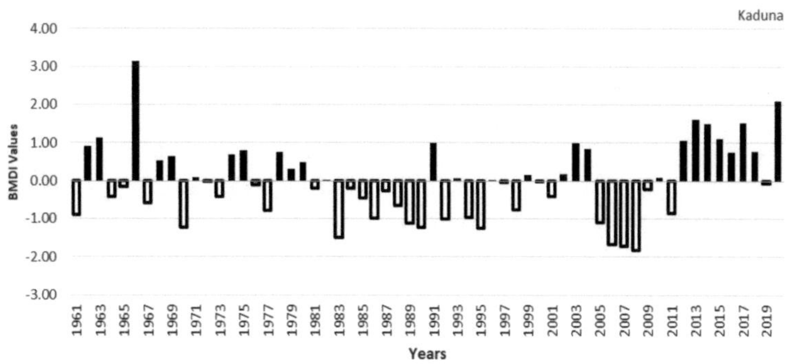

Figure 10.
Annual fluctuations in Bhalme and Mooley drought index for Kaduna.

climate variability is further supported by the fact that throughout the study period, the intensity of drought fluctuated between mild and moderate and never increased to severe or extreme.

4.2 Occurrences of droughts in the sub-periods

4.2.1 Decades

The occurrences of drought within decades varied during the period of study from one sub-area to another. This means the drought percentages of occurrence are not the same. Kano and Nguru experienced drought in four decades of the six decades considered, that is 66.66% drought to 33.33% of wetness. The decades 1961–1970, 1971–1980, 1981–1990, and 2011–2020 were those of droughts with that of 1981–1990 being significant. Other decades were wet with 2001–2010 being significant (**Table 3**). In Nguru, the first and last decades of the study were wet,

Sub-period	Yel	Sok	Gus	Kad	Kat	Kan	Ngu	Pot	Mai
1961–									
1970	−0.62	0.20	0.69	1.35	1.95	−0.31	1.77	1.14	1.11
1971–								—	
1980	−1.31	−1.49	0.35	0.39	−0.01	−1.58	−0.19	0.68	0.21
1981–	—	—	—		—	—	—	—	—
1990	1.99*	2.13*	1.62	−1.73	2.09*	2.16*	2.29*	0.85	2.43*
1991–					—				
2000	2.03*	0.78	1.49	−0.89	2.26*	1.84	−1.46	0.75	0.52
2001–									
2010	1.03	1.44	0.72	−1.56	1.65	2.39*	−0.73	0.27	1.42
2011–			—					—	
2020	1.23	1.59	1.45	2.20*	1.32	−0.65	2.47*	0.67	0.00

**Significant at $p < 0.05$.*
Source: Section 3.2.
Yel, Yelwa; Sok, Sokoto; Gus, Gusau; Kad, Kaduna; Kat, Katsina; Kan, Kano; Ngu, Nguru; Pot, Potiskum; Mai, Maiduguri.

Table 3.
Decadal analysis of drought.

with that of the last decade being significant. Other decades were those of drought with that of 1981-1990 being significant.

In some sub-areas such as Yelwa, Kaduna, Katsina, and Potiskum, the drought decades constituted about 50% the same as the wet decades (50%), although the decades were not the same in all the sub-areas of study. In Yelwa three consecutive decades of 1961–1970, 1971–1980, and 1981–1990 were those of drought with the value for 1981–1990 being significant. Other decades were wet with that of 1991–2000 being significant. Katsina sub-area also experienced the same situation, and the decades with droughts were 1971–1980, 1981–1990, and 1991–2000 with the last two being significant. These decades were also those of drought in the Potiskum sub-area, although without any being significant. Other decades were wet and were not significant in both sub-areas. In Kaduna, the drought decades were 1981–1990, 1991–2000 and 2001–2010 although with values that were not significant. Other decades in this sub-area were wet with the last, 2011–2020 being significant.

The situations experienced in Sokoto and Gusau sub-areas were slightly different from the above. Drought dominated in 1971–1980 for Sokoto and 2011–2020 for Gusau, while the decade 1981–1990 was a drought decade in both sub-areas although it was significant in Sokoto but was not in Gusau. Other decades were wet for both sub-areas. This indicates that the drought decades constituted about 33.33% while the wet decades constituted about 66.66%. Maiduguri sub-area had 1981–1990 as a significant drought decade. Other decades were wet although not significant. This indicated 16.66% of the decades were drought-dominated. The above percentages of drought and wetness do not suggest a permanent shift in

Stations	Sub-periods	Mean	SD	t_d
Yelwa	1961–1990	897.75	172.14	3.63*
	1991–2020	1082.78	213.43	
Sokoto	1961–1990	588.82	139.29	3.40*
	1991–2020	714.02	141.30	
Gusau	1961–1990	852.84	146.30	0.59
	1991–2020	885.01	256.73	
Kaduna	1961–1990	1251.80	241.01	0.14
	1991–2020	1259.96	216.14	
Katsina	1961–1990	577.71	163.12	0.21
	1991–2020	586.47	153.83	
Kano	1961–1990	743.80	226.64	4.12*
	1991–2020	1059.35	344.66	
Nguru	1961–1990	431.57	117.25	0.91
	1991–2020	459.83	120.26	
Potiskum	1961–1990	646.75	186.90	0.26
	1991–2020	656.94	88.88	
Maiduguri	1961–1990	553.37	141.46	1.57
	1991–2020	613.93	152.00	

**Significant at $p < 0.05$.*

Table 4.
Values for student's t-test (30 years non-overlapping).

climate. It shows more of an alternation between drought and wetness and therefore climate variability.

4.2.2 Sub-periods (30 year non-overlapping)

Values of the student's t-test show a significant deviation in trend from the sub-period, 1961–1990 to the sub-period, 1991–2020 (**Table 4**). These values show a significant deviation and shift towards a positive and increasing trend in the series of Yelwa, Sokoto, and Kaduna. For these stations/sub-areas, the sub-period of 1991–2020 appears to be wetter than the preceding sub-period of 1961–1990.

4.3 The trend in temperature of study area

4.3.1 Decadal

The decades 1961–1970, 1971–1980, and 1981–1990 were those of decrease of temperature in Yelwa with that of 1981–1990 being significant. These decreases were in the early decades of the study period. However, in Kaduna, the decreases were in the decades of 1981–1990, 1991–2000, and 2001–2010 without any being significant. The decreases occurred mainly within the mid-decades of the study period. The decreases in Katsina temperature occurred in the decades 1971–1980, 1981–1990, and 1990–2000 with the last two being significant. Also, Potiskum was with a decrease in average decadal temperature in 1971–1980, 1981–1990, and 2011–2020 without any being significant. These three decades constitute 50% of the study period. This indicates a shift in the climate of these sub-areas particularly of the

Stations	Sub-periods	Mean	*SD*	t_d
Yelwa	1981–2010	27.99	0.46	−1.31
	1991–2020	27.83	0.50	
Sokoto	1981 2010	29.03	0.43	−1.94
	1991–2020	28.72	0.74	
Gusau	1981–2010	26.82	0.52	−0.61
	1991–2020	26.74	0.53	
Kaduna	1981–2010	25.85	0.58	−1.51
	1991–2020	25.58	0.75	
Katsina	1981–2010	26.75	0.55	−0.67
	1991–2020	26.65	0.57	
Kano	1981–2010	26.81	0.60	−1.17
	1991–2020	26.63	0.54	
Nguru	1981–2010	27.97	0.47	−0.95
	1991–2020	27.85	0.47	
Potiskum	1981–2010	27.26	0.53	0.80
	1991–2020	27.36	0.48	
Maiduguri	1981–2010	27.90	0.43	1.10
	1991–2020	28.01	0.37	

Table 5.
Student's t-test (t_d) *analysis for temperature (30-years overlapping).*

element of temperature. Sokoto and Gusau had two decades (about 33.33% of decades) each with a decrease in temperature; 1971–1980 and 1981–1990 for Sokoto and 1981–1990, 2011–2020 in Gusau with only that of 1981–1990 in Sokoto being significant. The decadal value of 1981–1990 for Maiduguri was also significant and was the only decade (about 16.66% of decades) with a decrease in the sub-area during the study period.

All other decades in the stations/sub-areas indicated increases in temperature with the following being significant; 1991–2000 (Yelwa), 2011–2020 (Kaduna), 2001–2010 (Kano), and 2011–2020 (Nguru). These situations indicate fluctuations in the average decadal temperature of the study area and back up the assertion of climate variability being dominant as opposed to change in the climate of the study area.

4.3.2 Sub-periods (30 year overlapping)

The values of the student's *t*-test (**Table 5**) show deviation and shift in trend from the sub-period, 1981–2010 to the sub-period, 1991–2020. They indicate a shift towards decreasing temperature (negative) in the series for most of the stations used during the study except for Potiskum and Maiduguri that indicated an increasing (positive) trend However, both increasing and decreasing trends were not significant.

5. Conclusion

This study's findings indicate that there is no consistent increase in the occurrence of droughts in the study area and period considered. Years of drought alternated with those of wet years in all the stations and sub-areas of study. This indicates that drought and its variability are a part of the climate of the area. The fact that the intensity of the drought remained between mild and moderate and did not become worse throughout the study period indicates that there was no permanent shift towards drought that can suggest a change in the climate of the area. Also, the changes observed in the temperature of the study area (both increases and decreases) were not significant to conclude that global warming with the attendant climate change is underway. To ascertain the true situation of the state of the climate in the study area, rainfall and temperature data, and where possible, of other climatic variables such as wind, insolation and relative humidity of several more years and decades would be required.

Author details

Ojonigu F. Ati[1], Kayode Aremu[2], Adewale F. Olatunde[3], Iliya B. Abaje[1*] and Emmanuel O. Oladipo[4]

1 Department of Geography and Regional Planning, Federal University, Dutsin-Ma, Nigeria

2 Department of Geography, Nigerian Defence Academy, Kaduna, Nigeria

3 Department of Geography, Federal University, Lokoja, Nigeria

4 Department of Geography, University of Lagos, Lagos, Nigeria

*Address all correspondence to: abajebest@yahoo.com

References

[1] IPCC (2012). Managing the risks of extreme events and disasters to advance climate change adaptation. In: C.B. Field, V. Barros, T.F. Stocker, D. Qiu, D. J. Dokken, K.L. Ebi, M.D. Mastrandrea, K.J. Mach, G.K. Platter, S.K. Allen, M. Tignor, and P.M. Midgley, editors. A Special Report of Working Groups I and II of the Intergovernmental Panel on Climate Change. Cambridge: Cambridge University Press.

[2] Ayoade, J.O. (1974). A statistical analysis of rainfall over Nigeria. Journal of Tropical Geography 29:11–23.

[3] Lamb, P.J. (1980). Large scale tropical Atlantic surface circulation patterns associated with Sub Saharan weather anomalies. Tellus 30:240–251.

[4] Kerr (1985). Fifteen years of African drought. Science 227:1453–1454.

[5] Ojo, O. (1986a). The Sahel droughts: 1941–1984. In: Proc. ISLSCP Conference, Rome, Italy. ESA. SP-248, pp. 411–415.

[6] Ojo, O. (1986b). Drought persistence in tropical Africa since 1969. In: Programme on Long-Range Forecasting Research WMO Report. Series No. 6; Vol. 1; WMO/TD 87. pp. 73–85.

[7] Nicholson, S.E. (1980). The nature of rainfall fluctuations in subtropical West Africa. Monthly Weather Review 108: 473–487.

[8] Nicholson, S.E. (1981). Rainfall and atmospheric circulation during drought period and wetter years in West Africa. J. Monthly Weather Review 109:1388–1391.

[9] Nicholson, S.E. (1985). Sub Saharan rainfall 1981–1984. J. Clim. Appl. Meteorol. 24:1388–1391.

[10] Adefolalu, A.D. (1986). Further aspects of Sahelian drought as evident from rainfall regime of Nigeria. Arch. Met. Geoph. Biol. Ser. B. 36:277–295.

[11] Oladipo, E.O. (1993a). Some aspects of the spatial characteristics of drought in Northern Nigeria. Natural Hazards 8: 171–188.

[12] Olaniran J.O. (1991a). Rainfall anomaly patterns in dry and wet years over Nigeria. Int. J. Climatology 11:177–204.

[13] Olaniran J.O. (1991b). Evidence of climatic change in Nigeria based on annual rainfall series of rainfall of different daily amounts, 1919–1985. J. Climate Change 19:319–341.

[14] Olatunde, A.F. (2013). Invisible drought in some stations above latitude 9°N of Nigeria. Journal of Sustainable Development Studies 2(1):69–90.

[15] Olatunde, A.F., and Aremu, J.K. (2013). Temporal analyses of drought intensities occurrence within recent decades in some stations in Northern Nigeria. Journal of Sustainable Development Studies 2(1):99–108.

[16] Olatunde, A.F., and Aremu, J.K. (2014). The extent and intensity of extreme drought in some parts of the Savanna region of Nigeria. Journal of Developing Countries Studies 4(9):1–7.

[17] Olatunde, A.F., and Aremu, J.K. (2016). Mild and moderate droughts in Savanna ecological zone of Nigeria. Advances in Social Sciences Research Journal 3(4):236–247.

[18] Aremu, J.K., and Olatunde, A.F. (2012). Drought intensities in the Sudano-Sahelian region of Nigeria. Journal of Sustainable Society 1(4):88–95.

[19] Abaje, I.B., Ati, O.F., Iguisi, E.O., and Jidauna, G.G. (2013). Droughts in

the Sudano-Sahelian ecological zone of Nigeria: Implications for agriculture and water resources development. Global Journal of Human Social Sciences, Geography, Geo-sciences and Environmental 13(2).

[20] Adedoyin, J.A. (1989). Global-scale sea surface temperature (SSTs) anomalies and rainfall characteristics in Northern Nigeria. Int. J. Climatol. 9:133–144.

[21] Adebayo, W.O. (1999). Spatial and Temporal Dynamics of Rainfall and Temperature Fluctuations over Nigeria. Unpublished Ph.D. Thesis. Ibadan: Department of Geography, University of Ibadan.

[22] Olatunde, A.F., and Momoh, M. (2019). Assessment of the relationship between urban growth and surface temperature in Abuja Municipal Area Council. FUDMA Journal of Sciences 3(3).

[23] Winstanley, D. (1973a). Recent rainfall trend in Africa, the Middle East, and Indian. Nature 243:464–465.

[24] Winstanley, D. (1973b). Rainfall pattern and general atmospheric circulation. Nature 245:190–194.

[25] Lamb, P.J. (1973). Rainfall in sub-Saharan Africa during 1941–1983. Z. Gletshcherk Glazialgeo 21:131–139.

[26] Landsberg, H.E. (1975). Sahel drought: Change of climate or part of climate? Achieves.

[27] Olatunde, A.F. (2019). A review of drought occurrences and its implications for agricultural planning and development in some parts of Northern Nigeria. In: Onovwakponoko, editor. Perspectives on Research and Sustainable Development in the 21st Century Nigeria. Lokoja: Faculty of Arts and Social Sciences, Federal University. pp. 260–280.

[28] IPCC (2007). Climate change 2007: Impacts, adaptation, and vulnerability. In: Parry, M.L., Canziani, O.F., Palutikof, J.P., Vander Linden, P.J., Hanson, C.E., editors. Contribution of Working Group II to the Fourth Assessment Report of the Intergovernmental Panel on Climate Change. Cambridge: Cambridge University Press.

[29] Mitchell, J.M., Dzerdzeenski, B., Flohn, H., Hofmeyer, W., Lamb, L., Rao, K.N., and Wallen, C.C. (1966). Climate Change, WTO Tech Note 79 (No 195. TP-100). Geneva: World Meteorological Organization. 79p.

[30] Olatunde, A.F., and Alaci, D.S.A. (2012). Is the trend and persistence of drought in the Sudan-Sahel region of Nigeria an indication of climatic change? In: Iliya, M.A., Abdulrahim, M. A., and Dankani, I.M., editors. Climate Change and Sustainable Development in Nigeria. Sokoto, Nigeria: A Publication of the Association of Nigerian Geographers (ANG) and the Department of Geography, Usmanu Danfodiyo University.

[31] Shuiabu, U., and Oladipo. E.O. (1993). A Bhalme-Mooley-type meteorological drought intensity index for Northern Nigeria. The Zaria Geographer 14:18–27.

[32] Odekunle, T.O., Andrew, O., and Aremu, S.O. (2008). Towards a wetter Sudano-Sahelian ecological zone in twenty-first century Nigeria. Weather 63(3).

[33] Abaje, I.B., Ati, O.F., and Iguisi, E. O. (2012). Changing climatic scenarios and strategies for drought adaptation and mitigation in the Sudano-Sahelian ecological zone of Nigeria. In: M.A. Iliya and I.M. Dankani, editors. Climate Change and Sustainable Development in Nigeria. Ibadan: Crown F. Publishers.

[34] Abaje, I.B., and Oladipo, E.O. (2019). Recent changes in the

temperature and rainfall conditions over Kaduna State, Nigeria. Ghana Journal of Geography 11(2):127–157.

[35] Abaje, I.B., Ndabula, C., and Garba, A.H. (2014). Is the changing rainfall patterns of Kano State and its adverse impacts an indication of climate change? European Scientific Journal 10(2):192–206.

[36] Amos, B.B., Musa, I., Abashiya, M., and Abaje, I.B. (2015). Impact of cement dust emissions on soils within 10 km radius in Ashaka area, Gombe State, Nigeria. International Journal of Environment and Pollution 4(1):0909–0917.

[37] Oladipo, E.O. (1993b). A comprehensive approach to drought and desertification in northern Nigeria. Natural Hazards 8:235–261.

[38] Abaje, I.B., Achiebo, P.J., and Matazu, M.B. (2018). Spatio-temporal analysis of rainfall distribution in Kaduna State, Nigeria. Ghana Journal of Geography 10(1):1–21.

[39] Odekunle, T.O. (2006). Determining rainy season onset and retreat over Nigeria from precipitation amount and number of rainy days. Theoretical and Applied Climatology 83: 193–201.

[40] Odekunle, T.O. (2010). An assessment of the influence of the inter-tropical discontinuity on inter-annual rainfall characteristics in Nigeria. Geographical Research 48(3):314–326.

[41] Ayoade, J.O. (2008). Introduction to Agroclimatology (2nd ed.). Ibadan: University Press Plc.

[42] Oladipo, E.O., and Salahu, S. (1993). Spatial and temporal variations in rainy days and mean daily rainfall intensity in Northern Nigeria. Mausam 44(1):85–92.

[43] Adefolalu, A.D. (2002). Climate. In: Okafor, S., Osaghae, E., Ikporukpo, C., and Abumere, S., editors. Africa Atlases: Atlas of Nigeria. Paris, France: Les Editions J.A. pp. 62–65.

[44] Midgley, S.J., Eldridge, K.G., and Doran, J.C. (1989). Genetic resources of Eucalyptus camaldulensis. Commonwealth Forestry Review 68(4): 295–308.

[45] Ayoade, J.O. (1982). Climate 4: Evaporation, evapotranspiration, and humidity. In: Babour, K.M., Oguntoyinbo, J. S., Onyemelukwe, J.O. C., and Nwafor, J.C., editors. Nigeria in Maps. Ibadan: Nigeria Publishers' Services Ltd. pp. 20–21.

[46] Oguntoyinbo, J.S. (1982). Climate 3: Precipitation (2) and radiation. In: Babour, K.M., Oguntoyinbo, J. S., Onyemelukwe, J.O.C., and Nwafor, J.C., editors. Nigeria in Maps. Ibadan: Nigeria Publishers' Services Ltd. pp. 18–19.

[47] Iloeje, N.P. (1982). A New Geography of Nigeria. Fifth Edition. Ikeja: Longman Nigeria Plc.

[48] Anyadike (1993). Seasonal and annual rainfall variations over Nigeria. International Journal of Climatology 13 (5):567–580.

[49] Lawson, M.P., Balling, R.C., Peters, A.J., and Rundquist, D.C. (1981). Spatial analysis of secular temperature fluctuations. Journal of Climatology 1: 325–332.

[50] Nigerian Meteorological Agency (NIMET) (2005). Hydrometeorological Bulletin: A Hydromet. Abuja, Nigeria: Nigerian Meteorological Agency. pp. 15–16.

Chapter 13

The Impact of Global Warming and Climate Change on the Development of Agriculture in the Northern Latitudes of the Eurasian Continent

Inga Ryumkina, Sergey Ryumkin, Anastasiia Malykhina, Dmitry Ursu and Andrey Khanturgaev

Abstract

In the northern regions of the Eurasian continent, an increase in the sum of active temperatures up to 1500–2000 °C degrees is observed, which creates more favorable conditions for growing crops. The study reveals the prerequisites for the development of crop production in the northern latitudes and analyzes the yield of crops according to the Doctrine of Food Security. Also considered the yields of the main crops in the northern countries of Europe. In the south of the European part of the Eurasian continent, a decrease in crop yields is expected due to climate change and severe aridity. At the same time, this process will have a more negligible effect in the central regions. Improving the thermal regime in the North of the Far East will also increase the yield of fruit and berry, and vegetable crops. In the northern part of the circumpolar belt, an improvement in the thermal regime with a relatively insignificant change in climate humidity will create conditions for increasing crops' productivity and growing a more comprehensive range of crops, especially in river valleys.

Keywords: Nordic farming, global warming, climate-smart breeding, food security, Eurasian continent, Northern latitudes

1. Introduction

The history of agriculture is inextricably linked to the account of humanity. Different theories of the origin of farming, although divergent in their views, are united in the fact that "the cultural force of alienation" [1] significantly influenced the development of humanity. And the whole domestication of plants went along the way of concentrating on their various organs, which are consumed for human consumption. The story of agriculture is directly related to the cultivation of plants, landscape, and climatic conditions. Concerning technological innovations in plant cultivation, these technologies appeared and improved in cases where humanity, for various reasons, had to grow plants in unfavorable conditions. Less fertile soils, less light, heat, and other vital factors forced people to improve agricultural practices. Some researchers [2] and [3] rightly note that all four ancient civilizations of the Old World were located in the temperate zone of the Northern Hemisphere in a narrow strip along the thirtieth parallel. The alluvial deposits of the flooding Tigris, Euphrates, Nile, Indus and Yellow River allowed highly developed civilizations entirely dependent on river irrigation.

According to one of the hypotheses of the development of agriculture, the increase in population was its consequence, thanks to which man colonized new lands, settling in different corners of the planet, involving in its use an increasing amount of land resources.

The next round of the industrial revolution led to the widespread use of fossil fuels, in all areas of human activity, including, for the most part, the agro-industrial complex. The result has been a disruption of the natural cycle and balance of greenhouse gases on the planet, resulting in a gradual increase in the average annual temperature of the earth (**Figure 1**). On a planetary scale, this will lead to irreversible consequences, and already all countries of the world are concerned about global warming.

In addition to changes in temperature, humans are applying various mineral fertilizers and crop protection products in large quantities to maintain crop yields. Unfortunately, this also has negative consequences, but for soil fertility in traditional crop-growing regions. **Figures 2** and **3** show the dynamics of demand for mineral fertilizers in the world.

Soil degradation results from the pursuit of profit and squeezing all the nutrients out of the earth in all traditionally agricultural regions. And actively stimulating the soil with nutrients using fertilizer application only worsens the situation because the concentration of fertilizers is so high that the ground is deprived of its top quality - fertility - for many years.

These two reasons make it necessary to look for new land to cultivate and organize agricultural production. The way out of this situation, scientists see in the transfer of agricultural land to the North. For example, researchers at Singapore's Rajaratnam School of International Studies (RSIS), Goh Tian and Jonathan Lassa, point out that the greenhouse effect will make Northern Canada, Russia and the

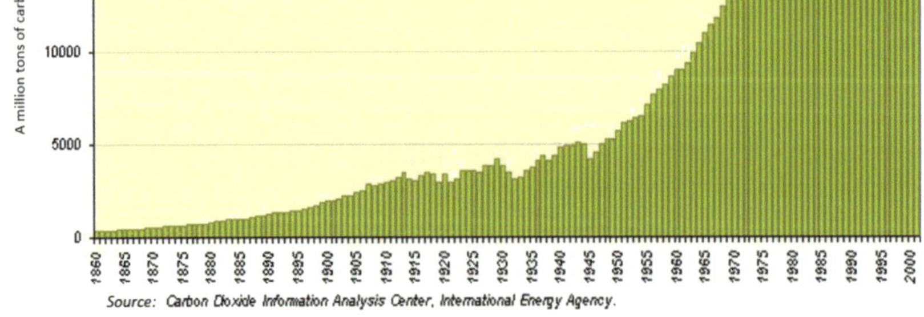

Figure 1.
Change in carbon dioxide emissions from human activities over the past 140 years [http://www.ladoga-lake.ru/pages/artcl-geology-ivashchenko-climat.php].

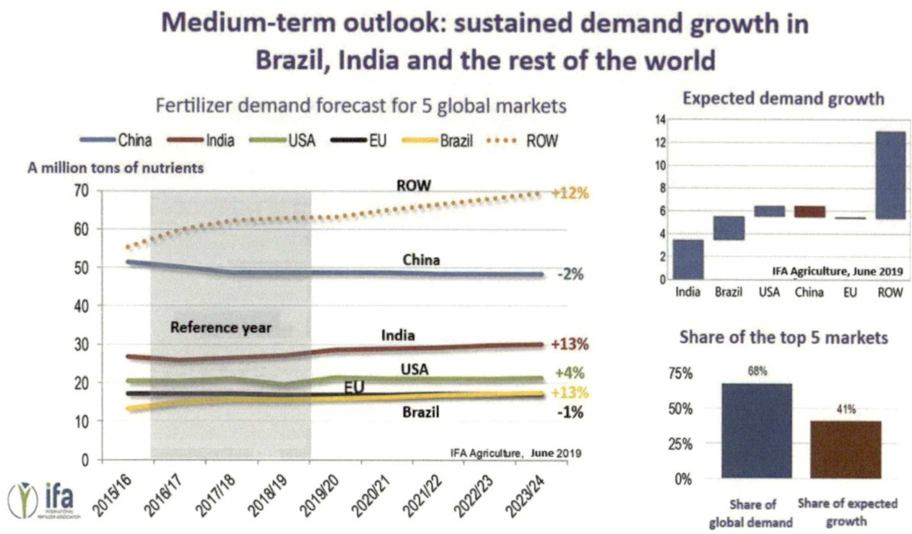

Figure 2.
Fertilizer demand trends in selected countries of the world [https://belchemoil.by/news/tehnologii-i-trendy/trendy-mirovogo-rynka-udobrenij].

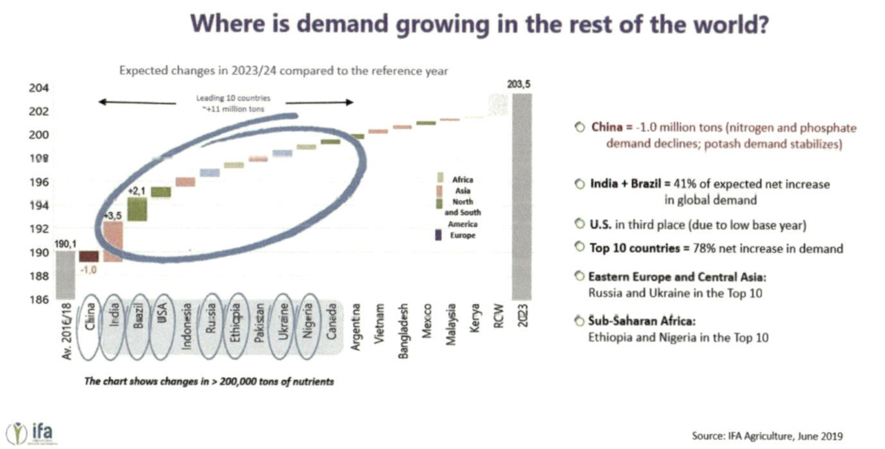

Figure 3.
Expected demand for fertilizers in selected countries of the world [https://belchemoil.by/news/tehnologii-i-trendy/trendy-mirovogo-rynka-udobrenij].

USA, southern Argentina mountainous tropics suitable for farming. "However, favourable conditions for some countries do not guarantee an increase in yields. The winners of climate change will be the countries that can take advantage of the opportunities presented by climate change. Above all, warming will create new regional agricultural centres to replace the old ones and reshape the market for producers - not only between exporters and importers but also between small and large companies" [https://penzanews.ru/analysis/92531-2015].

2. Materials and methods

Climate change has direct implications for the stability of food production systems. The increased frequency and intensity of extreme events, such as droughts and floods, will pose severe threats to the strength of both domestic and global food markets. In addition, the frequency and magnitude of food-deficit emergencies may increase as a result of the complex interaction between political conflict and migration, with increased competition for scarce resources.

One consequence of climate change is the growing scarcity of water resources. Water plays a crucial role in food production, both regionally and globally.

On the one hand, more than 80% of all agricultural land in the world is not irrigated; crop productivity depends on sufficient available moisture in these areas. On the other hand, in areas where this value is limited by climatic conditions, such as the arid and semi-arid regions of the tropics and subtropics and the Mediterranean-type regions in Europe, Australia and South America, agricultural production is very vulnerable to climate change.

On the other hand, global food production depends not only on moisture in precipitation but also on water resources for irrigation. With a warmer climate, the pressure on irrigation systems will increase, which will require additional resources and costs.

Under conditions of global climate change, the risk of competition for water resources increases, especially in regions with inter-boundary rivers and reservoirs, such as Central Asia [4].

In general terms, the adverse effects of global warming on agriculture can be as follows:

- loss of agricultural land fertility through erosion, compaction, desertification, salinization, waterlogging, soil contamination, insufficient mineral content in the soil;
- restructuring of soil biota, reduction of overall land productivity;
- a drop in the yield of cultivated crops due to exposure to high temperatures and dehydration;
- the complete death of plants during wintering;
- deterioration of livestock conditions, thermal stress on animals, leading to reduced productivity;
- lack of water supply, especially in arid areas;
- increased floods and inundations in water-abundant regions;
- the unprecedented spread of traditional crop pests and microorganisms, including in areas where they have not previously been found, the emergence of alien pest species [4].

Rising temperatures are not always uniformly bad; they can lead to increased crop productivity, particularly in the plateaus and high-altitude tropical areas or in northern latitudes where low temperatures limit crop growth.

3. Research results

The research subject is the northern part of the Eurasian continent, which starts from the coast of Norway in the western region and ends with the Kamchatka Peninsula in the eastern part.

The area is limited by the 70th parallel to the North, as further North is the far North where no crops can be grown despite warming. To the south, the 60th parallel.

On the whole Eurasian continent, only a few countries are included in this area: Iceland, Norway, Sweden, Finland and Russia [5, 6]. The main crops cultivated in these countries are shown in **Table 1**.

An important observation of the study is that only livestock production is practised in northern latitudes in all countries other than the Russian Federation. Crop production in northern latitudes is either undeveloped or underdeveloped in these countries. Therefore, further research will focus on the areas of the North of the Russian Federation.

In the context of increasing climate change, Russia, as a country with not only a vast territory but also a variety of different climatic zones, can have a significant impact on global food security, although it, like other regions of the world, will not escape the harmful effects of global warming.

According to many experts, climate change in Russia is already occurring and is often unfavorable for agriculture, the economy and the social sphere.

In studies of climate change across the country, all models without exception show substantial warming of the climate in Russia in the 21st century. Moreover, temperature changes are significantly more significant than the standard deviations throughout the area in question, even during the cold season when intrinsic temperature variability is exceptionally high.

In the scientific literature, various researchers show impressive results in studying the economic impact of climate change on agriculture, especially on the productivity of staple crops (cereals, forage).

Scientists estimated Yield Changes by the IPCC A1F1 Global Development Scenario, which assumes high economic growth with intensive use of fossil fuels. The data obtained were presented by the All-Russian Research Institute of Agricultural Meteorology - RRIAM.

For grain crops in Russia, yields are forecast to fall by up to 17% by 2050. In the Volga and Ural Federal Districts, the decline in grain yields will be catastrophic - by 30% and 38%, respectively. Likewise, the reduction in forage yields will be significant in the Southern and Volga Federal Districts, down 17% and 12%, respectively [5].

Similar conclusions are drawn by international research. In particular, the International Food Policy Research Institute (IFPRI) obtained estimates of yield changes of wheat and some other crops in Russia by 2050, based on which the experts came to several conclusions about the state of agricultural land in the future until 2050

- in the southernmost regions of Russia, a large area may cease to be used for wheat cultivation altogether;
- vast areas of the southwest will face a reduction in yields of more than 25%;
- yield reductions of less than 25% are expected in various areas of southern European Russia, the Southen Urals, Eastern and Western Siberia;
- an increase in climate-dependent wheat yields in the range of 5–25% may be observed in the regions bordering Kazakhstan and in the south of Western Siberia;
- the involvement of new land in agricultural turnover for wheat production is insignificant.

No.	Crops	Finland			Iceland			Norway			Russian Federation			Sweden		
		2017	2018	2019	2017	2018	2019	2017	2018	2019	2017	2018	2019	2017	2018	2019
1	Anise, badian, fennel, coriander	—	—	—	—	—	—	—	—	—	48488	10146	37578	—	—	—
2	Apples	6758	7200	8090	—	—	—	12664	14878	15593	1493600	1859400	1950800	22130	30610	22210
3	Apricots	—	—	—	—	—	—	—	—	—	53300	66300	69600	—	—	—
4	Barley	1460100	1353190	1701960	7400	3900	7900	574100	440900	581000	20628955	16991907	20489088	1635200	1094400	1546500
5	Bast fibers, other	—	—	—	—	—	—	—	—	—	50076	50061	50073	—	—	—
6	Beans, dry	87	90	60	—	—	—	1032	1131	1308	4368	5432	5715	—	230	260
7	Berries nes	303	130	130	—	—	—	122	84	156	3200	4000	4200	20	20	20
8	Blueberries		130	200	—	—	—	30	26	44	2700	3300	3500	70	80	90
9	Broad beans, horse beans, dry		24160	30170	—	—	—	—	—	—	7419	7600	8044	109400	34400	59500
10	Buckwheat	—	—	—	—	—	—	—	—	—	1524879	931713	785702	—	—	—
11	Cabbages and other brassicas	26213	23300	25400	290	160	276	37911	31613	32364	2687090	2495839	2623230	18290	19470	20470
12	Carrots and turnips	63823	66720	77340	750	520	900	51889	43290	49282	1438420	1408348	1558866	109080	92540	106730
13	Castor oil seed	—	—	—	—	—	—	—	—	—	117	81	319	—	—	—
15	Cauliflowers and broccoli	4288	3290	4050	55	47	77	11525	12157	11707	18744	23530	23237	10300	9330	8230
16	Cereals nes	3243		1020	—	—	—	—	—	—	13257	12807	12538	—	—	—
17	Cherries	—	—	—	—	—	—	623	956	689	37300	46400	48700	—	—	—
18	Cherries, sour	—	—	—	—	—	—	—	—	—	186500	232200	243600	100	90	90
19	Chickpeas	—	—	—	—	—	—	—	—	—	418646	620400	506166	—	—	—
20	Cucumbers and gherkins	50763	55330	56130	1800	1927	1924	16484	17462	18475	1504965	1604346	1626360	38100	35790	37900
21	Currants	1888	1420	1870	—	—	—	695	676	631	319800	398000	417600	360	290	290
22	Flax fiber and tow	—	—	—	—	—	—	—	—	—	38795	36715	38464	—	—	—

No.	Crops	Finland			Iceland			Norway			Russian Federation			Sweden		
		2017	2018	2019	2017	2018	2019	2017	2018	2019	2017	2018	2019	2017	2018	2019
23	Fruit, fresh nes	—	—	—	—	—	—	—	—	—	700	900	1000	—	—	—
24	Garlic	75	10	70	—	—	—	—	—	—	206074	211981	202120		40	50
25	Gooseberries	—	—	—	—	—	—	—	—	—	53300	66300	69600	—	—	—
26	Grain, mixed	28900	38890	46210	—	—	—	—	—	—	—	—	—	41100	22200	42300
27	Grapes	—	—	—	—	—	—	—	—	—	580077	627739	677997		60	70
28	Hemp tow waste	—	—	—	—	—	—	—	—	—	1240	1212	1187	—	—	—
29	Hempseed	—	—	—	—	—	—	—	—	—	1078	2117	2893	—	—	—
30	Hops	—	—	—	—	—	—	—	—	—	174	144	205	—	—	—
31	Leeks, other alliaceous vegetables	628	520	730	—	—	—	2717	3335	2702	—	—	—	4170	3180	3510
32	Lentils	—	—	—	—	—	—	—	—	—	197858	194726	116618			
33	Lettuce and chicory	14099	14350	13620	—	—	—	28541	27511	26576	—	—	—	30760	28410	32140
34	Linseed		500	400	—	—	—	—	—	—	611283	557888	658644	7800	4000	4300
35	Lupins	—	—	—	—	—	—	—	—	—	161684	136352	166271	—	—	—
36	Maize	—	—	—	—	—	—	156	142	101	13208095	11419020	14282352	—	4600	11300
37	Millet	—	—	—	—	—	—	—	—	—	316137	217200	439771	—		
38	Mushrooms and truffles	1320	1320	1330	580	580	560	—	—	—	16088	30686	47951	—	1200	1540
39	Mustard seed	—	—	—	—	—	—	—	—	—	98398	123507	164857	—	—	—
40	Nuts nes	—	—	—	—	—	—	—	—	—	17832	20044	19686	—	—	—
41	Oats	1013900	831520	1187480	—	—	—	282700	156600	233700	5456237	4719324	4424433	676400	363500	671200
42	Oilseeds nes	—	—	—	—	—	—	—	—	—	63438	39110	39806	—	—	—
43	Onions, dry	26252	23300	31440	—	—	—	27195	19034	29008	1794417	1642106	1670129	62800	52980	54870

No.	Crops	Finland			Iceland			Norway			Russian Federation			Sweden		
		2017	2018	2019	2017	2018	2019	2017	2018	2019	2017	2018	2019	2017	2018	2019
44	Oranges	—	—	—	—	—	—	—	—	—	84	83	75	—	—	—
45	Peaches and nectarines	—	—	—	—	—	—	—	—	—	29300	36500	38300	—	—	—
46	Pears		230	250	—	—	—	249	553	416	53300	66300	69600	2050	1740	1650
47	Peas, dry	9100	20150	34200	—	—	—	3800	3000	4000	3285500	2304432	2369479	82200	48330	68490
48	Peas, green	7096	6290	8860	—	—	—	3548	1941	4141	121768	79892	86599	12050	9580	21030
49	Plums and sloes	—	—	—	—	—	—	1859	1948	2546	133200	165800	174000	250	250	250
50	Potatoes	611900	600300	618900	9000	6020	8200	315500	326400	332200	21707645	22394960	22074874	852500	723000	846900
51	Pulses nes	—	—	—	—	—	—	—	—	—	9095	10573	8770	—	690	1190
52	Pumpkins, squash and gourds	1686	2350	2300	—	—	—	—	—	—	1165150	1191538	1195611	—	3980	4760
53	Quinces	—	—	—	—	—	—	—	—	—	5300	6600	7000	—	—	—
54	Rapeseed	91300	70900	41900	—	—	—	10300	7250	15000	1510324	1988697	2060320	362700	217700	381500
55	Raspberries	1071	950	1310	—	—	—	2994	2599	2408	133200	165800	174000	430	480	480
56	Rice, paddy	—	—	—	—	—	—	—	—	—	986620	1038222	1098660	—	—	—
57	Rice, paddy (milled rice equivalent)	—	—	—	—	—	—	—	—	—	658076	692494	732806	—	—	—
58	Rye	113500	42990	185260	—	—	—	50000	7900	50000	2548719	1916056	1428421	141800	88200	221300
59	Safflower seed	—	—	—	—	—	—	—	—	—	100885	25259	81189	—	—	—
60	Sorghum	—	—	—	—	—	—	—	—	—	103550	49128	98702	—	—	—
61	Soybeans	—	—	—	—	—	—	—	—	—	3621712	4026850	4359956	—	—	—
62	Spinach	1028	150	450	—	—	—	—	—	—				150	290	260
63	Strawberries	13861	15510	17820	—	—	—	8090	7964	9142	159900	199000	208800	15740	15640	16250
64	Sugar beet	430300	355400	501400	—	—	—	—	—	—	51913442	42065957	54350115	1963500	1698400	2028900

DOI: http://dx.doi.org/10.5772/intechopen.99392

No.	Crops	Finland			Iceland			Norway			Russian Federation			Sweden		
		2017	2018	2019	2017	2018	2019	2017	2018	2019	2017	2018	2019	2017	2018	2019
65	Sunflower seed	—	—	—	—	—	—	—	—	—	10480958	12755725	15379287	—	—	—
66	Tea	—	—	—	—	—	—	—	—	—	554	504	298	—	—	—
67	Tobacco, unmanufactured	—	—	—	—	—	—	—	—	—	4	6	9	—	—	—
68	Tomatoes	39386	39320	40450	1334	1213	1183	10574	12801	11311	2668993	2899664	3015010	14450	18230	16900
69	Triticale	—	—	—	—	—	—	—	—	—	500939	400651	355883	156500	66900	178300
70	Vegetables, fresh nes	12431	—	—	—	—	—	6887	9445	7629	2006655	2128047	2104172	40100	1180	1330
71	Vegetables, leguminous nes	—	—	—	—	—	—	—	—	—	375	342	332	—	—	—
72	Vetches	—	—	—	—	—	—	—	—	—	177459	156115	163163	—	—	—
73	Watermelons	—	—	—	—	—	—	—	—	—	1815022	1969954	1785277	—	—	—
74	Wheat	802000	501600	914180	—	—	—	400500	137700	459000	86002542	72136149	74452692	3298600	1620300	3476800

Table 1.
Gross harvest of the main cultivated crops in the Nordic countries of the Eurasian continent, tones [6].

Fluctuations in the production and supply of grain in the grain market caused by climate change strongly affect grain prices.

The food security of the Russian Federation [5, 7] in the long term largely depends not only on the readiness of agricultural systems to adapt to extreme climatic and weather events but also on the ability to adapt to these changes in the rest of the agro-industrial complex - logistics, agricultural processing and food consumption, as well as the social-economic sphere of the regions and the country as a whole.

In conclusion, the global community recognizes that climate change will make it challenging to produce enough food for the world's growing population and alter water resources' availability, quality, and mode of use. Avoiding over-intrusions into already stressed ecosystems will require countries to double the current rate of agricultural productivity growth while minimizing agricultural-related harm to the environment. This productivity will require the deliberate application of new or pre-existing technologies and practices, the development of crop varieties resistant to climatic shocks, the diversification of rural livelihoods, improved forest and fisheries management, investment in information technology and systems, and the active use of emerging computer technologies such as precision farming, GIS, etc.

International organizations including FAO [6], the International Food Policy Research Institute - IFPRI, the World Bank and others have developed recommendations for adapting world agriculture to global climate change.

Addressing this major challenge requires joint efforts and actions by all countries to effectively implement measures to mitigate climate change effects on agriculture and adapt the world food system.

The level of food security of the Northern regions of Russia depends on the local production of agricultural products and their regular import from favorable areas. And, in this regard, we have considered food security based on the gross harvest of the above indicators in the regions of the North [8].

The table shows that it is practically impossible to grow wheat (winter and spring) in the northern latitudes. Winter wheat is produced only in two subjects of the Federation out of 12 regions that we selected for the study (**Table 2**). According to the data, during the reporting period, winter wheat cultivation in the Tyumen region decreased by 38.04 thousand centners (or 24.9%), while in spring wheat, it decreased by 1037.2 thousand centners (or 11.6%). On the contrary, there has been a 15.6% increase in wheat yields in Russia's North.

The status of the Doctrine [7] indicator "potatoes" is more favorable than that of wheat. Potatoes are grown in 11 out of 12 regions. In the Krasnoyarsk region as a whole, potato self-sufficiency (6.182.97 thousand quintals) is high per capita (0.6 kilograms per day), i.e., exceeds 2.5 times the dietary standards. Still, in 2 autonomous Districts (Taimyr (Dolgano-Nenetskiy) and Evenk), potatoes are not grown due to difficult climate conditions and specific arctic [8]. Potatoes in these districts are not a staple food, unlike in Russia as a whole. In the other constituent entities of the Federation, potatoes are grown, albeit in small quantities. For example, in 3 regions (Yamalo-Nenetskiy Autonomous District, Magadan region, Chukotskiy Autonomous District), self-sufficiency in "potatoes" is low and equals less than 0.1% of the total in Russia. The Tyumen region (without districts), just as in wheat, shows a decrease in the gross output of potatoes by 1,031.79 thousand quintals. There is also an insignificant reduction in the gross harvest in 3 regions - the Republic of Sakha (Yakutia), the Magadan region and Chukotskiy Autonomous District. Three federal subjects (Khanty-Mansiysk Autonomous Area, Yamalo-Nenetskiy Autonomous Area, Kamchatka region) show increased potato cultivation. Still, only one part, Yugra, shows a significant growth rate of 140.84 thousand quintals.

No.	Regions of the Russian Federation	Wheat (thousand quintals)				Rate of increase/ decrease, in % in 2020 to 2017
		2017 year	The 2018 year	2019 year	The 2020 year	
Winter wheat						
1	Tyumen region (without districts)	152.66	115.52	127.64	114.62	↓ 24.9%
2	Krasnoyarsk region*	67.91	122.09	82.58	136.78	↑ 101.4%
Spring wheat						
3	Arkhangelsk region	22.6	17.8	15.7	16.1	↓ 28.7%
4	Tyumen region (without districts)	9 018.74	7 414.64	8 656.4	7 981.54	↓ 11.6%
5	Krasnoyarsk region*	12 560.18	11 352.66	13 274.28	16 983.47	↑ 35%
6	Republic of Sakha (Yakutia)	12.31	15.81	17.21	11.44	↓ 7%
7	Kamchatka region	0.07	0.07	—	—	—
8	**IN TOTAL**	**21 834.47**	**19 038.59**	**22 173.81**	**25 243.95**	↑ **15.6%**

**The entire Krasnoyarsk region, as no wheat is grown in the Taimyr (Dolgano-Nenetskiy Autonomous District) and Evenk Autonomous Districts.*

Table 2.
Gross harvest of wheat (winter and spring) in farms of all categories in the northern regions of the Russian Federation for the period 2017–2020 [5].

During the study period, potato yields increased from 24.5% in the Murmansk region to 104.1% in the Komi Republic. Therefore, self-sufficiency in the potato indicator in these regions is high.

The third Doctrine [7] indicator is "vegetables and gourds". These crops are more thermophilic than potatoes and are practically not intended for growing in the permafrost zone. Therefore, their cultivation rate is meager in this macro-region. The histogram (**Figure 1**) shows this clearly.

Despite the climatic conditions, the regions of Russia's northern latitudes are looking for opportunities to grow vegetables and gourds food crops. But the yield of the crops grown is low enough for the body to obtain all the vitamins it needs.

Melon crops are not grown in all regions of the northern latitudes due to the incompatibility of these crops' physiological and morphological characteristics and the difficult climatic conditions of the North. Thus, in 2020, according to federal statistics and the histogram, only three areas found it possible to supply themselves with minimal amounts of melon food crops, such as the Krasnoyarsk region (the entire region), the Republic of Sakha (Yakutia) and the Tyumen region (without districts) with a gross harvest of 3.56 thousand quintals, 3.04 thousand quintals and 0.17 thousand quintals respectively.

As for outdoor and indoor vegetables, small volumes of gross harvest are present in the regions of Northern Russia. But the values are so low that the histogram (**Figure 2**) does not show them. As a result, only two areas give a small visualization (Krasnoyarsk region at 1.543.02 thousand quintals and Tyumen region at 1.267.82 thousand quintals in 2020) (**Figure 5**).

The average harvest of outdoor and indoor vegetables in 2020 is recorded in three regions. For example, the Republic of Sakha (Yakutia) shows 263.63 thousand quintals, the Kamchatka region - 135.77 thousand quintals, and Khanty-Mansiysk Autonomous District - 267.64 thousand quintals.

And the lowest indicators for the cultivation of outdoor and indoor vegetables in 2020 are represented by three subjects of the Federation (Magadan region (44.05 thousand quintals), Chukotskiy region (2.7 thousand quintals) and Yamalo-Nenetskiy Autonomous District (1.36 thousand quintals).

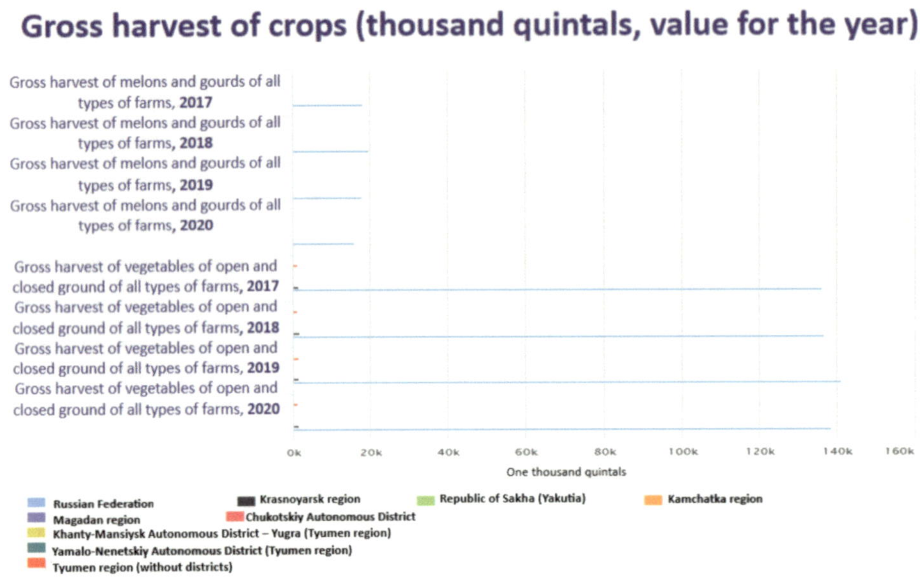

Figure 4.
*Gross harvest of vegetables and gourds in farms of all categories in the northern regions of the Russian Federation 2017–2020 [5]. *The entire Krasnoyarsk region, as in the Taimyr (Dolgano-Nenetskiy Autonomous District) and Evenk Autonomous Districts, no vegetables and melons are grown.*

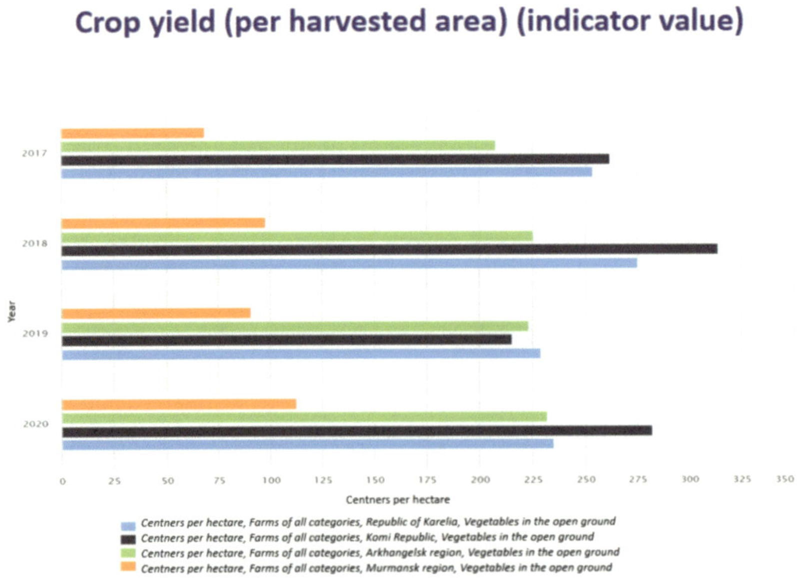

Figure 5.
*Gross harvest of outdoor vegetables in farms of all categories in the northern regions of the Russian Federation 2017–2020 [5]. *The entire Krasnoyarsk region, as in the Taimyr (Dolgano-Nenetskiy) and Evenk Autonomous Districts, no vegetables and melons are grown.*

The regions of the European North of Russia do not grow melons. But open field vegetables are grown in farms of all categories. The leader in growing vegetables among the European North of Russia regions is the Komi Republic in 2018 and 2020. The lowest yields are obtained in the Murmansk region due to the relatively small territory, climatic conditions and, of course, the specialization of the area itself. This region is an outpost of Russia in the North Seas.

Thus, we can conclude from this very Doctrine [7] indicator that man cannot change weather conditions in northern latitudes. Still, human engineering and technology of recent years show that it is possible to grow vegetable and food gourds even in extreme conditions. Besides, the facts recorded by meteorologists on global warming and climate change in northern latitudes over the past decades show that the sum of active temperatures (SAT) is observed more frequently than 35–50 years ago.

For example, in the Murmansk region over the last 35 years (**Figure 3**), there has been a significant upward trend in July temperatures from 15.2 °C in 1985 to 17.9 °C

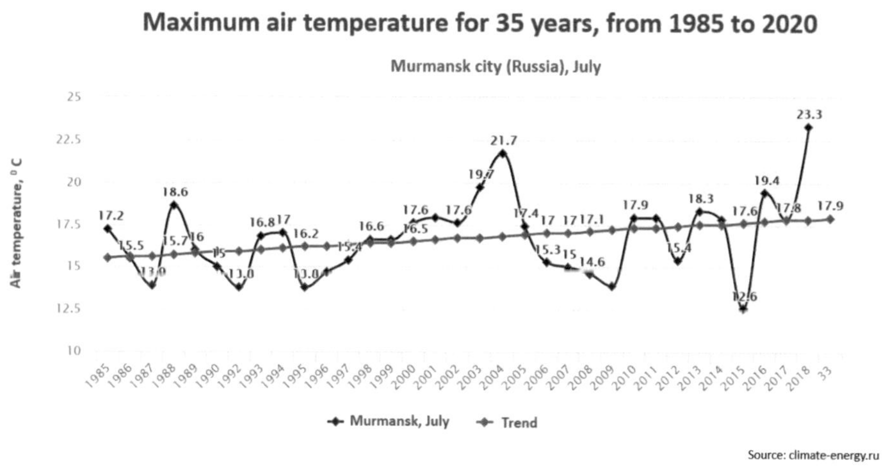

Figure 6.
Maximum air temperatures in Murmansk in July from 1985 to 2020.

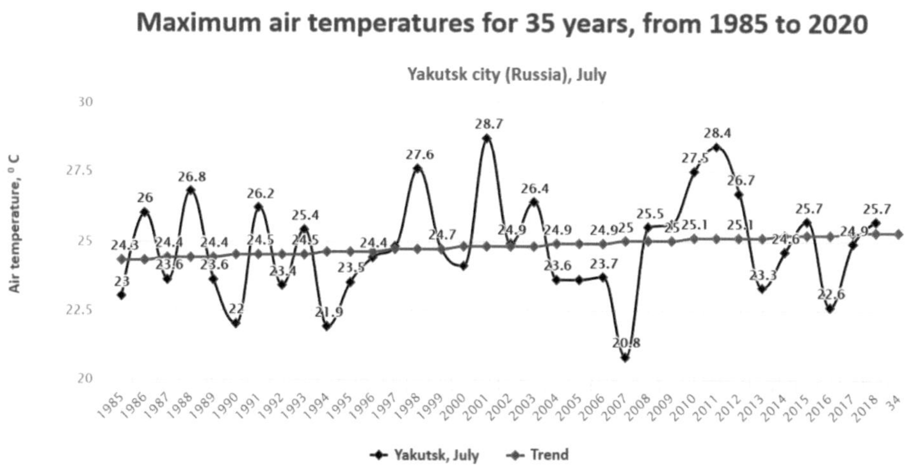

Figure 7.
Maximum air temperatures in Yakutsk in July from 1985 to 2020.

in 2020, which of course, shortly could affect the cultivation of even more vegetable and food crops to increase food security (**Figure 6**).

Scientists observe the same trend in the Republic of Sakha (Yakutia) over 35 years (**Figure 4**). In July, the temperature increased from 24.3 °C in 1985 to 25.5 °C in 2020, indicating an increase in the Russian North's range (**Figure** 7).

The next Doctrine [7] indicator, "fruit and berries", also depends on the sum of the active temperatures in the northern latitudes of the macro-region. **Table 3** proves this. The availability of fruit and berry plantations, including strawberries, strawberries, raspberries, currants, gooseberries, and other berries, does exist. Still, the gross yield is insignificant to supply the population of this macro-region.

These tables indicate the possibility of developing crop production in the regions of the North of Russia. However, food availability by the Doctrine indicator "Fruits" is high only in two areas (Tyumen and Krasnoyarsk regions) relative to other regions of the North. And, even these values of 69.36 thousand quintals and 64.27 thousand quintals do not cover the needs of these regions in fruit. For example, according to Order No. 614 of 09.08.2016 of the Ministry of Health of the Russian Federation, the norm for consumption of fruit and berries is 100 kg/year per capita. In the Krasnoyarsk region, fruit supplies are similar to those in the Tyumen Region. The per capita fruit deficit is 77.51 kg/per year per capita, without considering the increased consumption of fruits and vegetables for residents of the Far North by 15%. In other words, the current state of the gross fruit harvest is insufficient for the

No.	Regions of the Russian Federation	Potatoes (thousand quintals)				Rate of increase/ decrease, in % in 2020 to 2017
		2017 year	2018 year	2019 year	2020 year	
1	Republic of Karelia	133.2	156.2	163.7	167	↑ 25.3%
2	Komi Republic	88.7	143.7	120.5	181.1	↑ 104.1%
3	Arkhangelsk region	93.6	141.7	135.4	136	↑ 45.2%
4	Murmansk region	99.3	107.7	117.2	123.7	↑ 24.5%
5	Tyumen region (without districts)	4 248.69	4 352.6	4 101.38	3 216.9	↓ 24.2%
6	Khanty-Mansiysk Autonomous District – Yugra (Tyumen region)	682.83	667.04	596.27	823.67	↑ 20.6%
7	Yamalo-Nenetskiy Autonomous District (Tyumen region)	7.65	8.47	8.01	7.66	↑ 0.13%
8	Krasnoyarsk region*	—	—	—	—	—
9	Republic of Sakha (Yakutia)	756.1	829.35	814.82	711.78	↓ 5.86%
10	Kamchatka region	364.86	3558	400.88	396.71	↑ 8.72%
11	Magadan region	88.64	83.29	58.12	77.14	↓ 12.9%
12	Chukotskiy Autonomous District	1.26	1.24	1.27	1.04	↓ 17.4%
13	**IN TOTAL**	**6 564.83**	**6 847.09**	**6 517.55**	**5 842.7**	↓ **10.9%**

**Only the Taimyr (Dolgano-Nenetskiy Autonomous District) and Evenk Autonomous Districts.*

Table 3.
Gross potato harvest in farms of all categories in the northern regions of the Russian Federation for the period 2017–2020 [5].

self-sufficiency of the population of this region, let alone other territories of the Russian North.

As for fruit supply in Krasnoyarsk Region, the situation is similar to that in Tyumen Region. The per capita fruit deficit is 77.51 kg per year, and this does not include the 15% increased norms of fruit and vegetable consumption for residents of the Far North (**Table 4**).

No.	Regions of the Russian Federation	Fruit and berries (thousand quintals)				Rate of increase/ decrease, in % in 2020 to 2017
		2017 year	2018 year	2019 year	2020 year	
Fruit and berry plantations (strawberries, strawberries, raspberries, currants, gooseberries and other berries)						
1	Republic of Karelia	83.6	96.4	112.4	137.6	↑ 64.5%
2	Komi Republic	96.7	97.2	133.3	183.8	↑ 90.0%
3	Arkhangelsk region	122.0	130.6	143.1	148.3	↑ 21.5%
4	Murmansk region	20.7	24.2	25.4	40.9	↑ 97.5
5	Tyumen region (without districts)	345.36	368.37	293.95	253.05	↓ 26.7%
6	Khanty-Mansiysk Autonomous District - Yugra	41.38	47.28	68.84	71.96	↑ 73.9%
7	Yamato-Nenets Autonomous District	—	—	0.02	0.06	—
8	Krasnoyarsk region*	336.93	394.92	455.11	441.63	↑ 31.1%
9	Republic of Sakha (Yakutia)	0.97	0.94	2.57	6.35	↑ 554.6%
10	Kamchatka region	5.7	7.68	15.26	15.8	↑ 82.4%
11	Magadan region	2.57	2.52	2.85	2.22	↓ 13.6%
12	Chukotskiy Autonomous District	—	—	—	0.01	—
Seedlings (apple, pear, quince, etc.)						
1	Republic of Karelia	33.3	55.6	75.5	79.7	↑ 139.3%
2	Komi Republic	7.6	29.6	19.3	19.3	↑ 153.9%
3	Arkhangelsk region	41.5	40.6	43	39.9	↓ 3.85%
4	Tyumen region (without districts)	65.14	62.68	80.55	69.36	↓ 2.7%
5	Khanty-Mansiysk Autonomous District - Yugra	—	—	3.13	1.53	—
6	Krasnoyarsk region*	354	46.43	56.95	64.27	↑ 81.5%
7	Kamchatka region	—	—	—	0.09	—
8	**IN TOTAL**	**1 238.85**	**1 405.02**	**1 531.23**	**1 575.83**	↑ **27.2%**

**The entire Krasnoyarsk region, as in the Taimyr (Dolgano-Nenetskiy) and Evenki Autonomous Districts, no vegetables or melons are grown.*

Table 4.
Gross harvest of fruit and berries in farms of all categories in the northern regions of the Russian Federation for the period 2017–2020 [5].

For the Doctrine [7] indicator "berries", the situation, according to the table, is more optimistic, as there is no deficit in this position. Still, there is an oversupply of these crops per capita per year. For example, Khanty-Mansiysk Autonomous District - Yugra shows 42 kg/year per capita against the consumption norm of 7 kg/year per capita (not including the increase of consumption norm by 15% inhabitants of the Far North). Three other regions exceed the consumption norms: Tyumen region (without districts), Krasnoyarsk region (the entire region) and Kamchatka region.

We found that the gross berry harvest in four regions of the North of Russia (Yamalo-Nenetskiy Autonomous District, the Republic of Sakha (Yakutia), Magadan region, Chukotskiy Autonomous District) is lower than the average Russian consumption rate.

The growth rate for 2017–2020 in the regions of the European North for fruit and berry plantations and berries shows a steady increase in yield as these regions have significant stocks of non-timber products, which are in demand both domestically and internationally. Therefore, this area of crop production needs to be developed and maintained. Furthermore, because of the change in air temperature by 2.7 °C in these regions, it can be said that the growing season increases in days and, accordingly, some crops can adapt to the weather and climatic conditions of the North.

4. Discussion

There is considerable uncertainty in quantifying how climate change is expected to play out in the future and its impact on ecosystems, economic activity, and social processes in different countries and regions.

Studies on the impact of climate change on the economies of world regions and individual countries (including agriculture) have been carried out by various international organizations and national research centres. For example, the World Bank's Economics of adaptation to climate change synthesis report (2010) estimates that the costs of adapting to 2-degree global warming between 2020 and 2050 are in the range of $70 billion to $100 billion a year, depending on future climate change scenarios.

The same report notes that in addition to the financial costs of adapting the world's regions to climate change, the prices of mitigating the negative impacts on developing economies will rise, totaling US$265 billion to US$565 billion [4]. As a result, the cost of mitigating the effects of climate change would increase to a total of US$265 billion to US$565 billion [4].

It should be noted that even under constant climate conditions, as shown in the model studies, prices for the most critical crops will rise. By 2050, it is predicted that the cost of wheat could increase by 39%, rice by 62%, maize by 63% and soybean by 72%. Climate change will result in additional price increases: an average increase of 32–37% for rice, 52–55% for maize and 12–14% for soybeans, with the highest growth expected for wheat, ranging from 94 to 111% [4].

Adapting the food system to global climate change will require complex social, economic and biophysical adjustments to food production, processing and consumption. Such changes will be most difficult for the poorest and most vulnerable regions and populations. Furthermore, climate change modeling shows that the most severe impacts are likely to occur in the tropical drylands. Many of the poorest countries are located in these regions, so nations least able to adapt will be most affected.

Nevertheless, the food system is embedded in global processes and linked to other systems, which has both advantages and disadvantages. For example, economic shocks in one geographic region may spread rapidly to others. Still, shocks due to sharp reductions in the food supply in one area may be offset by output from the other areas. The global food system also affects the efficiency of food production by allowing parts with advanced production systems to export to lagging regions [4].

A global problem of the scale of climate change requires coordinated efforts at the international level. However, its solution depends on the actions taken by each country in its territory. Primarily, it is a question of reducing greenhouse gas (GHG) emissions into the atmosphere.

5. Conclusion

Based on the research, there is a strong dependency on crop production in the Northern regions of Russia, and the level of "food security" is relatively low. A significant reason for this, of course, is the climate. But even so, this factor is beginning to "melt away" over time in favor of growing more thermophilic crops. Thus, opening even more opportunities for the state and population to develop rural areas and agriculture in this macro-region.

We suggest using the maximum possible agro-technologies for crop production under global warming conditions, such as snow retention, reduction of unproductive evaporation, and expansion of drought-tolerant crops (primarily corn, sunflower, millet), and growth of winter crops (wheat).

We also suggest a need to develop climate-friendly crop breeding, which will cope with abiotic stress conditions as much as possible and adapt to climate change.

Today, the primary vector in solving the food problem and the development of rural areas is the development of productive forces of agriculture with a bias towards innovative and nature-based technologies [6], in providing the population with domestic functional foods, except for exotic foods and the formation of the necessary social infrastructure for the people of the Russian North.

As for the 2030 Doctrine, the overall food self-sufficiency indicators in Russia are more than ambitious, given the background of the food problem. And these difficulties and barriers can be seen in the North of Russia and everywhere, especially in the North Federal District and the Far East of Russia.

However, crop production can and should be developed despite the archaic climatic conditions, especially in the river valleys. This study shows that the "sprouts" of crop production already exist in the regions of the North of Russia. It is necessary to build a growing season for crops in northern latitudes, using agronomic techniques, plant breeding, and genetic methods to gain access to safe, functional food for a healthy lifestyle throughout the season [9]. And, of course, public-administrative and private-investment levers and support are indispensable.

Author details

Inga Ryumkina[1*], Sergey Ryumkin[1], Anastasiia Malykhina[1], Dmitry Ursu[2] and Andrey Khanturgaev[3]

1 Novosibirsk State Agrarian University, Novosibirsk City, Russia

2 Factor Ltd, Moscow, Russia

3 East-Siberian State University of Technology and Management, Ulan-Ude, Russian Federation

*Address all correspondence to: ingaryumkina@gmail.com

References

[1] Goncharov, N.P. et al., Methodological basis of plant breeding. *Academic Publishing House "Geo"*, 2009, pp. 427.

[2] Diamond J. Evolution, consequences and future of plant and animal domestication //*Nature*. 2002, Vol. 418, No. 6898, pp. 700-707. https://penzanews.ru/analysis/92531-2015 (Date of access: 03/25/2021).

[3] Wood M. In search of the first civilizations. Moscow: Capital Print, 2007, pp. 288.

[4] Paptsov A., Shelamova N. The global agro-food system and global climate change // *Agribusiness: Economics and Management*, 2017, vol. 11, pp. 81-94.

[5] Russian Federal State Statistics Service. http://www.gks.ru/ (Date of access: 04/30/2021).

[6] FAO statistics. http://www.fao.org/faostat/en/ (Date of access: 04/15/2021).

[7] "On the approval of the Doctrine of food security of the Russian Federation": *Decree of the President of the Russian Federation of January 21, 2020 N 20 //* https://www.garant.ru/products/ipo/prime/doc/73338425/ (Date of access: 03/18/2021).

[8] Ryumkin S.V. et al. Forecasting of Food Security and Sustainable Development of Rural Territories in Russia's North/ Chapter 25 in the book. "*Handbook of Research on International Collaboration, Economic Development and Sustainability in the Arctic*". / Under edition Vasilii Erokhin, Tianming Gao, Xiuhua Zhang/ DOI: 10.4018 / 978-1-5225-6954-1/ https://www.igi-global.com .

[9] Ryumkina I.N. et al. World Market of Organic Food Products. Chapter in the book. "Handbook of Research on Globalized Agricultural Trade and New Challenges for Food Security". / Under edition Vasilii Erokhin and Tianming Gao / DOI: 10.4018/978-1-7998-1042-1/ https://www.igi-global.com .

Chapter 14

Sustainable Water Flows in Era of Climate Change

Deepika Pandey

Abstract

The flow of water in rivers is of paramount importance to maintain supply of food and energy requirements to a great extent. The minimum flow in perennial rivers is subjected to groundwater availability, it is further replenished by the water added through precipitation. Climate change not only increases the melting of glaciers and sea level rise, but also influences the surface water flow and quality. As agriculture is directly affected by changing precipitation pattern, the reduction in water resources and untimely addition of water, both act havoc to the food production process. This interconnection makes agriculture even more vulnerable to the scenarios of global warming and climate change. Studies on food-energy-water nexus has opened new avenues of research in sustainable water management. The role of sustainable flow of water in rivers is highlighted which needs to be understood in era of climate change.

Keywords: surface water flows, ground water, climate change, sustainability

1. Introduction: Era of climate change

The surface water has been the cradle of civilisations since the beginning of human history. The social and economic well-being of human beings is directly affected by availability of fresh water. The use of water in domestic, agriculture as well as industrial sector has increased the stress on fresh water. The major fresh water sources, the rivers, are turning into water bodies unfit for proper use. The pollutants have also reached the groundwater and contaminated the clean water source. Climate change is another threat posed on the availability of fresh water. Change in the climate during the last few decades is described by the changes in variables of climate such as temperature, humidity, wind velocity etc. in different parts of the world. This climate change is caused by the warming of the earth's atmosphere as a result of increasing greenhouse gas (GHG) emissions, change in land-use pattern and emission of aerosols and other pollutants through various anthropogenic activities. Thus climate variability in a particular region may be observed at hourly and daily basis in addition to monthly, seasonally, annually and decadal variability [1].

Climate change will cause significant impact on our water resources since it is directly related to changes in patterns of precipitation. The effects of climate change are already visible in many parts of the world. Many countries in the world are negatively affected by the impacts which range from increased frequency and intensity of floods and droughts, severe water scarcity, increased intensity of erosion, retreating glaciers, sea level rise, decreasing snow cover, increased sedimentation in water bodies, deteriorating quality of water and damaged ecosystems [2]. The impact of climate change on surface waters, as studied for Moldova, East Europe, has shown increase in temperature and precipitation from 1945 to 2011. The annual average temperature was predicted by climate models to increase by 2°C between 2010 and 2039, and up to 3°C by next thirty years. However, the predictions for precipitations indicated that total amount of precipitation could decrease steadily or remain stable by the end of the century, with seasonal variability [3].

The human populations are exposed to various climate change related risks. The modifications of climate often result in unpredicted cyclones, storms, floods, droughts and heat waves and with increased intensity. The water availability in surface water bodies get altered as a consequence; modified river flow regime and altered ground water recharge play a significant role. Indirect effects of changing climate causes siltation of reservoirs, coastal floods and salt-water intrusion in coastal aquifers are most common occurrences (**Figure 1**). These various factors are interrelated and they affect the human population intensely and in most unpredictable way [1]. Various sectors which directly depend on water resources, such as agriculture, hydropower, navigation etc. are getting affected and have devastating societal impacts like effects on health of large scale of population and economy of many countries [2].

1.1 Extreme events

Extreme events have caused major disasters in many parts of the world in last couple of decades. A stronger link is detected between the changing weather patterns and warming of the atmosphere. Extreme heat and drought with changing seasons are being experienced in many tropical and sub-tropical regions. The number of flood disasters have increased several times [4]. Storms and flash floods have increased frequency in mountainous regions [4, 5]. In tropical and sub tropical regions, the event of extreme rainfall is anticipated on land because of increasing convective activity during summer monsoon. Warmer atmosphere facilitates heavy moisture bearing clouds and rainfall is intensified. Extensive and continuous rainfall have triggered disastrous effects such as landslides, subsidence and submergence under water. The annual flows would become more unstable, spring and flash floods will become more common and increase the peak flow [3]. The run-off is expected to increase in many parts of the world. Increased snowfall and frigid weather are also linked with moisture in the warmer atmosphere. The non-climatic factors of human actions have exposed and increased the vulnerability, making these regular events more hazardous. The geographical location and socio-economic dynamics play a very important role in turning extreme events into disasters.

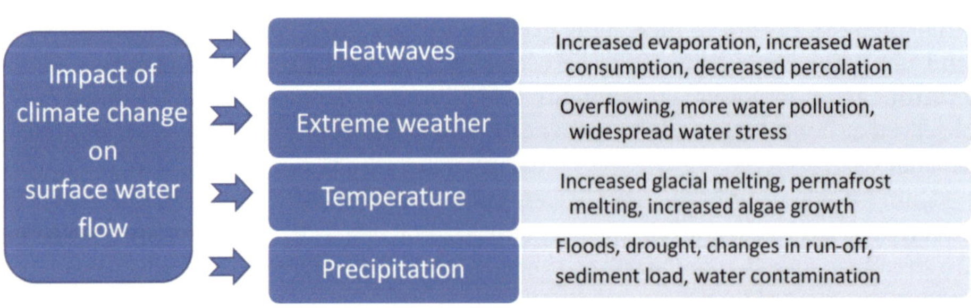

Figure 1.
Climate change impacts on flow of surface water (adapted from UNECE guidance document).

1.2 Effect on water quality

Surface water temperature increase will increase properties such as solubilisation, dissolution, and evaporation in the surface water bodies. Increased temperatures will also favour degradation of chemical and organic compounds, complex formation and their reactions, and bring about transformations in the chemical property of water to a much higher degree. As Arrhenius equation explains, kinetics of a given chemical reaction can be doubled for a temperature increase of 10°C, Increased temperature of water bodies due to high atmospheric temperature may alleviate dissolution of substances and decrease the dissolved gases in water. Depletion of the most important dissolved gases, oxygen, also affects the quality of water [6]. The increase in temperature and accommodating extreme inflow has a definitive effect on the quality of water due to mixing of pollutants, sediments, pathogens, pesticides, salt, organic carbon, nutrients etc. The surface water bodies are at the receiving end of the heavy contaminants through run-off which changes the quality of water. Extra heat released by anthropogenic causes, industrial processes, directed to water bodies, cause change in water quality called thermal pollution. Hence the quality of surface water is compromised due to increased heat as well as its calamitous effect on the physical and chemical properties of water and added contaminants. This negative impact is more pronounced to the aquatic ecosystem and the health of human population [7].

1.3 Effect on the flow of water

Climate change influences many characteristics of flowing surface water and is anticipated to have great impact on river regimes. The water levels of the rivers, flow velocity, residence times and overall hydraulic characteristics are affected as a consequence. This eventually brings changes in wetted areas of rivers, overflow patterns and affect habitat availability and connectivity [8]. Frequent and intensified flooding is expected as the run-off load and overall peak flow is increased. More intense rainfall and flooding is expected to increase the load of suspended solids and sediment yields [9, 10]. The impact of extreme events is severe on rivers in upper courses, leading to releasing heavier number of sediments which are carried downstream in suspended state and as fine sediments from land. Erosional capacity of heavy flow increases many times. Such changes have great influence on downstream habitats and ecology. Increased suspended load in the surface water is also a potential cause of bacterial (*Escherichia coli*) and harmful heavy metals [11] pollution. As studied in many parts of Europe, the key factor controlling the ecosystem behaviour is the hydrological and geomorphological processes in the region [12]. The carrying of sediments is beneficial for the upland stream ecology as it leads to wider range of habitats, meandering side channels and deeper sediment zones to support aquatic life. Enhanced habitat restoration due to varying flow regime of the rivers was found through extensive study of German river systems [13]. Sustained flow in rivers is difficult to maintain due to unpredicted extreme events of precipitation.

2. Sustainable water flows

Changes in the flow of river due to channel straightening, disconnect with flood plains, dam construction and loss of river ecology has a great impact on the overall ecology of the river, climate change is added problem. Fresh water ecosystems were studied through PRINCE (preparing for climate change impacts on freshwater ecosystems) project in UK which has shown that changes in climate have directly affected the aquatic ecosystems [14]. The changes in climate including changes in duration, frequency and magnitude of precipitation, especially extreme events, influences the aquatic ecosystems through episodic pulse effects. Combined with the changes in water velocity and dissolved oxygen content of the water bodies, the effect of temperature becomes more pronounced for fishes, amphibians, as well as birds. The more conspicuous effect would be on migratory or dispersal patterns between ecosystems such as between marine and freshwater, long-distance migrants or across watersheds. Hot summers with limited rainfall eventually decrease the amount of dissolved oxygen, particularly in standing waters, and rivers in middle and lower courses, where re-aeration in less. In addition, such water bodies also have high amount of nutrients resulting in profuse plant growth but leading to low oxygen levels which can be threat to fishes and invertebrates of water bodies. On the other hand, the high flow of storm water improves oxygen levels, but increases the contaminants such as pesticides and fertilisers. Such conditions take a toll on vulnerable sections such as ephemeral ponds, ditches and small water bodies which are most affected due to changing conditions. The sensitivity and resilience of the surface water ecosystem gets effected due to magnitude, variability, frequency and mistiming of the water flow and quality.

Surface water resources like lakes, rivers and reservoirs, as well as frozen water in glaciers and snow covers, the internal renewable water resources, receive water from total precipitation and upstream streamflow produced. The aquifer recharge is also dependent on the same. Ratio between total precipitation and total streamflow, runoff ratio, vary from country to country and is a factor of terrain, the mountainous region and steep terrain produce more run-off [15]. Hence total renewable water resources are the sum of internal renewable resources and surface run-off water. The upstream and island countries like Bhutan and Sri Lanka respectively have minimum run-off component whereas Bangladesh receives more than 90% [1]. The risks with water bodies gradually get worse with the climate change. Studies in tropical and sub-tropical regions have predicted increase in extreme flows due to climate change [16]. Land use induced by climate change may increase the erosion rates. Use and managing water resources which are differentially affected by modified rainfall patterns, temperature increase, and sea-level rise are very crucial in the times of climate change.

Hydrological impacts on river systems are aggravated due to climate change. The prime cause of reduced water flows is the increased water extractions, also a consequence of global warming. Overexploitation of both surface and ground water has affected the surface water flows more predominantly. The health of the rivers has been severely impacted due to consumption being overpowering. The per capita water availability has sharpy declined in many countries in the recent past [17]. With growing economy and ever-growing population, this is expected to become worse [18, 19]. The microscale studies on effect of climate change on different rivers or studies at global scale showing anthropogenic stresses on river systems, have not been thorough in understanding the situation in most countries, the meso-scale analysis of cross-continental basin-scale comparisons have revealed that water extractions from the rivers of past and present, have exceeded the projected impacts of climate change to much larger extent [20]. This finding realises the importance of effective water governance to reduce present water stresses and further deterioration because of decline in water flows and consequences of climate change. The current water-use practices consume water at the expense of natural water flows [21, 22]. With the threat of climate change, the policies to reduce consumptive use of water and maintain sustainable in-flow water uses for a longer time need to be encouraged. The water allocation management plan should share the variability equally between the users and the environment.

3. Climate change impacts

3.1 Floods

Floods are not only the excess flow of water in rivers during rainy season, but also the flash floods which are induced by changes in rainfall pattern and temperature rise. Such flash floods are expected more frequently in the mountainous regions as a consequence of convective activity during monsoon, collapse of natural ice or debris deposited due to landslides or sudden subsidence of glacial lake. The global warming has accelerated the melting of glaciers in Himalayas and number and size of glacial lakes have increased [5, 23]. The outburst of the glacial lake is anticipated due to moraine deposition, and flash floods play havoc to such conditions. It is disastrous for thousands of people living downstream and are capable of paralysing hydropower projects [24].

The more common riverine floods are still a disaster in many parts of the world and climate change has increased its probability. It has long-term effects on land-use, economy, and development of the effected region. Naturally, the floods are essential feature of a river life cycle and is highly required for the nutrients carried down through silt and water, which is beneficial for the crops like rice. However, the increased in the frequency and intensity of rainfall has resulted in longer flood duration and greater spatial extent. The frequent flooding damage crops, delays cropping cycles, spreads infectious diseases and damages properties and land.

The sea-level rise due to warming of atmosphere and increased frequency of storms in sea has aggravated the coastal floods making them more damaging. The low-lying lands get completely submerged everytime and the adjoining coastal lands face severe agricultural losses. The frequency and severity of coastal storms has increased, and cyclone have become more common, posing a serious threat to the coastal regions. The subsidence and retreating of delta are other consequences due to increase in silt load of the rivers in the coastal region [5, 18].

3.2 Droughts

Increasing water demand and increasing water scarcity in areas due to climate change have made regions even more vulnerable to droughts [24, 25]. The reduction in groundwater recharge is more prominent due to altered flow and changed rainfall patterns. The discharge in the rivers which are snow-fed will face huge changes in pattern as the peak season discharge will be smoothed throughout the year and extended in the early part of the year. Hence during the peak summer season less water will be available, along with increased rate of evaporation, for irrigation and hydropower projects [26]. This problem will increase as the temperatures increases, owing to enhanced evapotranspiration, increased water demand by crops and greater risk of drought stress, siltation of reservoirs and reducing storage capacities of the reservoirs [27].

3.3 Groundwater

Groundwater is the major source of freshwater, estimating up to one third of the total freshwater withdrawals, and is crucial for agriculture, industry as well as domestic requirement [28]. The groundwater also sustains the baseflow of surface waters such as rivers, lakes and wetlands and help them to survive during low or no rainfall period. On the other hand, the water table is maintained by percolation of rainwater and leakage from surface water such as lakes, streams and wetlands.

Hence, the groundwater recharge is highly dependent upon climate, land cover, terrain and geology of the region. The amount of precipitation and evapotranspiration is decided by the climate and vegetation cover, and the water percolation in the soil also depends upon the type of soil and underlying rocks [29]. It is the surplus water after run-off and evaporation, which enters the repository of ground water. At higher altitudes, the spatial distribution of snow and ice is largely affected by climate change, less accumulation of snow and earlier melting. In winter season this effect causes more precipitation and increased frequency of rainfall plus snow events. The overall impact is that the recharge regime of the groundwater is changed in this era of water scarcity and global warming, and it does not get replenished with sufficient water [30, 31].

Groundwater storage is the prime source of freshwater which has been stored since decades and is one of the cleanest sources of water. The reserves of groundwater act as buffer for the surface water supplies and provide a sustainable flow in surface water bodies under climate variabilities, however, there is a limit to their natural resilience due to withdrawal exceeding the recharge rate [32]. The recharge is in areas which allow inflow by infiltration of excess rainwater and surface water bodies, also includes irrigation practices in agriculture, water seepage of wastewater or storage water. Natural outflows include discharges in springs, wetlands and lagoons, most prominently extraction through borewells. The balance between inflow and outflow in groundwater determines the maintenance of the water table. The sustainability of groundwater depends upon the balance between its recharge and discharge; climate change would affect this balance in case of increasing aridity and overexploitation, decreasing the recharge rate. The lesser and more intensive rainfall may increase the water table of an area, even higher than their previous level, and damage the crop and property. It will also deplete the soil moisture, cause soil compaction, soil erosion and will further reduce the infiltration capacity and groundwater recharge.

Another important consequence is saline water intrusion in coastal aquifers and reduced flow in delta region. Many coastal countries are at the risk of climate-related groundwater contamination [1]. As the sea-level rise, the aquifers and wells are affected by saltwater intrusion and contaminate the drinking water source [33]. Extensive pumping of groundwater due to increase water demand in warm temperatures, accelerates the pollution of aquifers through lateral movements of saltwater or contaminants. The geogenic contaminants including arsenic, uranium, fluoride [34, 35] can be hazardous for living beings. Agricultural additives, fertilisers and pesticides also infiltrate to the groundwater and make it toxic. Domestic wastewater and other industrial effluents are increasingly reaching the groundwater and changing its quality. This will further deteriorate the situation as it will lead to excessive pumping from tube wells and water withdrawal from upstream rivers. This phenomenon of sea-level rise, depleting aquifers, increased water demand will accelerate the groundwater pollution and is already happening in many parts of the world [36, 37].

4. Indirect risks

The indirect risks of climate change phenomenon are closely interrelated and affect each other dynamically, creating a situation of environmental emergency. The extent of the risks and its effect on the human civilization may yet to be fully understood, as the stressors may have cumulative effect [38]. Hence feedback loops and synergies of indirect stressors need to be understood to safeguard against disasters. Some possible interconnections are as following:

4.1 Urbanisation

Land encroachment, high population and overexploitation of fresh water resources affect the environment most potently. Increasing industrial as well as other wastes and irresponsible dumping of wastes, add to the havoc of water contamination and increasing greenhouse gases, the cause of global warming and climate change.

4.2 Changes in land use pattern

Increasing temperature and changes in hydrology of a region bring about changes in the vegetation and land use pattern of a particular region. On the other hand, poorly planned land use is the arching factor behind increased incidences of disastrous conditions like flooding. Flooding in urban areas can be controlled by restricting urban development in low-lying flood-prone areas and floodplains of adjoining rivers, appropriate wastewater management released by high populated areas. Environmental degradation along with poor planning of land use can result in major disasters due to climate change.

4.3 Agriculture

Deterioration of soil due to extensive agriculture practices, unregulated irrigation and use of excessive fertilisers and pesticides have adverse effect on environment, particularly water bodies. Major changes in the water demand of irrigation due to climate change is expected. Practices such as clearing of forest for agriculture, cultivation on steep slopes, combined with overgrazing by cattle stock, unsustainable use of pastures and scrublands, have aggravated the conditions created by climate change. Soil degradation and erosion reduces the water absorbing capacity of soil and increases surface run-off. This will further increase the destructive effects of flash floods, siltation of downstream water reservoirs and also reduces groundwater recharge. Change in water quality and quantity as a result poses insecurity and the threat of food scarcity, hence increasing the vulnerability of poor farmers of arid and semi-arid regions.

4.4 Population growth and migration

Demographic changes due to increasing climate refugees increases the stress on the host land, adding to the problem of environmental degradation and making lives of people more vulnerable. Settlement along the rivers is more prone to disaster of floods, more damaging to the temporary or semi-temporary infrastructure for living. This also adds to the contamination of surface water and creates stress on dwindling freshwater resources. Social inequalities, less opportunities for livelihood, economic weakness, food insecurity and political instability causes fundamental threat to human life posing serious repercussions to the environment. Global demand of fresh water is sharply increasing primarily due to increasing population.

4.5 Growing energy consumption

It will trigger increasing demands of energy and stress on hydropower, thermal power and nuclear power to produce more energy. Inadequate infrastructure towards energy generation also comes with a risk of water stress, wastage of water, release of pollutants and major accidents. Increasing soil erosion and other extreme events such as flash floods, mud water flow has caused severe siltation of the reservoirs which will effectively reduce the capacity of hydro power generation. It also reduces the capacity of meeting water needs for irrigation and other activities during dry season.

5. Conclusion

IPCC Report 4 [5] has clearly emphasised on the linkage of atmospheric warming to the large-scale changes in the hydrological cycle, increase in the water vapour content, changing in the patterns of precipitation, increased frequency and intensity of precipitation, reaching extremes of dry season and flooding, reduced snow cover, enhanced melting of ice and reduced permafrost, changes in soil moisture content and increased runoff. A remarkable spatial and inter-decadal variability in the precipitation has been observed. The events of intensive precipitation have increased worldwide on one hand, and area classified as very dry land has increased to double on the other hand. The snow cover and water stored in glaciers have decreased considerably. This ice melting related phenomenon with snow-fed rivers, run-off in glaciers and increased number of glacier lakes have been observed in last few decades. Such changes have also been associated with disastrous flash floods, landslides, riverine floods and heavy mud flow. Sustainability of the surface water due to delicate hydrological cycle seem to be disrupted beyond cure.

Climate change models are predicting increased precipitation in higher latitudes and decreased precipitation in subtropical and lower mid-latitude regions, increasing disparity and creating extreme conditions in different parts of the world. As a result, increase in average river water availability and run off is predicted in the higher latitudes and a decreasing trend towards dry tropics. Reduction in the water supplies in glaciers causes reduced water availability in rivers, hence creating water stress conditions.

Freshwater resources are also influenced by many non-climatic drivers such as changes in the patterns of population, food consumption and production of food and other products. The economics of water pricing, technology and societal views towards value of freshwater ecosystems also plays a role [2]. Thus, climate change is one of the many factors causing future water stresses, effecting sustainable water flows. The socioeconomic and technological changes along with demographic changes can prove to be more effective towards water management in most regions of the world and on most time scales. Talking about the effect of extreme events, the impacts more often depend upon the socioeconomic vulnerability. The development pattern, wealth distribution, demographics and environmental conditions determine the severity of such disasters like extreme weather events and migration of disease vectors.

The water management practices presently in action are not enough to cope with the impacts of climate change and may not be able to assess and manage risks such as devastating floods, human health, food insecurity, damaged ecosystems and inadequate power supplies. The current variability in climate is impossible to predict completely and plan for long term adaptation strategies. The effect of climate change stress in tropical and sub-tropical regions is more pronounced and a threat to sustainable surface water flow. The indirect stressors such as population growth, urbanization etc., will aggravate the conditions. The challenges posed by changing climate in the past can be a good guidance for planning for the future. The management practices based on projected hydrological changes may help to avoid disastrous events and bring situation under control. The adaptation strategies for maintaining sustainable flow of surface water require integrated strategies for both demand and supply sides. Improving water-use efficiency, recycling of water, water harvesting, water conservation, water metering and water markets may be valuable efforts in this regard. Integrated water resource management, expanded use of economic incentives and providing framework adaptation strategies in administrative systems, across socio-economic barriers are the key factors. Integrated approaches must be at appropriate scales to be effective in practice and to provide total environmental solution. Such mitigation methods will not only reduce the impact of global warming but will also reduce the stress of adaptation to changing climate.

Author details

Deepika Pandey
Amity University Haryana, Gurugram, India

*Address all correspondence to: pdeeps@gmail.com

References

[1] Lacombe G, Chinnasamy P, Nicol A. Review of climate change science, knowledge and impacts on water resources in South Asia. Background Paper 1. Colombo, Sri Lanka: International Water Management Institute (IWMI). 2019. 73p

[2] Timmerman J, Gierk M, Sinisi L, Kalinin M, Agaltseva N, Pirumyan E, Teruggi AG, Van Schaik H, Koeppel S, Bernardini F (editors). The Guidance on Water and Adaptation to Climate Change. United Nations Publications; 2010

[3] EU. Assessing the impact of climate change on water supply sources and WSS systems in Moldova and inventory possible adaptation measures (Task 1). 2013

[4] IPCC. Summary for Policymakers. Intergovernmental Panel on Climate Change. 2014. pp. 1-161

[5] Monirul Qader Mirza M, Ahmad Q. Climate Change and Water Resources in South Asia. Taylor & Francis, CRC Press; 2005. 275p

[6] Delpla I, Jung AV, Baures E, et al. Impacts of climate change on surface water quality in relation to drinking water production. Environment International. 2009;**35**:1225-1233

[7] Bates BC, Kundzewicz ZW, Wu S and Palutikof, JP. Climate Change and Water: Technical Paper of the Intergovernmental Panel on Climate Change. Geneva: IPCC Secretariat; 2008. 210p.

[8] Brown LE, Hannah DM, Milner AM. Vulnerability of alpine stream biodiversity to shrinking glaciers and snowpacks. Global Change Biology. 2007;**13**:958-966

[9] Lane S, Tayefi V, Reid S, et al. Interactions between sediment delivery, channel change, climate change and flood risk in a temperate upland environment. Earth Surface Processes and Landforms. 2003;**32**:429-446

[10] Wilby RL, Dalgleish HY, Foster IDL. The impact of weather patterns on historic and contemporary catchment sediment yields. Earth Surface Processes and Landforms. 1997;**22**:353-363

[11] Longfield SA, Macklin MG. The influence of recent environmental change on flooding and sediment fluxes in the Yorkshire Ouse basin. Hydrological Processes. 1999;**13**: 1051-1066

[12] Whitehead PG, Wilby RL, Battarbee RW, et al. A review of the potential impacts of climate change on surface water quality. Hydrological Sciences Journal. 2009;**54**:101-123

[13] Hering D, Schmidt-Kloiber A, Murphy J, et al. Potential impact of climate change on aquatic insects: A sensitivity analysis for European caddisflies (Trichoptera) based on distribution patterns and ecological preferences. Aquatic Sciences. 2009; **71**:3-14

[14] Conlan K. Preparing for climate change impacts on freshwater ecosystems (PRINCE). Environment Agency:Great Britain. 2007

[15] Brown S, Nicholls RJ. Subsidence and human influences in mega deltas: The case of the Ganges-Brahmaputra-Meghna. Science of the Total Environment. 2015;**527-528**:362-374

[16] Eccles R, Zhang H, Hamilton D. A review of the effects of climate change on riverine flooding in subtropical and tropical regions. Journal of Water and Climate Change. 2019;**10**:687-707

[17] Shiklaomanov I, Rodda JC, editors. World Water Resources at the Beginning

of the Twenty-First Century. UK: Cambridge University Press; 2004. DOI: 10.5860/choice.41-4063

[18] Molden D. Comprehensive Assessment of Water Management in Agriculture. Water for Food, Water for Life: A Comprehensive Assessment of Water Management in Agriculture. London: Earthscan, and Colombo: International Water Management Institute; 2007

[19] United Nations world water development report 4: Managing water under uncertainty and risk. 2012

[20] Grafton RQ, Pittock J, Davis R, et al. Global insights into water resources, climate change and governance. Nature Climate Change. 2013;**3**:315-321

[21] National Water Commission 2011, The National Water Initiative—securing Australia's water future: 2011 assessment, NWC, Canberra

[22] National Water Commission. Impacts of Water Trading in the Southern Murray-Darling Basin between 2006-07 and 2010-11. 2012. Available from: http://archive.nwc.gov.au/library/topic/rural/impacts-of-trade-2012

[23] Bolch T, Kulkarni A, Kääb A, Huggel C, Paul F, Cogley JG, Frey H, Kargel JS, Fujita K, Scheel M, Bajracharya, S, Stoffel M. The state and fate of Himalayan glaciers. Science; 2012;**336**(6079):310-314

[24] Shrestha S, Shrestha M, Babel MS. Assessment of climate change impact on water diversion strategies of Melamchi Water Supply Project in Nepal. Theoretical and Applied Climatology. 2017;**128**:311-323

[25] Dai A. Increasing drought under global warming in observations and models. Nature Climate Change. 2013; **3**:52-58

[26] King BM, Sturtewagen B. Making the Most of Afghanistan's River Basins: Opportunities for Regional Cooperation. EastwestNgo. 2010. p. 27

[27] Rosenzweig C, Elliott J, Deryng D, et al. Assessing agricultural risks of climate change in the 21st century in a global gridded crop model intercomparison. Proceedings of the National Academy of Sciences of the United States of America. 2014;**111**: 3268-3273

[28] Döll P, Hoffmann-Dobrev H, Portmann FT, et al. Impact of water withdrawals from groundwater and surface water on continental water storage variations. Journal of Geodynamics. 2012;**59-60**:143-156

[29] Taylor RG, Scanlon B, Döll P, et al. Ground water and climate change. Nature Climate Change. 2013;**3**:322-329

[30] Taguc C, Grant GE. Groundwater dynamics mediate low-flow response to global warming in snow-dominated alpine regions. Water Resources Research. 2009;**45**W07421:1-12

[31] Sultana Z, Coulibaly P. Distributed modelling of future changes in hydrological processes of Spencer Creek watershed. Hydrological Processes. 2011;**25**:1254-1270

[32] IAH. Global Change & Groundwater. IAH Strategic Overview Series. 2016. p. 6

[33] Tabrez A, Inam A, Rabbani G. The State of the Environment of the Pakistan Coast. 2008. Available from: https://www.researchgate.net/profile/Asif_Inam/publication/259485497_The_State_of_the_Environment_of_the_Pakistan_Coast/links/55cffa0808aee19936fd9a9f/The-State-of-the-Environment-of-the-Pakistan-Coast.pdf

[34] Gosselin RE, Smith R, Hodge H. Clinical Toxicology of Commercial

Products. Baltimore, MD: Williams & Wilkins; 1984. ix + 2002p

[35] Heiken A. Arsenic contamination of irrigation water, soil and crops in Bangladesh: Risk implications for sustainable agriculture and food safety in Asia. 2006

[36] Laattoe T, Werner AD, Simmons CT. Seawater Intrusion Under Current Sea-Level Rise: Processes Accompanying Chapter Coastline Transgression. In: Groundwater in the Coastal Zones of Asia-Pacific. Netherland: Springer; 2013. pp. 295-313

[37] Gunaalan K, Asanthi HB, Gamage TPD, et al. Geochemical variations of groundwater quality in coastal and Karstic chapter aquifers in Jaffna Peninsula, Sri Lanka. In: Management of Water, Energy and Bio-Resources in the Era of Climate Change: Emerging Issues and Challenges. Springer International Publishing; 2013. pp. 51-61

[38] Zimmerman JB, Mihelcic JR, Smith J. Global stressors of water quality and quantity. Environmental Science & Technology. June 15 2008:4247-4254

Chapter 15

Recent Climate Shocks in the Sahel: A Systematic Review

Terence Epule Epule, Driss Dhiba and Abdelghani Chehbouni

Abstract

In Africa, the Sahel is increasingly susceptible to climate shocks such as droughts, sandstorms (winds), and floods. Through a systematic review this work tracks the frequency with which these shocks are reported in the literature during the period 1975–2020. This work examines trends to identify which shocks are most reported, documenting their spatial distribution and evaluating the impacts of climatic and non-climatic drivers. In general, 388 shocks were reported in 164 relevant peer review papers. Southern Niger recorded 15.97% of all the shocks while Ethiopia and Senegal recorded 11.85% and 10.85% respectively. Also, West African Sahel saw about 49.97% of all shocks followed by East African Sahel with 29.89% and Central African Sahel with 12.11%. Generally, droughts (n = 219), appear to be the most frequently reported shocks followed by floods (n = 123) and winds (n = 46). The 1975–1985 decade recorded the most shocks (n = 207), followed by the 1997–2007 decade which saw (n = 80) shocks while between 1986 and 1996 a total of 52 shocks were recorded. 52% of the shocks are driven by climatic factors while 47% are driven by non-climatic drivers.

Keywords: Climate change, Shocks, Droughts, Floods, Winds, Climatic, Non-climatic, Regional, Country-level, Sahel

1. Introduction

Global environmental changes cannot be excluded from the debates on development and environmental protection in Africa. The latter is true because ecosystems respond to climate change and climate change also determines the pace of global development in general and African development in particular [1, 2]. The influence of climate change in sub-Saharan Africa is even more evident as in the last three and a half decades the region has witnessed temperatures increasing in the range of 0.2–2.0° C amidst declining precipitation [3]. The projections of future temperatures right up to 2100 shows that it will cumulatively exceed the range of 0.2–2.0° C that was observed in the last three and a half decades. The fifth assessment report (AR5) of the IPCC further supports this assertion by observing that mean temperatures will potentially increase on a decade basis by 1.8° C in 2020, by 1.9° C in 2030, by 1.8° C in 2040, by 1.7° C by 2050 and by 1.2° C by 2100 [4]. The impacts of the above observations are enhanced vulnerability of cropping systems and amplified poverty, low adaptive capacity, and more frequent climate shocks [5, 6]. Despite the importance of climate factor in the region, non-climatic drivers such as agricultural expansion, deforestation, occasioned by rapid population growth have tilted the debate on which of climatic and non-climatic factors drives the shocks more [7].

Three main climate shocks (droughts, floods, and sandstorms (winds)) are dominant across Africa and the Sahel [8–15]. The Sahel is well known for its droughts and as a result, the region is one of the most susceptible to droughts globally [16–18]. The causes of these droughts are either human related (human induced climate change) or climatically driven (sea surface temperature, effects of vegetation, CO_2 emissions and land degradation, and dust feed backs) [11]. It has been reported that hydrological droughts are becoming frequent in the Sahel due to observations of enhanced discharge in rivers. Since the 1970s, river Niger for example has witnessed increase discharge of its tributaries. It is indeed surprising that while temperatures are observed to be rising in the Sahel and projections of the same are pessimistic, some rivers in the region are witnessing increase discharge in what has been termed the "Sahel paradox" [19–23]. Furthermore, the Sahel is increasingly impacted by airborne dust or sandstorms which are also impacting the climate of the region [3]. Satellite sensors have shown that Sahelian dust are not only regionally restricted but also globally distributed [14, 24]. In fact, reports hold that most of the dust outbreaks around the world are linked to the Sahel [25–27].

Climate change stakeholders in the Sahel have been active seeking ways and means of addressing these climate shocks. In this context, adaptations actions, policies and programs have been designed to help address the surge [28–30]. For example, the United Nations Reductions of Emissions from Deforestation and Forest Degradation (REDD+); has been designed to support reforestation project across the world [12, 31–37]. We also have the United Nations Framework Convention on Climate Change (UNFCCC) [38] which enhances research and adaptive capacity in developing countries. The African Development Bank (AfDB) has created the African Climate Fund which focuses on enhancing increase to climate finance in Africa [39]. Additionally, the Pan African Agency for the "Great Green Wall (GGW)" has been involved in planting trees across the Sahel [40].

Even though there is an increase in the proliferation of studies on climate shocks in the Sahel, there is still a shortage of studies that provide updated and holistic information on the dominant shocks, their distribution, across the Sahel and their drivers. In fact, it is still unclear which climate shocks are most reported in the scientific literature over time. This enhances the ability to monitor existing gaps in the scientific scholarship and to pave the way for the future. In addition, it provides a holistic picture of the regional and country level variations in the dominant shocks impacting the Sahel and further provides and opportunity to leverage our understanding in shaping adaptation actions.

2. Country and regional distribution of climate change shocks and affiliation of authors

A total of 388 climate shocks (droughts, floods, and winds) were documented in 164 peer reviewed papers. The results show further that in terms of countries, Southern Niger recorded the highest frequency and percentage of reports of climate change shocks (n = 62, 15.97%) between 1975 and 2020. Ethiopia which is second recorded 46 or 11.85% of the shocks, Senegal next with 41 or 10.56% of the shocks and Kenya with 33 or 8.5% of the climate shocks (**Figure 1**). Regionally, West African Sahel recorded the highest number of shocks (n = 193, 49.97%), then second was East African Sahel (n = 116, 29.89%) and Central African Sahel with (n = 47, 12.11%) (**Figure 2**).

A total of 164 peer review papers were recorded for this chapter. Of this number, 55 authors are affiliated in the USA, 19 in France, 17 in the United Kingdom and 10 in Germany. These countries are countries that have colonial ties in Africa and

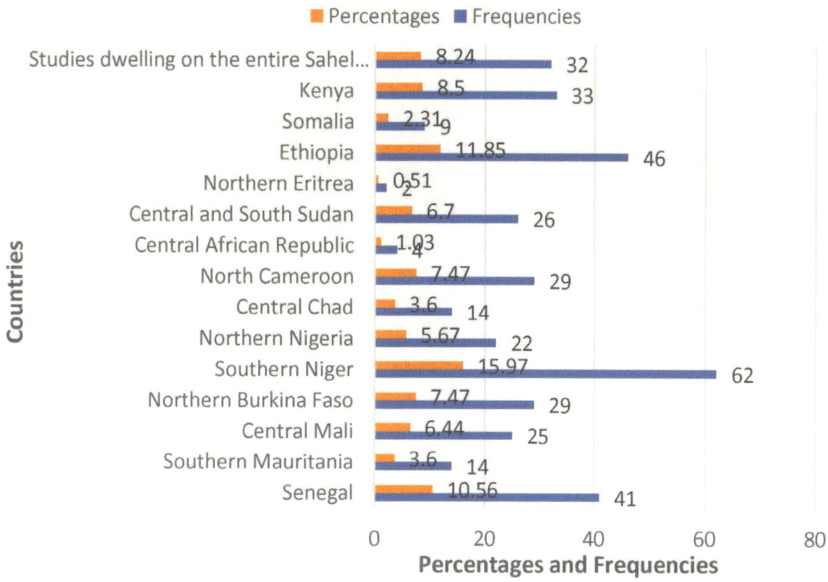

Figure 1.
Percentages and frequencies per country of reported/tracked climate shocks in the Sahel from 1975 to 2020.

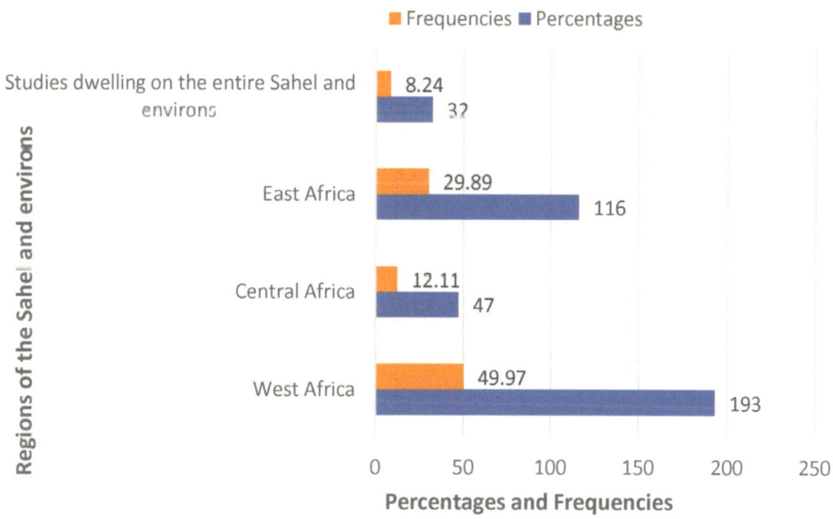

Figure 2.
Percentages and frequencies of reported/tracked climate shocks in various regions of the Sahel from 1975 to 2020.

are leveraging huge research grants on the Sahel troubled region. In the context of African authorship 9 authors are from Nigeria, 6 from Ethiopia and Niger and about 4 authors from Cameroon. In general, we observe that most of the authors involved in this genre of research are affiliated out of Africa.

3. Frequently reported climate change shocks in the Sahel

This study has found out that three key shocks are recorded in the Sahel of Africa. These shocks include droughts, floods, and winds/sandstorms. Droughts are reported as the most frequent among these shocks (n = 219) (**Figures 3** and **4** and

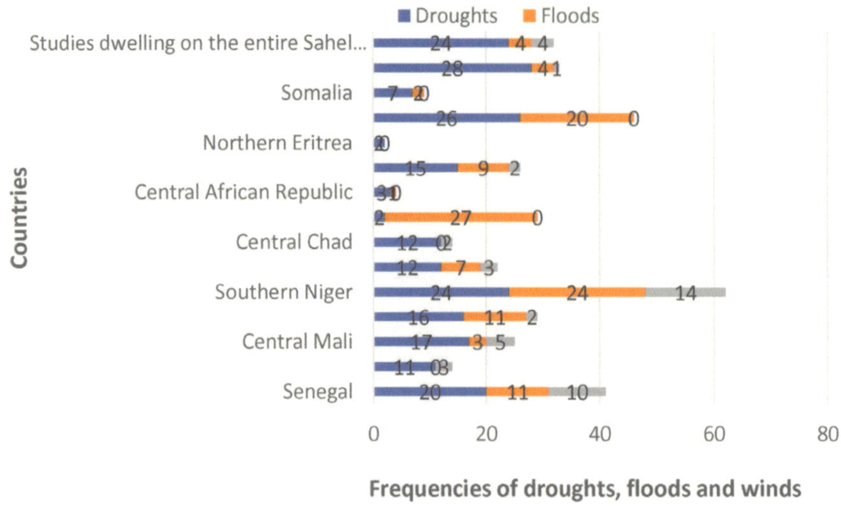

Figure 3.
Composite frequencies per country of reported/tracked climate shocks in the Sahel from 1975 to 2020.

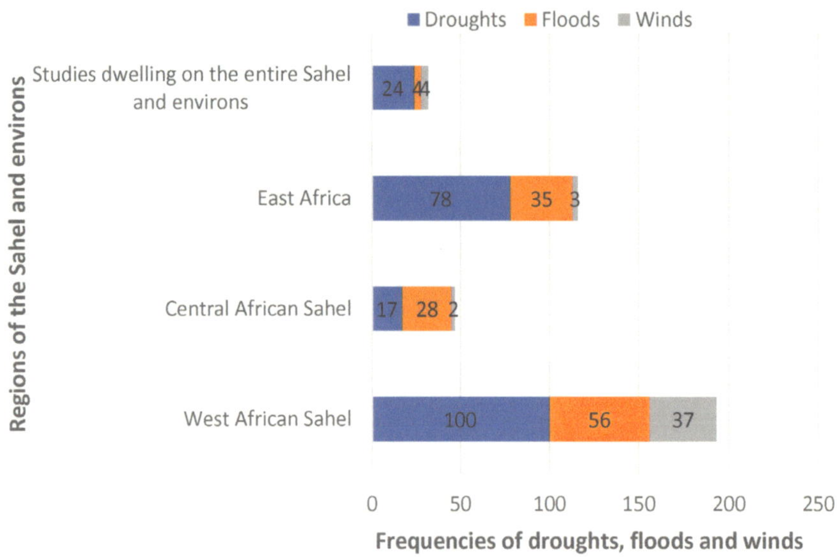

Figure 4.
Composite frequencies per region of reported/tracked climate shocks in the Sahel from 1975 to 2020.

Table 1). The other shocks in order of importance are floods (n = 123) (**Figure 2**) and winds/sandstorms (n = 46) (**Figures 3** and **4**). It can be inferred that the Sahel is a zone of climatic extremes with droughts topping the chart during the dry periods and with floods taking over during periods of prolonged precipitation which confirms the fact that one of the consequences of climate change is the amplification of extreme events.

In terms of the distribution of these shocks in various countries, the following countries recorded the highest number of shocks in order of importance: in term of droughts, Kenya recorded 28, Ethiopia recorded 26 and Niger recorded 24. In the domain of floods, Cameroon was the highest with 27 shocks, Niger with 24 shocks, and Ethiopia with 20 shocks. In the case of winds/sandstorms Niger observed the most shocks with the highest of 14 shocks, Senegal recorded 10 shocks and Mali recorded 5 shocks (**Figure 3**).

Shocks	Examples of studies
Droughts: WAS	
Senegal	Faure and Gac [39]; Agnew and Warren [8]
Niger	Boyd et al. [40]; Reenberg [41]
Mali	Nicholson [42]; Hiernaux et al. [43]
Burkina Faso	Nicholson [44]; Prospero and Nees [16]
Mauritania	Le Houerou [45]; Nicholson [42]
Northern Nigeria	Adefolalu [46]
Droughts: EAS	
Somalia	Boyd et al. [40]; Hitchcock and Hussein [47]
Ethiopia	Deressa et al. [48]; Turton and Turton [49]
Kenya	Epule et al. [12]; Nicholson [44]
Eritrea	Huho et al. [50]; Keller [51]
Sudan	Agnew and Warren [8]; Epule et al. [30]
Droughts: CAS	
Central African Republique	Nicholson [42]; Epule et al. [30]
Northern Cameroon	Nicholson [42]; Nicholson [44]
Chad	Nicholson [42]; Ozer et al. [52]

WAS: West African Sahel; EAS: East African Sahel; CAS: Central African Sahel.

Table 1.
Examples of Sahel countries with drought studies.

Regionally, West African Sahel witnessed the highest number of droughts of about 100 cases while East African Sahel had 78 shocks. As concerns floods, West African Sahel recorded the highest of about 56 cases while East African Sahel recorded 35 shocks and Central African Sahel recorded 28 shocks. As concerns winds/sandstorms, West African Sahel witnessed 370 shocks, while East and Central African Sahel recorded East 3 and 2) shocks respectively (**Figure 4**).

4. Decade with the highest number of shocks and the drivers of climate change shocks in the Sahel

At the decade scale, the 1975–1985 decade witnessed the largest number of shocks which 207 shocks recorded. The decade 1997–2007 recorded 80 shocks while the decade 1986–1996 recorded 52 shocks and the decade 2008–2016 recorded 49 shocks (**Figure 5**). In addition, the decade 1975–1985 witnessed the greatest number of droughts with 182 shocks. The decade 1997–2007 recorded the largest number of floods and winds of about 50 & 23 respectively (**Figure 5**). There is a decrease in the number of reports on climate shocks in general over time. We observe for example that the number of droughts reported has steadily decline over time from 182 in the decade 1975–1985 to 9 in the decade 2008–2016. Floods on the other hand witnessed increased reports of cases from about 12 in the decade 1975–1985 to 39 in the decade 2008–2016. Winds/sandstorms are irregular in the scientific publications and no concrete established trends have been recorded (**Figure 5**).

The most important drivers of the observed climate shocks are climatic with about 341 or 52% of the shocks. The non-climatic shocks account for about 314 or 47% of the shocks. When combined, these shocks account for about 655 shocks

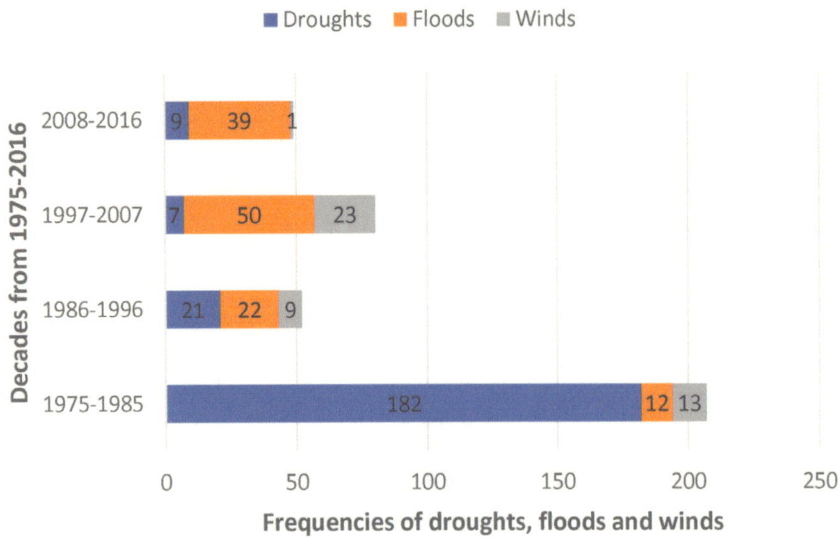

Figure 5.
Composite distribution of shocks per decade from 1975 in the Sahel.

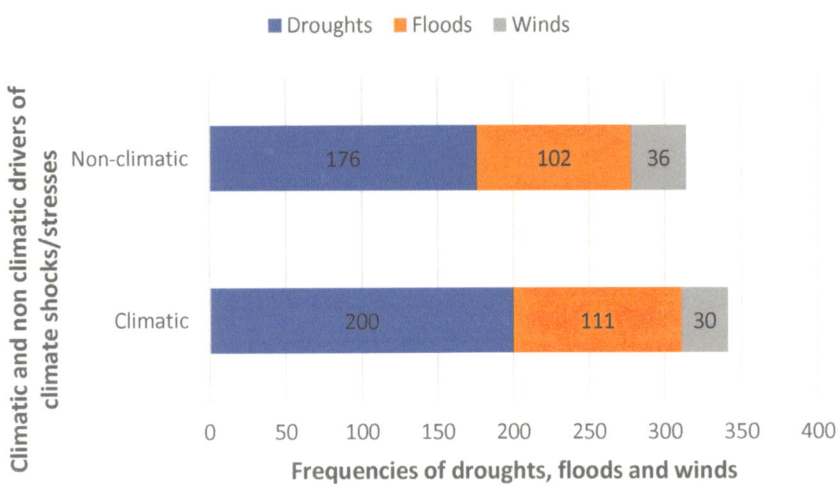

Figure 6.
Composite frequencies of reported/tracked climate shocks and their relationships with climatic and non-climatic variables.

which are more the 388 which are the total number of shocks reported as some studies reported both climatic and non-climatic drivers (**Figure 6**). Climatic drivers account for about 200 droughts, 111 floods and 30 winds/sandstorms. Non-climatic drivers accounted for 176 droughts, 102 floods and 36 winds/sandstorms. Climatic drivers account for most of the climate shocks but non-climatic drivers are increasingly becoming important. Some shocks such as droughts, floods, and winds (sandstorms) can result from a combination of climatic and non-climatic factors.

5. Discussion and implications

In this chapter, we examine the occurrence of climate shocks across the Sahel based on reports in the primary peer review literature. The work also examines the role of climatic and non-climatic drivers of the climate shocks and their evolution over time. These results provide a basis for assessing where shocks are occurring and to design adaptation actions based on the latter.

An analysis of where climate change shocks are occurring, what climate change shocks researchers are focusing on and their evolution over time goes a long way in providing insights that can enhance the ability to cope with the shocks. Already, this work has shown that in the Sahel, current adaptation strategies must be tailored to cope against droughts which are the most frequently reported climate shock in the Sahel. Invariably, droughts should be followed by floods and winds in any policy strategies aimed at cubbing shock and enhancing climate resilience in the Sahel. Even though floods have gained importance in the scientific scholarship over time, droughts remain the dominant climate forcing in the Sahel around which most adaptation actions should be drawn. Basing adaptations on site specific drivers is consistent with the systematic approach described by [53] and [11] describing the drivers of forest degradation in Bolivia and Cameroon respectively and using these drivers to propose policy options. This approach provides room for sustainability and accuracy and is a departure from the approach that is based on generalizations the copy pasting of what has worked elsewhere. This chapter provides a snapshot of the frequency in which climate change shocks are reported in the scientific literature, the affiliations of the authors, when and where the shocks occurred and the relative contributions of climatic and non-climatic drivers of these shocks.

Furthermore, this chapter has found out that Southern Niger and West African Sahel recorded the highest number climate change shocks in the Sahel. Kenya however recorded the highest number of reported droughts, while Cameroon recorded the highest number of floods. On the other hand, Niger recorded the highest number of winds. Researchers, and other decision making stakeholders must prioritize these dominant shocks. Most of the peer review studies reported here argue both climatic and non-climatic drivers are driving these climate shocks in the Sahel. For a long time, the scientific literature has noted that many of the environmental problems faced across Africa were caused by climate change. However, this current chapter notes that non climatic drivers are gaining importance. There is scientific evidence across Africa and the Sahel that illustrates that most of the environmental issues facing the region are driven by non-climatic drivers such as deforestation, population growth, wars and unsustainable approaches of land tenure and use like shifting and slash and burn cultivation inter alia while climate change only playing a reinforcing role [1, 7, 44, 54–58].

The decade 1975–1985 recorded the highest number of shocks in the Sahel. This surge has been attributed to the ravaging droughts that impacted most sub-Saharan African countries in the 1970s and early 1980s [50, 59–64]. At the dawn of the mid 1980s, the droughts that ravaged the continent had reduced in intensity giving rise to a surge in other climate shocks such as floods. Recent normalized difference vegetation index (NDVI) studies across the Sahel show that from the mid-1980s onwards, the Sahel has witnessed increasing precipitation and greening of the landscape [7, 54, 60, 61, 65–71]. This work is unique in that it is the first to use the systematic approach to verify the temporal and spatial distribution of climate shocks in the Sahel as well as their drivers. This approach has previously been used in tracking adaptations around the world as reported in the following studies, [1, 41, 43, 63, 64, 67–72]. It is important to caution that, this work provides a proxy-based approach offering a baseline from which climate change can be evaluated and monitored in the Sahel.

As this work is a first attempt to track climate change adaptations in the Sahel, going forward, the following recommendations are made: 1. similar systematic reviews across larger scales across Africa such as African continent. 2. Inclusion of both peer review and gray literature to better capture the shocks that might have been reported in non-standardized studies 3. Providing further insights into the leverage of research in the context of who is researching what and where are they affiliated. This option will help is encouraging a higher leverage of research funding for the region. Though this work is based on proxies, it has also been verified that through ground truthing that these results are generally valid as ground truthing has not resulted in any new information that can contradict these findings.

Acknowledgements

We would like to thank all the authors whose works we consulted. Special thanks go to UM6P for providing the funding for this research.

Conflict of interest

"The authors declare no conflict of interest."

Author details

Terence Epule Epule[1*], Driss Dhiba[1] and Abdelghani Chehbouni[1,2]

1 Mohammed VI Polytechnic University (UM6P), International Water Research Institute (IWRI), Benguerir, Morocco

2 CESBIO UMR (CNES-CNRS-UT-INRAe-IRD), Toulouse, France

*Address all correspondence to: epule.terence@um6p.ma

References

[1] Giannini, A., Saravanan, R., Chang, P. Oceanic forcing of Sahel Rainfall on interannual to interdecadal timescales. *Science,* 2003, 302:1027-1030.

[2] Reynolds, J.F., Stafford, S.D.M., Lambin, E.F., Turner, B.L., Mortimore, M., Batterbury, S.P., Downing, T.E., Dowlatabadi, H., Fernandez, R.J., Herrick. J/E., Huber-Sannwald. E/, Jiang. H., Leemans, R., Lynam, T., Maestre, F.T., Ayarza, M., Walker, B. Global desertification: building a science for dryland development. *Science,* 2007, 316: 847-851.

[3] IPCC. Climate Change 2007: Impacts, Adaptations and Vulnerability. Contributions of the Working Group 2 to the Fourth Assessment Report of the IPCC. Cambridge University Press, Cambridge, 2007, pp. 976.

[4] IPCC. Climate change 2013: the physical science basis. Contributions of the Working Group 1 to the Fifth Assessment Report of the IPCC. Cambridge University Press, Cambridge, 2013, pp. 979.

[5] Epule, T.E., Ford, J.D., Lwasa, S., Lepage, L. Vulnerability of maize yields to droughts in Uganda. *Water*, 2017a, 29, 181. http://dx.doi.org/10.3390/w9030181.

[6] Epule, T.E., Ford, J.D., Lwasa, S. Projections of maize yield vulnerability to droughts and adaptation options in Uganda. *Land Use Policy*, 2017b, 65, 154-163.

[7] Giannini, A., Biasutti, M., Held, I.M., Sobel, A.H. A global perspective on African climate. *Clim Chang*, 2008, 90:359-383. Doi: 10.1007/s10584-008-9396-y.

[8] Agnew, C.T., Warren, A. A framework for tackling drought and land degradation. *Journal of Arid Environments*, 1996, 33, 309-320.

[9] Agnew, C.T., Chappell, A. Drought in the Sahel. *Geojournal,* 1999, 48, 299-311.

[10] Vanmaercke, M., Zenebe, A., Poesen, J., Nyssen, J., Verstaeten, G., Deckers, J. Sediments dynamics and the role of flash floods in sediment export from medium-sized catchments: a case study from the semi-arid tropical highlands in northern Ethiopia. *Journal of Soil Sediments*, 2010, 10, 611-627.

[11] Epule, T.E., Peng, C., Lepage, L., Chen, Z. The causes, effects, and challenges of Sahelian droughts: a critical review. *Reg. Environ. Change*. 2014a, 14, 145-156.

[12] Epule, T.E., Peng, C., Lepage, L., Chen, Z. Policy options towards deforestation reduction in Cameroon: an analysis based on a systematic approach. *Land Use Policy*, 2014b, 36, 405-415. Doi: 10.1016/j.landusepol.2013.09.004.

[13] Mamadou, I., Gautier, E., Descroix, L., Noma, I., Moussa, I.B., Maiga, O.F., Genthon, P., Amogu, O., Abdou, M.M., Vandervaere, J.P. Exorheism growth as an explanation of increasing flooding in the Sahel. *Catena,* 2015, 131, 130-139.

[14] Karam, D.B., Flamant, C., Knippertz, P., Reitebuch, O., Pelon, J., Chong, M. Dabas, A. 2008. Dust emissions over the Sahel associated with West African monsoon intertropical discontinuity region: a representative case study. *Q.J.R Meteorol. Soc*. 2008, 134, 621-634.

[15] Hastenrath, S., Polzin, D., Mutai, C. Diagnosing the droughts and floods in Equatorial East Africa during Boreal Autumn 2005-08. American Meteorological Society, 2009, 813-818. Doi: 10.1175/2009jcli3094.1.

[16] Prospero, J., Nees, R. Dust concentration in the atmosphere of the

equatorial north Atlantic: possible relationship to the Sahelian drought. *Science,* 1976, 196, 1196-1198.

[17] Sivakumar, M.V.K., and Wallace J.S. Soil water balance in the Sudano-Sahelian zone: need, relevance, and objectives of the workshop. In: Sivakumar M.V.K., Wallace J.S., Renard C. and Giroux C. Soil water balance in the Sudano-Sahelian zone. IAHS Publication 199, Wallingford, 1991, pp. 3-10.

[18] Tarhule, A. Damaging rainfall and flooding: the other Sahel hazards. *Climate Change*, 2005, 72, 355-377.

[19] Amogu, O., Descroix, L., Yero, K.S., Le Breton, E., Mamdou, I., Ali, A., Vischel, T., Bader, J.C., Moussa, I., Gautier, E., Boubkraui, S., Belleudy, P. Increasing river flows in the Sahel? *Water,* 2010, 2, 170-199. Doi: 10.3390/w2020170.

[20] Albergel, J. Sécheresse, désertification et ressources en eau de surface-Application aux petits bassins du Burkina Faso. The Influence of Climate Change and Climatic Variability on the Hydrologic Regime and Water Resources (Proceedings of the Vancouver Symposium, August 1987). IAHS Publ.1987, no. 168 Available online at: http://iahs.info/redbooks/A168/iahs_168_0355.pdf.

[21] Lu, J., Delworth, T.L. Oceanic forcing of the late 20th century Sahel drought. *Geophysical Research Letters,* 2005, 32: L22706. Doi: 10.1029/2005GL023316.

[22] Mertz, O., Mbow, C., Reenberg, A., Diouf, A. 2009. *Environmental Management*, 43, 804-816.

[23] Lopez-I-Gelats, F., Fraser, E.D.G., Morton, J.F., Rivera-Ferre, M.G. What drives the vulnerability of pastoralist to global environmental change? A qualitative meta-analysis. *Global Environmental Change,* 2016, 39, 258-274.

[24] Anuforom, A. Spatial distribution and temporal variability of Harmattan dust haze in sub-Sahel West Africa. *Atmospheric Environment*, 2007, 41, 9079-9090.

[25] Jankowiak I, Tanre D. Satellite climatology of Saharan dust outbreaks: Method and preliminary results. *J. Climate, 1992,* 5: 646-656.

[26] Alpert, P., Ganor, E. A jet stream associated heavy dust storm in the western Mediterranean. *J. Geophys. Res. 1993,* 98, 7339-7349.

[27] Tengberg, A. Nebkha dunes as indicators of wind erosion and land degradation in the Sahel zone of Burkina Faso. *Journal of Arid Environments*, 1995, 30, 265-282.

[28] Tall, R., Mason, S.J., Van Aalst, M., Suarez, P., Ait-Chellouche, Y., Diallo, A.A., Braman, L. Using seasonal climate forecast to guide disaster management: the Red Cross experience during the 2008 West African floods. *International journal of Geophysics*, 2012, 986016, Doi: 10.1155/2012/986016.

[29] Lwasa, S. A systematic review of research on climate change adaptation policy and practice in Africa and South Asia deltas. *Regional Environmental Change*, 2015, 15, 815-825. Doi: 10.1007/s10113-014-0715-8.

[30] Epule, T.E., Peng, C., Lepage, L. Environmental refugees in sub-Saharan Africa: a review of perspectives on trends, causes, challenges and way forward. *Geojournal,* 80, 79-92.

[31] Dixon, R., Smith, J., Guill, S. Life on the edge: vulnerability and adaptation of African ecosystems to global climate change. *Mitigation and Adaptation Strategies for Global Change, 2010,* 8, 93-113.

[32] Drees, L.R., Manu, A., Wilding, L.P. Characteristics of Aeolian dust in the Niger, West Africa. *Geoderma,* 1993, 59, 213-233.

[33] Eklundh, L., Olsson, L. Vegetation index trends for the African Sahel 1982-1999. *Geophysical Research Letters*, 2003, 30, 8, 1430. Doi: 10.1029/2002GL016772.

[34] Mulitza, S., Heslop, D., Pittauerova, D., Fischer, H., Meyer, I., Stuut, J.B., Zabel, M., Mollenhauer, G., Collins, J.A., Kuhnert, H., Schulz, M. Increase in African dust flux at the onset of commercial agriculture in the Sahel region. *Nature Letters*, 2010, 466, 226-228.

[35] UNREDD+. Executive summary: 2015 annual report of the UN-REDD Program Fund. 2015, Available online at: http://www.unredd.net/documents.html?view=browse&customtags=172&startdate=&enddate=&dmlang=

[36] United Nations (UN). Paris agreement. United Nations, New York. 2015, Available online at: http://unfccc.int/files/essential_background/convention/application/pdf/english_paris_agreement.pdf.

[37] African Climate Change Fund. Supporting African countries to access international climate finance: enabling a transition towards climate resilient, low-carbon development. The African Development Bank. 2016, Available online at: https://www.afdb.org/fileadmin/uploads/afdb/Documents/Project-and-Operations/Africa_Climate_Change_Fund__Supporting_Africa_countries_to_access_international_climate_finance.pdf.

[38] O'Connor, Ford, J.D. Increasing the effectiveness of the "Great Green Wall" as an adaptation to the effects of climate change and desertification in the Sahel. 2014, *Sustainability*, 6, 7142-7154. Doi: 10.3390/su6107142.

[39] Faure, H., Gac, J.Y. Will the Sahel drought end in 1985? *Nature*, 1985, 291, 475-478.

[40] Boyd, E., Cornforth, R.L., Lamb, P.J., Tarhule, A., Lele, M.A., Brouder, A. Building resilience to face recurring environmental crisis in African Sahel. *Nature Climate Change*, 2013, 3, 631-637.

[41] Reenberg, A. Land-use dynamics in the Sahelian zone in eastern Niger-monitoring change in cultivation strategies in drought prone areas. *Journal of Arid Environments*, 1994, 27, 179-192.

[42] Nicholson, S.E. The west African Sahel. A review of recent studies on the rainfall regime and its interannual variability. *ISRN Meteorology*, 2013, 453521, Doi.org/10.1155/2013/453521.

[43] Hiernaux, P., Diarra, L., Trichon, V., Mougin, E., Soumaguel, N., Baup, F. Woody plant population dynamics in response to climate change from 1984 to 2006 in Sahel (Gourma, Mali). *Journal of Hydrology,* 2009, 375, 103-113.

[44] Nicholson, S.E. Sub-Saharan rainfall 1981-84. *Journal of Climate and Applied Meteorology,* 1985, 24, 1388-1391.

[45] Le Houerou, H.N. Climate change, drought, and desertification. *Journal of Arid Environments*, 1996, 34, 133-185.

[46] Adefolalu, D.O. 1986. Further aspects of Sahelian drought as evident from rainfall regime of Nigeria. *Arch. Met. Geoph. Biocl*, 36, 277-295.

[47] Hitchcock, R., Hussein, H. Agricultural and non-agricultural settlements for drought-affected pastoralists in Somalia. *Disasters,* 1987, 11, 1, 30-39.

[48] Deressa, T., Hassan, R., Ringler, C., Alemu, T., Yesuf, M. Determinants of farmers' choice of adaptation methods

to climate change in the Nile Basin of Ethiopia. *Global Environmental Change,* 2009, 19, 248-255.

[49] Turton, D., Turton, P. Spontaneous resettlement after drought: an Ethiopian example. *Disasters*, 1984, 8, 3, 178-189.

[50] Huho, J.M., Ngaira, J.K.W., Ogindo, H. *Educational Research*, 2011, 2, 1, 779-789.

[51] Keller, E. Drought, war and the politics of famine in Ethiopia and Eritrea. *The Journal of Modern African Studies*, 1992, 30, 4, 609-624.

[52] Ozer, P., Erpicum, M., Demaree, G., Vandiepenbeeck, M. The Sahelian drought may have ended during the 1990s. Hydrological Sciences Journal, 2003, 48, 3, 489-492. Doi: 10.1623/hysj.48.3.489.45285.

[53] Anyamba, A., Tucker, CJ. Analysis of Sahelian Vegetation dynamics using NOAA-AVHRR NDVI data from 1981-2003. *Journal of Arid Environments,* 2005, 63: 595-614.

[54] Nicholson, S.E. Rainfall and atmospheric circulation during drought periods and wetter years in West Africa. *American Meteorological Society*, 1981, 109, 2191-2208.

[55] Maranz, S. Tree mortality in the African Sahel indicates an anthropogenic ecosystem displaced by climate change. *Journal of Biogeography*, 2009, 36, 1181-1193.

[56] Anyamba, A., Eastman, J. Inter-annual variability of NDVI over Africa and its relation to El Niño/Southern Oscillation. *International Journal of Remote Sensing*, 1996, 17, 2533-2548.

[57] Nielsen, J. Ø., Reenberg, A. Cultural barriers to climate change adaptation: A case study from Northern Burkina Faso. *Global Environmental Change, 2010, 20*(1), 142-152. Doi: 10.1016/j.gloenvcha.2009.10.002.

[58] Nielsen, J.Ø., D'Haen, S., Reenberg, A. Adaptation to climate change as a development project: A case study from Northern Burkina Faso. *Climate and Development*, 2012, 4(1), 16-25.

[59] Muller, R., Pistorius, T., Rohde, S., Gerold, G., Pacheco, P. Policy options to reduce deforestation based on a systematic approach of drivers and agents in lowland Bolivia. *Land Use Policy*, 2013, 30, 895-907.

[60] Birkett, C., Mutugudde, R., Allan, T. Indian Ocean climate event brings floods to East Africa's lakes and Sudd Marsh. *Geophysical Research Letters*, 1999, 26, 8, 1031-1034.

[61] Samimi, C., Fink, A.H., Paeth, H. The 2007 floods in the Sahel: causes, characteristics, and its presentation in the media and FEWS NET. *Natural Hazards and Earth System Sciences,* 2012, 12, 313-325.

[62] Billi, P., Alemu, Y.T., Ciampalini, R. Increased frequency of flash floods in Dire Dawa, Ethiopia: change in rainfall intensity or human impact? *Natural Hazards*, 2015, 76, 1373-1394.

[63] Nka, B.N., Oudin, L., Karambiri, H., Paturel, J.E., Ribstein, P. Trends in floods in West Africa: analysis based on 11 catchments in the region. *Hydrology and Earth System Sciences*, 2015, 19, 4707-4 719.

[64] Ologunorisa, T., Adeyemo, A. Public perception of floods hazard in the Niger Delta, Nigeria. *The Environmentalist*, 2005, 25, 39-45.

[65] Nicholson, S.E., Tucker, C.J., Ba, M.B. 1998. Desertification, drought, and surface vegetation: an example from the west African Sahel. *Bulletin of the American Meteorological Society*, 1998, 79, 5, 815-829.

[66] Nicholson, S.E. A detailed look at the recent drought situation in the

Greater Horn of Africa. *Journal of Arid Environments*, 2014, 103, 71-79.

[67] Moreno, T., Querol, X., Castillo, S., Alastuey, A., Cuevas, E., Hermann, L., Mounkaila, M., Elvira, J., Gibbons, W. 2006. Geochemical variations in Aeolian mineral particles from the Sahara-Sahel dust corridor. *Chemosphere,* 2006, 65, 261-270.

[68] McLeman, R. Settlement abandonment in the context of global environmental change. *Global Environmental Change*, 2011, 21, 108-120.

[69] Ochola, S., Eitel, B., Olago, D. Vulnerability of schools to floods in Nyando River catchment, Kenya. *Disaster,* 2010, 34, 3, 732-754.

[70] Noble, I., Huq, S. Adaptation Needs and Options (New York: Cambridge University Press) chapter, 2014, 14.

[71] Fernandez, A., Najafi, M., Durand, M., Mark, B.G., Moritz, M., Jung, H.C., Neal, J., Shastry, A., Laborde, S., Phang, S.C., Hamilton, I.M., Xiao, N. Testing the skill of numerical hydraulic modeling to simulate spatiotemporal flooding patterns in the Logone floodplain, Cameroon. *Journal of Hydrology*, 2016, 539, 265-280.

[72] Kaly, F., Marticorena, B., Rajot, J.L., Janicot, S., Niang, A., Yahi, A., Thiria, S., Zakou, A., Coulibaly, B.S., Kone, I., Traore, S., Diallo, A., Ndiye, T. Variability of mineral dust concentrations over West Africa monitored by the Sahelian Dust Transect. *Atmospheric Research, 2015,* 164, 165, 226-241.

Chapter 16

Recent Climate Change Adaptation Strategies in the Sahel: A Critical Review

Terence Epule Epule, Abdelghani Chehbouni and Driss Dhiba

Abstract

Climate change adaptation continues to be central on the agenda of most African countries. Current understanding of the state of adaptation is limited in Africa. The Sahel is selected because of persistent declines in precipitation and rising temperatures. Here, we examine the status of adaptation actions across the Sahel by reviewing the primary peer review literature. A total of 70 peer review papers that document 414 discrete adaptations provide a snapshot of adaptations developed between 1975 and 2020. From a country-to-country perspective, Kenya has the highest number of reported adaptation actions (75 or 18.1%). From a regional standpoint, West Africa recorded about 261 or 18.1% of all adaptation actions reported. Income diversification of livelihoods, and water harnessing were reported as the most used adaptation actions in the Sahel. Based on categories, technically based adaptation actions are the most used options. The period 2008–2016 registered 65.2% of all adaptations. 98% of adaptation actions are reported to be driven by climate while non-climatic drivers account for 95% of adaptation actions. The findings presented here are proxies of climate change adaptation; some relevant information might be found in gray literature which not used because gray literature is less standardized because it is not subject to peer review.

Keywords: Climate change, Adaptation, West Africa, Sahel, Income diversification, Technical adaptation

1. Introduction

It is now clear that the world is experiencing climate change and the Sahel of Africa is no exception to this [1–7]. In the past thirty years the Sahel has witnessed temperatures in the range of between 0.2–2.0° C [8]. At the same time, the precipitation gradient has witnessed a decline from the south to the north of the Sahel. The effects of the above precipitation and temperature nexus in the Sahel have been declining crop yield, water resources and degradation of forest species. Specifically, they include: (i) increase mortality and dieback of various tree species in parts of the Sahel such as Mauritania, Chad, Mali, Burkina Faso, Senegal, and Niger, as seen in observations of the last quarter of the 20th century [9]; (ii) enhanced stress on food systems, with about 50% of the 60 million people living in the Sahel believed to be facing food insecurity linked to climate change [10–12], with the region likely projected to potentially experience about 250 million tons of food deficits by 2020 [12–14]; (iii) enhanced occurrence of malaria and diarrheal diseases, [15–17]; (iv) with more frequent water shortages also documented [18]. These impacts, in turn, are believed to have increased the number of climate refugees in the Sahel [19–22]. Climate change may also present opportunities, including increasing food production through better water management, irrigation, rainwater harvesting [3], and potential increase in crop productivity due to increased aerial fertilization by carbon dioxide [23].

Climate change Adaptation is important as it reduces the damages caused by climate change and takes advantage of new opportunities in-light of the rapid climate change already occurring and expected future impacts [24–31]. Governments, indigenous people, farmers, non-governmental organizations, donor organizations, the African Development Bank, the World Bank, and United Nations Environment Program have been active making suggestions through several policies, programs, and adaptations. For example, actions include those of the United Nations Reductions of Emissions from Deforestation and Forest Degradations (REDD+) which supports reforestation efforts in the Sahel [32, 33], and adaptation funding programs established under the United Nations Framework Convention on Climate Change (UNFCCC) [34] and at the regional level, the African Development Bank (AfDB) is now masterminding the African Climate Change Fund which has as objective to increase access of African countries to international climate finance [35]. Despite the importance of climate change and adaptations in the Sahel, a lot remains unknown with respect to how adaptation actions have occurred over time in various part of the Sahel. This knowledge gap reduces the ability to characterize key gaps in adaptations. Furthermore, this underscore the necessity is create adaptation metrices across Africa that can assist in evaluating progress and monitoring the status of adaptation.

This chapter provides a better understanding of the status of climate change adaptation actions across various Sahel regions and countries, showing how adaptations have changed over time as well as the relative contributions of climatic and non-climatic drivers. This is based on what has been recorded in the scientific peer review literature, underscoring a general and baseline overview of adaptation in the region. This work contributes to an important gap in the literature, with most studies examining the state of adaptation focusing on developed nations e.g., [31, 36, 37], or focused on other regions of Africa e.g., [38, 39]. This chapter is vital because there are currently insufficient studies that adopt the holistic approach fostered by this chapter. Prior to this study, the peer review data on climate change adaptation in the Sahel was found in several scattered publications which evidently did not provide a holistic approach. Furthermore, this chapter is vital because the Sahel remains a territory witnessing persistent shocks that need to be better addressed [40–42] that calls for urgent adaptation actions to enhance resilience. Therefore, a holistic picture on where the peer review literature has so far focused goes a long way in improving our understanding on where the emphasis is and where attention needs to be given.

2. The Sahel: geography, livelihood and overview of climate change

The Sahel is the semiarid strip of land located between the tropical rainforest in the south and the arid north of Africa and covers an area of about 3.053×10^3 km^2 and has about 60 million inhabitants [43]. The Sahel is located between latitude 10° and 20° north and extends from about 5000 km around northern Senegal around the west towards southern Mauritania, central Mali, northern Burkina Faso, south-western Niger, northern Nigeria, central Chad,

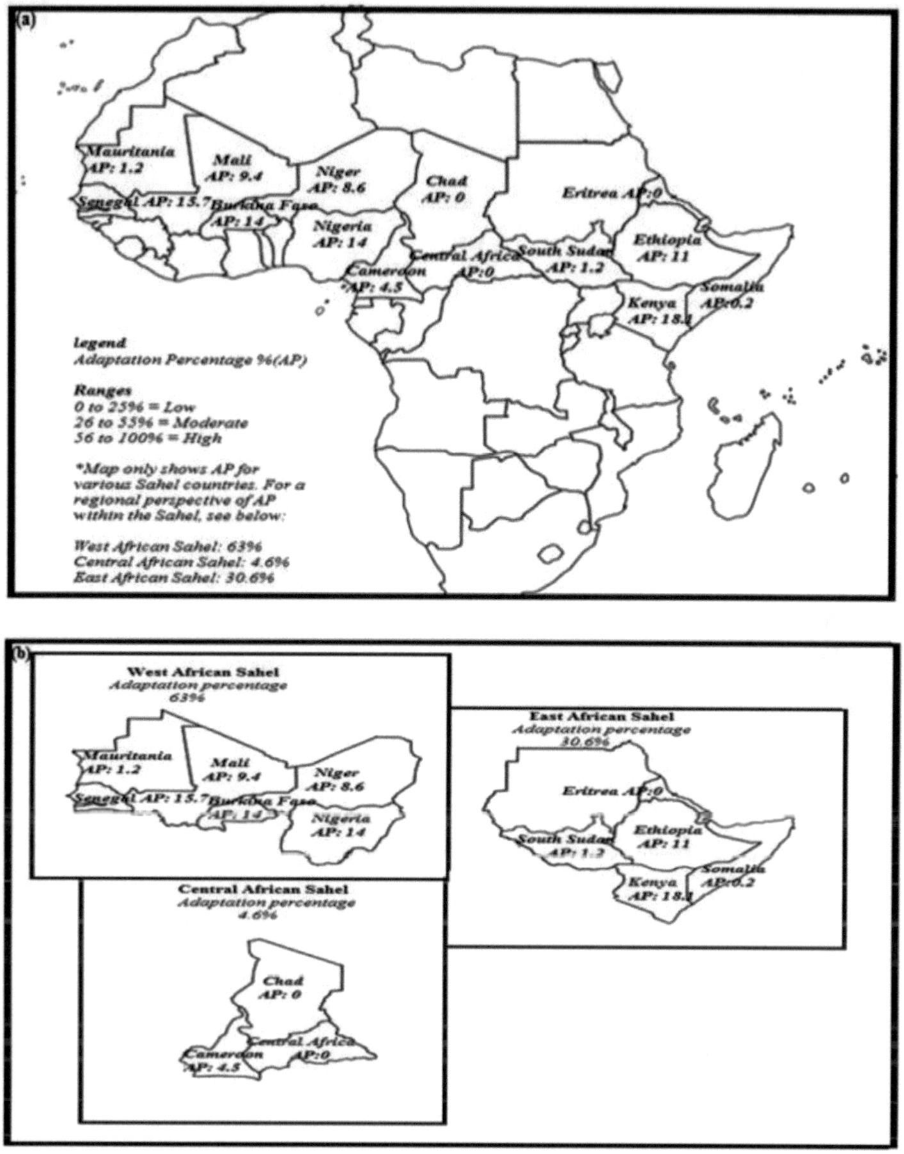

Figure 1.
Location of the various (a) countries and (b) regions of the Sahel.

north of Cameroon, Central African Republic, central Sudan and southern Sudan, northern Eritrea, extreme north of Ethiopia, to Somalia in the east and south east of the Sahel into Kenya (**Figure 1**).

In the Sahel, the vegetal landscape is covered by open Acacia shrubs and grassland. The Sahel represents a transition between the humid savanna in the south and the desert in the north [44, 45]. In terms of rainfall, the Sahel experiences declining rainfall with increasing latitude. At the southern border of the Sahel, about 450–500 mm of rainfall are recorded yearly while towards the higher latitudes less than 200 mm of rainfall are recorded yearly [46, 47]. Between 1930 and 1965 and 1966 and 2000, the Sahel recorded about 100 mm of rainfall per year [48]. The rainfall pattern in the Sahel is tied to the migration of the Inter-tropical Convergence Zone (ITCZ) [44, 46, 47, 49]. In the Sahel, a rainfall gradient of between 250 and

300 mm between the southern and northern spheres of the region is recorded. At 17° latitude north less than 200 mm of rainfall are recorded annually while southwards at about 15° latitude north (southern boundary of the Sahel), more rainfall of about 450–500 mm is recorded annually [44, 46, 47].

3. Countries and regions where adaptations are occurring in the Sahel?

In the Sahel, climate change adaptation actions can be categorized into four main categories (**Figure 2**). These include technically based, economically based, indigenous based and socially based (**Figure 2**). A total of 414 adaptation actions were reported in 70 peer review journal articles. An analysis of the articles illustrates that Kenya had recorded the highest number of adaptation actions of about 75 or 18.1% during the period 1975 and 2015. This was followed by Senegal with 65 or 15.7% of all recorded adaptations (**Figure 3**). Even though these countries recorded the highest rates of adaptation actions, it is worth noting that the overall picture in term of adaptation actions in the Sahel appears to be generally low (**Figure 1**).

Regionally, with a frequency and percentage of about 261 or 63%, West African Sahel witnessed the highest adaptation actions as reported by the peer review literature. East Africa recorded 127 or 30.6% and therefore is the second most adapted region in the Sahel (**Figure 4**). Furthermore, it can be said that when a region has more adaptation actions, the same is said of its adaptive capacity (**Figure 1**).

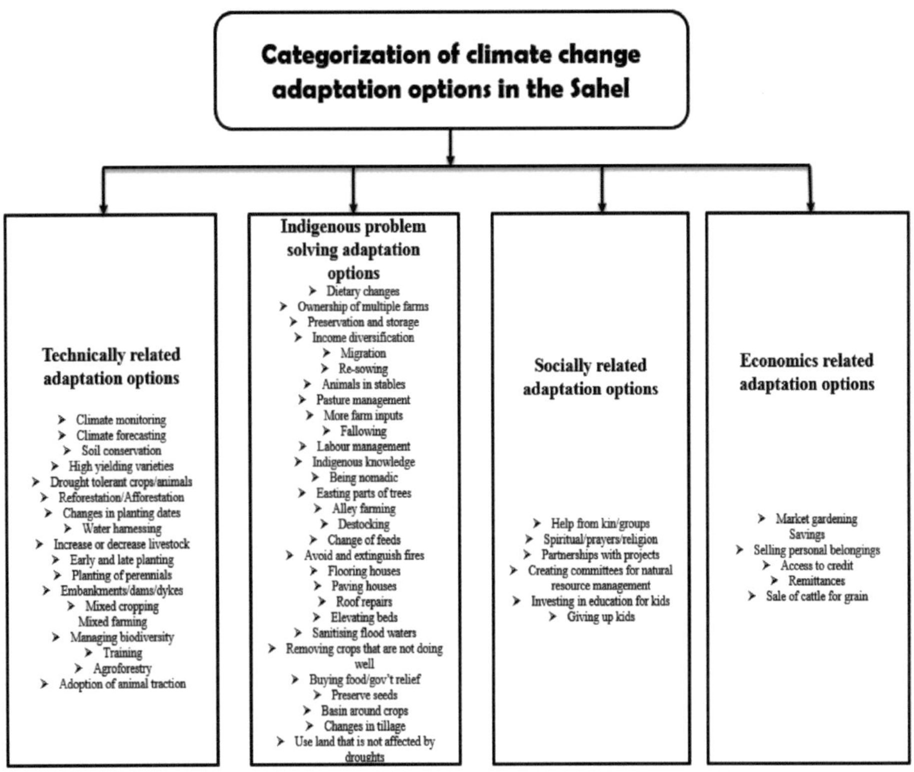

Figure 2.
The four categories of climate change adaptation actions in the Sahel.

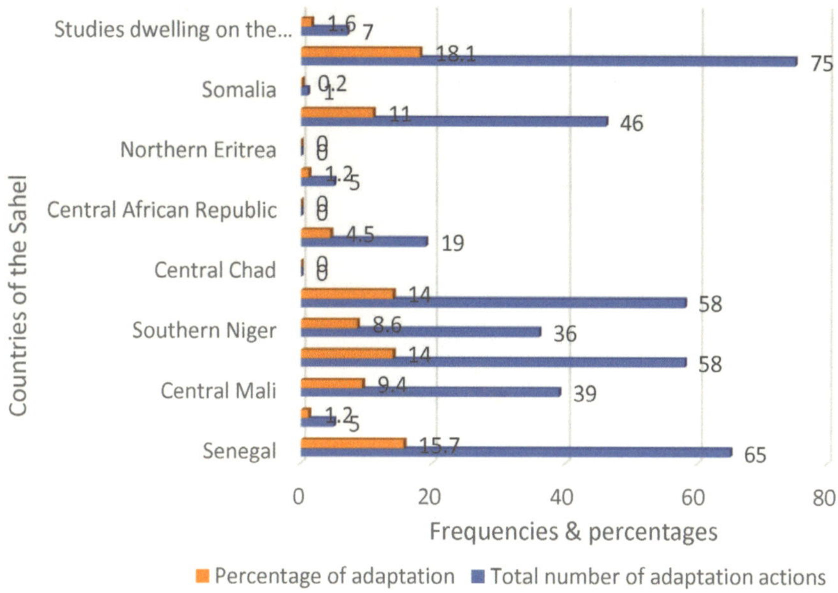

Figure 3.
Number of adaptation options and calculated percentage reported in the peer review literature published between 1975 and 2015 for various Sahel countries.

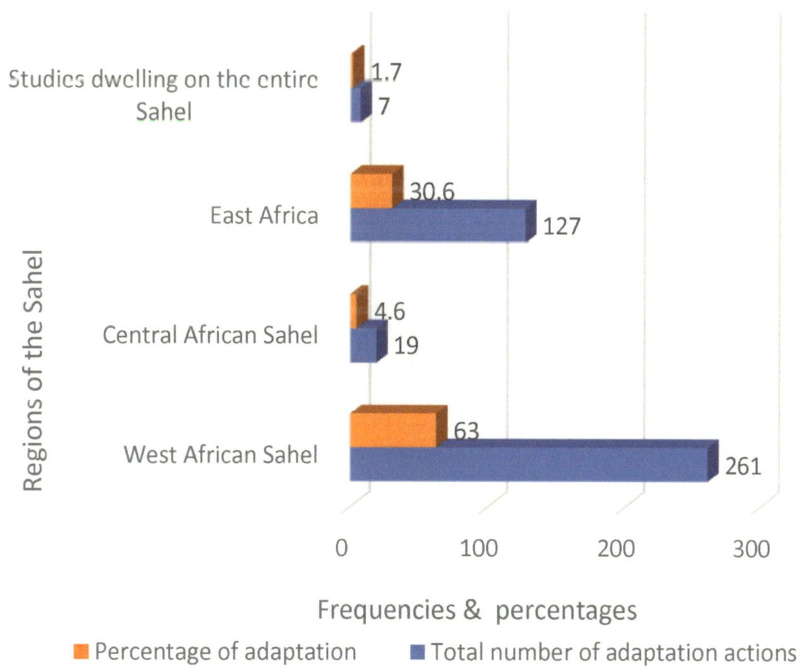

Figure 4.
Number of adaptation options and calculated percentage reported in the peer review literature published between 1975 and 2015 for various regions of the Sahel.

4. Actual adaptation actions in the Sahel

Based on this systematic review, income diversification adaptation actions recorded the highest frequencies in the scientific peer review literature. This is seen as they stood at a frequency of 53 or 12.8% and tailed closely by water harnessing actions which recorded a frequency of 48 or 11.5%. Adaptation actions like soil conservation, farm inputs, and planting high yielding varieties recorded 6.5%, 6.2% and 5.5% respectively (**Figure 5**, **Table 1**). From these results, we note

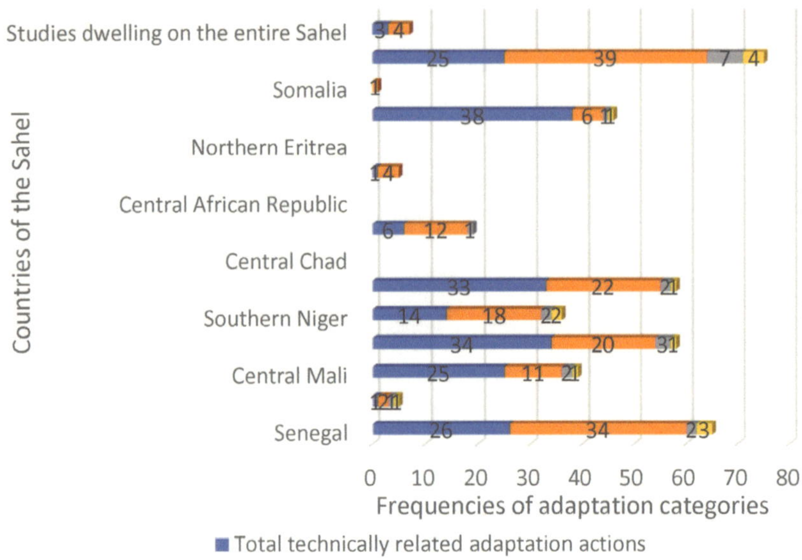

Figure 5.
Composite distribution of categorized number of adaptations options reported in the peer review literature published between 1975 and 2015 for various countries of the Sahel.

Adaptation options	References
Income diversification actions	
Off farm employment	Fleuret [50]; Mertz et al. [6]; Elmqvist and Olsson [51]; Morand et al. [52]
Working for development projects	Nielsen and Reenberg [53]; Moretimore [54]
Small-scale commerce	Nielsen and Reenberg [53]
Non-farm income	Mertz et al. [55]; Reardon et al. [56]; Burnham and Ma [57]
Selling of personal belongings	Schaer [58]; Brockhaus et al. [59]; Opondo [60]
Supplemental occupation	Epule et al. [61]; Opiyo et al. [62]
Hunting	Mosberg and Eriksen [63]
Water Harnessing	
Rainwater harvesting	Bryan et al. [64]; Zampaligré et al. [65]; Barbier et al. [66]
Irrigation	Bryan et al. [64]; Deressa et al. [67]; Fleuret [50]
Water management	Rockström [68]; Douxchamps et al. [69]; Burnham and Ma [57]
Boreholes	Mbow et al. [70]
Construction of dams and drainage systems	Magistro and Medou [71]
Use of water pumps	Schaer [58]
Use of green and blue water	Recha et al. [72]

Adaptation options	References
Soil conservation	
Mulching	Burnham and Ma [57]
Change in tillage and rotation	Bryan et al. [64]
Half-moon and stone dykes	Zampaligré et al. [65]
Planting shade trees	Burnham and Ma [57]; Deressa et al. [67]; Bryant et al. [61]
Fallowing	Gebrehiwot and Van der Veen [73]; Nyong et al. [74]
Soil erosion control/stabilization dunes	Okoye [75]; Brockhaus et al. [59]; Burnham and Ma [57]
Farm inputs	
Manure	Mertz et al. [6]; Marenya and Barrett [76]; Wood et al. [77]
Fertilizers	Mertz et al. [6]; Mertz et al. [55]; Croppenstedt et al. [78]
Inorganic fertilizers	Marenya and Barrett [76]; Wubeneh and Sander [79]
Adoption of compost	Somda et al. [80]
Organic fertilizers	Epule et al. [61]
High yielding crops and animals	
High yielding crop varieties	Bryan et al. [64]; Deressa et al. [67]; Fleuret [50]
Improved varieties of potatoes	Thuo et al. [81]; Burnham and Ma [57]; Abebe et al. [82]
Improved varieties of sorghum	Adesina and Baidu-Forson [83]
Improved varieties of maize	De Groote et al. [84]
High biomass variety of maize	De Groote et al. [84]
Improved animal husbandry	Zander et al. [85]

Table 1.
Adaptation actions in the Sahel in order of importance.

that emphasis is placed on employment diversification actions or on providing water harnessing options. While Kenya and Senegal recorded the highest cumulate amounts of adaptation actions with income diversification actions appearing to record the highest frequencies.

5. Categorization of actual adaptation actions in the Sahel

In terms of broad adaptation categories, four key categories of broad adaptation actions were identified. All other actions fall within any of these four categories, these include: (i) technically based actions; (ii) indigenous problem-solving based; actions; (iii) socially based actions and (iv) economics-based actions (**Figure 5**).

Technically based adaptation actions recorded the highest number of adaptation actions with a frequency of 206 or 49%. The other broad-based categories in order of importance are: Indigenous problem-solving based options with a frequency of 173 or 41.7% (**Table 2**), socially related based actions with a frequency of 21 or 5% and economics-based actions with frequencies of 14 or 3.3%. Though the technically based actions are leading, Kenya has the highest number

Adaptation options	References
Technically related adaptation options	
Water Harnessing	Zampaligré et al. [65]; Barbier et al. [66]; Recha et al. [72]
Soil conservation	Burnham and Ma [57]; Deressa et al. [67]; Bryan et al. [64]
Climate monitoring	Boyd et al. [86]; Huq et al. [87]; Reenberg [88]
Early harvesting	Tambo and Abdoulaye [89]; Burnham and Ma [57]; Deressa et al. [67]
Mixed farming	Moretimore [54]
Indigenous problem-solving adaptation options	
Income diversification	Schaer [58]; Brockhaus et al. [59]; Opondo [60]
Migration/relocation	Zampaligré et al. [65]; Scheffran et al. [90]; Burnham and Ma [57]
Farm inputs	Marenya and Barrett [76]; Wubeneh and Sanders [79]
Pasture management	Brockhaus et al. [59]; Burnham and Ma [57]; Mertz et al. [6]
Buying food/gov't relief	Epule et al. [61]; Oluoko-Odingo [91]
Socially related problem-solving options	
Help from kin/social organizations	Mortimore [54]; Epule et al. [61]; Oluoko-Odingo [91]; Fleuret [50]
Spiritual/prayers/religion	Fleuret [50]; Mertz et al. [6]; Burnham and Ma [57]
Partnership between local people and development project	Nielsen et al. [92]
Creating local committees for natural resources management	Brockhaus et al. [59]
Investing in education of kids	Opondo [60]
Economics related adaptation options	
Market gardening	Mertz et al. [55]; Mertz et al. [6]; Nielsen et al. [92] and Reenberg [88]
Savings	Schaer [58]
Selling of personal belongings	Schaer [58]
Access to credits	Ebi et al. [93]
Remittances	Fleuret [50]

Table 2.
Four categories of adaptation actions in the Sahel.

of total adaptation actions. The indigenous problem-solving based actions recorded 39 or 9.4%. Similarly, in Senegal, indigenous problem-solving adaptation recorded a frequency of 34 or 8.2%. When all the categories are put together, the frequencies for the latter category, does not outbid the former (**Figure 5**). Regionally, the West African Sahel recorded the highest number of adaptation actions and tallied highest in the context of technically based adaptation actions with a frequency of 133 or 32.1%. In terms of other regions, East Africa records the second highest frequency as technically based adaptation actions recorded 64 or 15.4% (**Figure 6**).

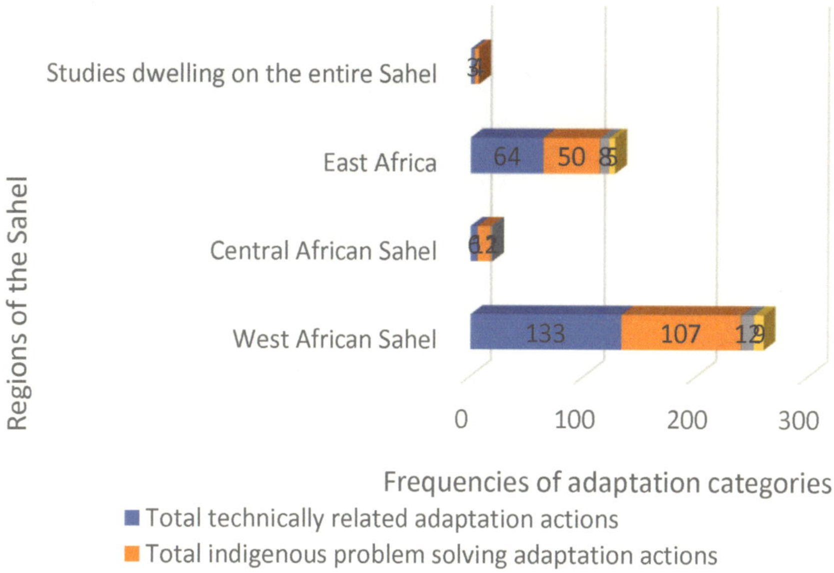

Figure 6.
Composite distribution of categorized number of adaptations options reported in the peer review literature published between 1975 and 2015 for various regions of the Sahel.

6. Temporal variations in adaptation actions and drivers of adaptation in the Sahel

The decade 2008–2016 recorded a total of 270 or 65.2% adaptation actions which represent the highest number of adaptation actions in the Sahel. The subsequent decades witnessed adaptation actions as follows: 1997–2007: (n = 58, 14%); 1986–1996: (n = 29, 7%); 1975–1985: (n = 57, 13.7%). In summary, we end up having a total of 414 adaptation actions which is like the total number of adaptation actions (**Figure 7a**). It can be observed that there has been an increase in the number of adaptation actions over the decades. A review of the drivers shows that most of the adaptation actions are anchored on climatic factors. The climatic drivers recorded 406, or 98% while the non-climatic drivers recorded 395 or 95% (**Figure 7b**). Though the climatic drivers are dominant, an interesting observation is the increase prominence of non-climatic drivers. Some potential climatic drivers include increase precipitation, decreasing precipitation, temperature increase, temperature decline, sea surface temperature changes, trade winds, El Niño, increase solar radiation, atmospheric circulation, and the prevalence of winds/sandstorms like the equatorial westerlies. Non-climatic drivers of adaptation cited in the literature encompass agriculture, population growth, settlements, poor urban planning, pastoralism, and cattle rearing, over exploitation of resources, grazing of animals, deforestation and land degradation, wildfires, insects, economic fluctuations and socio-political.

In the Sahel, some adaptation actions have gained prominence over time. Income diversification which is the highest used adaptation action recorded a frequency of 21 during the decade 1985–1996. Th decade 2008–2016; also recorded 18 adaptation actions. Here also, water harnessing actions were dominant during the decade 1985–1996 in which they recorded a frequency of 27. The decade 2008–2016, witnessed 15 adaptation actions. On the other hand, soil conservation related actions recorded their highest frequency during the decade 1985–1996. Migration had a frequency of

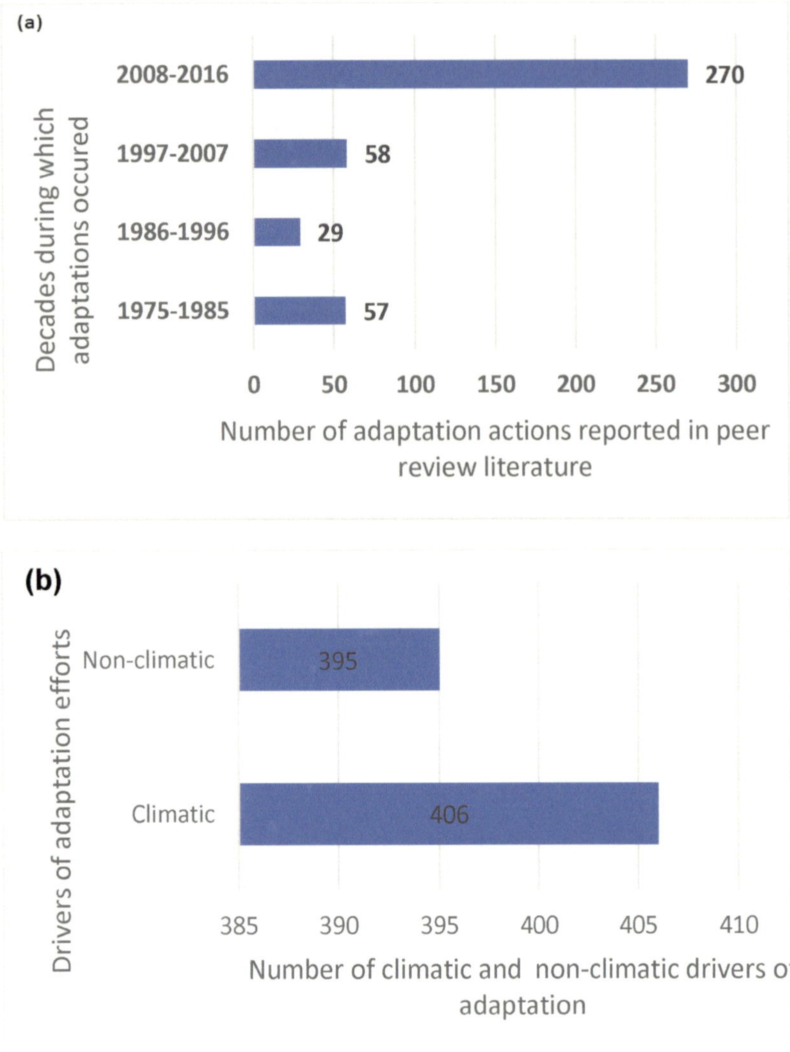

Figure 7.
(a) Variations in adaptation action in various decades in the Sahel. (b) Climatic and non-climatic drivers of adaptation actions in the Sahel.

8 and this was during the decade 2008–2016. High yielding varieties also recorded a frequency of 13 adaptation actions during the decade 2008–2016. Generally, during the decade1975–1985, two adaptation actions were most frequently used. These included *income diversification* and *migrations*. Help from *kin* and *network* such as friends and family options were second (**Figure 8**). The decade 1985–1996 also had as the most important adaptation option *water harnessing* related options with a frequency of 27; soil conservation recorded the second with a frequency of 19 while income diversification was third with a frequency of 18 (**Figure 8**). During the decade 1997–2007, *income diversification* led with the highest frequency of 8 while more *farm inputs* came second with a frequency of 6 and migration recorded a frequency of 3 (**Figure 8**). During the decade 2008–2016, *income diversification* was still the most used adaptation option with a frequency of 21; water harnessing came second with a frequency of 15 while high yielding varieties were third with a frequency of 13 (**Figure 8**).

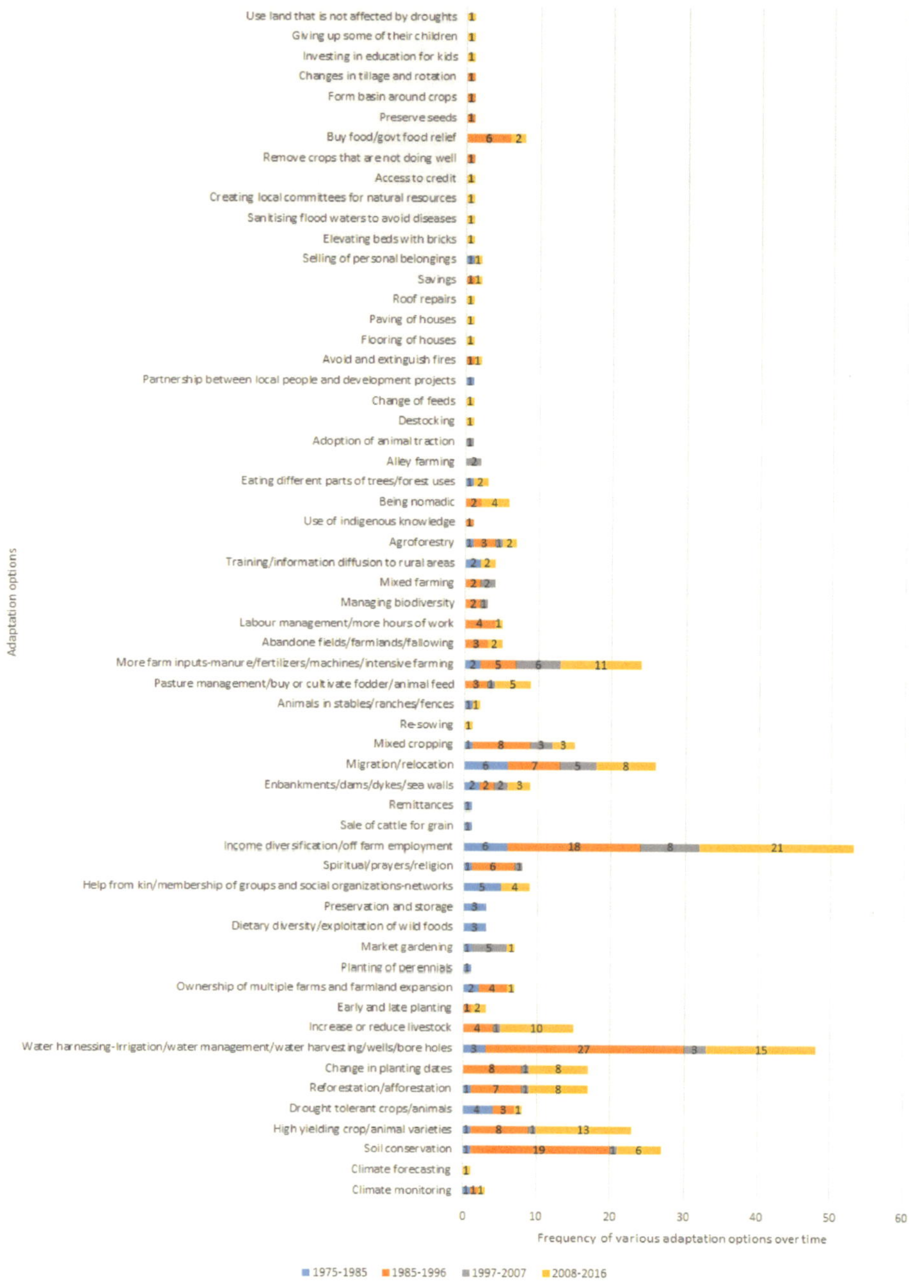

Figure 8.
Composite distribution of variations in different adaptation options reported in the peer review literature published over time for various countries of the Sahel.

7. Discussion

Climate change adaptation is currently receiving a lot of global attention as governments, NGOs, International organizations, civil society groups inter alia are now placing climate change adaptation as a priority on their political and environmental agenda [94, 95]. Even though we have agreed that climate change adaptation is vital in establishing resilience in the Sahel, stakeholders are still in disagreement with respect to what approach climate change adaptation actions represent [96, 97]. This chapter attempts at improving our understanding of the regional and country level adaptation actions in the Sahel as well as their drivers and evolution through several decades. The approach used in this study is similar to those used by Ford et al. [30, 98]. Systematic reviews enhance the ability to compare how adaptation is taking place across the Sahel [30, 98, 99]. The results from this study should therefore be considered as proxies for the level of adaptation in the Sahel as they are based on what exist in the scientific literature and therefore provide a baseline.

These findings can be weakened by the fact that several adaptation actions in the Sahel might be beyond the scope covered by this current study. For example, this work does not consider the gray literature and studies beyond 2020. However, the fact that this study covers a time frame that goes beyond 40 years makes these results valid. The decade 1975–1985 recorded the highest number of adaptation actions while the decade 1985–1996 recorded fewer adaptation actions. The surge in adaptation actions recorded during the decade 1975–1985 can be attributed to the surge in climate change shocks in the form of droughts that dominated this decade [43, 48]. These shocks triggered a lot of actions from governments, international and non-governmental organizations to enhance adaptations [7, 46]. From 1985, the droughts that affected the Sahel had reduced and so did the number of adaptation actions due to 'adaptation fatigue' of the previous decade. The adaptation actions of the 1970s and 1980s have already produced positive results in the Sahel. Recent studies show that the Sahel has become greener with a parallel increase in precipitation. Normalized difference vegetation index (NDVI) and precipitation across the Sahel show a strong correlation between NDVI and greening [3, 53, 100, 101]. Prince et al. [23] has argued that the increase precipitation and greening recorded in the Sahel can be attributed to rising aerial fertilization effect of carbon dioxide. The decades 1997–2007 and 2008–2016 recorded 14% and 65.2% adaptations, respectively. These decades are consistent with an increase in the number of adaptation initiatives over time. It is difficult to argue that the climate change adaptations reported in this study are mainly driven by the adaptation initiatives presented in the peer review literature. In addition, with a total of 414 discrete adaptation actions reported in this study, it remains clear that the level of adaptation is still relatively low in the Sahel. Regionally, West African Sahel dominates in terms of adaptation actions. However, in term of countries, Kenya in East Africa records the highest frequency of 18.1%. According to Ford et al. [95], during the period 2006–2012, Kenya recorded a total of 34 adaptation actions and thus was the highest in terms of adaptation actions from a set of African and Asian countries. The fact that West African Sahel has a higher adaptation rate from a regional perspective can be linked to the "Great Green Wall of the Sahara and the Sahel initiative" (GGWSSI) [102].

The finding that income diversification is the most used adaptation action in the Sahel is consistent with other previous studies that report that to better reduce the effects of climate shocks in the Sahel, it is important to resort to a diversification of livelihoods [103, 104]. Water management actions on the other hand which are second in terms of use are vital as the regions keeps facing recurrent droughts and therefore there is an urgent need for adaptation actions to remedy the situation. With respect to the broad categories of adaptation, technically based adaptations are dominant. This in other words shows that adaptation actions anchored on science and technology are those that are mostly reported in the scientific literature. This is without doubts as the response of technical field experts in the Sahel is often in the form of technically based solutions at the expense of indigenous knowledge, economic based and socially based actions. Examples of these technically based actions are drought resistant species, rainwater harvesting and high yielding varieties.

In terms of policy, it has become evident that income diversification and water harnessing actions have gained prominence over time. This provides an opportunity to further leverage these actions and to investigate how other actions can be further enhanced to improve adaptations. The fact that technically based adaptation actions are more frequently reported shows that policies that enhance both technically based adaptations and others that are less cited should be leveraged in other to enhance the level of adaptation actions in the Sahel. It is important to emphasize the fact that no single adaptation action can single handedly revamp the adaptation landscape in the Sahel because several actions indeed work together better. Those that have been ignored this far need to be researched into to verify what the indigenous people think about them before concrete suggestions are made.

Most of the studies presented here showed that most of the adaptations are driven by climatic variables. However, we also observe an increase attribution of the drivers to non-climatic drivers. Non-climatic drivers are becoming increasingly vital as there are many cases when adaptations are not just taken to adapt to climate change but also in response to other issues such as health, poverty, literacy, and socio-cultural disparities that affect adaptive capacities [95, 96]. The evidence across Africa and the Sahel shows that the problems that Africa and the Sahel are facing ranging from food crisis, deforestation, water scarcity, pandemics and epidemics are mainly caused by non-climatic variables such as deforestation, population movements, wars, and land degradation inter alia with climate change only playing a reinforcing role [1, 3–5].

These results tend to be similar and different from some previous studies based on results and methodology. A study by Ford et al. [95] researched into the status of climate change adaptation in Africa and Asia based on a systematic approach and showed that Kenya had the highest number of adaptation actions. When compared with other vulnerable regions, as the Arctic, there are more reports of adaptation actions in the Sahel. Also, another study by, Ford et al. [30], reported about 157 adaptation actions between 2003 and 2013 while this current study reports 414 adaptation actions for the Sahel for the period 1975–2020. The major areas of differences are that the Ford et al. [95] study focused on both peer review and gray literature while this chapter is based on peer review literature. The Ford et al. study is based on Africa and Asia and covers studies published from 2006 while this current study is based on the Sahel and covers studies from 1975 to 2020. Also, Berrang-Ford et al. [97] used the systematic approach and focused on the peer review literature only, an approach like that adopted by this current study. The results from the latter are also consistent with this in that they report that considerable research has been done with a greater focus on intentions than groundwork. Also, non-climatic drivers are increasingly being presented as the drivers of climate change adaptation [38].

8. Conclusions

In this chapter, we investigated the state of climate change adaptation in the Sahel. The findings show that Kenya recorded the most adaptation actions while regionally, West Africa reported the highest. The approach used in this study provides a proxy-based perspective and baseline from which climate change can be monitored and tracked over time. It is important that various stakeholders come together using this information to further enhance climate change adaptation in the Sahel. This is important as the results from this study show that at country level adaptations to climate change are still at infancy judged by the low frequencies and percentages recorded at the country level.

Due to the existence of a lot that is unknown about climate change adaptation across the Sahel, it becomes important that moving forward attention is given to evidence-based research. It is suggested that more systematic reviews that cover most other regions of Africa be considered. Additionally, gray literature and a wider time scale should be considered. Advancing more research into other regions will go a long way in providing a bigger picture of the status of climate change adaptation across the continent of Africa. Looking at the gray literature will help in brining to the lamplight other potential options not found in the peer review literature. Finally, it is also important to better understand the climatic and non-climatic shocks that are driving adaptation actions in the Sahel and Africa in general. This will enhance the policy process aimed at triggering better adaptation options based on the actual drivers of change.

Acknowledgements

We would like to thank all the authors whose works we consulted. Special thanks to UM6P for providing the funding for this research.

Conflict of interest

The authors declare no conflict of interest.

Author details

Terence Epule Epule[1]*, Abdelghani Chehbouni[1,2] and Driss Dhiba[1]

1 Mohammed VI Polytechnic University (UM6P), International Water Research Institute (IWRI), Hay Moulay Rachid, Benguerir, Morocco

2 CESBIO UMR (CNES-CNRS-UT-INRAe-IRD), Toulouse, France

*Address all correspondence to: epule.terence@um6p.ma

References

[1] Giannini, A., Saravanan, R., Chang, P. Oceanic forcing of Sahel Rainfall on inter-annual to inter-decadal timescales. Science, 2003, 302, 1027-1030.

[2] Giannini, A., Saravanan, R., Chang, P. Dynamics of the boreal summer African monsoon in the NSIPP1 atmospheric model. Climate Dynamics, 2005, 25, 517-535. Doi: 10.1007/s00382-005-0056-x.

[3] Giannini, A., Biasutti, M., Held, I.M., Sobel, A.H. A global perspective on African climate. Climate Change, 2008, 90, 359-383. Doi: 10.1007/s10584-008-9396-y.

[4] Reynolds, J.F., Stafford, S.D.M., Lambin, E.F., Turner, B.L., Mortimore, M., Batterbury, S.P., Downing, T.E., Dowlatabadi, H., Fernandez, R.J., Herrick, J.E., Huber-Sannwald, E., Jiang, H., Leemans, R., Lynam, T., Maestre, F.T., Ayarza, M., Walker, B. Global desertification: building a science for dryland development. Science, 2007, 316, 847-851.

[5] Lu, J., Delworth, T.L. 2005. Oceanic forcing of the late 20th century Sahel drought. *Geophysical Research Letters,* 2005, 32: L22706. Doi: 10.1029/2005GL023316.

[6] Mertz, O., Mbow, C., Reenberg, A., Diouf, A. Farmers' perceptions of climate change and agricultural adaptation strategies in rural Sahel. Environ Manage, *2009, 43*(5), 804-816. Doi: 10.1007/s00267-008-9197-0.

[7] Epule, T.E., Peng, C., Lepage, L., Chen, Z. The causes, effects, and challenges of the Sahelian droughts: a critical review. Regional Environmental Change, 2013a, 14, 145-156. Doi: 10.1007/s10113-013-0473-z.

[8] IPCC. Climate change 2007: impacts, adaptations, and vulnerability. Contributions of the working group 2 to the fourth assessment report of the IPCC (Cambridge: Cambridge University Press). 2007, p976.

[9] Gonzalez, P., Tucker, C.J., Sy, H. Tree density and species decline in the African Sahel attributable to climate. Journal of Arid Environments, 2012, 78, 55-64. Doi: 10.1016/j.jaridenv.2011.11.001.

[10] Clover, J. Food security in sub-Saharan Africa. African Food Review, 2010, 12(1), 5-15. Doi:10.1080/10246029.2003.9627566.

[11] Verpoorten, M., Arora, A., Stoop, N., Swinnen, J. 2013. Self-reported food insecurity in Africa during the food price crisis. Food Policy, 2013, 39, 51-63. Doi: 10.1016/j.foodpol.2012.12.006.

[12] Nyariki, D.M., Wiggins, S. Household food insecurity in sub-Saharan Africa: lessons from Kenya. British Food Journal, 1997, 99(7), 249-262.

[13] Battisti, D., Naylor, R. Historical warnings of future food insecurity with unprecedented seasonal heat. Science, 2009, 232, 240-244.

[14] Sissoko, K., Keulen, H.V., Verhagen, J., Tekken, V., Battaglini, A. Agriculture, livelihoods, and climate change in the West African Sahel. Regional Environmental Change, 2011, 11, S119-S125.Doi:10.1007/s10113-010-0164-y.

[15] UNAIDS. Report on the global HIV/AIDS epidemic 2002. 2002, UNAIDS, Geneva.

[16] Costello, A., Abbas, M., Allen, A., Ball, S., Bellamy, R., Friel, S., Grace, N., Johnson, A, Kett, M., Lee, M., Levy, C., Maslin, M., McCoy, D., McGuire, B., Montgomery, H., Napier, DD., Pagel, J.,

Puppin de Oliveira, J.A., Redclift, N., Rees, H., Rogger, D., Scott, H., Stephenson, J., Twigg, J., Wolff, J., Patterson, C. Managing the health effects of climate change. The Lancet Commission, 2009, 373, 1693-1733.

[17] Watts, N., Adgar, W.N., Agnolucci, P., Blackstock, J., Byass, P., Cai, W., Chaytor, S., Colbourn, T., Collins, M., Cooper, A., Cox, M.P., Depledge, J., Drummond, P., Ekins, P., Galaz, V., Grace, D., Graham, H., Grubb, M., Haines, A., Hamilton, I., Hunter, A., Jiang, X., Li, M., Kelman, I., Liang, Lu., Lott, M., Lowe, R., Luo, Y., Mace, G., Maslin, M., Nilsson, M., Oreszcyn, T., Pye, S., Quinn, T., Svensdotter, M., Venevsky, S., Warmer, K., Xu, B., Yang, J., Yu, C., Zhang, Q., Gong, P., Montgomery, H., Costello, A. 2015. Health and climate change: policy responses to protect public health. The Lancet Commissions, 386, 1861-1914.

[18] Rohr, J., Dobson, A., Johnson, P.T.J., Kilpatrick, A.M., Paull, S.H., Raffel, R.R., Ruiz-Moreno, D., Thomas, M.B. Frontiers in climate change-disease research. Trends in Ecology and Evolution, 2011, 26(6), 270-277.

[19] Myers, N. Environmental refugees: Our latest understanding. Philosophical Transactions of the Royal Society B, 2001, 356, 16.1-16.5. Doi:10.1098/rstb.2001.0953.

[20] Myers, N. Environmental refugees: a growing phenomenon of the 21st century. Philosophical Transactions of the Royal Society, 2002, 357(1420), 609-613. Doi: 10.1098/ rstb. 2001.0953.

[21] Myers, N., Kent, J. Food, and hunger in Sub-Saharan Africa. The Environmentalist, 2001, 21(1), 41-69.

[22] Epule, T.E., Peng, C., Lepage, L. Environmental refugees in sub-Saharan Africa: a review of perspectives on trends, causes, challenges and way forward. Geojournal, 2015, 80, 79-92. Doi: 10.1007/s10708-014-9528-z.

[23] Prince, S.D., Brown, D., Coulston, E., Kravitz, L.L. Evidence from rain-use efficiencies does not indicate extensive Sahelian desertification. Global Change Biology, 1998, 4, 359-374.

[24] Ford, J., Pearce, T., Smit, B., Wandel, J., Allurut, M., Shappa, K., Ittusujurat, H., Qrunnut, K. Reducing Vulnerability to Climate Change in the Arctic: The Case of Nunavut, Canada. *Arctic, 2007,* 60,768 150-166.

[25] Verchot, L., van Noordwijk, M., Kandji, S., Tomich, T., Ong, C., Albrecht, A., Mackensen, J., Bantilan, C., Anupama, K., Palm, C. Climate change: linking adaptation and mitigation through agroforestry. Mitigation and Adaptation Strategies for Global Change, 2007, 12, 901-918.

[26] Ford, J.D., Pearce, T. What we know, do not know, and need to know about climate change vulnerability in the western Canadian Arctic. Environmental Research Letters, 2010, 5. Doi: 10.1088/1748-765 9326/5/1/014008.

[27] Newton, J., Paci, C.D.J., Ogden, A. Climate Change and Natural Hazards in Northern Canada: Integrating Indigenous Perspectives with Government Policy. Mitigation Adaptation Strategies of Global Change, 2005, 10, 541-571. Doi: 10.1007/s11027-005-0060-9.

[28] Pearce, T., Ford, J.D., Duerden, F., Smit, B., Andrachuk, M., Berrang-Ford, L., Smith, T. Advancing adaptation planning for climate change in the Inuvialuit Settlement Region (ISR): a review and critique. Reg. Environ. Change 2011, 11, 1-17. Doi:10.1007/s10113-010-0126-4.

[29] Berrang-Ford, L., Ford, J.D., Lesnikowski, A., Poutianinen, C., Barrera, M., Heymann, S.J. What drives national adaptation? A global assessment. *Climate Change*. 2014, Doi: 10.1007/s10584-014-1078-3.

[30] Ford, J.D., McDowell, G., Jones, J. The state of climate change adaptation in the Arctic. *Environmental Research Letters*, 9, 2014, 104005. Doi: 10.1088/1748-9326/9/10/104005.

[31] Lesnikowski, A.C., Ford, J.D., Berrang-Ford, L., Barrera, M., Berry, P., Henderson, J., Heymann, S.J. National-level factors affecting planned, public adaptation to health impacts of climate change. Global Environmental Change, 2013, 23, 1153-1163. Doi: 10.1016/j.gloenvcha.2013.04.008.

[32] Epule, T.E., Peng, C., Lepage, L., Chen, Z. Policy options towards deforestation reduction in Cameroon: an analysis based on a systematic approach. Land Use Policy, 2014, 36, 405-415. Doi: 10.1016/j.landusepol.2013.09.004.

[33] UNREDD+. Executive summary: 2015 annual report of the UN-REDD Programme Fund. 2015. Available online at: http://www.unredd.net/documents.html?view=browse&customtags=172&startdate=&enddate=&dmlang= (Accessed March 31, 2021).

[34] United Nations (UN). Paris agreement. United Nations, New York. 2015. Available online at: http://unfccc.int/files/essential_background/convention/application/pdf/english_paris_agreement.pdf (Accessed March 31, 2021).

[35] African Climate Change Fund. Supporting African countries to access international climate finance: enabling a transition towards climate resilient, low-carbon development. The African Development Bank. 2016. Available online at: https://www.afdb.org/fileadmin/uploads/afdb/Documents/Project-and-Operations/Africa_Climate_Change_Fund_-_Supporting_Africa_countries_to_access_international_climate_finance.pdf. (Accessed March 31, 2021).

[36] Ford, J.D., Berrang-Ford, L., Patterson, J. 2011. A systematic review of observed climate change adaptation in developed nations. Climate Change Letters, 106, 327-336.

[37] Austin, S.E., Ford, J.D., Berrang-Ford, L., Araos, M., Parker, S., Fleury, M.D. Public Health Adaptation to Climate Change in Canadian Jurisdictions. *International Journal of Environmental Research and Public Health,* 2015, 12, 691 623-651. Doi: 10.3390/ijerph120100623.

[38] Lwasa, S. A systematic review of research on climate change adaptation policy and practice in Africa and South Asia deltas. Regional Environmental Change, 2015, 15, 815-825. Doi: 10.1007/s10113-014-0715-8.

[39] Lesnikowski, A., Ford, J., Biesbroek, R., Berrang-Ford, L., Heymann, S.J. National-level progress on 816 adaptation. Nature Climate Change, 2016, 6, 261-264. Doi: 10.1038/nclimate2863.

[40] Agnew, C.T., Chappell, A. Drought in the Sahel. *Geojournal,* 1999, 48, 299-311.

[41] Mamadou, I., Gautier, E., Descroix, L., Noma, I., Moussa, I.B., Maiga, O.F., Genthon, P., Amogu, O., Abdou, M.M., Vandervaere, J.P. Exorheism growth as an explanation of increasing flooding in the Sahel. Catena, 2015, 131, 130-139.

[42] Karam, D.B., Flamant, C., Knippertz, P., Reitebuch, O., Pelon, J., Chong, M. Dabas, A. Dust emissions over the Sahel associated with West African monsoon intertropical discontinuity region: a representative case study. *Q.J.R Meteorol. Soc*. 2008, 134, 621-634.

[43] Anyamba, A., Tucker, CJ. Analysis of Sahelian Vegetation dynamics using NOAA-AVHRR NDVI data from 1981-2003. Journal of Arid Environments, 2005, 63: 595-614.

[44] Nicholson, S.E. Sahel, West Africa. Encyclopedia of Environmental Biology, 1995, 3:261-275.

[45] Nicholson, S.E. Climatic and environmental change in Africa during the last two centuries. Climate Research, 2001, 17, 123-144.

[46] Zeng, N. Drought in the Sahel. *Science,* 2003, 302:999-1000.

[47] Wang, B., Ding, Q., Fu, X., Jin, K., Shukla, J., Doblas-Reyes, F. Fundamental challenge in simulation and prediction of summer monsoon rainfall. *Geophysical Research Letters*, 2005, 32, 15711. Doi: 10.1029/2005GL0 22734.

[48] Maranz, S. Tree mortality in the African Sahel indicates an anthropogenic ecosystem displaced by climate change. Journal of Biogeography, 2009, 36, 1181-1193.

[49] Sinclair, A.R.E., Fryxell, J.M. The Sahel of Africa: Ecology of a Disaster. Canadian Journal of Zoology, 1985, 63, 987-994.

[50] Fleuret, A. Indigenous responses to drought in sub-Saharan Africa. Disaster, 1986, 10(3), 224-229.

[51] Elmqvist, B., Olsson, L. 2007. Livelihood diversification: continuity and change in the Sahel. GeoJournal, 2007, *67*(3), 167-180. Doi: 10.1007/ s10708-007-9043-6

[52] Morand, P., Kodio, A., Andrew, N., Sinaba, F., Lemoalle, J., Béné, C. Vulnerability and adaptation of African rural populations to hydro-climate change: experience from fishing communities in the Inner Niger Delta (Mali). Climatic Change, 2012, *115*(3-4), 463-483. Doi: 10.1007/s10584-012-0492-7.

[53] Nielsen, J. Ø., Reenberg, A. Cultural barriers to climate change adaptation: A case study from Northern Burkina Faso. Global Environmental Change, *2010, 20*(1), 142-152. Doi:10.1016/j.gloenvcha. 2009.10.002.

[54] Moretimore, M. Adapting to drought in the Sahel: lessons for climate change. WIREs Climate Change, 2010, 1, 134-143.

[55] Mertz, O., Mbow, C., Reenberg, A., Genesio, L., Lambin, E. F., D'Haen, S.... Sandholt, I. Adaptation strategies and climate vulnerability in the Sudano-Sahelian region of West Africa. Atmospheric Science Letters, *2011, 12*(1), 104-108. doi:10.1002/asl.314.

[56] Reardon, T., Malton, P., Delgado, C. 1988. Coping with household-level food insecurity in drought-affected areas of Burkina Faso. *World Development*, 1988, 16(9), 1065-1074.

[57] Burnham, M., Ma, Z. Linking smallholder farmer climate change adaptation decisions to development. Climate and Development, *8*(4), 2015, 289-311. Doi:10.1080/17565529.201 5.1067180.

[58] Schaer, C. Condemned to live with one's feet in water? International Journal of Climate Change Strategies and Management, *2015, 7*(4), 534-551. Doi: 10.1108/ijccsm-03-2014-0038.

[59] Brockhaus, M., Djoudi, H., Locatelli, B. Envisioning the future and learning from the past: Adapting to a changing environment in northern Mali. Environmental Science & Policy, *2013, 25*, 94-106. Doi: 10.1016/j.envsci.2012. 08.008.

[60] Opondo, D, O. Erosive coping after the 2011 floods in Kenya. *International journal of Global Warming*, 2013, 5(4), 452-466.

[61] Epule, T., Bryant, C. Small-scale farmers' indigenous agricultural adaptation options in the face of

declining or stagnant crop yields in the Fako and Meme divisions of Cameroon. Agriculture, *2016, 6*(2), 22. Doi:10.3390/agriculture6020022.

[62] Opiyo, F., Wasonga, O., Nyangito, M., Schilling, J., Munang, R. Drought Adaptation and Coping Strategies among the Turkana Pastoralists of Northern Kenya. International Journal of Disaster Risk Science, *2015, 6*(3), 295-309. Doi: 10.1007/s13753-015-0063-4.

[63] Mosberg, M., Eriksen, S. H. Responding to climate variability and change in dryland Kenya: The role of illicit coping strategies in the politics of adaptation. Global Environmental Change, *2015, 35*, 545-557. Doi: 10.1016/j.gloenvcha.2015.09.006.

[64] Bryan, E., Deressa, T., Gbetibouo, G., Ringer, C. 2009. Adaptation to climate change in Ethiopia and South Africa: options and constraints. Environmental Science and Policy, 12, 413-426.

[65] Zampaligré, N., Dossa, L. H., Schlecht, E. Climate change and variability: perception and adaptation strategies of pastoralists and agro pastoralists across different zones of Burkina Faso. Regional Environmental Change, *2013, 14*(2), 769-783. Doi: 10.1007/s10113-013-0532-5.

[66] Barbier, B., Yacouba, H., Karambiri, H., Zorome, M., Some, B. Human vulnerability to climate variability in the Sahel: farmers' adaptation strategies in northern Burkina Faso. Environ Manage, *2009, 43*(5), 790-803. Doi: 10.1007/s00267-008-9237-9.

[67] Deressa, T. T., Hassan, R. M., Ringler, C., Alemu, T., Yesuf, M. 2009. Determinants of farmers' choice of adaptation methods to climate change in the Nile Basin of Ethiopia. Global Environmental Change, 2009, *19*(2), 248-255. Doi:10.1016/j.gloenvcha.2009.01.002.

[68] Rockström, J. Resilience building and water demand management for drought mitigation. Physics and Chemistry of the Earth, Parts A/B/C, *2003, 28*(20-27), 869-877. Doi:10.1016/j.pce.2003.08.009.

[69] Douxchamps, S., Van Wijk, M. T., Silvestri, S., Moussa, A. S., Quiros, C., Ndour, N. Y. B.,... Rufino, M. C. Linking agricultural adaptation strategies, food security and vulnerability: evidence from West Africa. Regional Environmental Change, *2015, 16*(5), 1305-1317. Doi: 10.1007/s10113-015-0838-6.

[70] Mbow, C., Mertz, O., Diouf, A., Rasmussen, K., Reenberg, A. The history of environmental change and adaptation in eastern Saloum–Senegal-Driving forces and perceptions. Global and Planetary Change, *2008, 64*(3-4), 210-221. Doi:10.1016/j.gloplacha.2008.09.008.

[71] Magistro, J., Medou, L. Historical and human dimensions of climate variability and water resource constraint in the Senegal River valley. Climate Research, 2001, 19, 133-147.

[72] Recha, J. W., Mati, B. M., Nyasimi, M., Kimeli, P. K., Kinyangi, J. M., Radeny, M. Changing rainfall patterns and farmers' adaptation through soil water management practices in semi-arid eastern Kenya. Arid Land Research and Management, *2016, 30*(3), 229-238. Doi:10.1080/15324982.2015.1091398.

[73] Gebrehiwot, T., & van der Veen, A. Farm level adaptation to climate change: the case of farmer's in the Ethiopian highlands. Environ Manage, *2013, 52*(1), 29-44. Doi: 10.1007/s00267-013-0039-3.

[74] Nyong, A., Adesina, F., Osman Elasha, B. The value of indigenous knowledge in climate change mitigation and adaptation strategies in the African Sahel. Mitigation and Adaptation Strategies for Global Change, *2007, 12*(5), 787-797. Doi: 10.1007/s11027-007-9099-0.

[75] Okoye, C.U. Comparative analysis of factors in the adoption of traditional and recommended soil erosion control practices in Nigeria. Soil and Tillage Research, 1998, 45, 251-263.

[76] Marenya, P. P., Barrett, C. B. Household-level determinants of adoption of improved natural resources management practices among smallholder farmers in western Kenya. Food Policy, *2007, 32*(4), 515-536. Doi: 10.1016/j.foodpol.2006.10.002.

[77] Wood, S. A., Jina, A. S., Jain, M., Kristjanson, P., DeFries, R. S. Smallholder farmer cropping decisions related to climate variability across multiple regions. Global Environmental Change, *2014, 25*, 163-172. Doi: 10.1016/j.gloenvcha.2013.12.011.

[78] Croppenstedt, A. Demeke, M., Meschi, M.M. Technology adoption in the presence of constraints: the case of fertilizer demand in Ethiopia. Review of Development Economics, 2003, 7(1), 58-70.

[79] Wubeneh, N. G., Sanders, J. H. Farm-level adoption of sorghum technologies in Tigray, Ethiopia. *Agricultural Systems, 2006, 91*(1-2), 122-134. Doi: 10.1016/j.agsy.2006.02.002.

[80] Somda, J., Nianogo, A.J., Nassa, S., Sanou, S. Soil fertility management and socio-economic factors in crop-livestock systems in Burkina Faso: a case study of composting technology. Ecological Economics, 2002, 43, 175-183.

[81] Thuo, M., Bravo-Ureta, B., Hathie, I., Obeng-Asiedu, P. Adoption of chemical fertilizer by smallholder farmers in the peanut basin of Senegal. *African Journal of Agricultural Resource Economics,* 2011, 6(1). Available online at: http://purl.umn.edu/156960. (Accessed March 31, 2021).

[82] Abebe, G. K., Bijman, J., Pascucci, S., Omta, O. Adoption of improved potato varieties in Ethiopia: The role of agricultural knowledge and innovation system and smallholder farmers' quality assessment. Agricultural Systems, *2013, 122*, 22-32. doi:10.1016/j.agsy.2013.07.008.

[83] Adesina, A., Baidu-Forson, J. Farmers' perceptions and adoption of new agricultural technology: evidence from analysis in Burkina Faso and Guinea, West Africa. *Agricultural Economics, 1995,* 13, 1-9.

[84] De Groote, H., Dema, G., Sonda, G. B., Gitonga, Z. M. Maize for food and feed in East Africa-The farmers' perspective. Field Crops Research, *2013, 153*, 22-36. Doi:10.1016/j.fcr.2013.04.005.

[85] Zander, K. K., Mwacharo, J. M., Drucker, A. G., Garnett, S. T. Constraints to effective adoption of innovative livestock production technologies in the Rift Valley (Kenya). Journal of Arid Environments, *2013, 96*, 9-18. Doi:10.1016/j.jaridenv.2013.03.017.

[86] Boyd, E., Cornforth, R.J., Lamb, P.J., Tarhule, A., Lele, M.S., Brouder, A. Building resilience to face recurring environmental crisis in African Sahel. Nature Climate Change, 2013, 3, 631-637.

[87] Huq, S., Reid, H., Konate, M., Rahman, A., Sokona, Y., Crick, F. Mainstreaming adaptation to climate change in Least Developed Countries (LDCs). Climate Policy, *2004, 4*(1), 25-43. Doi:10.1080/14693062.2004.9685508.

[88] Reenberg, A. Land-use dynamics in the Sahelian zone in eastern Niger-monitoring change in cultivation strategies in drought prone areas. Journal of Arid Environments, 1994, 27, 179-192.

[89] Tambo, J. A., Abdoulaye, T. Smallholder farmers' perceptions of and adaptations to climate change in the

Nigerian savanna. Regional Environmental Change, *2012, 13*(2), 375-388. Doi:10.1007/s10113-012-0351-0.

[90] Scheffran, J., Marmer, E., Sow, P. Migration as a contribution to resilience and innovation in climate adaptation: Social networks and co-development in Northwest Africa. Applied Geography, *2012, 33*, 119-127. Doi: 10.1016/j.apgeog.2011.10.002.

[91] Oluoko-Odingo, A. A. Vulnerability and Adaptation to Food Insecurity and Poverty in Kenya. Annals of the Association of American Geographers, 2011, *101*(1), 1-20. Doi:10.1080/00045608.2010.532739.

[92] Nielsen, J.Ø., D'Haen, S., Reenberg, A. Adaptation to climate change as a development project: A case study from Northern Burkina Faso. Climate and Development, 2012, 4(1), 16-25.

[93] Ebi, K. L., Padgham, J., Doumbia, M., Kergna, A., Smith, J., Butt, T., McCarl, B. 2011. Smallholders' adaptation to climate change in Mali. Climatic Change, 2011, *108*(3), 423-436. Doi: 10.1007/s10584-011-0160-3.

[94] Noble, I., Huq, S. Adaptation Needs and Options 2014, (New York: Cambridge University Press) chapter 14.

[95] Ford, J.D., Berrang-Ford, L., Bunce, A., McKay, C., Irwin, M., Pearce, T. The status of climate change adaptation in Africa and Asia. Regional Environmental Change, 2015, 15, 801-814.

[96] Dupuis, J., Biesbroek, R. Comparing apples and oranges: the dependent variable problem in comparing and evaluating climate change adaptation policies *Global Environmental Change*, 2013, 23, 1476-1487.

[97] Berrang-Ford, L., Pearce, T., Ford, J.D. Systematic review approaches for global environmental change research. Regional Environmental Change, 2015, 15, 755-769.

[98] Ford, J. D., Berrang-Ford, L., Lesnikowski, A., Barrera, M., Heymann, S. J. 2013. How to track adaptation to climate change: a typology of approaches for national-level application. Ecology and Society, 18 40.

[99] Mannke, F. Key themes of local adaptation to climate change: lessons from mapping community-based initiatives in Africa. In: Leal Filho W (ed) Experiences of climate change adaptation in Africa, Springer, New York, 2011, pp. 17-32.

[100] Anyamba, A., Eastman, J. Inter-annual variability of NDVI over Africa and its relation to El Niño/Southern Oscillation. International Journal of Remote Sensing, 1996, 17, 2533-2548.

[101] Eklundh, L., Olsson, L. Vegetation index trends for the African Sahel 1982-1999. Geophysical Research Letters, 2003, 30(8), 1430. Doi: 10.1029/2002GL016772.

[102] O'Connor, Ford, J.D. Increasing the effectiveness of the "Great Green Wall" as an adaptation to the effects of climate change and desertification in the Sahel. Sustainability, 2014, 6, 7142-7154. Doi: 10.3390/su6107142.

[103] Le, H.D., Smith, C., Herbohn, J., Harrison, S. More than just trees: assessing reforestation success in tropical developing countries. Journal of Rural Studies, 2012, 28(1), 5-19.

[104] Epule, T.E., Peng, C., Lepage, L., Chen, Z. Enabling conditions for successful greening of public spaces: the case of Touroua, Cameroon based on perceptions. *Small Scale Forestry*. 2013b, Doi: 10.1007/s1182-013-9246-2.

Chapter 17

Transformation Action to Combat Desertification: A Direct Carbon Saving Mechanism in South Syria for Post-Conflict Management

Fidaa Fawwaz Haddad

Abstract

Climate change has been visible through observed deterioration in the environment. Despite mitigation policies, greenhouse gas emissions (GHG) have increased over the last decade. Undeniably, war and conflict have the potential to further exacerbate inequalities and put major stress on meeting the sustainable development goals (SDGs). Despite the 10 year civil war in Syria, carbon emissions have declined by 1.5 percent since 2015 due to the economic slowdown, resulting in approximately 13 million Syrians in need of humanitarian assistance. As a consequence of the protracted crisis, families have suffered significant losses of assets and income generation opportunities, in both rural and urban areas, which has increased the vulnerability of the land, fostered illegal over-pumping of irrigation water. Investing in renewable energy, such as installing solar panels to pump water for irrigation, or operating medium-small factories, can help community resilience in response to climate change. The research investigated humanitarian initiatives on solar panels targeted at households and small/medium industrial uses. It also addresses this intervention as an approach that might have potential for transnational water sharing for humanitarian, development, and peace nexus. This process helps in laying the basis for climate change resilience, and combatting land desertification as part of a humanitarian, development and peace nexus.

Keywords: resilience, carbon trade, humanitarian, nexus, post-conflict

1. Introduction

Countries of the Middle East and North Africa (MENA Region) face major challenges in terms of food supply and food security. High dependency on food imports combined with ongoing conflicts and reoccurring climate shocks make the region particularly vulnerable to food crises. Among the 10 countries with the most people affected by food insecurity (IPC/CH Phase 3 or above), three are located in the Middle East including Yemen, Afghanistan, and Syria [1]. In Syria, approximately 9.3 million people were food insecure and needed food, humanitarian, and livelihood assistance between October and December 2020 [1].

IntechOpen

1.1 Syria double-headed problem: conflict and drought

Over the past decade, MENA countries have experienced severe land degradation which has led to reduced productivity, increased water scarcity, and loss of arable land [2].

The agriculture sector in South Syria relies heavily on irrigation, but ever since the start of the conflict the agricultural irrigation infrastructure has been severely impacted by the total damage to agricultural infrastructure and assets [19]. Farmers in the south governorates are still linked to their lands. To irrigate their lands, the wells used generators running on diesel to pump water, due to the depth of the ground water table. However, due to limited income, fuel, and power, many wells are non-operational as a result. Most of the KII respondents highlighted that solar-powered pumps should be installed in locations where wells are of an appropriate size and have a sufficient catchment area. Their main constrains from installing solar systems are that wells with a high volume of flow and outputs require large solar arrays to power water pumping. Arrays that are too large may be targeted for air strikes.

A country like Syria is facing a double-headed problem [3]: Syria is going into its 10th year of conflict while facing impacts of climate change, such as droughts. In 2006, at the beginning of the current drought in Syria, rainfall declined by 66 percent across the country [4]. As rainfall declined and temperatures rose, the rate of evaporation of groundwater also increased through the years, which has caused widespread evaporation to remarkably decrease the level of groundwater [5]. Much of the water used in Syria comes from the Euphrates River, which originates in Turkey; at the same time, Syria is sharing the Al-Yarmouk transboundary basin with Jordan.

The Syrian case illustrates the risk multiplier effect of climate change. Since the 2007/2008 agriculture season, around 75 percent of agriculture-dependent households suffered total crop failure, leading 1 million people into food insecurity and forcing many to join urban centers. The intense stress upon urban resources resulting from the increased demographic pressure is likely to have aggravated an already tense socioeconomic and political context [2]. In addition, the conflict has increased the cutting and burning of forests and caused agricultural land and rangelands pollution because of irregular oil refining activities. Between 2000 and 2015, forest land cover decreased by 8.04 percent; shrubs, grassland, sparsely, vegetated areas by 14.2 percent, and arable lands by 0.08 percent. On the other hand, artificial area land cover has increased by 50.10 percent, bare lands by 4.06 percent, and water bodies/ wetlands by 0.75 percent [6].

1.2 Strengthening preparedness for improving communities' resilience

The Sendai framework for Disateter Risk Reduction 2015-2030[1] prioritizes reducing social, economic, and environmental vulnerability, through improving early warning systems and strengthening preparedness for effective response, as a tool for enhancing disaster-resilient societies [7]. The impact of conflict in terms of weak governance and absence of governmental legislation and policies, will increase the risk of climate change, such as food insecurity and loss of livelihoods assists [8]. According to a CARE International livelihoods assessment report published in 2016, it was indicated that from the consequence of the protracted crisis in Syria, families have suffered from significant losses of assets and income generation opportunities,

[1] The Sendai Framework is the successor instrument to the Hyogo Frame for Action 2005- 2015. It ecognizes that the State has the primary role to reduce disaster risk but that responsibility should be shared with other stakeholders including local government, the private sector and other stakeholders.

in both rural and urban areas [9]. The losses in gross domestic product (GDP) in 2011 are four times more compared to 2010, which has increased vulnerability and poverty of the population across Syria accordingly [10].

Therefore, improving the resilience of Syrian communities, development and humanitarian actors should ensure having mutual benefits that are focused on improving the ability of the communities to anticipate, cope with, and recover from the double impact of climate change and war [11].

For example, due to electricity shortages, Syrians in the south governorates were forced to rely on diesel power generators to meet domestic, industrial, and agricultural needs. However, diesel is proving not to be the most stable source of energy, especially with its expensive cost, which impacted livelihood opportunities, and depleted communities' productive assets [12]. On the other hand, using diesel power as a source of fuel increases the emission of carbon dioxide "CO_2" in the atmosphere [13].

Therefore, shifting attention to the solar energy sector has potential for communities in South Syria to mitigate the impact of climate change and enhance their livelihoods resilience [14]. As for Jordan, this will provide it with more than enough energy to function, allowing for a decrease in dependency on the ever-shrinking water resources in the region [15].

The research was conducted in June 2018, prior to the conciliation agreement between the Syrian regime forces and opposition groups in the south. It aimed to investigate how the implementation of solar panel systems, as a source of renewable energy at the household level, or a small scale of enterprises, can increase the carbon saving for the direct measures of GHG emissions, caused by a specific activity [16]. This could lead to the strengthening communities' resilience. Accordingly, the research findings will discuss the importance of renewable energy for humanitarian livelihoods' resilience initiatives in South Syria. And lastly, the research will recommend the potential mechanisms that would empower communities, to participate more actively in community resilience.

1.3 Research objectives

The research assessed solar panel initiatives, implemented by humanitarian aids in South Syria, as a case study to highlight the importance of using renewable energy technologies for climate change mitigation and how such interventions could be articulated toward improving community resilience in South Syria. The research objectives are as follows: (i) to identify and critically review the importance of renewable energy resources for the humanitarian livelihood resilience initiatives in South Syria governorates; (ii) to infer the viability of using renewable energy (such as solar panels at a small scale of households and industrial uses) toward improving the communities' resilience in South Syria; (iii) in addition, to estimate the effectiveness of carbon trading in South Syria for post-conflict management, and climate change mitigation; and (iv) to propose potential mechanisms to empower communities in laying the basis for climate change mitigation and resilience.

1.4 Research questions

The research focuses on the following questions:

- What are the main challenges and opportunities to adopt renewable energy initiatives in the Syrian conflict context along with transboundary water resources management considerations?

- What percentage of households have benefitted from solar panel systems, reporting reduced reliance on negative coping strategies, and an improvement in their income in comparison with their results before using solar panel systems?

- What is the percentage of medium-small factories, which have benefitted from the solar panel system, reporting reduced electricity cost even under the impact of the crisis, compared with their results before using the solar panels?

- What is the reduction in CO_2 emitted as a result of households and medium-small factories, that have benefitted from solar panel system, compared to others still using fuel as a source of energy?

- How might CO_2 trading contribute to post-conflict management for Syrian communities?

1.5 Research methodology

The research used the remote management[2] modality (Cross- boarder humanitarian program) in collecting the data needed, local organizations and enumerators worked inside Syria during the reconciliation agreements between the opposition groups and the Syrian army on 2018. Humanitarian agencies that worked to support the most vulnerable people in need in South Syria. The analysis of data for the survey was carried out in two stages—univariate and bivariate analyses in order to assess the single variables, and the relation between two variables for the research demographic and objectives. At the univariate stage of the analysis, variations within demographic variables were explored, to support the analysis of carbon footprint for each installed system at 13 households and six rehabilitated factories, in terms of using renewable energy source, such as the solar panel to mitigate climate change.

The second stage of the analysis was the bivariate level, where relationships between sociodemographic characteristics and the resilience indicators were assessed, for the targeted household and factories, with main consideration to the gender-disaggregated data.

A flexible approach to the research was adopted, due to the security context in Syria as well as the remote management of this research. A purposive sampling approach was used, in order to ensure the survey finding is reflecting the realities faced by whom is using the solar panels for shallow aquifer wells, rather than the population at large [17]. The research used the sample as a case study for potential investigation, and due to the challenges that faced the data collection, more focus discussions were conducted in small focus groups with the enumerators, remotely for further understanding.

1.6 Research sample demography

The research is based on a sample of 30 households (HHs) using solar panels and 21 households who did not install any. Of the HHs studied, 23.5 percent of HH heads are female and 76.5 percent are males. Women representing 17% of the total factories'

[2] "Remote Management" refers to a set of adapted procedures put in place because the access to field offices and/or field sites is limited – for security, practical or logistical reasons. International Rescue Committee's Global Remote Management Guidelines 2016.

owners. With regard to the level of education, 32 percent of HHs had a high level of education, while 69.2 percent only attained a secondary level of education. Among the female respondents, 33.3 percent had no education, and a total of 16.7 percent of households that were headed by women completed a primary level of education.

The average household (HH) income was 345,782 SYP per month (~ 1,609.00 USD). Informal work, such as farming activities or cash for work, accounted for 80.4 percent of household's income. About 13.7 percent of woman-headed households' earned money from home production and handicrafts, while 5.9 percent of the total sample used their savings. As shown in **Figure 1**, the factories' average monthly production quantity increased after installing the solar panels (2016-2018). Dairy products, cheese products, and dry dairy represented 72.6, 12.2, and 5.7 percent of the production, respectively.

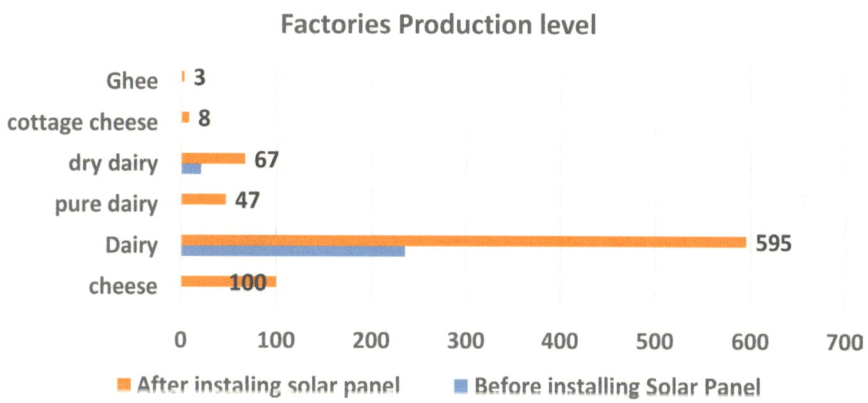

Figure 1.
Highlights of the production before and after installing solar panels for factories' operation.

2. Analysis and results

2.1 Importance of renewable energy resources for humanitarian-livelihoods initiative in South Syria

The primary data research was planned to investigate the impact of using renewable energy as one of humanitarian livelihood initiatives. Consequently, a number of questions were asked in order to identify the HHs' coping strategies to secure food and to what extent they used these strategies. According to the vulnerability analysis mapping unit at WFP (VAM), the Coping Strategy Index (CSI) is often used as a proxy indicator of household food insecurity. It can highlight the HH consumption behaviors, in terms of quantity and sufficiency of food, and can measure the adjustment the HH make in livelihood options and assets [18].

The first group consisted of those who installed solar panels—they showed positive behavior in their coping strategies, the second group consisted of HHs without solar panels who are still using the electrical pump for water pumping—they applied low coping strategy index.

Further disaggregated analysis of HH savings showed that the average saving for male respondents between those who installed solar panels and those who did not is 259,226 SYP (12,017 USD), with 25.8 percent as a positive increase of savings between the ones with solar panels and the ones without. Female respondents showed better improvement than male respondent: 169,444 SYP (788.7 USD) with a 40.4 percent improvement for families with solar panels. Women confirmed that last season, they used productions of their small yards to feed their children and to preserve food for winter.

Regarding the factories' owners, it is conveyed in the analysis that they showed positive behavior in their coping strategy after using renewable energy as source of electrical supply for their small factories, with 29.49 percent as mild the Reduced Coping Strategies Index (rCSi)[3] and 16.7 percent as low rCSi. The only female owner showed a sufficient ability in reducing the coping strategy index, which is a good indication that a woman can support her family financially and adapt to different situation.

2.2 Viability of using solar panels as the electric supply for the pumping irrigation of water at a household level in South Syria

The escalating of the conflict in South Syria increased food insecurity, as communities depended on agricultural and livestock activities for their livelihoods. The cost of food, fuel and transportation are usually linked with agricultural seasons [19]. The analysis below highlights the importance of using solar panels—as a source of electrical supply- to pump water to households to irrigate their small yards, which can be used for daily food supplies.

The analysis focused on the cost of HH electricity before and after installing the solar panels (2016-2018), and then linked these costs with the impact on HH incomes. The analysis was done by using the Statistical Package for the Social Science (SPSS) to extract arithmetic and standard deviations for the 30 HH samples on electrical consumption and income changes, before and after installing solar panels. About 100 percent of the sample showed that the fuel cost decreased in summer and winter after installing the solar panels. The electricity bill had an almost 1.5 percent difference between the before and after periods of solar panel installations.

Similarly, the decrease in fuel consumption and cost to irrigate the small yards reflected positively on the HH income. All sampled HHs decreased their fuel consumption in summer and winter compared to their annual income expenses, and mainly from the source of seasonal heating activities. This explains the decrease of negative coping mechanisms which were adopted before installing the system. In contrast, female-headed household showed interest in their responding to the use of solar panels, but at the same time, worried about using them. Most of the women's concerns were on the solar panel's maintenance and sustainability, as they have to support their families' needs and income within limited time; the security issue is adding a critical burden on their mobility and safety. Furthermore, 80 percent of women interviewed that they need ongoing training and follow-up during early summers and winters, to make sure that they fully understood how to deal with the panels' maintenance, to ensure their productivity.

2.3 Viability of using solar panel as the electric supply for the industrial activities in South Syria

Conflict and war can weaken the market system. However, improving the medium-small enterprises can support local market adaptability, so that communities can secure their food needs [20].

[3] The Coping Strategies Index (CSI) measures behavior, in particular the things people do when they cannot access enough food, which is converted into a single index. ***(IPCC Glossary)***

Focus group discussions, with small and medium food factory owners in South Syria, indicated that high fuel and unavailability of skilled labor are the main obstacles to market participation under the recent conflict. 83 percent of the interviewed owners had confirmed the mentioned obstacles. The research assumes that applying the use of renewable energy as a source of electricity can improve the local Small and Medium Enterprises (SMEs) productivity and can improve market system overtime of resilience. The analysis showed that the average income increased for the factories' owners. Similarly, for the factories' maximum productivity level, the average significantly increased by 31 percent. On the other hand, the working hours for factories, rehabilitated by using solar panels, were reduced by 37 percent in average. This finding highlighted the assumptions of the research, regarding the link of using renewable energy to improve the income resilience of the affected communities in South Syria.

2.4 Estimated carbon reduction to mitigate the climate change, if solar panels were used across the targeted areas

Using solar panels will reduce carbon emissions [21]. The research used the survey data and the key informant interviews (KII), to calculate the amount of kilowatts of energy per month/ year that solar panels produce, in order to estimate the carbon footprints—if adapted as a source of renewable energy, by small factories and irrigated wells at the household level.

However, the analysis of the full life cycle assessment (LCA) for solar panels, including manufacturing, transportation, installation, and so on, must be considered to estimate the carbon emissions [22]. Additionally, considering the wastes produced by solar panels—especially its manufacturing materials will help refine the calculation to estimate the clear reduction of carbon from the installed solar panels [23].

Presently, this research used a simple theory of calculation, based on the estimation of the energy produced by installing solar panels, multiplied by the remaining households with irrigation wells (50 200 m) and small factories/ workshops in South Syrian governorates, taking the estimated percentage for solar panel production into consideration.

2.4.1 Amount of energy produced at household level if installing solar panel for irrigation wells

The energy produced by solar panels depends on the location and shading, along with operational hours per day. The energy produced differs between summer and winter, based on sun exposure [24]. According to the Syria National Centre for Energy Research, the radiance period is 312 days/year. However, the USAID water resource assessment in 2016—conducted in South Syria—indicated that Syrians are relying on diesel power generators to meet domestic, industrial, and agricultural needs [25], but diesel is not the most stable source of energy. It is expensive which depletes communities' assets [26].

Generally, the calculation of the amount of energy produced in kilowatt hour (kWh) is done by a simple equation—of multiplying the watts of solar panel by the number of operational hours per day, with consideration of the 312 radiation days per year for Syria, to calculate the energy produced in 1 year [27].

The calculations for 30 HHs indicated that the average energy produced by households to pump irrigated water from wells—with depth between 50 and 200 m to support their small yard for basic food needs—is **5002.16 Kwh/year.**

2.4.2 Amount of energy produced if installing solar panel at medium-small factories

The amount of solar energy used for operating the electricity of factories' sample is calculated here by converting Amber hours to Watt hours, and out of the electrical grid.

$$\text{Watt hours} = Amp \times \text{Battery Voltage}$$

Then, the same equation of calculating the energy—produced by multiplying it with operational hours per day—was used,

$$\text{Energy produced } per \text{ year} = (\text{Battery power} \times \text{operation hours}) \\ \times 312 \text{ days of radiation } per \text{ year}$$

The calculations for 6 factories indicated that the average energy produced is almost **3876.6 Kwh/year.**

2.4.3 Calculation of estimated carbon reduction at household level and small factories

Assuming that the recent average value of carbon emission for Syria—which was only available up to 2014 on the World Bank data system—is 1.6 tonnes of CO_2 with 312 days of radiation per year. The 1.6 tonnes of CO_2 were also adopted in 2016 for the world development indicators database; in this section, the calculation of carbon footprint will be using the IBRD–IAA [28]:

$$\text{Energy produced } per \text{ year} \times 1.6 \text{ tonnes of } CO_2$$

The calculation does not take into account the CO_2 from other parts of the solar panel life cycle, such as the manufacturing process or the transportation. The simple theory of CO_2 reduction calculation showed that the total of 240,103.47 tonnes of CO_2 can be reduced from operating solar panel at HH irrigation wells, and 37,215,360 tonnes of CO_2 by factories that were rehabilitated by using solar panel, as a source of electrical energy, out of electrical grid and the potential carbon.

2.4.4 Calculation of estimated carbon reduction if solar panels were installed at household irrigation wells, with 50–200 m small factories for electricity purposes in South Syria

Primary data mapping conducted for South Syria—which targeted wells—showed that in total, there were 69 wells with almost the same depth and capacity specifications. Thirty out of 69 wells were rehabilitated by solar panels; the calculation assumed that the average carbon reduction of the 30 wells can give an overall indication of how much—in total—the carbon reduction in South Syria governorates per year will be and prevented pollution that occurs when using diesel.

A total of **552,238.05** tonnes of CO_2 could be reduced by the rehabilitation of 69 wells in South Syrian governorates.

Correspondingly, a total of **291,520,320** tonnes can be reduced by the rehabilitation of 47 factories/ small workshops that were potentially mapped in South Syrian governorates in May 2018.

This means that there is an obvious carbon reduction from such solar panel initiatives with a total of **843,758.37** tonnes of CO_2, from both irrigation wells and small factories, in two of the South Syrian governorates. Particularly, the government of Syria can save up to **843,758.37** tonnes of carbon per radiation year (312 days) and can have opportunity to trade this carbon market globally.

DOI: http://dx.doi.org/10.5772/intechopen.101462

Currently, this amount of carbon reduction does not take into account the life cycle cost of solar panel, as mentioned in Section 2.4.3. However, it must be noted that this approximation is speculative and needs thorough calculations of the energy inputs and emission outputs, throughout the whole production chain of solar panels, including its transportation and installation process.

Arguably, most research harmonizes the life cycle GHG emission from PV solar panels materials, in order to determine the average emission intensity. Production stage is significantly contributing with 30 percent of the total of GHG emission [22]. Hence, some research did not take into account the end cycle—such as recycling the system, which is contributing with 4 percent of the total GHG emission [29].

According to the report published by the World Nuclear Association in 2011 [30]—on the Comparison of Lifecycle greenhouse gas emissions of various electricity generation sources, the range between different studies for solar panels is 85 tonnes CO2e/Kwh, which specified that the manufacturing process is becoming more efficient compared to other sources of renewable energy.

Therefore, if the approximate 30 percent of the total carbon from solar panel lifecycle is considered, then it should be deducted from the first year of installation. Finally, the actual carbon saved from the solar panel initiative in South Syria will be calculated as below:

$$834{,}758.37 \times 30\,\text{percent} = 253{,}127.51\ \text{tonnes of}\ CO_2$$

$$\text{Carbon saving for year is} = 834{,}758.37 - 253{,}127.51$$

Based on the result, this initiative will save a total of **590,630.86 tonnes of CO_2**, which can be traded in year one. Starting from year 2, it will approximately be a total of **834,758.37 tonnes of CO_2**.

2.5 Potential impacts on transboundary water resources between Jordan and Syria

Both Syria and Jordan rely on different water sources to generate renewable energy. Precisely, the Yarmouk river basin is shared by both Jordan and Syria, and both of them receive energy from the Unity Dam [31]. With the decrease in precipitation and overall water levels, the dependency on hydroelectricity needs to shifted [32]. According to the Millennium Development Goals Report in 2014, water is becoming a scarcer resource in the region and is overall becoming a less reliable resource. This means that there will be a decrease in hydroelectricity coming from the dam, along with other water resources in the region [5]. This necessitates the need to find a new power resource [33].

Some of the main tools developed for this research are the focus group discussion (FGD) and the key informant interviews (KIIs). Due to the limitation and constrains during the data collection from South governorates in Syria, one FGD was conducted with the farmers group, and two KIIs were conducted with water engineers—who used to support the WASH[4] sector's actors in South Syrian governorates. On the other side of the Yarmouk basin—in Jordan, four KIIs were

[4] Water, Sanitation and Hygiene is the focus of the first two targets of the Sustainable Development Goal 6 – targets 6.1 and 6.2 aim at equitable and accessible water and sanitation for all. "Access to WASH" includes safe water, adequate sanitation and hygiene education.

conducted with related decision makers on water basin management, as well as wash cluster members, who used to support the people in need and the displaced people in South Syria with WASH assistance, such as sanitation, hygiene, and water access.

KIIs respondents mentioned that the wash assistance was focused on providing solar panels for households or villages, in order to support their access to water. Therefore, most actors need to consider the safety issues associated with arrays, and also need to have a clear discussion with the targeted communities before the confirmation of solar sites, and management of their operational modalities. Involving them will ensure the ownership, thereby, decreasing the likelihood of damage or theft of installation.

Most respondents from FGD and KIIs complained about the cost recovery practice that was introduced by different WASH actors. They complained about the inability of beneficiaries to afford the cost of the system's maintenance, due to their low land productivity, and the malfunction of surrounding market. However, they were interested in having cash vouchers or start-up kits to support the operation and maintenance of the system in the first 3 months, along with intensive capacity building on system operation and maintenance so that the communities will manage to run it efficiency and effectively.

Interestingly, all respondents believe that using solar panels is the most viable and appropriate solution for their water pumping. Respondents believe that using this system at the level of school and health care facilities will minimize electricity shortages. Hence, ensuring the working hours for these services are sufficiently increased, to support their communities. Unfortunately, although the drought seasons were mentioned by the respondents, no one linked using the solar panels of the potential mitigation activity to the impact of climate change on their country; this emphasizes the need to raise the communities' awareness on the biggest threat on their livelihoods assets, which is climate change. Moreover, there is an urgent need to bridge the gap between humanitarian aids - resilience and development activities to plan, monitor and track the different interventions to design and DO NO HARM to social and environmental interventions.

In Jordan, the other side of the Yarmouk basin has a different perspective on using the renewable energy technologies for communities that live in the basin areas. Jordan and Syria have a long history of nonsatisfaction on several treaties over the years, prior to the conflict in Syria. Both countries are facing drought, where the average precipitation dropped significantly [34, 35].

The main concern of the Ministry of Water and Irrigation is losing control of the measurement of the water table and flow, which will occur if such solar panel system becomes widely used in Syria. The ministry considered the electricity bill as the only enforcement measure that can control the illegal and over pumping of Yarmouk ground water. As a result, the Jordanian enforce limitations on trade the solar panels' items, with aims to control the flow level of water in their catchment areas as much as possible.

Additionally, the Ministry of Agriculture restricted the trade of seeds or seedlings for two reasons: The first is to avoid deficiency of native seeds in Jordan, such as wheat and barley, that already differ from the ones used to plant in Syria. The second reason is to avoid planting around drainage area, thus avoiding the low water table, changes in the land use in catchment areas, and water allocations to Jordan.

What also contributed to this problem was the agricultural lands inside of Syria, which led to decreases in the level of water, and increases in the level of salt by digging local wells [36]. This calls for both governments in Syria and Jordan to approach a more diplomatic way of ensuring the effective transboundary water management, by addressing the political and hydrological conditions of the basin [37].

WASH actors are also facing challenges in managing the water resources and needs in south governorates of Syria. The Syrian Human Needs Overview 2017 [26] identifies 14.9 million people in urgent need of support with WASH, in addition to the extensive needs across the sub-sectors of water in order to move toward increasing the resilience of affected population and ensure the sustainable access to safe water.

The main challenges facing the WASH actors are the inefficiency of piped water networks, which have been damaged, destroyed, or non-operational due to the lack of power and/or fuel. The 2017 WASH assessment conducted by the cluster members highlighted that the water networks currently operate at less than 20 percent of its real capacity. This added an extra burden on communities, mainly women and girls, due to the spending of more time in collecting and saving water during the day.

The Syrian communities recognize the importance of water resources and its safety; the cost of the fuel and irrigation systems is very high compared to the average monthly income. Most of the community members, mainly the medium or small businesses, cannot afford the high cost of fuel, which led to the non-functional Syrian market.

In 2018, the WASH and food Security clusters proposed multiple responses to provide multi-sectorial assistance to vulnerable people. The clusters' members highlighted that farmers and herders started complaining about the drought seasons—mainly the limited grazing lands—due to the highly displaced movement in south Syria governorates, and also the conflict which changed the access to land and water. This indicates that multi-clusters must address the potential impact of climate change, and include the environmental measures in their mandate and interventions, to ensure food security and reduce negative coping strategies for the affected communities in South Syria.

3. Research discussion

Renewable energy advanced rapidly in 2017 with the highest growth rate according to the Global Energy and CO_2 status report [38]. Solar panel technology is one of the green technologies that prove its efficiency among other technologies, such as wind, hydroelectric, nuclear, and biomass [30]. Although the improvement in global efficiency decreased to 1.7 percent in 2017 compared to 2.0 percent in 2016, the Arab countries recorded a 55 percent improvement in energy efficiency, whereas it was 35 percent in 2015 [39]. However, the demand of electricity and transportation is increasing due to higher temperatures [40], which will burden the energy efficiency performance especially with the conflict and political instability in the region [41].

In comparison, 99 percent of the total installed capacity in Yemen are from fossil fuels, according to the Country Profile Energy Efficiency in Yemen 2012. About 40 percent of the total population still remains without electricity [42]. About 2 percent of the total installed energy capacity in Yemen is from renewable energy, reflected in 22 Million Mt. of CO_2 emission from the consumption of energy [14]. This is less than in Iraq, of 137 Million Mt. of CO_2 emission—from consumption of energy—compared to 87.3 percent of total installed energy capacity, which are from fossil fuels and 0 percent from renewable energy.

Prior to the conflict in Syria, two renewable energy master plans were developed in 2002 covered 10 years aimed at maximizing the use of renewable energy to increase the contribution of the in grid connected electricity generated, in order to reduce gas consumption [43]. The 2010–2030 scenarios of the Syrian master plans for Energy Efficiency and Renewable Energy projects prioritized solar energy for replacing fossil fuel in households, industry, and service sectors; it showed that 22 Mtoe[5] of conventional fuels will be substituted by using renewable energy. Therefore, 97 Mtoe of total energy will be saved.

However, due to the drop of the energy intensities in 2009, instability, and the lack of political commitment, the master plans failed to be implemented [44]. In view of these proposed scenarios, the research argument reveals that solar energy can be a smart source of electricity and support the country's master plan to increase the energy demand from renewable energy from 0 to 4.31 percent by 2030 [43].

Numerical validation of greenhouse gas emitted from using the renewable energy at household's level or small scale of enterprises to mitigate the climate change impact and can lead to strengthen communities' resilience need to be calculated. Evidently, the simplified approach that was adopted to calculate the data available presented that more than 12 percent adapted sufficient coping mechanisms, between families who installed solar panel and families that did not. In addition, between 25.8 –and 40.4 percent improved their savings. As for factories' owners, approximately 53 percent were able to profit from installing solar panel systems to operate their factories.

Monetarily, the average August 2018 carbon price, according to the carbon emission future price (*https://www.investing.com/commodities/carbon-emissions*), is approximately 19.30USD per 1 tonne of CO_2. This means that the Syrian government can trade the carbon and get:

$$590{,}630.86\,\text{tonnes of}\ CO_2\ \text{X}\,19.30\,USD = 11{,}399{,}175.5\,USD$$

$$\text{in year}\ \mathit{one}\ \text{after deducting the life cycle}\ GHG\ \text{emission}$$

$$\text{While the actual trading starting from year 2 will}\ \mathit{be}\ 16{,}284{,}536.54\,USD\,/\,\text{year}$$

If the lifetime for the solar panels of 30 years is considered, this means that over the 30 years of horizon, the Syrian government can trade with a total of **472,251,559.59 USD** at country level from two south governorates, keeping in mind that the calculations were done after deducting the production cost of the solar panel for the first year.

Having the recent conflict in Syria, the demand and price of electricity doubled, and increased by five times for Industrial and commercial tariffs, during the conflict years between 2014 and 2017, according to the recent Global Energy and CO_2 status report. Presently, based on the analysis, it is projected that each household with an irrigated well that uses an average of 16.03 Kwh per day can trade money with **154,466.56 USD** annually as out grid source. Likewise, the factories with average of 12.425 Kwh per day can trade with roughly **119,709,408 USD** annually. Therefore, as the economic burdens deepen with the recent conflict in Syria, communities increasingly adopt negative coping mechanisms, which pose protection risks, such as child labor or early marriage [12].

Such initiative can support the communities' resilience capacity and reduce the negative coping mechanisms, which are shown for families who installed solar

[5] Millions of tonnes of oil equivalent.

panel systems. For example, The International Organization for Immigration in Yemen reduced 400 tonnes of carbon emissions every year, and saved 121,000USD as a result of using solar energy projects, other countries have installed solar energy projects to serve water scarcity, which are a result of internal and external conflicts. For instance, Iraq started to recover from war and is now operating its main streets by solar powered lamps—around six thousand solar lamps light up the street in Baghdad. Moreover, Iraq is looking to subsidize solar panels through carbon trading, but still needs to establish internal resolutions and laws to promote and allow the trading of carbon [45].

The recent studies done by Intelligence 2017 highlighted that due to the instability, the energy revenue declined in Syria leading to high fuel costs and fluctuation in Syrian currency [46]. The primary data demonstrated such impacts on the communities' coping strategy, electricity costs, and industrial activities, to support the currency in the market. The purposive sample showed insights on potential solutions to such declines. For instance, the productivity level of targeted rehabilitated factories increased by 37 percent, which ensured the availability of their products in the market. Communities living around these factories expressed their satisfaction on availability of items during the week, compared to the usual demand and increased number of displaced folks, from time to time during the recent conflict in the south governorates. In contrast, a total of 8878.76 Kwh/year out of electricity grid can be produced from a small sample of HH and factories, which was intended to be much more, if applied to the whole HHs and small factories or workshops in south governorates.

On the other hand, the changes at the political border level between Syria and Jordan showed the over abstraction variated on water-related records [47]. The two countries should review their agreements on including the impact of climate change on water flow, and to ensure the role and impact of communities' activities and behaviors on transboundary villages [48].

Interestingly, the effect of using renewable energy, such as solar panels at a small scale of households and factories, would have a remarkable impact on the national footprint, eventually mitigating the impact of climate change on the environment [49].

4. Conclusion and recommendations

With temperatures in drylands currently twice the global average, there is an urgent need for adaptation to mitigate climate change. The level of downside climate impacts is directly correlated with the level of development. Climate change is expected to have significant impacts on extreme poverty by 2030, affecting communities already coping with livelihood limitations and inequalities, resulting in increased resource competition and conflict.

The recent 2017 Arab climate change assessment report underlined the droughts in Syria of 1998–2000 and 2007–2010, impacting substantial economic losses and the displacement of more than 1 million people [40]. Therefore, seeking alternative sources of subsistence and income, and building the resilience of the communities will enable them to adapt to the impact of climate change, and to deal with shocks, such as security conflicts, to improve their livelihoods opportunities and assists [9].

Despite the challenges that faced this research, the recommendations have been structured into the level of transboundary management, national efforts, and communities' behaviors and equity. The interaction of these levels will contribute to avoid further degradation and help in post-conflict recovery plans:

Complementary assessment of the transboundary water management for the Yarmouk basin is a must to improve the regional understanding of climate change impacts on transboundary water basins [40]. Particularly after 10 years of crisis and leakage of joint treaties between Jordan and Syria, so as to investigate on the way to optimize its infrastructure for saving energy costs and reduction in carbon emissions [37].

Initiating the joint projects between communities that live on the transboundary in Jordan and Syria, such as implementing joint projects to support the water demand and legal use according to international water law, which could promote peace building and hence effective climate change mitigation policies.

Bridging the gap between humanitarian aids and early recovery stages in Syria will support the communities' resilience and inclusiveness. Considering future projections of climatic impact on livelihoods assets, in terms of environmental resources, will foster the understanding of vulnerability and capacity to design DO NO HARM social and environmental interventions' components that deal with current and future risks to prevent food insecurity. It will encourage communities to reduce cutting the trees and using sustainable practices to avoid further land degradation.

Investment in renewable energy as a source of reducing vulnerability, and improving the local economy through supporting the renewable energy subsidies. This can be done by promoting the public and private cooperation, in order to decrease installation costs of solar panels. A country like Syria needs to re-build the trust between the government and communities, post 10 years of conflict; building such solar panel initiatives as proved in this case study can assist the country in entering the global carbon markets, and support its re-construction, respectively. Moreover, environmental agencies can help both Jordan and Syria to apply different schemes of payment of water ecosystem services "PES," which stand for incentives offered to stewards of ecosystem services [50] including public and private scheme that can be legally bound by a specific legal framework of transboundary basin.

Recognizing the link between climate change adaptation and mitigation under security conflict could entirely reduce the human impact on carbon emissions. During the conflict, prioritizing the climate change adaptation activities to build the communities' resilience will improve the country's mitigation targets in the long run. The case presented by this research showed the improvement in income and coping mechanisms for communities, which used solar panels to reduce fuel cost, to pump irrigation water and feed their families from planting small yards; this adaptation technique impacted positively on saving energy out of the electrical grid and reduced carbon emissions.

Ensuring gender equality and women economic empowerment could achieve resilient livelihoods. Dealing with women as "agents of change" [51] based on their skills, capacities, and opportunities to engage with such initiatives could strengthen their role in households and communities, in more economically productive parts, therefore, securing their food to absorb the shocks.

Improving the communities' adaptation of climate smart technologies could help in reducing CO_2 emissions. The recent publication by The Center for Behavior & the Environment [52] showed that dealing with solar energy resources can support a decrease of 24.6–40.3 GtCo2- eq in 2020–2030. This means that although in Syria the conflict increased the carbon pollutions [53], the government should focus on building the knowledge and skills for communities in adapting to climate smart technologies, to support achieving the primary renewable energy targets or updating the country's target.

Strengthening governance of the energy sector could support the government in identifying how much electricity can be generated to feed into the grid is thought to improve the policy with international practices, to support trading the carbon, thus using the revenue for country re-construction and resilience activities.

Approaching the post-conflict from the lens of complementarity between humanitarian and development activities presents a transforming action toward peace and turning environment into an asset for food security. Hence, there is an urgent need to promote innovative, integrated approaches, shift paradigms, and share the best practices to build back better and to reduce the impact of conflict on desertification.

Acknowledgements

This research would not have been accomplished without the tremendous support I had in surpassing the difficulties faced undergoing the implementation of its research. My particular appreciation goes to CARE International, for facilitating the data gathering. Without their support, the research would have been impossible to conduct.

Author details

Fidaa Fawwaz Haddad
CARE international, Amman, Jordan (Jan 2017 - March 2019)

*Address all correspondence to: f2hadad@hotmail.com

References

[1] World Food Programme. Global report on Food Crises 2021. Report. Rome: WFP; 2021

[2] OECD/FAO. OECD-FAO Agricultural Outlook 2018-2027. Rome: OECD Publishing, Paris/Food and Agriculture Organization of the United Nations; 2018

[3] Smith D, Vivekananda J. A Climate of Conflict: The Links Between Climate Change, Peace and War. Philippine ed. London: International Alert; 2007. p. 44

[4] Wilson S. Climate change and the rise of terrorism in northeastern syria. ESSAI. 2017;**15**:127. Available from: https://dc.cod.edu/essai/vol15/iss1/37

[5] Gleick PH. Water, drought, climate change, and conflict in Syria. Weather, Climate, and Society. JSTOR. 2014;**6**(3):331-340. Available from: www.jstor.org/stable/24907379

[6] Government of the Syrian Arab Republic Ministry of Local Administrations and Environment. Land Degradation Neutrality Target Setting Programme. Syrian Arab Republic: In cooperation with the Global Mechanism - United Nations Convention to Combat Desertification; 2020. Available from: https://knowledge.unccd.int/sites/default/files/ldn_targets/2020-08/Syria%20LDN%20TSP%20Final%20Report%20%28English%29.pdf

[7] United Nations Office for Disaster Risk Reduction. Sendai Framework for Disaster Risk Reduction 2015-2030. Geneva; 2015

[8] EJF. Beyond Borders: Our Changing Climate – Its Role in Conflict and Displacement. UK; 2017. Available from: https://ejfoundation.org/reports/beyond-borders

[9] CARE International. WOMEN, WORK & WAR: Syrian women and the struggle to survive five years of conflict. Research Study. 2015;**2016**:4-5

[10] World Bank. 2017. The Toll of War : The Economic and Social Consequences of the Conflict in Syria. World Bank, Washington, DC. © World Bank. Available from: https://openknowledge.worldbank.org/handle/10986/27541 License: CC BY 3.0 IGO

[11] Paton D, Johnston D. Disasters and communities: vulnerability, resilience and preparedness, disaster prevention and management. An International Journal. 2001;**10**(4):270-277

[12] Humanitarian Situation in Dara'a, Qunaitra and As-Sweida Governorates Situation Report No. 6 (3-16 August 2018) [en/ar] - Syrian Arab Republic [Internet]. ReliefWeb. Available from: https://reliefweb.int/report/syrian-arab-republic/syrian-arab-republic-humanitarian-situation-dara-qunaitra-and-sweida. [cited 2021 May 20]

[13] Intergovernmental Panel on Climate Change. Summary for policymakers. In: Climate Change: Mitigation of Climate Change. Contribution of Working Group III to the Fifth Assessment Report of the Intergovernmental Panel on Climate Change. Cambridge, United Kingdom and New York, NY, USA: Cambridge University Press; 2014

[14] World Resources Institute. Climate Data Explorer. CAIT. 2015. Available from: http://cait.wri.org/profile/Syria [Accessed: 19 September 2018]

[15] United Nations Development Program (UNDP). Human Development Report: Human Development for Everyone. 2016. Available from: http://hdr.undp.org/en/content/human-development-report-2016 [Accessed: May 2018]

[16] East AJ. What is a Carbon Footprint? An Overview of Definitions and

Methodologies. Sydney: Horticulture Australia; 2008

[17] Babbie ER. The Practice of Social Research. 2016. Available from: https://openlibrary.org/works/OL1903244W/The_practice_of_social_research [Accessed: 2 May 2018]

[18] CARE, WFP, Feinstein International Center, and TANGO. The Coping Strategies Index: Field Methods Manual. Second ed. USA: Copyright © 2008 Cooperative for Assistance and Relief Everywhere, Inc; 2008. Available from: https://www.fsnnetwork.org/

[19] FAO, IFAD, IOM, and WFP. The Linkages between Migration, Agriculture, Food Security and Rural Development. Rome; 2018. Available from: http://www.fao.org/3/CA0922EN/CA0922EN.pdf. [Accessed: 26 July 2018]

[20] CARE International. The Forgotten South: Food Security and Livelihoods in Syria. Jordan: CARE International; 2016

[21] Intergovernmental Panel on Climate Change. In: Metz B, editor. and othersMitigation. Contribution of Working Group III to the Fourth Assessment Report of the Intergovernmental Panel on Climate Change. Cambridge, United Kingdom: Cambridge University Press; 2007b

[22] Hsu DD, O'Donoughue P, Vasilis M. Fthenakis, Garvin A. Heath, Hyung Chul Kim, Pamala Sawyer, Jun-Ki Choi and Damon Economy Turney. "Life Cycle Greenhouse Gas Emissions of Crystalline Silicon Photovoltaic Electricity Generation: Systematic Review and Harmonization; 2012

[23] Ayres R, Cornis LVD, Bassi A, Jeroen BVD, Bing X, Brent A, et al. Manufacturing: Investing in Energy and Resource Efficiency. Kenya, Nairobi: Copyright © United Nations Environment Programme; 2011. pp. 245-285

[24] Rapp D. The Earth's heat balance and the greenhouse effect. In: Assessing Climate Change. Cham: Springer; 2014

[25] United States Agency for International Development. Syria Essential Services II – Annual Activity Report, Southern Syria Governorate. Sponsoring USAID Office USAID/Jordan: Implementing Partner(s) International Relief and Development, Inc.; 2017. Available from: https://pdf.usaid.gov/pdf_docs/PA00TKFR.pdf [Accessed: 21 November 2017]

[26] United Nations Office for the Coordination of Humanitarian Affair. Humanitarian Needs Overview: Syrian Arab Republic. New York. 2016:39 and 42. Available from: Https://hno-syria.org/#severity-of-needs [Accessed: 25 September 2018]

[27] Mathew A, Biju B, Mathews N, Pathapadu V. Design and stability analysis of solar panel supporting structure subjected to wind force. International Journal of Engineering Research and Technology. 2013;2:559-565

[28] IBRD –IAA. Management's Discussion & Analysis and Financial Statements. 2016. Available from: https://openknowledge.worldbank.org/bitstream/handle/10986/24985/210852v2.pdf?sequence=4&isAllowed=y. [Accessed: 2 August 2018]

[29] Frankl P, Menichetti E, Raugei M, Lombardelli S, and Prennushi G. Final Report on Technical Data, Costs and Life Cycle Inventories of PV Applications. Deliverable n° 11.2 - RS 1a of the NEEDS. Ambiente Italia (NEEDS project consortium); 2005

[30] World Nuclear Association. Report on Comparison of Lifecycle Greenhouse Gas Emissions of Various Electricity Generation Sources. London: World Nuclear Association; 2016

[31] Hussein H, Grandi M. Dynamic political contexts and power asymmetries: The cases of the Blue Nile and the Yarmouk Rivers. International Environmental Agreements. 2017;**17**:795-814

[32] Verner D. Adaptation to a Changing Climate in the Arab Countries: A Case for Adaptation Governance and Leadership in Building Climate Resilience. Washington DC: World Bank; 2012

[33] Fischhendler I. Legal and institutional adaptation to climate uncertainty: A study of international rivers. Water Policy. 2014;**6**:15-35

[34] Ministry of Environment. Jordan's Third National Communication to the United Nations Framework Convention on Climate Change. Jordan: The Hashemite Kingdom of Jordan The Deposit Number at the National Library (2014/11/5178); 2014

[35] Mohtadi S. Climate Change and the Syrian Uprising. Chicago: Bulletin of the Atomic Sentients; 2013. Available from: http://thebulletin.org/web-edition/features/climate-change-and-the-syrian-uprising [Accessed: 15 September 2018]

[36] Saleeby S. Sowing the Seeds of Dissent: Economic Grievances and the Syrian Social Contract's Unravelling. 2012. Available from: http://www.jadaliyya.com/pages/index/4383/sowing-the-seeds-of-dissent_economic-grievances-an [Accessed: 24 September 2018]

[37] UEA. Hydropolitical Baseline of the Yarmouk Tributary of the Jordan River. Norwich USA: Water Security Research Centre of the University of East Anglia; 2018

[38] International Energy Agency. Publications and Statistics. 2017. Available from: https://www.iea.org/ [Accessed: 15 September 2018]

[39] AFEX EE. Remarkable Progress in the Arab Energy Efficiency Markets Report. Cairo, Egypt: The Regional Center for Renewable Energy and Energy Efficiency (RCREEE); 2017

[40] United Nations Economic and Social Commission for Western Asia. Arab Climate Change Assessment Report – Main Report. Beirut: United Nations Economic and Social Commission for Western Asia; 2017

[41] El HediArouri M, AdelBen Y, Hatem M, Rault C. Energy consumption, economic growth and CO_2 emissions in Middle East and North African Countries. Energy Policy. 2012;**2012**(45):342-349

[42] Regional Center for Renewable Energy and Energy Efficiency. Energy Conservation Indicators in Southern Mediterranean Countries: Country Report for Syrian Arab Republic. RCREEE. 2012. Available from: https://www.rcreee.org/content/energy-efficiency-indicators-southern-and-eastern-mediterranean-countries [Accessed: 15 September 2018]

[43] Syrian Energy Balance for the Years 2003-2009. Syria: National Energy Research Center. Available from: www.nerc.org.sy [Accessed: 15 September 2018]

[44] Regional Center for Renewable Energy and Energy Efficiency Arab Future Energy Index™ (AFEX). RCEREE. Cairo, Egypt: Regional Center for Renewable Energy and Energy Efficiency; 2018

[45] Al Waeli M. Change must begin with Iraq's political parties. In: 1001 Iraqi Thoughts. Baghdad, Iraq: Muhammad Al-Waeli is an Iraqi Commentator on Political and Social Issues; 2018

[46] Shaffer B. A guide to the application of energy data for intelligence analysis. Washington, DC Studies in Intelligence.

2017;**61**(4). Available from: https://dokument.pub/energy-as-analytic-tool-flipbook-pdf.html

[47] Hobler M, Margane A, Almomani M, and Subah A. Contributions to the hydrogeology of Northern Jordan. Ministry of Water and Irrigation (MWI) and Federal Institute for Geosciences and Natural Resources (BGR). Report Vol. 4. In: Groundwater Resources of Northern Jordan. China: Scientific Research Publishing Inc.; 2001

[48] United Nations Economic and Social Commission for Western Asia. Inventory of Shared Water Resources in Western Asia. UN-ESCWA and German Federal Institute for Geosciences and Natural Resources. 2013. Available from: http://www.escwa.un.org/information/pubaction.asp?PubID=1362

[49] Ürge-Vorsatz D, Eyre N, Graham P, Harvey D, Hertwich E, Jiang Y, et al. Energy end-use: Buildings. In: Global Energy Assessment: Toward A Sustainable Future. Cambridge: Cambridge University Press; 2012

[50] Greiber T. Payments for Ecosystem Services. Gland, Switzerland: Legal and Institutional Frameworks. IUCN; 2009

[51] Aguilar L. How a Changing Climate Impacts Women. New York: WEDO-CWWL; 2007

[52] Williamson K, Satre-Meloy A, Velasco K, Green K. Climate Change Needs Behavior Change: Making the Case For Behavioral Solutions to Reduce Global Warming. In: Corinn W, Kyla T, editors. Harold M. & Adeline S. Morrison Family Foundation Editing and Design. Arlington, VA: Rare; 2018. Available from: rare.org/center

[53] Lelieveld J, Beirle S, Hörman C, Wagne T. Abrupt recent trend changes in atmospheric nitrogen dioxide over the Middle East. Science Advances. 2015;**1**(7). Available from: https://www.science.org/doi/10.1126/sciadv.1500498

Chapter 18

Droughts and Floodings Implications in Agriculture Sector in Rwanda: Consequences of Global Warming

Muneza Lydie

Abstract

Global warming driven either by human activities that emit greenhouse gases in the atmosphere or by natural shifts in weather patterns results in climate change. Short as well as long climate changes have impacts on the people's livelihoods worldwide through the intensification of natural desert, increased temperatures and precipitations, lack of access to (clean) water among others. This description is true in Rwanda as climate change affects agricultural sector which is proven to be a cornerstone of sustainable development of the country. However, it is possible to minimize the negative effects of climatic changes by reducing the vulnerabilities to it. This requires sound adaptation measures and mitigation strategies. Using literature review, this chapter shed light on ambitious mechanisms developed by the government of Rwanda to combat and to adapt to the effects of climate change, especially in its agricultural sector.

Keywords: Droughts, flooding, global warming and agricultural sector

1. Introduction

The degradation of the environment is undoubtedly one of the most serious threats to the people's livelihood across the world today, and Climate change happens to be one of principal causes and consequences of environmental degradation. Originally, climate change was considered as an environmental challenge, but now is being recognized as a political, social and sustainable development issue. In this views, Antwi [1] explains that adaptation related policies have been viewed in different contexts over the years. Antwi points out that climate change is now regarded as a major threat to human security with focus on different aspects. Martin [2] argues that climate change severely affects people's livelihoods worldwide through the intensification of natural disasters; global warming which leads to increased temperature and changes in precipitations resulting in droughts and flooding. These hazards affect agricultural production and access to clean water. The Stern Review [3] points out that with a greater amount of warming, global food production is likely to be seriously affected; especially in Africa, declining crop yields could leave hundreds of millions without the ability to produce or purchase sufficient food.

In addition, Misra [4] mentions that the impacts of climate change such as the depletion of water resource and the rise in global average temperature have created conditions for a decline in agricultural production which, in turn, leads to food price inflation globally as well as food shortages in developing countries.

However, it is crucial to shed light on the complex link between climate change and land use in order to capture, the impact of global warming on agricultural sector in general, and in Rwanda particularly. Land use practices (anthropogenic activities) affect the global concentration of greenhouse gases. While Land use practices constitute an important driver of climate change, a changing climate change on the other hand can lead to changes in land use. Thus, for example, farmers might change their customary crops to crops that are climate smart.

In Rwanda, agriculture is an important sector, because it largely contributes to crops production, national and household income, and food security to mention few [5]. Despite its important contribution to the Rwanda's economy, agriculture in Rwanda is highly sensitive to climate change. High temperatures eventually reduce yields of desirable crops and tend to encourage weed and pest proliferation, long periods of sun provoke desertification in some areas, while erratic rainfall leads to floods, erosion, landslides and loss of suitable arable soil. That is why the government of Rwanda aligns its self in the implementation of mitigation and adaptation strategies recommended by international protocols such as Sustainable Development Strategies (SDGs).

2. Description of Rwanda in terms of population and climate

2.1 Relief of Rwanda

The topography of Rwanda constitutes itself, an element of vulnerability. The relief profile of Rwanda is hilly and mountainous with an altitude varying between 900 m and 4507 m above sea level. The components of that relief are: Congo Nile Ridge Overlaying Lake Kivu with an altitude between 2500 m and 3000 m, it is dominated in the Northwest by the volcanic ranges consisting of five volcanic massifs of which the highest is Karisimbi itch 4507 m. The central plateau presents a relief of hills with an altitude ranging between 1500 m and 2000 m. The lowlands of the East are dominated by a depression characterized by hills with a more or less round top and 1000 to 1500 m in altitude. And the lowlands of the South-West in the Bugarama plain with an altitude of 900 m are part of the tectonic depression of the African Rift Valley [6].

2.2 Climate of Rwanda

Rwanda is located in tropical temperature climate due to its high altitude. The average annual temperature ranges between 16°C and 20°C, without significant variations. Rainfall is abundant although it has some irregularities. Winds are generally around 1–3 m/s in the high regions of the Congo Nile ridge, average temperatures ranges between 15°C and 17°C and the rainfall is abundant. The volcanic region has much lower temperatures that can go below 0°C in some places. In areas with intermediary altitude, average temperatures vary between 19°C and 21°C and the average rainfall is around 1000 mm/year. Rainfall is less irregular, and sometimes causes periods of drought. In the lowlands (East and Southeast), temperatures are higher and the extreme can go beyond 30°C in February and July–August. The absolute temperature of 32.8°C was recorded in the Southeast by Karama-plateau station on September 1980. Thermal constraints are more considerable in lowlands than in the remaining parts of the country. Rainfall is also less abundant in that region with around 700-970 mm/year. Rwanda has two rainy periods that are from mid-September to mid-December and from March to May, other periods are characterized by dry seasons [7].

2.3 Population of Rwanda

Rwanda's population density is the highest in Africa and it is estimated to be 13.11 million in 2021, 17.5% lives in urban centres, while 82.5% lives in rural areas [8]. About 72% of the population that is economically is engaged in agricultural sector [9]. FAO further argues that the majority of people who lives in rural areas relies on agriculture as the main source of income.

3. Impact of climate hazards on agricultural sector in Rwanda

Climate change has the potential to threaten agriculture through its effects on soil properties and processes [10]. Thus, the soil degradation implies decline in the quality and capacity of soil to perform its function. Qafoku [11] reveals that climate change, i.e., high atmospheric carbon dioxide concentration (≥400 ppm), along with increasing air temperature (2–4°C or greater) that persist for an extend period of time, significantly affect soil proprieties and fertility, thus agriculture sector fails in providing food quantity and quality. Furthermore, Islam and Wong [12] points out that, changes in the carbon dioxide concentration in soil affects the organic matter content of soil and soil quality, resulting in soil degradation. In return, Soil degradation reduces the output of agriculture and the efficiency of inputs.

It is crucial to consider the the impact of changing climate on soil in Rwanda, in order to cope with it. According to Famine Early Warning Systems Network [13] Rwanda is ecologically diverse with highland mountain forests, savannah grasslands, and lowlands. The western part of the country is characterized by highland mountain landscapes and gentle slopes leading toward the lowland areas in the east.

Rwanda is divided into twelve zones excluding Kigali city which is considered as an urban area and they offer a form of economic geography adapted to food security analysis. These livelihood zones are defined on the basis of local economies and livelihood opportunities, National Institute of Statistics of Rwanda [14], and are regularly used to analyze food security in Rwanda [13].

According to the 2018 analysis of food security and vulnerability in Rwanda conducted by NISR, the three eastern livelihood zones, namely, Bugesera Cassava Zone; Eastern Agro-Pastoral zone and Eastern Semi-Arid Agro-Pastoral zone are areas prone to drought conditions which prevent farmers from performing well in their farming activities. Besides, there is East Congo-Nile highlands subsistence farming zone that is at risk of prolonged dry spells as well as severe flooding. This zone is located in a high altitude mountainous area of Rwanda Southern Province. As the Eastern Semi-Arid zone, East Congo-Nile Highlands zone suffer from poor roads conditions and steep terrain which impede access to markets during the rainy seasons, the period when landslides occur more frequently. These conditions prevent farmers from selling their products.

Moreover, Ministry of Foreign affairs [15] notes that the mountainous west of the country constitutes a subject to erosion, part of central north and south experience severe floods, and the east and south suffer from droughts and desertification.

The consequences are not only seen on production but also on the demand of labour. Especially, poor households are at risk of food insecurity since they rely on their own production which is done in small plots of land, in-kind payment and market purchase mainly through agricultural activities/labour to meet most of their annual food needs.

3.1 Vulnerability of agricultural sector in Rwanda

The vulnerability of the agriculture sector in Rwanda is found in the characteristics of the country's topography. The western part of the country is dominated by highland mountain landscapes and gentle slopes while the eastern is consisted of lowland areas. These conditions make the whole country's arable land susceptible to severe landslides, erosion and floods. Another challenge is the demography of the country, which constitutes one of the most densely populated in the region. In addition, there is limitation in ability to cultivate with climate smart related options coupled with small size plot hold by farmers.

According to REMA [5], agriculture is highly sensitive to climate change and extreme weather conditions such as droughts, floods and severe storms. This is so in Rwanda, since all farming activities are dominantly rain-fed. In line with the observation of REMA [5] several factors account for the direct connection between climate change and agriculture productivity. These include average temperature increase, changes in rainfall amount and patterns; rising atmospheric concentrations of CO_2; pollution levels such as tropospheric ozone; climatic variability and increase in frequency and intensity of extreme events such as droughts, floods, intense and destructive rains, as well as other related events such as landslides.

As observed by Mutabazi [16] rainfall behavior shows that the rainy seasons in Rwanda are becoming shorter with higher intensities. As a result, there are events such as floods and landslides in areas experiencing heavy rains; and drought in dry areas. This tendency leads to decrease in agricultural production. For instance, heavy rains that have been observed in the northern and the western provinces coupled with a loss of ecosystems services due to the deforestation have resulted in soil erosion, rock falls, landslides and floods which destroy crops. On the other hand, the eastern province has been experiencing rainfall deficits over the decade of 2000–2010, leading to droughts.

The impact of climate change on agriculture and allied sectors in Rwanda is presented in **Figure 1**.

As REMA [5] observes, there are some climatic conditions which have increased crop growth vulnerability. First of all, the shifts in rainy seasons (September–December) and (March–May) and short or prolonged dry seasons in some regions distort growing seasons. This usually confuses farmers on critical decisions such as planting dates, which affects the timing of field preparation and planting in turn. These also affect crop growth, cause intensification of crop diseases and pests and result in lower crop yields. For instance, the shift in rainfall patterns leads to reduction in the amounts of rain water harvested therefore affecting both hillside and valley irrigation projects, through either decreased water levels in ponds/dams or high amount of water destroying dam/pond embankments and causing erosion and silting. Sometimes, the high amount of rainfall destroys the quality of the soil and affects the growth of crops.

For instance, REMA [5] highlights the floods observed in the Northwest and in the marshes of the River Valleys of Nyabugogo and Akanyaru in Rwanda resulted in loss of food production, property destruction which usually leave the farmers homeless and without food. Besides, they lead to serious soil erosion and destruction of irrigation infrastructure both on the hill slopes and down the valleys.

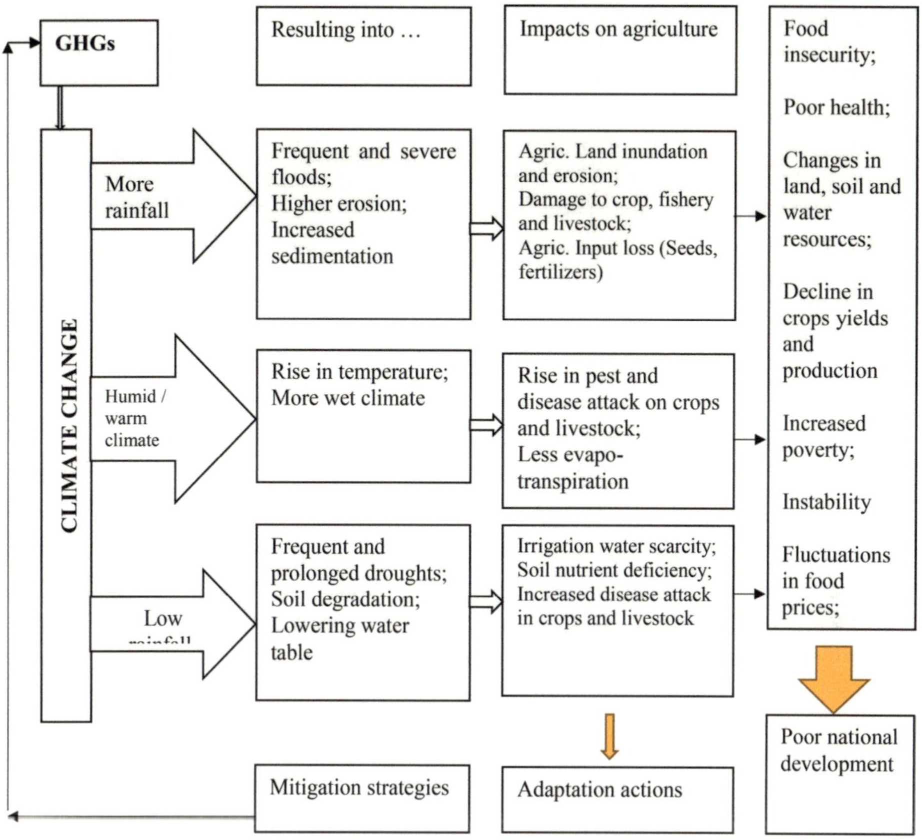

Figure 1.
Climate change effects and related impacts in the agricultural and allied sectors. Source: Adapted from [5].

Another major factor of vulnerability that REMA [17] showed is the prolonged and cyclical droughts. This drought is encountered in the Southeast of the country resulting from decreasing annual rainfall amounts (from 1000 mm to 700 mm) which leads to food insecurity and displacement of communities. Moreover, this calls for relief supplies and resettlement which would have budgetary implications. On the other hand, REMA reported that since 2011, the intense rains in the Northwest and the droughts and erratic rainfall in the districts of Bugesera, Nyanza, Gisagara, Huye, Rusizi-Nyamasheke have affected over 60–90% households. These climate hazards do not only affect food production, but also results in food and nutrition insecurity.

3.2 Droughts

As a result of global warming, Rwanda's drought vulnerability is observed to be high. Mutabazi [16] notes that drought; irregular rains and dry spells were the most commonly reported shocks for households. For instance, drought made the Southeast region vulnerable where decrease in annual rainfall as well as prolonged and cyclical droughts led to food insecurity.

Drought mostly affects the agro-bioclimatic regions of the East and Southeast of the country and sometimes some zones of the Central plateau. For example, Bugesera used to receive rainfall amounts between 700 and 800 mm/year before the 1990s but in the 2000s; it only receives 300 mm/year which is more that 70% of water deficit. As a result, the production of cereals and leguminous plants such as maize and beans has become almost impossible in this region. Another consequence of drought is the proliferation of parasites like caterpillars on sweet potatoes and predators of beans [18]. These parasites prevent crops from growing normally leading to low production which increases the risk of food insecurity in the country.

For example, NISR [14] shows that severe drought in the main growing seasons (A and B see **Figure 1**) impact between 28,500 and 157,700 people in all seven districts of the Eastern Province (Kayonza, Gatsibo, Kirehe, Nyagatare, Rwamagana, Ngoma and Bugesera). Moreover, the analysis highlights that about 62,000 MT and 157,700 MT of major crops are vulnerable to severe drought in season A and season B, respectively. Among these major crops such as banana, cassava, and Irish potato are the most vulnerable.

3.3 Flooding

The Global Facility for Disaster Reduction and Recovery (GFDRR) associated with the World Bank [19] state that floods are most frequent disasters in Africa. This may be a result of number of transboundary river systems located in Africa in addition to 160 major fresh water lakes and several small rivers, consequently this situation leads to frequent seasonal floods. Moreover, the report further notes that the impacts of natural disasters such as floods and landslides among others are further exacerbated by the weak flood protection, insufficient urban drainage systems, and increased run off caused by land degradation and poorly planned development.

Concerning landslides, Kervyn *et al.,* [20] highlights that landslides constitute a major hazard in many mountainous and hilly regions. Obura [21] weigh in by advancing the view that in countries with hilly terrain and high levels of rainfall, landslides risk is high due to widely prevalent soil erosion, deforestation and unsustainable land management which constitute the main causes of landslides in Africa. Rwanda is a good example in these case. Rwanda is in the African Great lakes region and is highly elevated; its geography is dominated by mountains in the west and savanna to the east, with numerous lakes throughout the country, where does the nickname "country of a thousand hills' come from.

Ministry of Environment [6] describes the relief of profile of Rwanda as follow: Rwanda is hilly and mountainous with an altitude varying between 900 m and 4507 m above sea level. The components of that relief are: Congo Nile Ridge Overlaying Lake Kivu with an altitude between 2500 m and 3000 m, it is dominated in the Northwest by the volcanic ranges consisting of five volcanic massifs of which the highest is Karisimbi itch 4507 m. The central plateau presents a relief of hills with an altitude ranging between 1500 m and 2000 m. The lowlands of the East are dominated by a depression characterized by hills with a more or less round top and 1000 to 1500 m in altitude. And the lowlands of the South-West in the Bugarama plain with an altitude of 900 m are part of the tectonic depression of the African Rift Valley [6].

According to MINITERE [18], in Rwanda, the most frequent direct risks to which agriculture is exposed are related to flooding and landslides. Erosion affects at least 50% of all farmers with a 30% decline in farm productivity while deterioration of soil is detrimental to Rwanda's food security for over 90% of the people who depend solely on agriculture in the country [16]. Soil erosion contributes to the decrease of agricultural productivity, causing therefore food insecurity problems. Floods, landslides and collapses are acute in regions with high altitude. These are dominantly in the west, southwest and north and central regions of the country.

4. Rwanda's specific climate change related actions in agricultural sector

Rwanda seeks to transform the agricultural sector from subsistence farming to a sustainable, value-creating, market- oriented food sector with high contributions to national output and household food security [22]. This makes the government of Rwanda to be ambitious concerning the mitigation of and adaptation to climate change. The basis for actions, by the government of Rwanda, on climate change is enshrined in international and national legal provisions. Among them we can mention United Nations Framework Convention on Climate Change (UNFCCC), the Millennium Development Goals (MDGs) targets that guided actions of countries. MDGs largely succeeded in capturing the popular imagination and reflecting shared priorities [23]. In 2015 UN, member states adopted an ambitious new agenda "transforming our world": the 2030 Agenda for Sustainable Development (Agenda 2030), establishing 17 Sustainable Development Goals (SDGs) to be achieved by all countries and stakeholders by 2030 to finish the unfinished mission of MDGs and transform the economies and societies of all countries. It has been proven that achieving all the SDGs will be much more challenging without urgent climate action (SGD 13).

At the regional level, Rwanda subscribes to EAC Climate Change Policy (EACCCP) that has been developed for the EAC region to engage a more strategic and cooperative approach to responding to the impacts of climate change and maximizing any potential benefits of the changing climate. Moreover, Rwanda's commitment to combating climate change impacts on the agricultural sector is set out in its Nationally Determined Contributions (NDCs) reports under the 2015 Paris Climate Change Declaration that embody its efforts to reduce national emissions and adapt to the impacts of climate change.

Specifically, in Rwanda climate change is recognized at the highest level of government as a potential threat to productivity and sustainability of the agricultural sector and livelihoods. This is more so because the agricultural sector constitutes the backbone of the economy of the country and contributes up to the third of GDP of the country. As a result, the government of Rwanda developed and is implementing a number of policy initiatives including: mechanisms to implement provisions of international climate change provisions, for instance the implementation of National Adaptation Programmes of Actions (NAPA); reforming public institutions by including climate change management functions; and mainstreaming climate change within policy processes.

4.1 Land use consolidation

The agricultural land in Rwanda is fragmented and mainly located at hilly slopes [22]. In order to optimize the use of available farmland, the government focused on the agriculture land utilization systems. In this regard, the Ministry of Agricultural and Animal Resources (MINAGRI) initiated the Crop Intensification Program in 2007 in which the main pillar was the Land Use Consolidation policy of 2008. According to NISR [14], land consolidation is a reallocation of parcels of land to overcome the effects of fragmentation. Farmers in the same area, adjacent to each other choose a single crop and use of a single cultivation regime (inputs, schedule of planning...). through this approach, the boundaries and rights of parcels remain intact and the government provides subsidized inputs for farmers in a given area with closed parcels to grow the same priority crops on a minimum size area of 5 hectares in a synchronized manner and hence improves the productivity.

4.2 Soil protection and land husbandry

Analysis of rainfall trends shows that the rainy seasons are becoming shorter with high intensity rains which lead to decrease in agricultural production. Moreover, MINAGRI [22] notes that 90% of agriculture land is on slopes. In this regard, the putting in place of effective land management structures is of a great importance to improving productivity. Thus the government of Rwanda has introduced a number of measures which comprise radical terraces accompanied with soil fertility management and soil erosion control. For farmlands on gentle slopes, progressive terracing and agroforestry have proven success in reducing soil erosion.

NISR [14] highlights that an average of 67% of agricultural households in Rwanda is engaged in land conservation practices. These practices include terracing and agroforestry that minimize soil erosion and promote water conservation. According to Gasheja and Gatabazi [24], PSTA I and PSTA II programmes were largely successful in terms of progressive and radical terraces, accompanied by soil fertility management and soil erosion control.

4.3 Irrigation and water management

Around 22% of agricultural households have land under the land consolidation programme and 9% of them have a part of their land irrigated [14]. Irrigation was identified as a key strategic activity in PSTA II in concordance with the commitment of Rwanda under CAADP compact which establishes Land and Water management in its pillar I. Researches show that irrigation is important in increasing agricultural productivity by allowing multiple cropping, and reducing vulnerability to weather shocks. By signing the CAADP, the government of Rwanda promised to allocate 2% of public funds for irrigation development.

In fact, irrigation allows farmers to move from rain-fed agriculture to diversified high value crops, thus increased cropping intensity and land productivity as well as resilience to climate change [22].

4.4 Input in farming

The provision of farming inputs has also contributed to increased crop productivity in a number of areas in Rwanda. Since 2007, there has been an increase in fertilizer use. Gasheja and Gatemberezi [24], reveal that the application of fertilizers in crop intensification programme areas has reached an annual average of 29 kg/ha/year in 2011–2012 compared to a national average of 4.2 kg/ha/year from 1998 to 2005 which increased crops yields. For example, maize yields increased from 0.65 MT/ha in 2000 to 2.5 MT/ha in 2010 while wheat yields increased by 2.5 times during the same period.

4.5 Improved seeds

As noted by Huggins [25], the efforts exist to make the agriculture in Rwanda more climate-smart. MINAGRI [26] observes that availability, accessibility and optimal quality seeds enhance crop yields and their subsequent contribution to food security, balanced nutrition, value of the product in the market, and economic growth. In the line with CIP policy, farmers, where it is applicable for now, must use improved seeds for selected crop to increase the production to ensure food security and to increase incomes [27].

4.6 Agricultural mechanization

Agricultural mechanization is for efficiency and expending production scope. In Rwanda, mechanization of the agriculture sector consists of different forms of machinery which performs different farm operation: land development and land preparation, planting, crop treatment, harvesting, post-harvesting and agro-processing.

Mechanization of agriculture has many benefits. Gasheja and Gatemberezi [24], argue that mechanization contributes to improving productivity of cultivated land and facilitates expansion of cropping areas, thus improving overall food security. MINAGRI [22] emphasized that agricultural mechanization is among the key technological pillar to boost the yield and production by minimizing the harvest and post-harvest losses while saving human and capital investment farm operation.

MINAGRI has facilitated the ownership of farm machinery to interested farmers, through a lease agreement with them. Actually, MINAGRI, in partnership with private service providers, provides the necessary equipment, tools and training with the aim of accelerating the mechanization of agriculture.

4.7 National post-harvest staple crop strategy

The National Adaptation Programme of Action and the National Strategy of Rwanda on Climate Change and Low Carbon Development highlight improved post-harvest management as a key climate change adaptation priority.

In Rwanda, climate change is causing floods and droughts, increasing the incidence of thunderstorms and associated high winds. It is also changing traditionally dry seasons, increasing the incidence of rainfall in the midst of higher temperature. These changes affect existing rural infrastructure and pose a threat to traditional practices in harvest storage [28]. Besides, the success of the Crop Intensification Program, launched in 2007 with the objective of increasing productivity in selected food crops while improving food security and self-sufficiency, increases in crop yields resulted in unanticipated surpluses in key staple grains and cereals [29]. In handling the problem of crop yield loss, under MINAGRI, the government initiated the National Post-Harvest Staple Crop Strategy in 2011 with the goal of assisting and strengthening farmers with the harvesting, post-harvest handling, storage, trade and marketing within staple crop value chains in Rwanda in the effort to reduce post-harvest losses. The fundamental vision of this strategy is to reduce food insecurity through an efficient post-harvest private sector system delivering staple food to the people of Rwanda [29]. This framework has been guided by Vision 2020 of the Government of Rwanda, EDPRS and PSTA.

5. Conclusion

Climate change is a reality. It has much more impacts in regions where vulnerability is high. That is why countries, regions and even at international level, concerned stakeholders must look on how to reduce vulnerabilities. Based on literature this study noted that achievement is seen in land management using soil erosion mitigation strategies such as terracing and irrigation schemes. The land consolidation model has been key to raising production.

The government of Rwanda developed and is implementing numerous strategies and programmes across the country. Since 1992, the country started to ratify and sign international and regional treaties and protocols which it domesticated according to its needs. In this sense, Rwanda started its journey with Vision 2020 which the objective was transforming the country from the status of a low income economy into a middle-income economy and a mission of overcoming poverty. Because of successful adaptive strategies in farming activities, a significant growth in agricultural productivity has been one of the main rivers of growth and poverty reduction. The performance of agricultural sector has been impressive after the introduction of the Rwanda national agricultural policy and strategies such as the Land use consolidation programme, improved seed, inputs like fertilizes and use of pesticides, water management, terracing, agricultural mechanization and improved post-harvest storage.

Author details

Muneza Lydie
Pan African University, Institute of Governance, Humanities and Social Sciences (PAUGHSS), Cameroon

*Address all correspondence to: mnlydie@gmail.com

References

[1] Antwi, A. (2013). Climate Change and Food Security: An Overview about the Issue. Friedrich *Eberto stiftung Foundation*, p. 13.

[2] Martin, S. F. (2009). Managing environmentally induced migration. In F, Laczko & C. Aghazarm (Eds.), Migration, environment and climate change: Assessing the evidence (pp. 353-384). Geneva, Switzerland: IOM.

[3] Stern, N. (2006). What Is the Economics of Climate Change? The Stern Review: On *the Economics of Climate Change.* HM Treasury, London.

[4] Misra, A. K. (2014) Climate change and challenge of water and food security, Int. J. Sustain. Built Environ., 3, 153-165.

[5] REMA (2011). Guidelines for mainstreaming climate change adaptation and mitigation in the environment and natural resources sectors. CRA 9-11.

[6] Ministry of Environment (2018). Third national communication under the UNFCCC.

[7] MINITERE (2006). National Adaptation Programmes of Action NAPA to Climate Change - Report. Ministry of Lands, Environment, Forestry, Water and Mines (MINITERE), Kigali.

[8] Kemp, S. (2021) Digital 2021: Rwanda.

[9] FAO (2020). Rwanda food and agriculture policy monitoring review 2020. Policy report.

[10] Brevik, E. C. (2013). Soil and human health – An overview. In E. C. Brevik, L. C. Burgess, and FL. R. Boca (Eds.). *Soils and Human Health.* (pp. 29-56). CRC Press, USA.

[11] Qafoku, N.P. Climate change effects on soils: Accelerated weathering, soil carbon and elemental cycling Adv. Argon. 2015. 131, 111-172.

[12] Islam, M. S., and Wong, A. T. (2017). Climate change and food in/security: A critical nexus. Environment. https://doi:10.3390/environments4020038.

[13] FEWSNET (2012). Rwanda: Livelihood zone and descriptions June 2012.

[14] NISR (2018). Comprehensive Food Security and Vulnerability and Nutrition Analysis Survey of 2018 (CFSVA).

[15] Ministry of Foreign Affairs (2018). Climate change profile: Rwanda.

[16] Mutabazi, A. (2011). Rwanda Country Situational Analysis. *CAMCO.*

[17] REMA (2018). The Assessment of Climate Change Vulnerability in Rwanda.

[18] MINITERE (2005). Initial national communication. Under the united nations framework convention on climate change.

[19] GFDRR and World Bank (2015). Starving toward disaster resilient development in sub-Sahara Africa. Strategic framework *2016-2020.*

[20] Kervyn, M. Jacobs, L., Maes, J. ... Mertens, K. (2015). Landslide resilience in Equatorial Africa: Moving beyond problem identification! *Revue Belge de geographie.* https://doi.org/10.4000/belgeo.15944.

[21] Obura, D. (2006). Impacts of the 26 December tsunami in eastern Africa. Ocean and Coastal Management,

49(11), 873-888. http://doi.org/101016/j.ocecoaman.2006.08.04.

[22] MINAGRI (2019). Annual Report 2018-2019, Kigali.

[23] UNDP (2016). From the MDGs to sustainable development for all. Lessons from 15 years of practice. New York, NY, 10017 USA.

[24] Gasheja, F and Gatemberezi, P. (2017). An assessment of Rwanda's agricultural production, climate change, agricultural trade and food security. *KIPPRA Working paper series* No.23, Kigali.

[25] Huggins, C. (2017). "Climate Change Adaptation in Rwanda's Agricultural Sector: A Case Study from Kirehe District, Eastern Province". African Centre for Technology Studies.

[26] MINAGRI (2017). National Agriculture Policy (NAP).

[27] IRDP (2018). Crop Intensification Programme (CIP) Satisfaction Survey-2017.

[28] IFAD (2013). Smallholders, Food Security, and the Environment. Rome: International Fund for Agricultural Development.

[29] MINAGRI (2011). National Post-Harvest Staple Crop Strategy. Kigali.

Chapter 19

Evaluation of Climate Change-Induced Impact on Streamflow and Sediment Yield of Genale Watershed, Ethiopia

Tufa Feyissa Negewo and Arup Kumar Sarma

Abstract

In the 21st century, changes in induced climate can significantly affect the water resources system in the watershed. Understanding climate change disrupts hydrological processes can facilitate sustainable water resource strategies to resilient impacts of global warming. The hydrological response of watersheds will be accelerated by climate change, altering the rainfall, magnitude & timing of runoff, and sediment yield. The study investigates climate change aspects on the hydrological responses using Soil and Water Assessment Tool (SWAT) model interfaced with Geographical Information System (GIS) of Genale Basin, Ethiopia. The calibrated SWAT was applied to simulate the impact of climate, and the SUFI II algorithm was used for parameter optimization & finalization. The change of climate scenarios was built using the outcomes bias-corrected CORDEX RCM daily precipitation, min/max temperature for Ethiopia under RCP 4.5 and RCP 8.5. The average monthly change of streamflow from −16.47% to 6.58% and − 3.6% to 8.27% under RCP4.5 and RCP8.5, respectively (2022–2080). The monthly average sediment yield change was −21.8% to 6.2% and − 5.6% to 4.66% for the RCP4.5 and RCP8.5 scenarios, respectively, over 2022–2080. It implies that the climate change-induced impacts on sediment yield are more significant than streamflow and suggest substantial adaptive management in watershed systems.

Keywords: Climate change, SWAT, Genale watershed, Sediment yield, RCM, Streamflow, Hydrological impacts evaluation

1. Introduction

Water is a unique resource given to humankind from nature impacted by induced climate change [1]. The atmospheric scientists suggest that the Earth is warming as a global temperature increase the hydrological cycle more actively. Greenhouse gases (GHGs) increase in the surroundings is a significant concern for global warming & climate changes. These changes may influence natural water resources in the catchment [2]. The Intergovernmental Panel on Climate Change (IPCC) appraisal report stated that global mean precipitation, surface temperature, droughts, and floods had changed significantly, and the changes are expected to continue [3]. Principally, developing country like Ethiopia is now facing severe climate change effects on water and agriculture sector. Currently, it is of great importance to evaluate the consequence of climate change on the regional and local water resources. The rise in surface earth air temperature and precipitation patterns are prominent features of change in climate that directly impact almost all other hydrological responses [4]. A temperate climate will accelerate the hydrological process, altering rainfall patterns and the magnitude & timing of streamflow. Climate changes are also expected to have remarkable impacts on the soil type since rainfall and runoff are the factors governing soil erosion and sediment yield/transport within landscapes [5].

The information derived from Global Climate Models (GCMs) is currently the most applicable in evaluating both past and possible future changes in climate scenarios. This climate data is then used as input to drive the hydrologic process. Long-term locally-observed climate data are also needed to validate climate model outputs to capture local settings [6]. However, direct implementation of GCM outputs to any hydrological model for subsequent evaluation of impact is despondent in climate studies because of coarse resolution issues. The simulation of GCMs runs on large scales to consider various grids across the globe, and GCM typically takes about 2.80 x 2.80 longitude and latitude resolution. To tackle the problems downscaling is assumed, a process of bringing down the climate information from GCM to regional & local hydrologic scales to produce outputs of the more acceptable resolution, which are more realistic with the local scale before estimating the risks associated with the future hydrologic scenarios [7, 8].

Different downscaling techniques have been advanced over the past two decades, deriving from two major blueprints; dynamic downscaling and statistical downscaling approaches. The dynamic approach is often viewed as a mini-GCM because it stimulates regional climate variables by decreasing the horizontal area covered (typically around 25 by 25 km) using the same boundary conditions as the evolving GCM. Because they produce high-resolution climate data, they have not been extensively accepted because of the complexities and costs involved in running this type of technique to capture regional-scale climate variables. Statistical downscaling approach, involving weather typing procedures, transfer functions, and stochastic weather generators, are the most known methods used in climate change studies nowadays [9]. They give future climate scenarios based on a statistical relationship between climate variables at one or more GCM grid points at a particular station. They are adopted because they are relatively economical to apply and give point climate data at a specific site of interest [5, 7].

The changes in streamflow and sediment yield characteristics resulting from climate change depend on individual watershed aspects. Decisive evaluations of the quantity and rate of runoff and sediment yield are needed to help decision-makers develop catchment management plans for better soil & water conservation measures [10, 11]. The SWAT-Soil and Water Assessment Tool model simulates the climate change-induced impacts for the San Jacinto River basin in Texas [12]. The effect of climate change on catchment hydrology is typically evaluated by characterizing climate change scenarios to a hydrological model based on the futuristic GHGs [1, 5, 4].

Streamflow modeling is essential to know sediment concentration in the stream, whereas peak streamflow rate is vital for hydraulic structure, watershed management practices, and flood protection. Different studies used empirical, statistical, and simulation methods to resolve the impacts of climate change on hydrological responses [13]. Recent studies recommended that SWAT is widely used as a capable model to evaluate environmental and hydrological changes with varying land types and climate conditions [14]. Additionally, the output components incorporated in the SWAT model are found to address various water-related systems in the watershed. The study highlighted that an increase in the concentration of CO_2 has a notable effect on streamflow, sediment yield, evaporation, and water yield. Carbon emission scenarios are the main driving forces in climate models. Scenarios are images or pictures of how the world is likely to emerge in the future in terms of greenhouse gasses (GHGs). In the recent study, we use the latest scenarios, called Representative Concentration Pathways (RCPs), which have rarely been applied in the study catchment. The IPCC characterizes a set of RCP scenarios (2.5, 4.5, 6.0, and 8.5) for projection of future climate based on Coupled Model Intercomparison Project (CMIP5) [15]. These four RCPs consolidate one alleviation scenario priming a low driving level (RCP2.6), two stabilization (medium) scenarios (RCP4.5 and RCP6), and one with a high GHGs emissions scenario (RCP8.5). These emissions scenarios are emerged based on the driving force such as socio-economic development, population growth, and GHGs [16]. Based on the IPCC report, by the end of the 21st century, global warming/temperature may increase by 1–5°C. Climate change scenarios for the Global Climate Model (GCM) or simple analog models are sometimes adapted to investigate climate change impacts on hydrology [17].

Nevertheless, their spatial resolutions are extremely coarse for regional climate study and need to downscale it. Therefore, either through statistical or dynamic regional climate models, the downscaling approach is required to convert GCM data into acceptable resolution before using for any hydrological study [18, 19]. Limited reports address the climate change analysis using Regional Climate Model (RCM) on streamflow and sediment concentration in the region. Nevertheless, most studies have used coarse-resolution GCM data, which are not favored for watershed hydrological modeling. The SWAT model was selected for this study because of its ability & wide range of applications, demonstrating that the model is a flexible and robust tool that can simulate various regional water flow at a watershed scale provide effective results [20].

This study contributes to investigate the effects of future climate change projection on the streamflow and sediment yield of Genale catchment using the calibrated/validated SWAT model under baseline and future two emissions and offers baseline information for adaptive soil and water resource management in a changing climate region. For the SWAT input, the future climate projection (2022–2080) statistically downscaled Regional Climate Model (RCM) Bias-corrected Coordinated Regional Climate Downscaling Experiment (CORDEX) precipitation, max/min temperature for Ethiopia, under RCP 4.5 and RCP 8.5 emissions scenario was used with historical data of (1990–2013). The climatic model data for the hydrologic modeling tool (CMhyd) is used to extract and bias-correct the climate variables obtained from RCM-CORDEX.

2. Materials and methodology

2.1 Description of the study area

The surface of the Earth has three main climate zones: tropical (hot & higher humidity zones), temperate (moderate between tropical & polar), and polar (floating and pack ice). Ethiopia is placed in the tropical climate zone lying between the Equator and the Tropic of Cancer. The latitude, longitude, & altitude of Ethiopia is given as 9^0 8′ 53” N, 40^0 29′ 35″ E, & 1343 m respectively. Based on elevation, the country has three different climate zones: Tropical zone (Dega, Weyna Dega, and Kola), with an average annual temperature of about 27°C and annual rainfall of about 510 millimeters. The study area is located on the Genale watershed with 54,941.583 Km^2 of the part of Genale Dawa River, situated in the South-Eastern part of Ethiopia and joins with Dawa River at the border with Somalia (Dolo Ado) (4° 16'N, 42° 04′E) to become the Juba River. In the Genale Basin, a total of 464 HRUs were created and scattered among 25 sub-basins. The annual mean of precipitation experienced in the area 810 mm distribution of rainfall in the watershed is 300 to 1302 mm per year. The daily max and min temperatures are 34.5 °C and 8.6 °C, respectively, with a daily average of 19 °C (**Figure 1**).

2.2 Description of the SWAT model

The SWAT was advanced in the 1990s by the United States Department of Agriculture (USDA). It is a mechanism-based and spatially semi-scattered hydrological model or flexible tool in different parts of the world, designed to calculate and route water, sediments, management practices, and nutrient-point sources of pollution from individual sub-basins through the mainstream watersheds towards its outlet resulting from changes in land use/cover in the river basins [21]. In general, SWAT simulates the hydrological cycle and water balance in the catchment using equation (1).

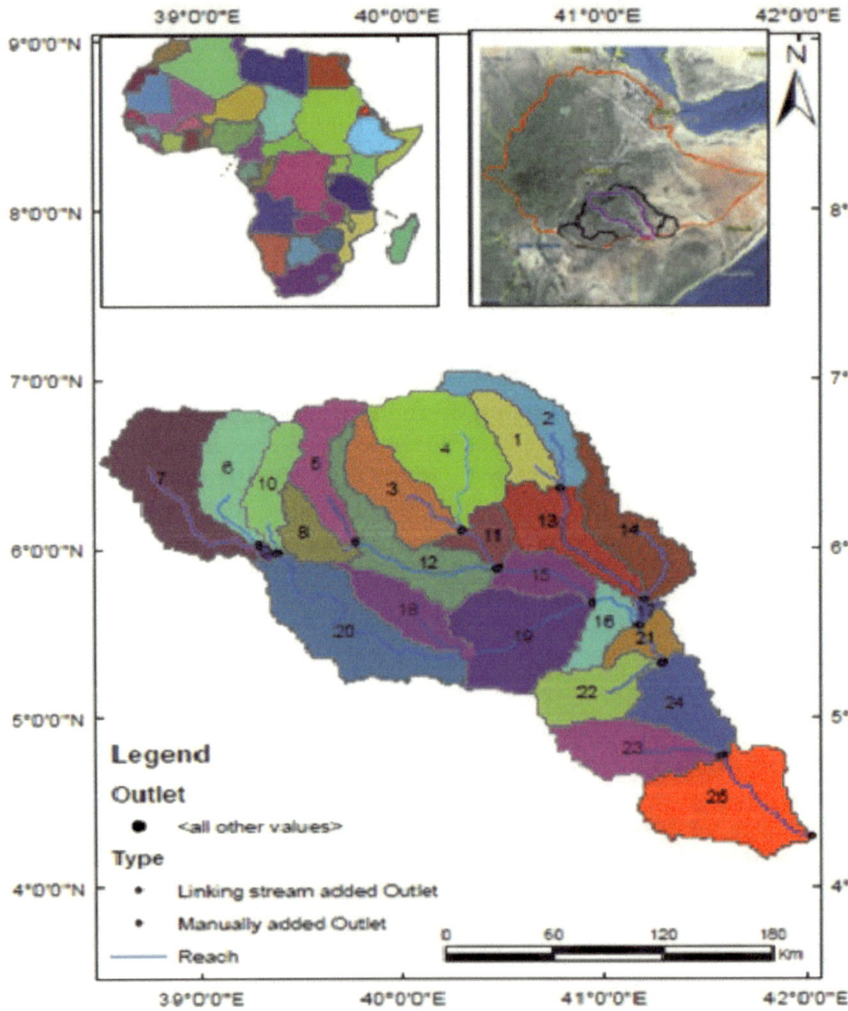

Figure 1.
Shows the delineated watershed of the study area extracted from the Africa, Ethiopia map.

$$SW_t = SW_o + \sum_{i=1}^{t}\left(R_{day} - Q_{surface} - E_a - W_{seep} - Q_{gw}\right) \quad (1)$$

Where; SWt = Final soil water content on a day i (mm/day), SWo = Initial soil water content on day i (mm/day), t = time in days, Rday = amount of precipitation on day i (mm/day),Qsurface = amount of surface runoff on day i (mm/day), Ea = amount of evapotranspiration on day i (mm/day), Wseep = amount of water entering the vadose zone from the soil profile on day i (mm/day), Qgw = amount of return flow on day i (mm/day). The SWAT uses the soil conservation service curve number (SCS-CN) approach to evaluate surface runoff, illustrating runoff to soil type, land use/cover, slope classes, and management practices, and is computationally effective [22]. The model estimates the streamflow in the sub-basins as a result of the total daily rainfall SCS- using the Soil Conservation Service curve number (CN) method as follows:

$$Q_{Surface} = \frac{\{R_{Day-0.2S}\}2}{\{R_{Day} + 0.8S\}} \quad (2)$$

The retention parameter(S) and prediction of lateral flow by SWAT model expressed as;

$$\mathrm{S} = 25.4(1000/\mathrm{CN} - 10) \quad (3)$$

Where; S = drainable volume of soil water per unit area of a saturated thickness (mm/day), CN = curve number.

The model's water yield within a watershed has been evaluated based on the equation; (Negewo & Sarma, 2021).

$$W_{YLD} = Q_{Surface} + BF - T_{Loss} = Q_{Surface} + Q_{GW} + Q_{LAT} - T_{Loss} \quad (4)$$

Where; W_{YLD} = water yield (mm), $= Q_{Surface}$ = surface runoff (mm), $+Q_{LAT}$ = lateral flow contribution to stream(mm), $+Q_{GW}$= groundwater contribution to streamflow (mm), and T_{Loss} = the transmission losses (mm) from tributary in the HRU through the bed.

For individual HRU, the sediment losses attributed to the surface runoff were evaluated based on the Modified Universal Soil Loss Equation (MUSLE) [23]. The MUSLE formula of sediment yield in the sub-basin roughly estimates the gross soil erosion caused by sheet, rill, and rain splash but does not include the erosion caused by landslides and gullies.

$$\mathrm{QSED} = 11.8 * \left(Q_{Peak} * Q_{Surface} * \mathrm{A_{hru}}\right) 0.56 * \mathrm{K} * \mathrm{C} * \mathrm{P} * \mathrm{LS} * \mathrm{CFRG} \quad (5)$$

Where; QSED = Sediment loss/Sediment yield(ton/ha/day) from individual HRU, $Q_{Surface}$ = surface runoff associated to HRU (mmH2O/ha/day), A_{hru} = Area of HRU in(ha), Q_{Peak} = peak flow rate(m^3/s), K_{USLE} = soil erodibility factor, C_{USLE} = Cover and management practice factor, P_{USLE} = Conservation support practice factors of land use, LS_{USLE} = Topographic factor, hill slope steepness factor/ the length slope factor, CFRG = coarse fragment factor.

Typically, the application of the SWAT contained five mains: (a) watershed delineation and streams network generation, (b) combination of DEM, soil data, and land use/cover data and create slopes classes, (c) creating HRU

(Hydrological response unit) definition, (d) combination of climate data (e) run the simulation (**Figure 2**).

2.3 Future climate change data

The statistically downscaled Regional Climate Model (RCM) Bias-corrected Coordinated Regional Climate Downscaling Experiment (CORDEX) precipitation, min/mean/max temperature for Ethiopia, under RCP 4.5 and RCP 8.5, downloaded from(https://dataservices.gfzpotsdam.de/pik/showshort.php?id=escidoc:3124935) is provided as input data for hydrological modeling of this study. This dataset contributes bias-corrected daily precipitation, min/mean/max air temperature of ten CORDEX RCM runs covering the country of Ethiopia for historical (1970–1999) and over the 21st century for RCP 4.5 and RCP 8.5 [24]. For this study, daily rainfall and maximum & minimum temperature data of historical (1990–2013) of eight climatic stations were obtained from the Meteorological Agency of Ethiopia, and other eight climatic stations were from the global database of Climate Forecast System Reanalysis (CFSR) after filling missing data, consistency, and outlier checked [1].

Accordingly, all the bias correction has improved the simulation of precipitation and temperature before using CORDEX-RCM outputs for any climate impact modeling. The study used climate model data for hydrological modeling CMhyd to extract CORDEX-NetCDF and bias correction of precipitation, minimum and maximum temperature to predict climate change-induced temperature changes in the Genale catchment.

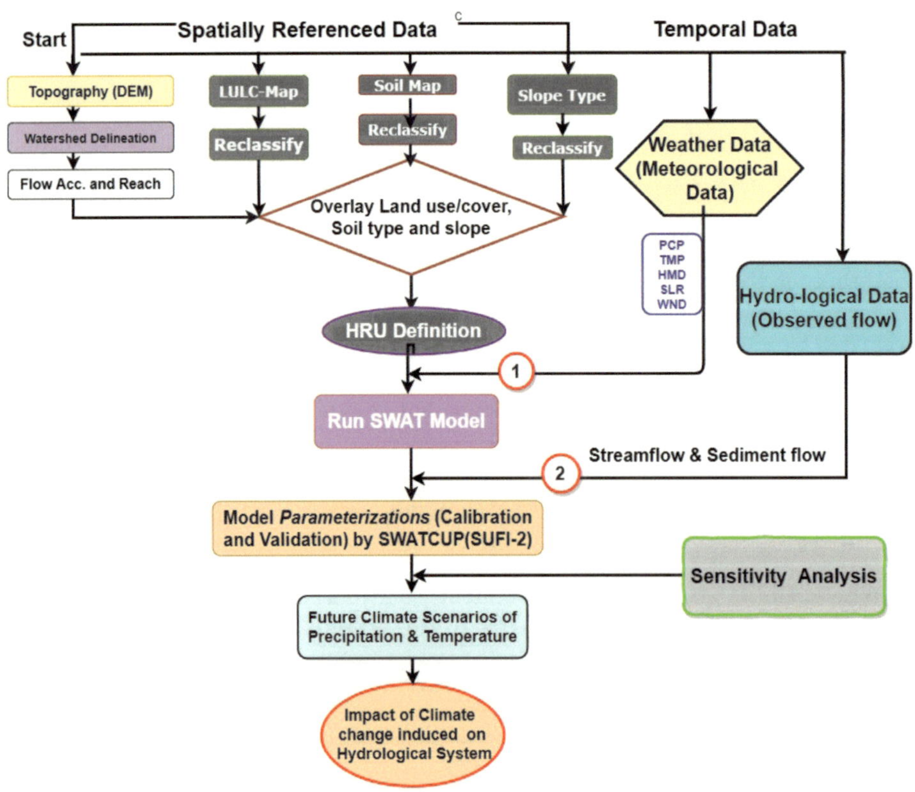

Figure 2.
Steps of the implementation of the SWAT model for the study area. Analysis of SWAT input data.

SWAT model was used to simulate water yield using the RCM under the future emission scenarios of two representative concentration pathways (RCPs) (medium emission scenario (RCP4.5) and high emission scenario (RCP-8.5)). Climate data for different periods are input into the SWAT model with the other components unchanged. The period of 1990–2013 is set as the baseline period.

Digital elevation model was downloaded from USGS Earth Explorer (http://earthexplorer.usgs. gov/) SRTM (Shuttle Radar Topography Mission) 90 m*90 m and used for watershed delineation, sub-basin, slope calculation/ classification, and extract stream networks. The spatial land use/cover collected from the Ethiopian Ministry of Water, Irrigation, and Electricity (MoWIE) GIS department of the year 2013 used for SWAT input, and the dominant land use/cover is range brushland (RNGB) accounts for about 71% of the area. This study's soil map/type is from the Food and Agricultural Organization (FAO) Digital Soil Map of the World (http://www.fao.org/geonetwork/srv/en/metadata) the scale of 1/5000000 for 2007. The soil data that integrated into the SWAT model are: the available water content, the texture, the hydraulic conductivity, the apparent density of the different soil layers, and the dominant soil type in the study watershed was Rc19-bc-204 (Calcaric Regosols), and it accounts about 40% of the catchment.

The climate data required for this paper has been taken from the National Meteorological Agency of Ethiopia, http://www.ethiomet.gov.et/etms. These data subsist of precipitation, max and min temperatures, wind energy, solar radiation, and relative humidity daily and covered the period from 1990 to 2013 for sixteen stations. The discharge data from the Ethiopian Ministry of Water, Irrigation, and Electricity (MoWIE) Hydrology department Genale @ Halwen gauging station a bit upstream of the outlet, and then transferred to the outlet, and arranged for SWAT language for the period from 1990 to 2013.

Figure 3, shows different 16 (sixteen) meteorological stations were distributed in the watershed, hydrological gauging station, watershed outlet, stream reach, and basin mark of the study area. The stations which were designated as; GMS1-Gridded Meteorological station-1, GMS2- Gridded Meteorological station-2, GMS3-Gridded Meteorological station-3, 4,5,6,7, & 8 respectively (**Figure 3**).

Figure 4 shows the distribution of rainfall in the study area for the selected different gauge stations.

From **Figure 5b**, the details of maximum, average, and minimum yearly temperature of the study area were pinpointed as 24.6, 19, & 12.93 °C, respectively, for 1990–2013 (**Figure 6**).

2.4 SWAT-CUP(SUFI-2) description

The automatic calibration and validation adjustment in the SWAT model achieved using the SWAT-CUP (SUFI-II) public user software developed by [25]. The SWAT-CUP has interfaced with five algorithms: (1) sequential uncertainty fitting (SUFI-2, (2) generalized likelihood uncertainty estimation (GLUE), (3) parameter solution (ParaSol), (4) Markov chain Monte Carlo (MCMC), and (5) particle swarm optimization (PSO) [26].

In this study, the analysis of uncertainty, calibration, validation was conducted using the SUFI-2 optimization algorithm; this algorithm needs less simulation number, faster, and one of the most used in the automatic calibration of model for several basins the semi-arid region like Genale Basin.

Assessment of performance criteria for the model is; Nash-Sutcliffe Efficiency (NSE), PBIAS, and Coefficient of Determination (R^2) has been used as the efficiency criteria to evaluate the performance of models in the Genale watershed.

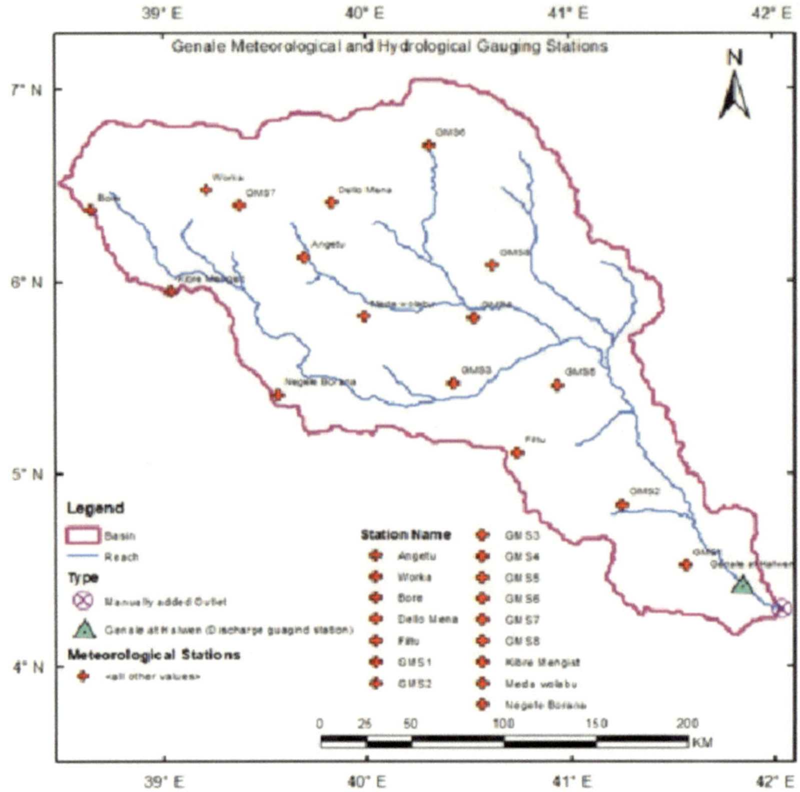

Figure 3.
Meteorological and hydrological gauging station distribution sub-basin wise for Genale River.

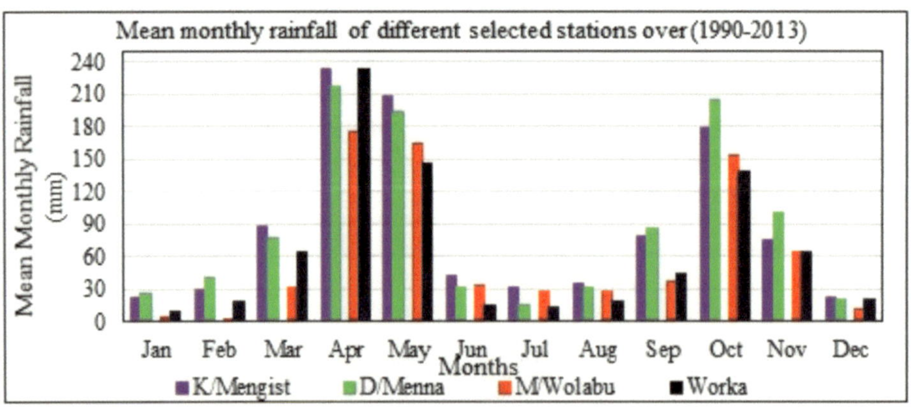

Figure 4.
Mean monthly rainfall for selected stations in the study area over 1990–2013.

The first three objective functions are mainly used for daily and monthly streamflow /sediment calibration–validation uncertainty analysis.

Coefficient of Determination (R^2)

$$R^2 = \frac{\left[\sum_{i=1}^{n}(Q_{si} - Q_{sm})(Q_{oi} - Q_{om})\right]^2}{\sum_{i=1}^{n}(Q_{si} - Q_{sm})^2 \sum_{i=1}^{n}(Q_{oi} - Q_{om})^2} \tag{6}$$

DOI: http://dx.doi.org/10.5772/intechopen.98515

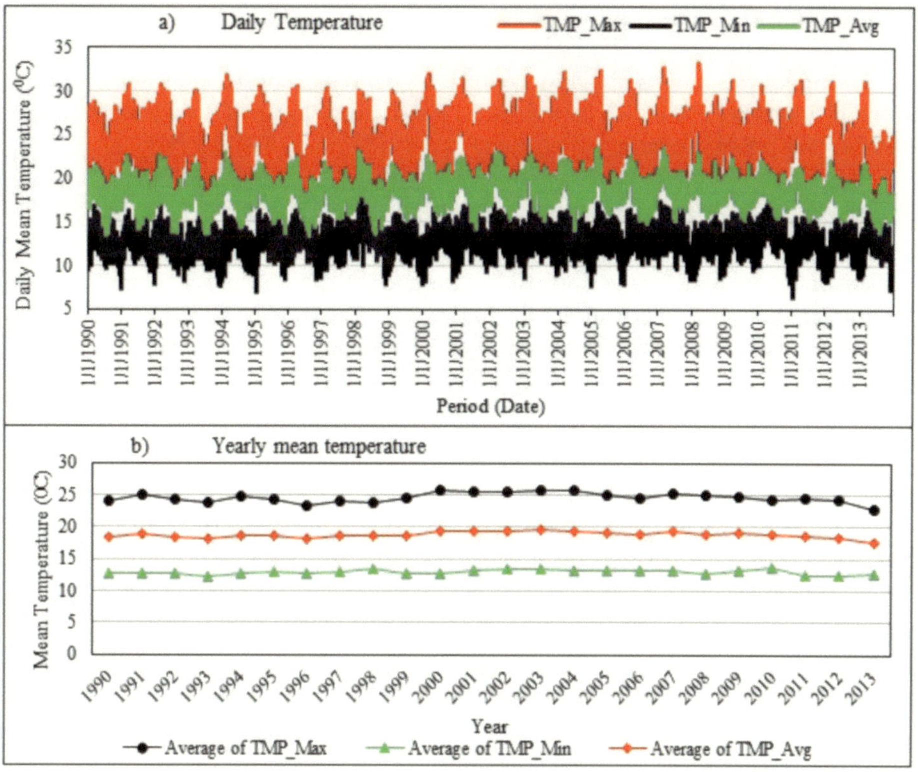

Figure 5.
Daily and yearly average maximum, minimum, and average daily temperatures in the study area, respectively.

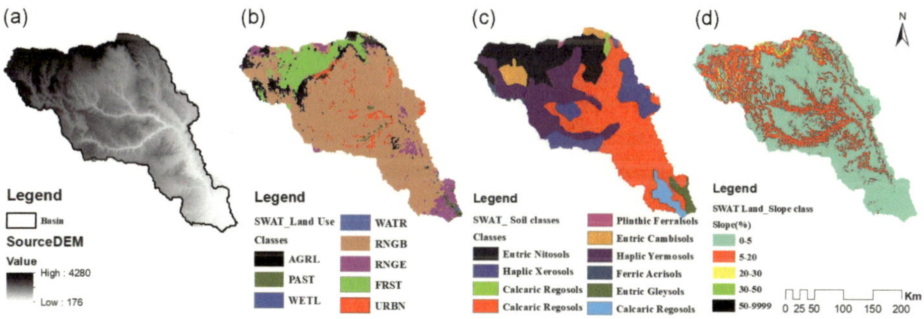

Figure 6.
DEM, LULC, soil type, and slope classes of the Genale watershed, respectively.

Where, Q_{si} is the simulated value, Q_{oi} is the measured value, Q_{om}, is the average observed value and Qsm is an average simulated value? Nash Sutcliffe Efficiency (NSE).

$$NSE = 1 - \frac{\sum_{i=1}^{n}(Q_{oi} - Q_{si})^2}{\sum_{i=1}^{n}(Q_{oi} - Q_{om})^2} \tag{7}$$

where, Q_{oi} is the observed, Q_{si} is the simulated, and Qom is the observed discharge.

$$\text{Percent Bias (PBIAS)}; PBIAS = 100\% \times \frac{\sum_{i=1}^{n}(Q_{oi} - Q_{si})}{\sum_{i=1}^{n}Q_{oi}} \tag{8}$$

3. Results and discussions

The model was built with DEM, soil classes, land use/cover, and slope types for the Genale watershed, which contained 25 sub-basins, 464 HRUs with a catchment area 54,942Km2 at the outlet.

3.1 Model parameter sensitivity analysis

Sensitivity analysis was performed to navigate the calibration action and pinpoint parameters that significantly affect the discharge and sediment flow. In a sensitivity analysis of the model, SCS curve number (CN2.mgt), an available water capacity of the soil layer (SOL_AWC.sol), and saturated hydraulic conductivity (SOL_K.sol) are the most sensitive parameters for runoff estimation. However, model efficiency is also influenced by the reliability of spatial and temporal data.

Sediment sensitivity analysis was carried out for three years warm-up period 1987 to 1989- and 16-years calibration period 1990 to 2005- and 8-years validation period 2006 to 2013. Based on the p-value and t-stat results obtained from sensitivity analysis, the ranks of parameters were finalized. The simulated sediment was sensitive to the amount of sediment re-entrained during channel sediment routing (SPCON.bsn), (SOL_AWC.sol), CN2, etc., respectively.

A parameter with a larger absolute value of t-stat is more sensitive to flow. The p-value gives the relevance of the sensitivity. Thus, when the p-value is close to zero, then the sensitivity of the parameter is a priority (**Table 1**).

3.2 Calibration/validation

Calibrated parameters and the fitted values are final notes for the modeler from the calibration process used for the required objectives. Calibration of discharge and sediment flow was performed with several iterations of 500 simulations number; each was carried out for the calibration period of 1990–2005 monthly.

Validation is required to verify whether the calibrated parameters also work for other data of different years within the watershed. Validation time (2006–2013) results revealed a satisfactory performance, as statistical measures are in the acceptable range for discharge and sediment. **Table 2** shows the acceptable range for the model's performance in light of the calibration and validation process.

The results show satisfactory and well responded to calibration and validation process (**Table 3**).

The calibration was done from 1990 to 2005 & the validation period from 2006 to 2013, and the model performance shows satisfactory agreement between the observed and simulated flow (**Figure** 7). The calibrated/validated model also responded to the rainfall with the respective months.

As indicated, the simulated and observed sediment load agreed and showed a satisfactory performance during the calibration and validation action (**Figure 8**).

3.3 Impact of climate change in the watershed

3.3.1 Climate change impacts on temperature and precipitation

The climate change impact on hydrology was evaluated by driving the calibrated/validated SWAT model with the bias-corrected RCM-CORDEX weather corresponding to the present-day historical data and future emission scenarios. The analysis was executed on a monthly basis for streamflow and sediment yield.

Process	Parameter name	Description of the parameter	Range value	Fitted value	p-value	t-stat	Rank
Streamflow	CN2.mgt	SCS runoff curve number	35–98	−0.17	0.0	−42	1
	SOL_AWC.sol	Available water capacity of the soil layer	0–1	1.0	0.004	2.9	2
	SOL_K.sol	Saturated hydraulic conductivity	0–2000	0.566	0.12	−1.5	3
	SOL_BD.sol	Moist bulk density	0.9–2.5	0.984	0.20	1.2	4
	ALPHA_BF.gw	Baseflow alpha-factor (days).	0–1	0.570	0.21	−1.2	5
	REVAPMN.gw	Threshold depth of water in a shallow aquifer for "revap" to occur (mm)	0–500	408.6	0.308	−1.0	6
	GW_REVAP.gw	USLE support practice factor	0–1	1.2	0.49	−0.6	7
	ESCO.hru	Soil evaporation compensation factor	0–1	0.27	0.65	−0.4	8
	HRU_SLP.hru	Average slope steepness	0–1	0.578	0.72	0.34	9
	SURLAG.bsn	Surface runoff lag time	0.05–24	0.072	0.96	−0.05	10
Sediment	SPCON.bsn	The max amount of sediment that can be retrained during channel routing.	0.0001–0.01	0.0002	0.0	−29.5	1
	SOL_AWC(.) sol	Available water capacity of the soil layer	0–1	0.639	0.0	14.2	2
	CN2.mgt	SCS runoff curve number	35–98	−0.24	0.0	−10	3
	SOL_K(..).sol	Saturated hydraulic conductivity	0–2000	0.845	0.0	7.18	4
	SPEXP.bsn	Exponent parameter for calculating sediment retrained in channel sediment routing.	1–1.5	1.156	0.0	−5.63	5
	CH_COV1.rte	Channel erodibility factor.	−0.05-0.6	0.78	0.145	−1.44	6
	USLE_K(..)sol	USLE equation soil erodibility (K) factor.	0–0.65	0.012	0.57	−0.6	7
	USLE_P.mgt	USLE equation support parameter	0–1	0.029	0.73	0.35	8

Table 1.
Fitted values and rank of parameters used in the SWAT model calibration/validation (1998–2012).

The statistically downscaled Regional Climate Model (RCM) Bias-corrected Coordinated Regional Climate Downscaling Experiment (CORDEX), precipitation, min/mean/max temperature for Africa-Ethiopia, under RCP 4.5 and RCP 8.5. The average annual rainfall in the study climate stations during the baseline 24-years period (1990–2013) was 810 mm, and the maximum and minimum yearly rainfall accounts were 1,303 mm and 300 mm, respectively.

The monthly temperature of the catchment varies from 14.5°C to 24.6°C, with an average of 19.5°C. We predicted the long-term average precipitation with the historical data for two climate emission scenarios. As shown in **Figure 9**, significant changes occur in the dry season (December, January & February).

p-factor	r-factor	R2	NSE	PBIAS		RSR	Rating
				Flow	Sediment		
0.7–1	<1, (close to 0)	0.75–1	0.75–1	<±10%	<±15%	0–0.5	very good
		0.65–0.75	0.65–0.75	±10–15%	±15–30%	0.5–0.6	good
		0.5–0.65	0.5–0.65	±15–25%	±30–55%	0.6–0.7	satisfactory
Close to 0	>1, (infinite)	<0.5	≤0.5	> ± 25%	> ± 55%	>0.7	unsatisfactory

Table 2.
SWAT statistical performance index acceptable range [25, 26].

Types of assessment		p-factor	r-factor	R2	NSE	PBIAS	RSR	Rating
Flow	Calibration	0.51	0.78	0.87	0.81	−2.1%	0.50	good
	Validation	0.54	0.86	0.85	0.78	−0.5%	0.52	good
Sediment	Calibration	0.48	0.37	0.84	0.79	3.8%	0.61	satisfactory
	Validation	0.43	0.39	0.82	0.75	3.9%	0.67	satisfactory

Table 3.
Actual index value for SWAT output during calibration/validation process (1990–2013).

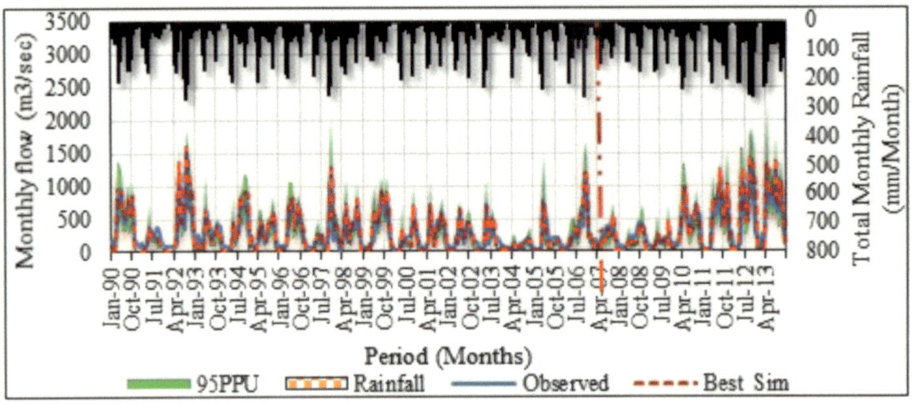

Figure 7.
Monthly calibration and validation of streamflow (1990–2013) for Genale River basin at Genale Halwen.

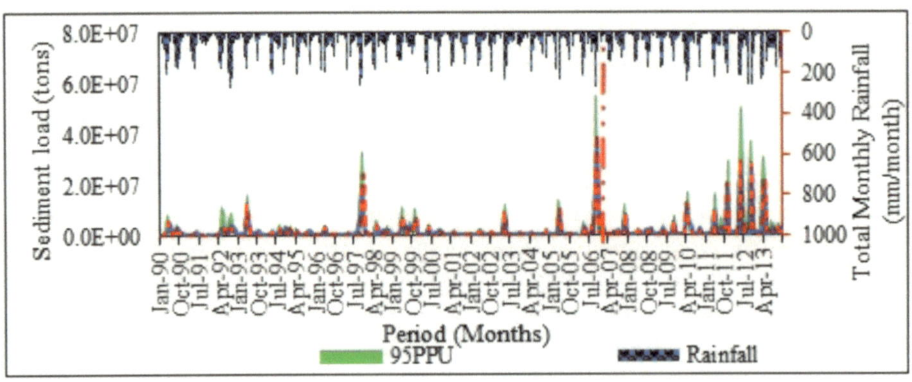

Figure 8.
Monthly observed and simulated sediment load plots for the calibration (1990–2005) and validation (2006–2013).

Figures 9 and **10** show the climate changes in the average monthly precipitation and the maximum and minimum air temperatures over the catchment between the historical and future periods (2022–2080) for the two emission scenarios. Generally, the climate change over the Genale basin will likely become warmer, especially in autumn and spring, considering the higher emission scenario (**Figure 10**).

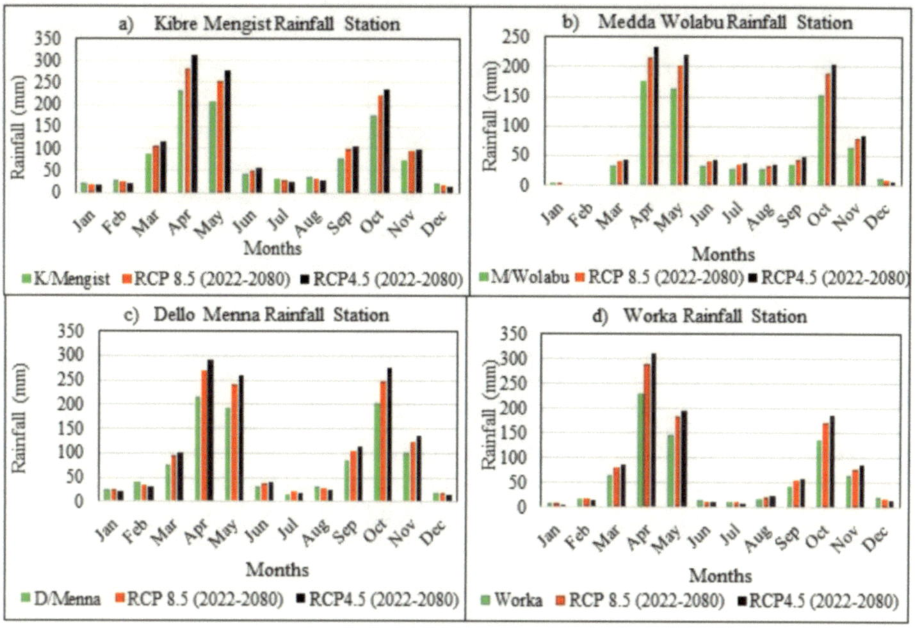

Figure 9.
Comparison of average observed monthly precipitation for baseline condition, RCP4.5, & RCP8.5 scenarios of four stations in the catchment.

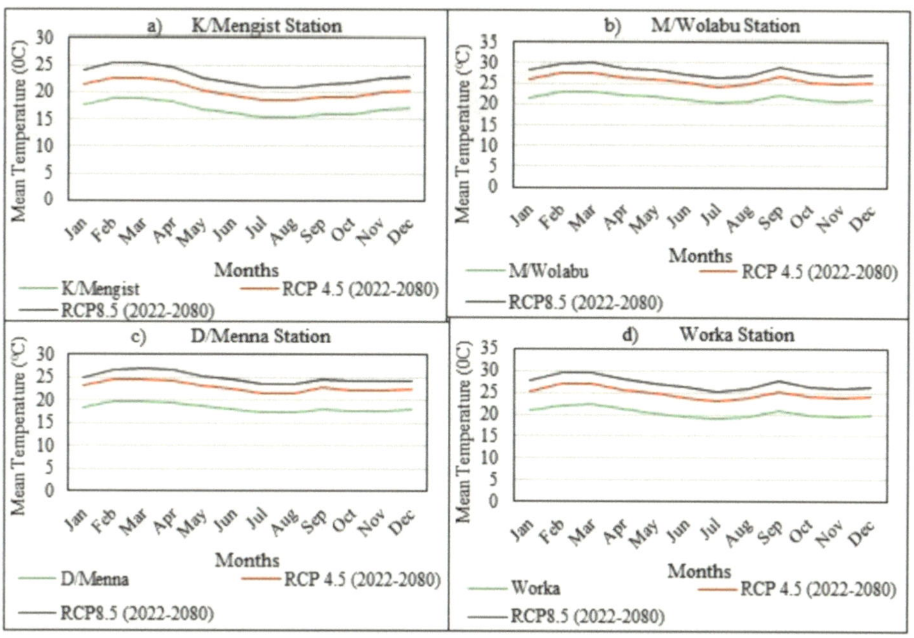

Figure 10.
Comparison of mean temperatures for historical data, RCP4.5, & RCP8.5 scenarios of four stations in the catchment.

Indistinct, the maximum temperature increase is somewhat higher than that of the minimum temperature in the region.

Figure 10 shows an average of mean monthly changes in temperatures in the study watershed, and it is increasing under emission scenarios of RCP4.5 and RCP8.5.

3.3.2 Climate change impact on streamflow and sediment

The bias-corrected rainfall and maximum/minimum temperature outputs were used as inputs to the calibrated/validated SWAT model to examine the Genale catchment streamflow and sediment yield responses in the future years. The climate-induced discharge changes are understood by assessing differences produced by the SWAT model when driven by future scenarios and present-day climates. A similar study by Negewo & Sarma (2021) for the Genale watershed revealed that the mean annual quantity of water resources is possible to increase under RCP4.5, but variations are substantial for individual sub-basins and HRUs. The study results reflected that climate change might increase the high flows in the catchment in the Autumn season (April, May, June) and Spring season (September, October & November) (**Figure 11**).

Monthly discrepancy showed that the increase in discharge is more pronounced in March, April, May, August, September under RCP4.5, and the decrease is more pronounced in the same months under RCP8.5 scenarios (**Figure 11**). The average monthly change of streamflow for the RCP4.5 and RCP8.5 was running from −16.47% to 6.58% and − 3.6% to 8.27%, respectively, of 2022–2080(**Figure 12**).

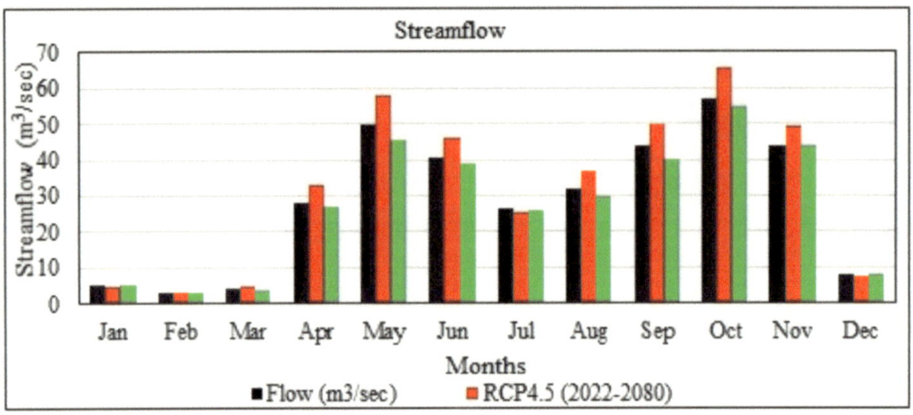

Figure 11.
Streamflow variations under historical data and two emissions scenarios.

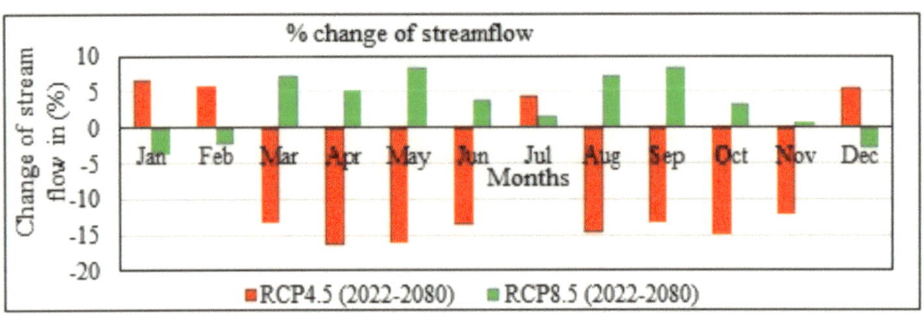

Figure 12.
Predicted relative changes (percent of baseline levels) in monthly streamflow by RCP4.5 & RCP8.5.

The change in monthly streamflow is consistent with the predicted changes in rainfall and temperature patterns in the future period.

The change patterns of sediment yield follow that of streamflow in the region. Prediction of RCM showed an increase in sediment yield for RCP4.5 and slightly decreased for the RCP8.5 scenario (**Figure 13**). Monthly variation showed that the magnitude changes in sediment yield in March, April & June was the highest with values of −21.8%, −15.0%, −15.0% and − 13.7% respectively for RCP4.5 scenario, and slightly lower in January, March, November, September & December with values of −3.5%, 1.5%, 2.1%, 2.1% & 2.2% respectively for RCP8.5 scenario (**Figure 14**). It should be recognized that the maximum increase in heavy rainfall and extreme events was also predicted in the respective months. Hence, the corresponding change predicted by the model is reasonable. The monthly average changes in sediment yield for the RCP4.5 and RCP8.5 scenarios were − 21.8% to 6.2% and − 5.6% to 4.66%, respectively, over 2022–2080 (**Figure 14**).

The increase in change (percentage) of sediment yield is more significant than discharge, implying that the sediment concentration in the Genale catchment will likely increase in the future periods under the RCP4.5 scenario.

From the spatial distribution of the sediment yield for the baseline & two emissions scenarios periods (**Figure 15**), the high-sediment-yield regions are mainly

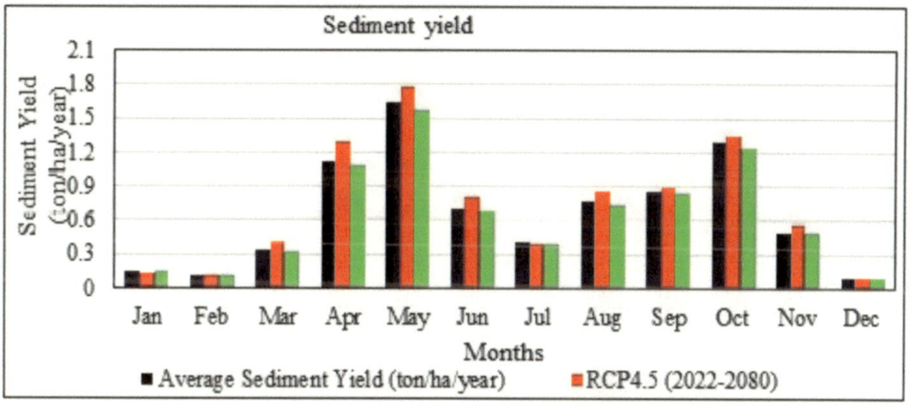

Figure 13.
Sediment yield variations for baseline and two GHGs scenarios.

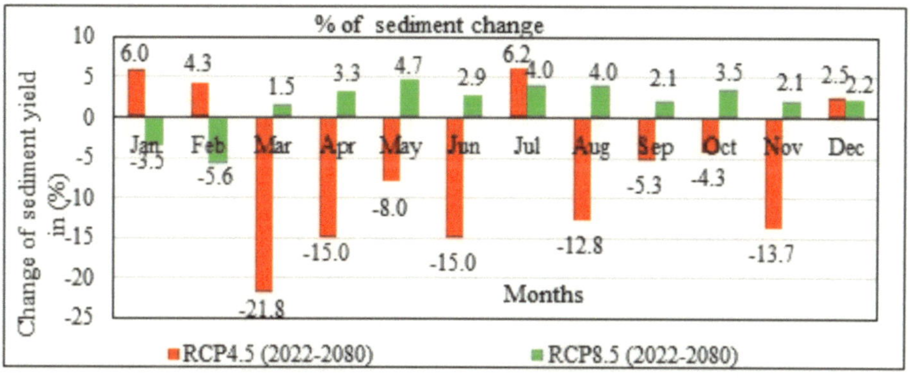

Figure 14.
Predicted relative monthly changes in sediment yield in future periods for two emission scenarios compared to baseline.

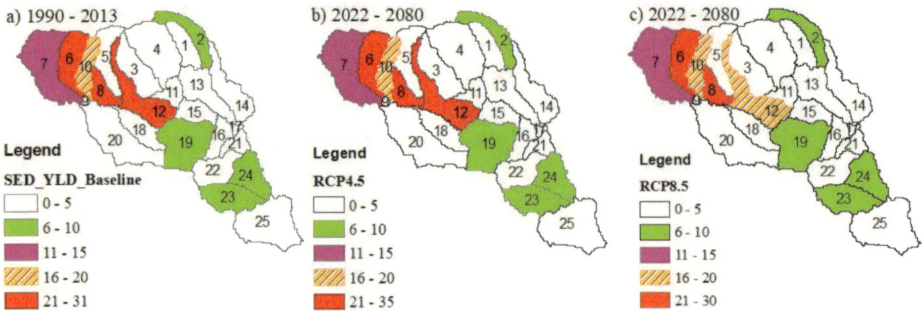

Figure 15.
Annual average spatial distribution of sediment yield (ton/ha/year) at sub-basin scale in historical data & future periods under baseline, RCP4.5& RCP8.5.

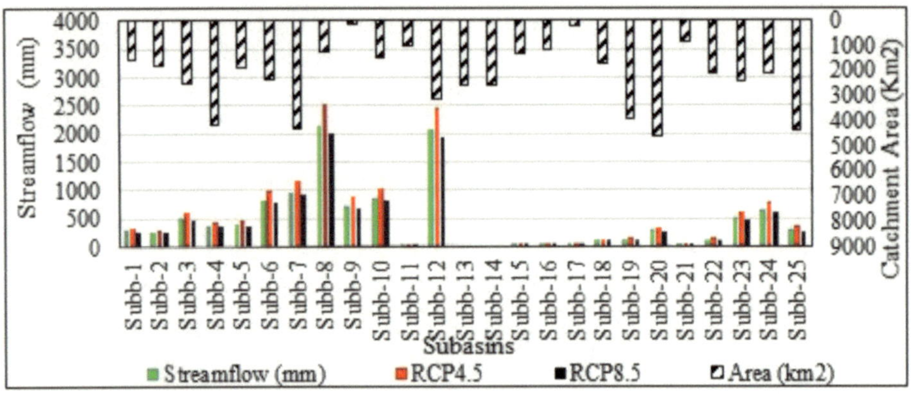

Figure 16.
Annual average spatial distribution of streamflow (mm) at sub-basin scale in historical data & future periods under baseline and two emission scenarios.

located in the upstream regions of the catchment for all cases, in which the sediment yield varies from 0 to 31 (baseline condition), from 0 to 35 (under RCP4.5) & from 0 to 30 (under RCP8.5) over 1990–2013, 2022–2080, 2022–2080 respectively.

Irrespective of the catchment area, the spatial distribution of streamflow at the sub-basin level follows the trend of sediment yield patterns in historical data (1990–2013) and future periods (2022–2080) under baseline conditions & two emissions scenarios (RCP4.5 & RCP8.5) (**Figure 16**).

From the results, a comparison of monthly stream discharge & sediment yield predictions for 2022–2080 indicated that the impact of climate changes induced on sediment yield is more significant than on streamflow under the two emission scenarios.

4. Conclusions

The study evaluated the impact of climate change-induced on the sediment yield and streamflow of the Genale catchment, Ethiopia, for the medium-future period 2022–2080 under the RCP 4.5 and RCP 8.5 emission scenarios. The SWAT hydrological model was applied to simulate discharge and sediment yield, and the SUFI-2 algorithm technique in the SWAT-CUP tool was used for parameterization. The process of uncertainty analysis, calibration (1990–2005), & validation (2006–2013) for both discharge and sediment were satisfactory. The sensitivity analysis enabled that the SCS curve number (CN2.mgt), an available water capacity of the soil layer (SOL_AWC.sol), and saturated hydraulic conductivity (SOL_K.sol) are the most sensitive parameters for runoff estimation.

The study used the change of climate scenarios built up using the outcomes bias-corrected CORDEX RCM daily precipitation, min/mean/max temperature for Ethiopia under RCP 4.5 and RCP 8.5 emission scenarios and fed them into the validated SWAT model to simulate future changes in streamflow and sediment yields due to change of climate. The average monthly change of streamflow for the RCP4.5 and RCP8.5 was running from −16.47% to 6.58% and − 3.6% to 8.27%, respectively, of 2022–2080. The monthly average changes in sediment yield for the RCP4.5 and RCP8.5 scenarios were − 21.8% to 6.2% and − 5.6% to 4.66%, respectively, over 2022–2080. The monthly average discharge varies significantly throughout the year and relatively high in March, April, May, August, September, and October. The monthly streamflow and sediment yield variations were more during the wet seasons (Autumn and Spring). The results revealed that the impact of climate changes induced on sediment yield is more significant than streamflow under the two emission scenarios for 2022–2080.

The results revealed that regional decision-makers and other stakeholders are helpful for the effective adaptive strategy, plan & management practices of soil and water resources improvement under changing climate.

Acknowledgements

The authors express special thanks to the Ministry of Water, Irrigation, and Electricity (MWIE), Ethiopia, Department of Hydrology, and National Meteorological Agency of Ethiopia for supporting streamflow, sediment concentration, and weather data of Genale catchment based on the request. Finally, we are grateful for the invitation of the IntechOpen publisher to contribute a chapter to "Global Warming and Climate Change," book, which extensively fills the knowledge gap of the readership.

Conflict of interest

The authors declare no conflict of interest.

Author details

Tufa Feyissa Negewo* and Arup Kumar Sarma
Department of Civil Engineering, Indian Institute of Technology Guwahati, Assam, India

*Address all correspondence to: tufa1886@gmail.com; tufa@iitg.ac.in

IntechOpen

References

[1] Negewo, T. F., & Sarma, A. K. (2021). Estimation of water yield under baseline and future climate change scenarios in Genale watershed, Genale Dawa River basin, Ethiopia, using SWAT model. Journal of Hydrologic Engineering, 26(3), 05020051.

[2] Dariane, A. B., & Pouryafar, E. (2021). Quantifying and projecting of the relative impacts of climate change and direct human activities on streamflow fluctuations. Climatic Change, 165(1), 1-20.

[3] IPCC, 2007. Climate Change 2007: Impacts, Adaptation, and Vulnerability. Contribution of Working Group II to the Fourth Assessment Report of the Intergovernmental Panel on Climate Change. Cambridge University Press, Cambridge, U.K.

[4] Azari, M., Moradi, H. R., Saghafian, B., & Faramarzi, M. (2016). Climate change impacts on streamflow and sediment yield in the north of Iran. Hydrological Sciences Journal, 61(1), 123-133.

[5] Nilawar, A. P., & Waikar, M. L. (2019). Impacts of climate change on streamflow and sediment concentration under RCP 4.5 and 8.5: A case study in Purna river basin, India. Science of the total environment, 650, 2685-2696.

[6] Chen, Y., Marek, G. W., Marek, T. H., Porter, D. O., Moorhead, J. E., Wang, Q., … & Brauer, D. K. (2020). Spatio-temporal analysis of historical and future climate data in the Texas High Plains. Sustainability, 12(15), 6036.

[7] Dlamini, N. S., Kamal, M. R., Soom, M. A. B. M., Mohd, M. S. F. B., Abdullah, A. F. B., & Hin, L. S. (2017). Modeling potential impacts of climate change on streamflow using projections of the 5th assessment report for the Bernam River basin, Malaysia. Water, 9 (3), 226.

[8] Worku, G., Teferi, E., Bantider, A., & Dile, Y. T. (2020). Statistical bias correction of regional climate model simulations for climate change projection in the Jemma sub-basin, upper Blue Nile Basin of Ethiopia. Theoretical and Applied Climatology, 139(3), 1569-1588.

[9] Gebrechorkos, S. H., Hülsmann, S., & Bernhofer, C. (2019). Statistically downscaled climate dataset for East Africa. Scientific data, 6(1), 1-8.

[10] Wang, X., Li, Z., & Li, M. (2018). Impacts of climate change on stream flow and water quality in a drinking water source area, northern China. Environmental Earth Sciences, 77(11), 1-14.

[11] Jasrotia, A. S., Baru, D., Kour, R., Ahmad, S., & Kour, K. (2021). Hydrological modeling to simulate stream flow under changing climate conditions in Jhelum catchment, western Himalaya. Journal of Hydrology, 593, 125887.

[12] Muttiah, R. S., & Wurbs, R. A. (2002). Scale-dependent soil and climate variability effects on watershed water balance of the SWAT model. Journal of hydrology, 256(3–4), 264-285.

[13] Immerzeel, W. W., Pellicciotti, F., & Bierkens, M. F. P. (2013). Rising river flows throughout the twenty-first century in two Himalayan glacierized watersheds. Nature geoscience, 6(9), 742-745.

[14] Tian, P., Lu, H., Feng, W., Guan, Y., & Xue, Y. (2020). Large decrease in streamflow and sediment load of Qinghai–Tibetan plateau driven by future climate change: A case study in Lhasa River basin. Catena, 187, 104340.

[15] Tan, M. L., Ficklin, D. L., Ibrahim, A. L., & Yusop, Z. (2014). Impacts and uncertainties of climate change on streamflow of the Johor River basin, Malaysia using a CMIP5 general circulation model ensemble. Journal of Water and Climate Change, 5(4), 676-695.

[16] Mahmoodi, N., Wagner, P. D., Kiesel, J., & Fohrer, N. (2020). Modeling the Impact of Climate Change on Streamflow and Major Hydrological Components of an Iranian Wadi System. Journal of Water and Climate Change.

[17] Fereidoon, M., & Koch, M. (2018). SWAT-MODSIM-PSO optimization of multi-crop planning in the Karkheh River basin, Iran, under the impacts of climate change. Science of the Total Environment, 630, 502-516.

[18] Jha, M., Pan, Z., Takle, E. S., & Gu, R. (2004). Impacts of climate change on streamflow in the upper Mississippi River basin: A regional climate model perspective. Journal of Geophysical Research: Atmospheres, 109(D9).

[19] Navarro-Racines, C., Tarapues, J., Thornton, P., Jarvis, A., & Ramirez-Villegas, J. (2020). High-resolution and bias-corrected CMIP5 projections for climate change impact assessments. Scientific data, 7(1), 1-14.

[20] Bekele, W. T., Haile, A. T., & Rientjes, T. (2021). Impact of Climate Change on the Streamflow of the Arjo-Didessa Catchment under RCP Scenarios. Journal of Water and Climate Change.

[21] Arnold, J. G., Srinivasan, R., Ramanarayanan, T. S., & Bednarz, S. T. (1998). Large area hydrologic modeling and assessment part II: Model application 1. JAWRA Journal of the American Water Resources Association, 34(1), 91-101.

[22] Arnold, J. G., Williams, J. R., & Maidment, D. R. (1995). Continuous-time water and sediment-routing model for large basins. Journal of Hydraulic engineering, 121(2), 171-183.

[23] Bonumá, N. B., Rossi, C. G., Arnold, J. G., Reichert, J. M., Minella, J. P., Allen, P. M., & Volk, M. (2014). Simulating landscape sediment transport capacity by using a modified SWAT model. Journal of Environmental Quality, 43(1), 55-66.

[24] Liersch, Stefan; Rust, Henning; Dobler, Andreas; Kruschke, Tim; Fischer, Madlen (2018): Bias-corrected CORDEX precipitation, min/mean/max temperature for Ethiopia, RCP 4.5 and RCP 8.5. GFZ Data Services. https://doi.org/10.5880/PIK.2018.009

[25] Abbaspour, K. C., Vejdani, M., Haghighat, S., & Yang, J. (2007, December). SWAT-CUP calibration and uncertainty programs for SWAT. In MODSIM 2007, an international congress on modeling and simulation, modeling, and simulation society of Australia and New Zealand (pp. 1596–1602).

[26] Moriasi, D. N., Arnold, J. G., Van Liew, M. W., Bingner, R. L., Harmel, R. D., & Veith, T. L. (2007). Model evaluation guidelines for systematic quantification of accuracy in watershed simulations. Transactions of the ASABE, 50(3), 885–900. Tical documentation version 2009. Texas Water Resources Institute.

Section 5

Climate Change, Humans and Insects

Chapter 20

The Energy and Climate Change Nexus in Uganda: Policy Challenges and Opportunities for Climate Compatible Development

Revocatus Twinomuhangi, Arthur Martin Kato and Adam M. Sebbit

Abstract

Although Uganda has abundant energy resources including hydropower, oil and gas, biomass, geothermal, and solar energy, energy poverty is still very high and constrains socio-economic transformation. Biomass energy accounts for approx. 88% of the energy mix and only up to 28% of the country population have access to electricity, and the two energy sources are climate sensitive. The reliance on biomass energy is a driver to deforestation and forest degradation that also reduces the country's resilience to climate hazards such as flooding, drought and landslides. Besides, deforestation is driver to greenhouse gas (GHG) emissions and adversely affects the delivery of ecosystem services. Uganda is also warming very first and rainfall patterns are becoming more variable. Coupled with increasing occurrence and severity of drought, intense rainfall, flooding and landslides, energy supply systems are becoming more vulnerable. While Uganda is currently not a major emitter of GHG, emissions will rise significantly in the future given the country's rapidly growing population and urbanization that are increasing demand for energy and exacerbated by ongoing oil and gas development. Therefore, as Uganda strives to attain a middle-income status country, building climate resilient and transiting to decarbonized energy systems is not only a necessity but transformational to reducing energy poverty, increasing access to clean and affordable energy services, spurring investment and economic growth, job creation, improved health and poverty reduction. In this chapter, we examine the nexus between energy and climate change in Uganda, focusing on energy as both a driver and victim of climate change while at the same time exploring opportunities for achieving enhanced access to affordable, reliable and clean energy as a contribution to sustainable, green and resilient development.

Keywords: energy, climate change, vulnerability, sustainable development, Uganda

1. Introduction

Energy and climate change are important topics in the sustainable development discourse globally and in Uganda. Indeed, the two touch on almost all sectors and achievement socio-economic and environmental sustainability. Excitingly, the two topics are closely linked to the extent that one cannot be exhaustively discussed without reference to the other. Energy is driver of climate change because it is the main source of greenhouse gases (GHG) that are responsible for global warming and climate change. But still energy is a victim of climate change because energy supply systems are sensitive to the impacts of climate change that affect the accessibility, reliability and affordability of energy services. Therefore, while increased investment in sustainable and clean energy systems can be a solution to the global climate change challenge, addressing climate risk can increase the reliability and resilience of energy supply systems.

While Uganda's contribution to global GHG emissions is still very low (given that fossil fuel consumption is low) the country is highly vulnerable to climate change, ranking 166th out of 181 countries in the 2019 ND-GAIN index[1] which makes it the 12th most vulnerable and 36th least ready country [1] to adapt to climate change. Energy poverty is high and the main energy related GHG emissions in the country are caused by overreliance on biomass energy which is also a driver to deforestation land degradation and in-door air pollution. However, the rising population, rapid urbanization and ongoing oil and gas development point to a significant increase GHG emissions in the future.

Uganda's long-term development agenda aims at socio-economic transformation and attainment of an upper middle-income country status by 2040 [2] and to achieve this the country needs increased access to clean, affordable, and reliable energy. But Uganda also seeks to 'pursue a green growth and resilient development path' in line with its national development agenda and international development commitments, and particular the implementation of the Paris Agreement on Climate Change but this will not be achieved unless increased investments are made in green and resilient energy systems. In this chapter, we explore the nexus between energy and climate change in Uganda and its implications to achieving green, resilient and sustainable development.

2. Uganda country profile

Uganda is a landlocked country in East Africa across the equator, located between longitudes 29° 34′ and 35° 0′ East, and latitudes 4° 12′ North and 1° 29′ South. It covers an area of 241,550 km^2 of which 41,743 km^2 (17.2%) is open water and swamps, and the land area is 199,807 km^2 [3]. The terrain is mostly plateau raging between 1000 and 2, 500 meters above sea level (a.s.l.) averaging about 1,200 meters a.s.l. The lowest point is 620 metres (in the Albert Nile) and the highest point is at 5100 meters (the peak of Mt. Rwenzori) [4]. The main mountain areas are Mt. Rwenzori, permanently snow-capped mountain located in Western Uganda marking the border with the Democratic Republic of Congo (DRC); Mt. Elgon (4,321 m a.s.l), an extinct volcano in the eastern part of the country at the border with Kenya, and Mt. Muhavura (4,127 m a.s.l.) in the south-west bordering Rwanda and D.R.C.

Uganda is generally fertile, and well-watered country with many lakes and rivers. The most important water bodies include the Nile River, one of the world's longest, and whose source is Lake Victoria, the largest lake in Africa. Other major lakes are Kyoga, Albert, Edward and George. Uganda has a pleasant tropical climate, with moderate temperatures and the rainfall is more regular with a bimodal

[1] ND-GAIN index summarizes a country's vulnerability to climate change and other global challenges in combination with readiness to improve resilience. https://gain.nd.edu/our-work/country-index/

rainfall distribution. The main rainy seasons are from March to May and October to December, although though the northern part has a single rainy season from March to mid-October. Mean daily temperatures are 28°C and the long-term mean near-surface temperature is 21°C. The highest temperatures are observed in the north, especially the north-east, while lower temperatures occur in the south, and these mild conditions make climate one of Uganda's most valuable natural resources [4, 5]. However, the climate is becoming more variable and changing i.e. temperatures are rising, rainfall is becoming more erratic and unreliable, and extreme weather events are on the rise. It is widely expected that the impacts of climate change will be felt in varying degrees across all the country's sectors and regions [6].

Uganda's population is still small, estimated to be 41 million (Uganda Bureau of Statistics - [7]) of which 51% are female and 49% are male, and the life expectancy at birth at 63.1 years having increased from 50.4 years in 2002 [8]. The highest proportion of the country's population (54%) is young, under 18 years, and 73% of population living in rural areas while only 27% is urban [7]. The challenge is that the population is growing very rapidly at 3% per annum, making it one of the world's fastest growing populations and [9], and the political leadership (government) seems to be happy with this population growth trend. At this growth rate, it is projected that Uganda's population will reach 57 million and 72 million by 2030 and 2049 respectively [10]. While the proportion of the urban population is still small, the country is also urbanizing very fast with one of the world's' highest annual urban population growth rates at more 5% [11]. Kampala, the capital is also largest city in the country, with the Greater Kampala Metropolitan Area accounting for more than 60% of the country's GDP.

Uganda is a Least Developed Country (LDC), and indeed a highly indebted poor country (HIPC), with per capita GDP of USD 794.3 [12], and annual growth

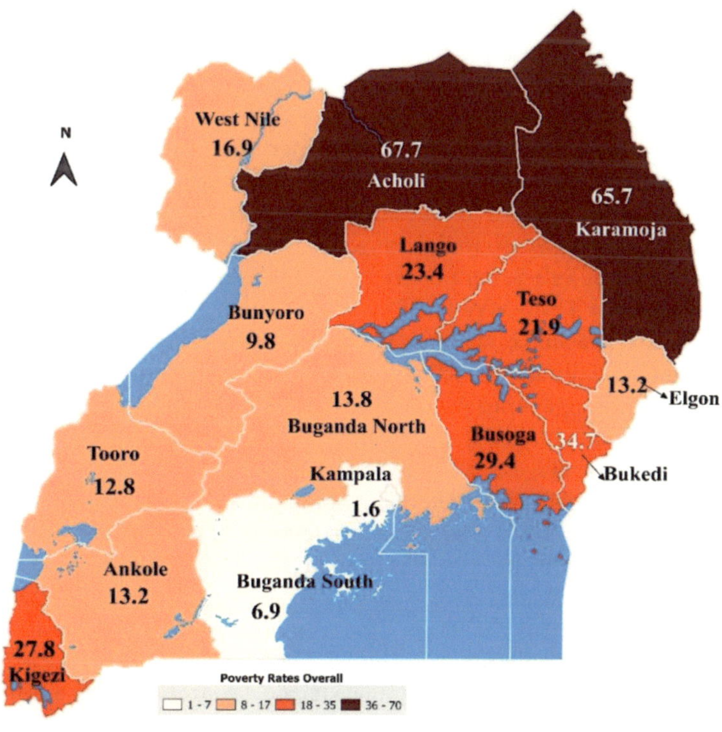

Figure 1.
Proportion of poor people in Uganda – Percentage (source: Uganda Bureau of Statistics - [7]).

was 2.9% in 2020, less than half the 6.8% recorded in 2019, due to the effects of the COVID-19 pandemic on the economy [13] and the debt to GDP ratio is very high 50.09% at in mid 2021. As at 2019, the population living below the poverty line was still high at 20.3%, with approx. 1.3 people living in absolute poverty [7]. The highest poverty rates are in northern Uganda in Acholi sub-region (67.7%) and Karamoja sub-region (65.7%), and the lowest is in the central region i.e. Kampala (1.6%) and Buganda (6.9%), (see **Figure 1**).

The country's long-term goal, articulated in the Uganda Vision 2040 [2], is 'transforming Ugandan society from a peasant to a modern and prosperous country within 30 years' and attainment of a middle-income country (MIC) status in which per capita income was envisaged to increase from USD 506 in 2010 to USD 908 in 2020, when the country would supposedly achieve a lower MIC status and then striving to achieve an upper MIC status by 2040 with a per capita income of USD 9,500. However as of 2021, the country had not yet achieved the MIC status. Agriculture is the Uganda's main economic activity employing around 64.3% of the country's population (47% employed in susbsistence farming), and acconted for 21.9% to the GDP in 2018/2019. Approx. 28% of the children are in child labour. However, the services and industrial sectors are also expanding and their contribution to GDP is increasing now standing at 51.1% and 18.7% respectively [7, 14, 15].

3. Energy profile and transitions

SDG 7 obligates member states to ensure access to affordable, reliable, sustainable, and modern energy for all. The Government of Uganda recognizes that energy is catalytic to economic and social development and that there is a direct relationship between per capita consumption of electricity to GDP and overall prosperity [16]. Recognizing, the link between energy and development, the Uganda Vision 2040 and NDP III (2020/21–2014/15) recognizes the critical role of electricity in the attainment of their key development priorities of economic growth, job creation and inclusive development. However, although Uganda is rich in energy resources, access to affordable, reliable electricity remains a major constraint to the country's socio-economic transformation and achievement of sustainable development. The energy resources potentials include renewables such as hydropower, biomass, solar, wind and geothermal, but also fossil fuels such as peat, and oil and gas, and these resources and are fairly distributed throughout the country.

Uganda has one of Sub-Saharan Africa's largest hydroelectric potential estimated to be well over 4,500 MW [2, 17] with a big portion of this potential lying along the Nile River, which increases dependency on the Nile. Renewable energy resource potential, estimated to be well over 7,500 MW, includes hydro, geothermal (450 MW), biomass cogeneration (1,650 MW), wind and peat (300 MW), as well as 460 million tons of biomass standing stock with a sustainable annual yield of 50 million tons, and an average of 5.1 kWh/m2 of solar energy [2, 17, 18]. As at 2019, the country's energy mix was dominated by biomass energy (88%) consumed mainly through firewood and charcoal, electricity (2%) and oil or fossil fuels at 10% [19], and energy consumption has been on rise over the years (see **Figure 2**).

The dominance of biomass energy in the country's energy mix remains socio-economic and environment challenge; hindering socio-economic transformation, a source of indoor air population with the associated health challenges, a driver to increased deforestation and forest degradation and a source of GHG emissions. However, limited availability of alternative affordable clean energy sources for cooking means that biomass will remain Uganda's most important source of cooking energy in the foreseeable future. Currently, Uganda depends on imported

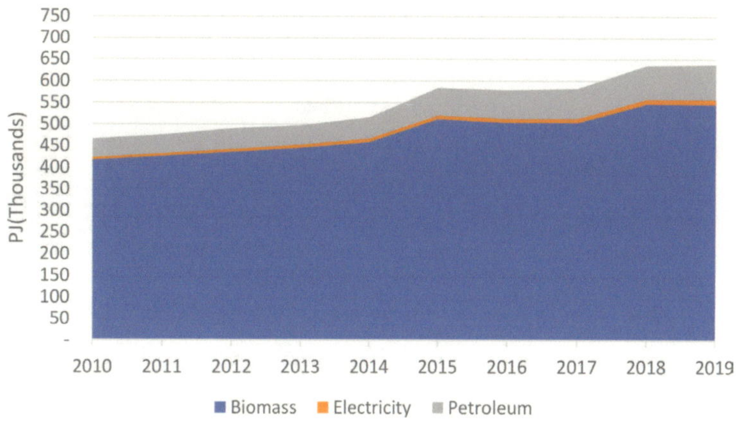

Figure 2.
The trends of contribution main energy by type (data sourced from MEMD and ERA 2010–2019).

oil/petroleum with the annual imports of petroleum products having increasing over the years e.g. from 1.4 billion liters in 2014 to 2.1 billion in 2018 [10]. Transport consumes 90% of the imported petroleum products, while 6% is consumed as kerosene in households consumes mainly for lighting [19].

Uganda depends on hydropower for its electricity accounting for more 80% of the country's electricity supply, the other sources being thermal (8%), co-generation (8%) and solar (4%). The grid installed electricity generation capacity has been increasing over the years from 183 MW in 1997 to 1,246 MW in 2019, while the energy losses in the distribution segment have reduced from 34% in 2008 to 16.4 in 2019% [16, 20]. Construction of Uganda's largest power plant (the Karuma hydropower plant) with 600 MW installed capacity is ongoing and could be commissioned by the end of 2021 or early 2022, and together with other small hydroelectricity plants that are under construction will increase the country's grid electricity generation to approx. 2,000 MW by 2025. However, the country's sustainable energy transition is still being hindered by the government's emphasis on large-scale hydroelectricity over other renewables [11], even when it is known that hydro-electricity is highly climate sensitive. Some interest in decentralized renewables exists, including solar and bioenergy, but they have not attracted the required state attention, and of late government attention is rather shifting to tap into oil and gas development for domestic power generation in the near future.

Nearly all the electricity generated in Uganda is consumed domestically, with households consuming the largest amount (61%), followed by commercial (14%), transport (13%), industry (8%), and others including export consuming only 4%. But then even at household level, the use of electricity is limited to lighting due to the high tariffs and for most households, charcoal and firewood remain the most accessible and affordable sources of energy for cooking. While there has been general increase in energy consumption in all the main sectors of economy over the last decade (see **Figure 3**), consumption of electricity in the industrial and commercial sectors remains low due to the low levels of industrialization, urbanization and overall economic development in the country.

Energy poverty is as electricity access remain low in Uganda with connectivity at only 28%, [21], which far lower than Sub-Saharan Africa average of 45% in 2017 [22]. In addition, electricity consumption per capita is approx. 100kWh [19], which is one of the lowest in the world. Besides the high costs of electricity, coupled with unreliable supply contribute to overdependence on biomass energy for cooking and the high cost of doing business which reduces the country's competitiveness [23].

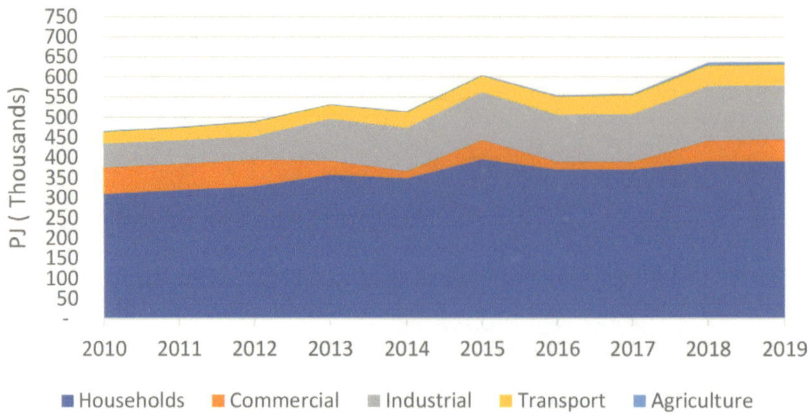

Figure 3.
Trends in energy consumption by sector (data sourced from UBOS 2010–2019; and MEMD, 2015–2020).

This makes it more than urgent for the country to increase access to modern energy in rural and urban so as to reduce deforestation, stimulate the growth of economies, and promote green and resilient development.

The government's target is to increase electricity connections to 60% by 2025 and 80% access to modern energy by 2040 [2]. An estimated US$1.6bn capital investment will required to meet the present plan of 60% electrification [16]. With the Uganda Vision 2040 foreseeing a country developing at a high to achieving an upper MIC status by 2040, Uganda will need much more modern energy with the energy demand is likely to outstrip the supply by 2030. By 2040, energy demand (including energy to propel industrial and commercial sectors) could reach 41,738 MW and capita electricity consumption could increase from the current 100 KWh to 3,668 kWh, indicating that therefore there is still a lot to be done in the energy sector.

4. Climate vulnerabilities of energy systems and adaptation

Uganda's climate is changing: temperatures are rising, rainfall is becoming more variable, and extreme weather events are increasing in frequency and severity. Between 1900 and 2010 temperatures in Uganda increased by 0.8°C - 1.5°C and are projected to increase by between 2-5° by 2100 and rainfall received in 2000–2010 was on average about 8% lower than rainfall received between 1920 and 1969rainfall was observed to be variable [24, 25]. Melting of the ice caps on Mt. Rwenzori, increased occurrence of droughts, rainstorms and flooding and landslides have been observed in the last three decades.

Drought and rising temperatures affect biomass availability, causing gradual drying up of biomass which will increasing hardship for the majority of Ugandan's who are dependent on wood fuel for their domestic energy needs. During extreme wet seasons, access to firewood is limited because of the reduction in the amount of available dry firewood, and the dangers associated with fetching firewood in such conditions. The scarcity and increasing prices of firewood greatly affects poor rural households that largely depend on biomass energy as they move long distances to collect firewood, a process that disproportionately affects women and girls exposing them to gender based sexual violence (GBSV), including sexual assault, rape and defilement. The projected change in climate means that droughts will occur which will increase scarcity of biomass. Even without climate change, the unmet demand for biomass energy will become extreme by 2050, estimated at 1,710 million tonnes over the period 2010–2050, and climate change will add a plausible loss of 5 to 10% of domestic wood between 2020 and 2050, which could increase the costs of inaction from US$123.6 bn to between US$130 bn and 136 bn [6]. Therefore, the current growth in demand for biomass is not sustainable and a sustainable solution is needed to address the predicted deficit [5]. Besides, the reliance on biomass energy is a major driver to widespread deforestation and land degradation across the country. Uganda's forest cover reduced from 24% percent in the 1990s to a mere to 12.4% currently, a two percent annual loss of forest cover, which is among the world's highest forest cover losses, driven by among others by increased demand for fire wood and charcoal, but also by expansion of farming and urban centres into forest land. But deforestation and forest degradation also results catchment degradation that indirectly affects hydro-electricity generation. i.e. it is a driver to soil erosion, run-off and flooding in the catchments which lead to siltation and sedimentation of lakes and rivers that in turn reduces water availability for hydro-power generation.

The impacts of climate change also significantly affect electricity supply. The Stern Report posits that that a 3-6°C increase in temperature in the next few years could result in a 30–50% water reduction [26] and this could disrupt hydro energy supply, alter energy demand patterns but also damage energy infrastructure [27]. With Uganda's temperature projected to increase by 2-5OC by 2100, the effect on energy will be significant, and indeed the functioning of electricity supply systems is already being adversely affected by the rising temperatures, increased rainfall variability, drought and flooding [28]. Hydro electricity generation, on which Uganda highly depends, is the most affected as droughts and higher temperatures often reduce water in lakes and rivers causing a reduction in electricity generation. For example, persistent droughts have in the past reduced water levels in Lake Victoria affecting hydropower generation at the Nalubale electric power generation dam in Jinja [29]. Electricity generation at hydro-power plant on Mpanga river in Western Uganda significantly declines during prolonged dry seasons or droughts [6]. A study conducted by MWE with support from the Climate and Development Knowledge Network (CDKN) predicts a reduction in rainfall that could reduce Uganda hydropower potential by about 26% by 2050 [5].

In **Figure 4**, the possible the impact of climate change on electricity generation is illustrated further. The loss in hydro electricity production is due to reduced runoff, a scenario where a drop in available capacity is between 0% in 2025 and 26% in 2050 [5]. The loss would be met by increasing generation through nuclear and renewables (geothermal and solar) where nuclear accounts for 95% of the difference of the reduced yield (35.4 TWh). However, global opposition to nuclear power generation is very high due to its environmental and political sensitivity, and besides nuclear power plants are very expensive to put up.

Besides, that is only part of the story as significant increase in rainfall amounts and intensity, and flooding as predicted could also seasonally increase water levels in lakes and rivers and also become barriers to energy generation. Between 2019 and 2020, the Lake Victoria basin received increased rainfall and run-off which increased the water level of Lake Vitoria to unprecedented levels, by 13.42 metres at Entebbe, which exceeded the 13.41 metres level recorded in 1964 [30][2]. This rise in water level created some challenges to electricity generation, with moving islands/floating vegetation blocking water flow to the hydro power stations on the Nile in Jinja. According to Eskom, Uganda's major power generation company, on April 14, 2020 a floating island docked at the Nalubale power dam blocking water for power

[2] https://storymaps.arcgis.com/stories/bd820937c06845faa86f7f8944d56f47

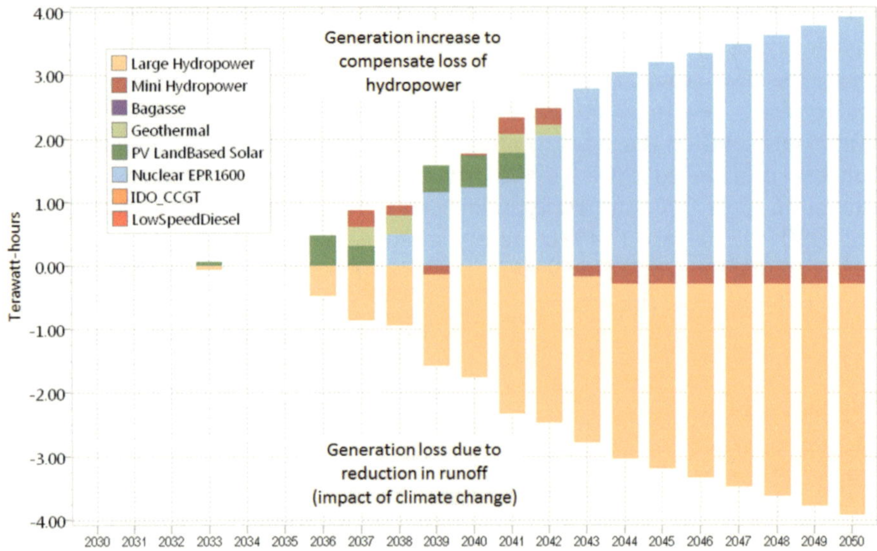

Figure 4.
Impact of climate change on electricity generation in TWh/yr. under the scenario of reduced runoff assuming high electricity demand with low supply capacity (DHSLWBC) vs. no impact of climate change (source [5]).

generation causing a nationwide power blackout.[3] However, Uganda's National Environment Management Authority (NEMA) also attributes the floating islands on the lake to environmental degradation in the lake basin. The old Maziba hydro-electric power station, in south western Uganda, was closed due persistent siltation that affected electricity generation and all these point to need to "climate-proof" the country's energy infrastructure.

Another major source of energy vulnerability is that the impacts of climate change disrupt the energy supply chain. The increased intensity and frequency of extreme weather events such as heavy rainfall storms and floods affect energy infrastructure - power plants, transmission lines and power lines - that disrupt energy supply resulting in power cuts and higher energy prices [31, 32]. Moreover, the heavy rains and flooding make road and rail transportation inaccessible. The disruption of rail and road transport indirectly affects energy supply, as fuel is typically imported into the country by train from the coast and distributed by trucks [23].

Adaptation to climate change for energy sector should that focus significantly reducing dependence on traditional biomass. A biomass supply deficit is already being experienced, with prices of charcoal rising each passing day, and indeed supply will not meet demand in the future even without climate change. It is also important to ensure the efficiency gains in biomass use (e.g. efficient cook stoves) are increased, and diversifying energy sources and promoting renewable energy that transition into more households switch to LPG, biogas, and electricity. It is critical to increase electricity connectivity, affordability, and reliability and de-risking the use of LPGs through tax incentives and supply in rural areas. The adaptation cost for replacing lost biomass is estimated at between $5 and $11 bn over the period 2020–2050 at a 10 percent discount rate [5].

The best use of hydropower needs careful management of the water resources and thus catchment management is also important. Afforestation and reforestation measures to protect watersheds that supply major hydroelectricity generating sources should be a key part of such protection, and Uganda is promoting

[3] Eskom Uganda Limited. Press Release. April 16, 2020.

catchment-based water resource management, and the country is divided into 17 catchments, for which catchment management plans are prepared to address the issues affecting water and other related resources (forests, biodiversity, wetlands, soils etc.) that could among others ensure sustained water availability for electricity generation.

Above all, coping with extreme events requires strengthening the structures of small dams to withstand increase in rainfall intensity or reduced water levels.

But still, Uganda government's preference for large hydro-power projects in areas of rich ecosystems and biodiversity, and eco-tourism sites is an environmental controversy, given the dependence on hydropower on natural systems (climate and water availability). The ongoing construction of Karuma hydro-electricity dam/station and the proposal to construction another hydro-electricity dam on Murchison Falls (and adjacent Uhuru Falls), all projects in Murchison Falls National Park, a protected area with rich biodiversity and eco-tourism potential is are cases in point. These energy projects remain controversial given that Uganda is not lacking alternative renewable energy resources – solar and geothermal potential is very high and there are many other alternative sites for hydro-electricity development. The larger questions are the levels of Uganda government's assessment of the 'economic benefits' versus 'environmental costs' of its energy development model. Besides, the dams on the Nile remain insecure to current and future climate risk i.e. fluctuating water levels and extreme weather events. But also, the Nile is a transboundary resource and water governance and security concerns are hot topic of debate and a potential source conflict in the Nile basin countries, especially with downstream neighbors (Sudan and Egypt), putting Uganda's dams at future risk of reduced flow.

5. Energy and GHG emissions in Uganda

Globally, energy is the main source of GHG emissions, with 70% of the global emissions emanating from fossil fuels combustion processes to generate electricity, in industry, buildings and transport, [27]. However, as an LDC, Uganda's energy consumption (especially fossil fuel consumption) is very low and electricity generation in Uganda is dominated by hydro-power which is a renewable energy source and thus the country'contribution to global GHG emissions and climate change is negligible. For example, Uganda accounts for only 0.01% of total world emissions its per capita CO_2 emissions is also low at only 0.13 t (2017)[4]. However, Uganda's GHG emissions have been on the rise, having slowly increased from 53 442 Gg CO2e in 2005 to 90 230 Gg CO2e in 2015 (see **Figure 5**).

The energy sector ranks third as a major source of GHG emissions in Uganda accounting for 10.7% of the country's total emissions, with dominant source being the Land Use and Land Use Change and Forestry (LULUCF) sector accounting for 59.5% (53 670 Gg CO2e) of the total emissions.

It needs to be recognized that energy is an indirect driver of the LULUCF emissions because the country's high dependency on biomass energy (charcoal and firewood) is a major driver to deforestation. Indeed, Zutari [37] observes that emissions from fuelwood and charcoal are the largest contributors to the CO2 emissions in Uganda but these emissions are not accounted for under energy emissions but rather the LULUCF sector as wood removal losses. The other major sources of GHG emission are agriculture (26.9%) and waste (2.3%).

Uganda has the basic infrastructure to manage GHG inventory system which includes software, a database to hold data from all the sectors and officers to

[4] https://ourworldindata.org/co2/country/uganda?country=~UGA)

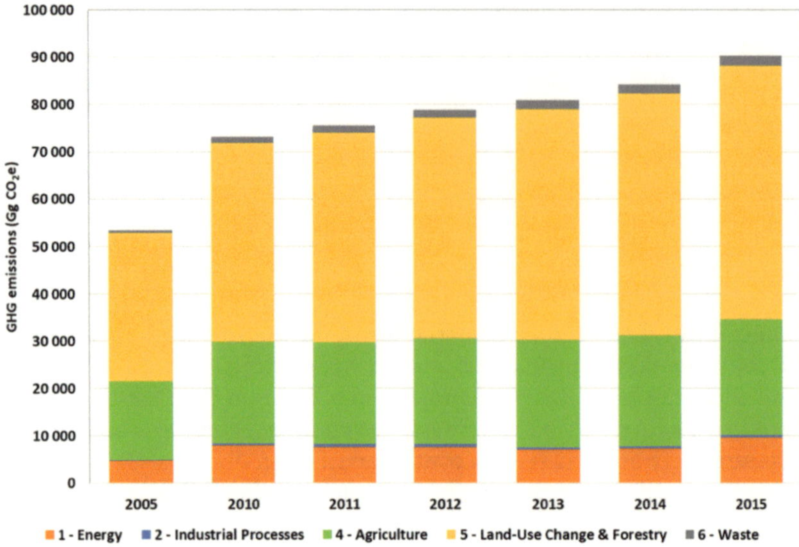

Figure 5.
Sectoral trends in GHG emissions for Uganda (source: [15, 33–36]).

manage the system, and the MEMD is one of the data provider sectors [14, 15]. In addition, Uganda has made efforts to mitigate climate change by deploying decarbonized development pathways as required by the UNFCCC. Uganda's first Nationally Determined Contribution (NDC) reveals that under the business-as-usual (BAU) scenario GHG emissions will increase to 77.3 Million tons of carbon dioxide equivalent per year (MtCO2eq/yr) by 2030 but the NDC commits to 22% emission reduction of national GHG emissions in 2030 compared to business-as-usual, through implementation of mitigation policies and measures, including those in energy. Various Nationally Appropriate Mitigation Actions (NAMAs) actions have been development and are implemented, with those that are energy relevant revolving around improved institutional cook stoves in Ugandan schools, integrated waste management and biogas production, vehicle fuel efficiency and bus rapid Transit (BRT).

6. Oil and gas: challenge for decarbonizing energy systems and the environment

Although Uganda is targeting a low carbon development path in line with its climate change policy and Nationally Determined Contribution to UNFCC, at the same time, there is ongoing development in oil and gas sector. There have been great expectations that oil and gas development will socioeconomically transform Uganda to MIC status through increased investments, revenue generation, infrastructure development and job creation along the value chains. For example, the oil and gas resources once fully developed could fetch up to US$2bn per annum for 20 to 30 years, and to contribute between 20% and 30% to the country's Gross Domestic Product [19]. However, the sub-sector has enormous challenges it poses to the environment, biodiversity and ecosystems and particularly for upstream and mid-stream activities, but besides oil development and use is a challenge for decarbonization efforts the country and achievement of carbon neutrality because it will greatly increase the country's GHG emissions. In this section, we provide an overview of Uganda's oil and gas sector and its likely impact on environment and naturals resources and Uganda's efforts to contribute to the achievement of the Paris Agreement on climate change.

6.1 Overview of the oil and gas sector in Uganda

Uganda has abundant oil resources, estimated at approx. 6.5 billion barrels of oil reserves of which 1.4 billion are said to be economically recoverable and the country's oil production potential is estimated at 200,000 barrels per day once the resources are fully developed [19]. The country's gas resources are estimated to be 700 billion cubic feet (bcf), with associated and non-gas gas estimated at 173 bcf and 500 bcf respectively.

Oil resources were first discovered in the 1890s MacKeinze *et al.* [38], but it was not until the 1920s that the country's oil and gas potential was evaluated [39]. Attempts on oil and gas exploration were first undertaken between 1936 and 1956; where the first deep well (Waki B1) was drilled in 1938 at Butiaba in Buliisa district, after which about 20 wells were also drilled in Kibiro and Kibuku, and several geological surveys were also undertaken the 1940s to 50s. These exploration efforts were halted due to the political conflicts and turmoil that ensued in the 1970s till the 1990s.

The economic viability of Uganda's oil and gas reserves was not ascertained till around 2006 [40, 41] and since then significant achievements have been made in the exploration and moving towards full development of the oil and gas resources in which various international companies have been licensed for oil exploration and development including Total E&P Uganda B.V, China National Offshore Oil Corporation Uganda Limited (CNOOC Uganda Ltd) and Tullow Uganda Operations Pty Limited together with Armor Energy Limited and Oranto Petroleum Limited [19]. By 2016, 21 oil and gas sites had been drilled, which is an 88% success rate of viable oil and gas resources. These oil and gas estimates are expected to increase as additional discoveries are made since only 40% of the total area with the potential for oil and gas reserves has been explored by 2021 [19].

Recognizing the socio-economic importance of the oil and gas sector, the GoU developed the National Oil and Gas Policy (2016) to operationalize and regulate the sub-sector. A number of laws have also been enacted including the Petroleum Exploration, Development and Production (PEDP) Act 2013 and the Petroleum (Refining, Conversion, Transmission and Midstream Storage) Act 2013 to replace the Petroleum Exploration and Production Act of 1985. The oil and gas policy and regulatory frameworks give MEMD an oversight role to guide and monitor work in the oil and gas sector and activities of the Directorate of Petroleum, Uganda Petroleum Authority [42], and the Uganda National Oil Corporation (UNOC) [43]. The Directorate of Petroleum is responsible for policymaking, licensing, and coordinating all developments in the sector. The Uganda Petroleum Authority [42] regulates and monitors actors by enforcing compliance of the operations of oil companies. The Uganda National Oil Corporation is mandated to handle business interests across the oil and gas value chain.

6.2 Effect of oil and gas exploration and development on vegetation and wildlife

Most of Uganda's oil and gas resources (and thus their exploration and development), are in the Albertine Graben region in the western part of the country, situated within the western arm of the East African Rift Valley at the border between Uganda and the Democratic Republic of Congo (DRC). The graben stretches for about 500 km with an average width of 45 km covering a total area of approx. 23,000 sq. km [41].

Unfortunately, the Albertine Graben region is not only one of Uganda's richest biodiversity hotspots [41, 44] but the eco-region is the most important forest system in Africa for biodiversity [24]. The eco-region ranks first out of the 119 distinct terrestrial eco-regions of continental Africa in terms of endemic species of birds, mammals, reptiles and amphibians, and second in terms of globally threatened species; there are around 567 endemic plant species" [45, 46]. Indeed, the graben reportedly has seven designated conservation (protect areas) and various eco-tourism attractions among which include the 45 m waterfall along the Victoria Nile River, 556 bird species, 144 different mammals that include endangered and rare species such as Rothschild's giraffe and tree climbing lions, 51 reptile species, 51 (28 and 23 known and unknown respectively) amphibian species.

The biodiversity in the area, and more especially wildlife, are very sensitive to oil and gas exploration and development. As observed by MacKeinze et al. [38], whereas oil exploration and development could take place within conservation areas (according to the Uganda Wildlife Statute 1996: Section 19(5)), the activities pose serious impacts on vegetation cover and wildlife in the Albertine Graben eco-region. Indeed, the Uganda Wildlife Authority (UWA), an agency mandated with management of conservation areas, the National Environment Management Authority, the agency mandated for environment management; and the communities in the region are cognizant of the negative impacts that up-steam and mid-stream oil and gas development activities have on wildlife [44, 47, 48].

One of the oil and gas exploration activities that affect biodiversity is the seismic surveying, that involves holes in the grounds into which dynamite or special vibrators that pump air or water into the ground. The associated seismic vibrations are reported to have caused migration of large mammals and fish, and impacted the breeding of wildlife [38, 40, 47, 49]. The most affected mammals are the elephants, giraffes, buffaloes and oribi [44] that have moved further (about 0.75–1 km) away from oil exploration (seismic drilling) sites (Wildlife Conservation Society of Uganda (WCS), [50]) showed that many big mammals have moved away to distances of about 750 to 1000 metres from seismic drilling, and seismic waves affect breeding and migratory patterns of fish. The migration of wildlife has impacts on tourism and fishing, which are prime socioeconomic activities for the communities in the Albertine Graben region.

The drilling and/or establishment of drilling pads is also a driver to wildlife migration (WCS, [50]) and is also linked to deforestation and the degradation of pristine vegetation cover and habitats [48, 50]. Habitat degradation is another driver to wildlife migration which has culminated into human-wildlife conflicts. Drilling pads cause habitat fragmentation which is another major factor leading to loss of biodiversity [48]. Besides, oil explorations is associated with increase noise resulting from drilling and road construction to the exploration sites that scares wildlife and forces it to migrate [40, 51–53], Indeed UNRA [53] recognizes the environmental impact of noise emanating from the construction of oil roads that negatively disrupted mammals and birds in the area.

The exploration and development processes require many workers that need to be accommodated which gives rise to settlements and other service provisioning. Therefore, oil companies have construction of workstations and more will come in future. These works stations have been constructed in remote and high vegetative areas which has caused deforestation and degradation, and Nnakayima [52] observes a decrease in tropical high and low forest, woodlands and grasslands in the areas caused up-stream oil and gas activities.

Oil spills from oil collection pits have been reported and more are expected as the oil development progresses [50]. These spills cause soil and water pollution there by affecting biodiversity and the livelihoods of many people in the region [40, 52].

Gas flaring is another activity that is linked to air pollution [54], with gases associated with gas flaring comprising carbon dioxide, methane, sulfur dioxide, nitrogen dioxides, dioxin (CJP/ERA 2005), some of which are GHG that cause global warming and climate change. Besides the increased dissolution of these gases could form acid rainfall, which could scotch vegetation. During gas flaring, hazardous compounds that ensue from unburned fuel components such as benzene, toluene, xylene, and hydrogen sulphide, and the seepage of these compounds into the environment results in soil pollution, destruction of vegetation and water contamination which could causing death of wildlife that drink the water.

6.3 Construction and infrastructure (pipeline, roads and airport)

In April 2021, Uganda and Tanzania signed an agreement to jointly construct a 1,440 km crude oil pipeline to connect Uganda's oil fields in the Albertine region to Tanzania's seaport of Tanga on the Indian ocean [42]. The US$ 3.55 bn East Africa Crude Oil Pipeline (EACOP) is needed to transport crude-oil from the Albertine for export. Once completed, the pipeline will be able to transport 216,000 barrels of oil per day, and will be the world's longest electrically heated crude oil pipeline. The construction of the pipeline will necessitate acquisition of large expanses of land for construction camps (172 acres); pipeline Right of Way (RoW) and orphan land (2,745 acres); and feeder roads covering 182 acres. The pipeline runs through forests, dense vegetation covers, and protected areas and will therefore cause deforestation and land degradation. Indeed, the EACOP environmental and social impact assessment (ESIA) shows that the pipeline construction would result in deforestation and forest degradation resulting in habitat fragmentation and wildlife disturbance and migration, and have significant impacts on wetlands that are vital sources of water for wildlife. For example, the proposed pipeline's right of way cuts through Bugungu Wildlife Reserve which is an important bird and key biodiversity area) in the Albertine Graben, [51, 55]. The deforestation and degradation of the vegetation cover which will result in loss of carbon sequestration potential, and thus will thus increase GHG emissions.

Although the new "oil roads" constructed in the Albertine region have improved access, jobs and incomes of communities [56], there have also been noticeable negative impacts. Road developments have increased access to pristine and remote high biodiversity hotspots, and caused land fragmentation that causes habitat fragmentation [57] and wildlife disturbance [40, 48]. Similarly, the increased accessibility has exacerbated illegal activities such as felling of trees, poaching and wildlife trafficking in the Albertine eco-region [52]. For example, the construction of the Kaiso-Tonya road has increased poachers' access to Kabwoya Wildlife Reserve as well as disturbance of wildlife, death of wildlife (knocked) along the road, as well as noise and air pollution [52].

7. Policy: challenges and gaps

Uganda is a signatory to the Paris Agreement on Climate Change that requires parties to develop climate change policies, strategies and plans that promote adaptation and mitigation. The Uganda Vision 2040 and the NDP III call for appropriate adaptation and mitigation strategies to ensure that Uganda is sufficiently cushioned from any adverse impact of climate change, while at the same time promoting low carbon intensive pathways. In 2015, Uganda unveiled both the National Climate Change Policy (NCCP) and Nationally Determined Contribution (NDC) that reference both adaptation and mitigation, but state that adaptation is the primary priority and mitigation is a secondary priority [15]. Both the NCCP and NDC prioritize climate sensitive energy services. The NCCP prioritizes sustainable energy access and utilization for increased climate change resilience, increased investment in renewable energy and clean energy technologies to reduce GHG. Similarly, the NDC priorities efficient biomass energy production and use, increased electricity access and energy efficiency, expansion of off grid solar systems and overall climate proofing of the energy sector.

Besides both the NCCP and NDC prioritize mainstreaming climate change in national, sectoral and local development plans, plans and budgets. In 2018, Uganda became the first country in Africa to sign the Partnership Plan for NDCs to achieve national climate goals as part of its obligations to the Paris Agreement. Under the NDCs, the country committed itself to reduce national emissions and adapt to the impacts of climate change. Currently Uganda is developing its Long-term Strategy for climate change and updating its NDC.

The Paris Agreement (Articles 2 and 4) also asks Parties to formulate and communicate a Long-term Low Emission Development Strategies (LT-LEDS) to strengthening global response to the threat of climate change i.e. "holding the increase in the global average temperature to well below.

2°C above pre-industrial levels and pursuing efforts to limit the temperature increase to 1.5°C above pre-industrial levels" [58]. Uganda is on track to develop its Long-Term Strategy for Climate change, ahead of COP 26 in in UK in December 2021. Only 29 parties have so far submitted their LTS for climate change, only two of which are from African i.e. Benin and South Africa[5], implying that Uganda will be one of the first leading countries worldwide to develop an LTS. The focus of the LTS is to identify one or more low-emission pathways in line with the country's development objectives and strives to identify emission pathways compatible with the temperature goal set by the Paris Agreement. Uganda's LTS will cover both mitigation and adaptation aspects, and will include all of the main outcomes from the major plans, policies and strategies that have been previously developed, including among others, the Vision 2020, NDPs, NCCP, Second National Communication to UNFCC, NDC of 2015, Uganda Green Growth Development Strategy, First Biennial Update Report (BUR) submitted in 2019, and the NDC submitted to UNFCCC 2015 and is currently being updated in the run-up to COP 26. A National Climate Change Act was passed by the Uganda parliament on 27th April 2021[6] and is now awaiting presidential assent before it becomes law. The new law provides for national participation in climate change, institutional arrangements for coordinating and implementing climate change response ansd for financing for climate change action, including for climate smart energy systems.

The main energy policies for Uganda are the National Energy Policy 2002 and **the National Renewable Energy Policy.** The aim of energy policy is to meet the energy needs of the Ugandan population for social and economic development in an environmentally sustainable way. The policy is b being revised and revised policy is place [19], aimed at "meet the energy needs of the Ugandan population by providing adequate and reliable energy supply for socio-economic growth and sustainable development". The renewable energy policy aims at increasing the use of modern renewable energy, from 4–61% by 2017. The energy policy framework addresses issues of with renewable energy technologies and energy efficiency devices which can translate into low GHG emissions and promoting low carbon development and provides the necessary framework for private sector investors in renewable energy projects to benefit from the available facilities in emissions trading. One of policy's

[5] https://www.climatewatchdata.org/lts-explore

[6] https://www.parliament.go.ug/news/5093/climate-change-bill-passed

strategies is to integrate biomass energy production and efficient utilization and its impacts on climate and health into the formal education system. The policy also recognizes the need for promoting the conversion of municipal and industrial waste to energy to reduce methane emissions.

The Ministry of Energy and Mineral Development (MEMD) is the main energy actor in Uganda, responsible for energy policy development and guidance. The Ministry has two departments that support climate friendly energy policy and practice i.e. the Renewable Energy Department and an Energy Conservation Department. However, the policy provides for the establishment of a National Energy Committee at the National level and District Energy Committee and District Energy Offices at the local Governments, to date, they are not yet in place. Small-scale renewable energy sources which would have been ideal the rural areas have remained marginal compared to large scale commercial energy and yet 85% of Uganda's population is rural. Taxes on energy saving products, efficient technologies and solar products for lighting, cooking and heating, water pumping (which are also climate change compliant) are still high. If these products continue to be highly priced due to high taxes, their future potential is at risk, given the low-income levels of the majority of Uganda's population. Consequently, renewable energy technologies have remained financially out of the reach of many Ugandans. Moreover, the high electricity tariffs make it unaffordable for cooking purposes and the alternatives remain charcoal and firewood. Besides, energy blackouts and outages remain frequent and many institutions, commercial enterprises and households' resort to use of thermal generators for energy as compared to use renewable energy like solar or biogas.

Following the discovery of commercial oil deposits, the Government of Uganda came up with the National Oil and Gas Policy for Uganda in February 2008 whose main policy goal is to use the country's oil and gas resources to contribute to early achievement of poverty eradication and create lasting value to society. Given that consumption of fossil fuels is the greatest contributor to GHG emission, the policy and utilization could prove a serious challenge to climate change and overall environmental management. Already, the country plans to use the oil reserves for thermal electricity generation to fill the electricity deficit, which will definitely increase GHG emissions. Although one of the policy priorities of the oil and gas policy is to ensure that oil and gas activities are undertaken in a manner that conserves the environment and biodiversity, the policy remains climate change not incorporating climate change issues.

8. Conclusion

Energy poverty remains high in Uganda amidst the high energy potential, but the current main sources energy (biomass and hydro-electricity) are extremely vulnerable to the impacts of climate change. With no viable and affordable alternative sources of energy is sight, dependence on biomass energy remains inevitable in the foreseeable future. This makes managing climate risks to energy systems an urgent necessity to ensure supply systems reliability and resilience, if Uganda is to achieve its development targets.

The main barriers to attaining climate smart and sustainable energy systems in Uganda financial constraints and weak technical ability, most which has to be obtained from external sources. But while the climate change vulnerability of the energy sector and need for increased resilience are highly recognized, international focus has been put on mitigating GHG emissions and greening energy, and much less on building climate resilience. This creates a conflict in policy for

Uganda because Uganda's primary climate change policy priority is adaptation, and mitigation is secondary. Still, the much-anticipated oil and gas development, compounded by rapid population and urbanization will inevitably increase Uganda GHG emissions, and thus what is needed is green energy innovations and solutions that can take the country towards carbon neutrality while deliver co-benefits for adaptation, health, economic growth and environmental sustainability. It will be essential for Uganda develops its oil and gas resources, but it also important that it is done in a way that protects ecosystems and biodiversity. It will be transformational if Uganda would invest a significant portion of its oil and gas revenues in green and renewable energy to enhance equitable and affordable energy services. With its huge renewables, future hydropower development should not cause ecosystem and biodiversity loss.

Author details

Revocatus Twinomuhangi[1,2*], Arthur Martin Kato[3] and Adam M. Sebbit[4]

1 Department of Geography, Geoinformatics and Climatic Sciences, Makerere University, Kampala, Uganda

2 Makerere University, Centre for Climate Change Research and Innovations (MUCCRI), Kampala, Uganda

3 Cities and Infrastructure for Growth (CIG) Uganda, Kampala, Uganda

4 Department of Mechanical Engineering, Makerere University, Kampala, Uganda

*Address all correspondence to: rtwinomuhangi@gmail.com

References

[1] ND-GAIN. 2019. Rank countries by ND-GAIN Country Index, Vulnerability and Readiness. https://gain-new.crc.nd.edu/ranking

[2] Government of Uganda. 2010. Uganda Vision 2040. National Planning Authority.

[3] Uganda Bureau of Statistics (UBOS). 2015. 2015 Statistical Abstract. Uganda Bureau of Statistics (UBOS).

[4] Government of Uganda. 2014. Second National Communication. Ministry of Water and Environment, Climate Change Department.

[5] MWE. 2015. Economic assessment of the impacts of climate change in Uganda. Final Report. MWE. Kampala Uganda

[6] Twinomuhangi, R. & Monkhouse C. 2015. Economic assessment of the impacts of climate change in Uganda: key results. Climate and Development Knowledge Network (CDKN).

[7] UBOS. 2021. Uganda National Household Survey 2019/2020. Report launched on June 2, 2021.

[8] UBOS. 2020. World Population Day Celebrations Saturday July 11, 2020 – brochure. Available at: https://www.ubos.org/wp-content/uploads/publications/07_2020WORLD-POPULATION-DAY-BROCHURE-2020.pdf. Accessed on May 24, 2021.

[9] UBOS. 2016. The National Population and Housing Census 2014 – Main Report. UBOS. Kampala, Uganda.

[10] UBOS. 2019. 2019 Statistical Abstract. UBOS, Kampala Uganda

[11] Government of Uganda. 2015. Second National Development Plan 2015/2016 – 2019/2020. National Planning Authority. Kampala Uganda.

[12] World Bank. 2019. World GDP per Capita Ranking 2019. Available at: https://data.worldbank.org/indicator/NY.GDP.PCAP.CD?locations=UG Accessed on May 24, 2021.

[13] World Bank. 2021. Uganda overview. https://www.worldbank.org/en/country/uganda/overview

[14] MWE. 2019a. Water and Environment Sector Performance Report 2019. MWE. Kampala Uganda.

[15] MWE 2019b. Uganda's first biennial update report to UNFCCC. Available at: https://unfccc.int/sites/default/files/resource/FBUR%20Final_2019.pdf. Accessed on May 25, 2021.

[16] MEMD. 2020. Roadmap for catalytic power sector transformation (RCPT) 2020-2025. MEMD, Kampala Uganda.

[17] MEMD (2015), Energy and Mineral Sector Development Plan 2015/15 – 2019/20. Ministry of Energy and Minerals. Kampala, Uganda

[18] WWF, 2015. Energy report for Uganda: A 100% renewable energy future by 2050. WWF Uganda Country Office. Kampala Uganda.

[19] MEMD. 2019. Draft National Energy Policy. MEMD, Kampala Uganda.

[20] Umeme, 2019. Umeme Annual Report 2019.

[21] Government of Uganda. 2020. Third National Development Plan (NDPIII) 2020/21-2024/25. National Planning Authority, Kampala.

[22] International Renewable Agency (*n.d*). Global Renewables Outlook: Energy transformation - Sub-Saharan

Africa. https://www.irena.org/-/media/Files/IRENA/Agency/Publication/2020/Apr/IRENA_GRO_R10_Sub-Saharan_Africa.pdf?la=en&hash=DB067EF85E0FDB6B8762833E77CC80F3975E46DC.

[23] Twinomuhangi R. 2018. Uganda national urban climate change profile. Ministry of Lands, Housing and Urban Development. Kampala, Uganda.

[24] Ministry of Water and Environment (MWE). (2015). Strategic Plan for the Northern Albertine Rift of Uganda 2011-2020. Ministry of Water and Environment, Kampala Uganda.

[25] Nimusiima, A., Basalirwa C.P.K, Majaliwa J.G.M., Kirya D., Twinomuhangi, R. 2018. Predicting the impacts of climate change scenarios on maize yield in the cattle corridor of Central Uganda. Journal of Environmental and Agricultural Sciences. 14: 63-78.

[26] Stern, N. 2007. The Economics of Climate Change: The Stern Review. Cambridge: Cambridge University Press.

[27] Ebinger, J., & Vergara W. 2011. *Climate impacts on energy systems: Key issues for energ sector adaptation.* The World Bank, Washington DC. http://documents.worldbank.org/curated/en/580481468331850839/Climate-impacts-on-energy-systems-key-issues-for-energy-sector-adaptation.

[28] Twinomuhangi, R., Sseviiri, H., Mulinde, C., Mukwaya P.I., Nimusiima, A., Kato A.M. 2021b. Perceptions and vulnerability to climate change among the urban poor in Kampala City, Uganda. Reg Environ Change **21,** 39 (2021). https://doi.org/10.1007/s10113-021-01771-5

[29] START. 2006. Assessing *the impacts of climate change and variability on water resources in Uganda: developing an integrated approach at the sub-regional scale.* Final Research Report. START.

[30] Lake Victoria Commission. 2020. Lake Victoria rising water levels. https://storymaps.arcgis.com/stories/bd820937c06845faa86f7f8944d56f47. Accessed on June 2, 2021.

[31] Office of the Prime Minister. 2019. National risk and vulnerability atlas of Uganda. https://www.necoc-opm.go.ug/NRVA/National%20Risk%20and%20Vulnerability%20Atlas%20of%20Uganda.pdf. Accessed on May 26, 2021.

[32] Twinomuhangi, R., Kato AM., Sebbit AM. 2021a. Umeme climate change study report. Cities and Infrastructure for Growth (CIG) Uganda.

[33] Electricity Regulatory Authority (ERA). 2010. Electricity Sector Performance Report: July 2009–December 2009, Electricity Regulatory Authority (ERA), Kampala.

[34] ERA. 2010. Transmission Energy Purchases. ERA. http://www.era.or.ug/TransmissionPurchases.php. Kampala. Accessed 30 June 2011.

[35] ERA. 2014. UETCL Total Volumes Purchased across All Suppliers of Electricity (MWh) 2006-2013,

[36] ERA. 2018. Energy Generated, Electricity Regulatory Agency. https://www.era.or.ug/index.php/stats/generation-statistics/energy-generated-by-utility, Kampala. Accessed June 30, 2018.

[37] Zutari. 2020. Metabolic assessment report: A circular approach to define a resource efficient and low-carbon future. Uganda climate promise project. UNDP and Government of Uganda. Kampala, Uganda.

[38] MacKenzie, C. A., Fuda, R. K., Ryan, S. J., Hartter, J. 2017. Drilling

through conservation policy: oil exploration in Murchison Falls Protected Area, Uganda. Conservation and Society, *15*(3), 322-333.

[39] Directorate of Petroleum Uganda. (*n.d.*). *Petroleum Exploration History*. Directorate of Petroleum., from https://www.petroleum.go.ug/index.php/who-we-are/who-weare/petroleum-exploration-history. Accessed on April 12, 2021

[40] Byakagaba, P., Mugagga, F., Nnakayima, D. 2019. The socio-economic and environmental implications of oil and gas exploration: Perspectives at the micro level in the Albertine region of Uganda. The Extractive Industries and Society, *6*(2), 358-366.

[41] Kassim, W. 2020. Land Conservation in the Albertine Graben Region of Uganda: A Critical Analysis of the Legal Regimes. In *Legal Instruments for Sustainable Soil Management in Africa* (pp. 79-99). Springer, Cham.

[42] Petroleum Authority of Uganda (PAU) (*n.d*). The East African Crude Oil Pipeline (EACOP) Project. Petroleum Authority of Uganda. https://www.pau.go.ug/the-east-african-crude-oil-pipeline-eacop-project/. Accessed on April 15, 2021.

[43] Mbabazi, P., Muhangi, M. 2017. Uganda's Oil Governance Institutions: Fit for Purpose? *Oil wealth and development in Uganda: prospects, opportunities and challenges*.

[44] Kamara, E. D., Nina, P. M., Ochieng, L. A. 2019. Effects of oil and gas exploration in Murchison Falls National Park on wildlife resources. African Journal of Environment and Natural Science Research, *2*(2), 48-57

[45] Plumptre, A.J., Behangana, M., Ndomba, E., Davenport, T., Kahindo, C., Kityo, R. Ssegawa, P., Eilu, G., Nkuutu, D. & Owiunji, I. 2003. The Biodiversity of the Albertine Rift. Albertine Rift Technical Reports No. 3. WCS, New York.

[46] Plumptre, A.J., Ayebare, S., Mugabe, H., Kirunda, B., Sekisambu, R., Mulondo, P., Mudumba, T. 2015. Biodiversity surveys of Murchison Falls Protected Area. New York, USA: *Wildlife Conservation Society*.

[47] National Environment Management Authority (NEMA), (2010). The Environmental Sensitivity Atlas of the Albertine Graben (2010): *2nd Edition*. NEMA. Kampala Uganda.

[48] Uganda Wildlife Authority (UWA). 2015. *Uganda Wildlife Authority Strategic Plan 2015 – 2020*. Kampala, Uganda.

[49] Sserwanga, I. 2018. Human wildlife conflicts and interaction: the impact of oil exploration and development in Buliisa, Uganda (Master's thesis, NTNU, Norway).

[50] World Conservation Society Uganda (WCS). (*n.d.*). Oil Development. Wildlife Conservation Society (WCS) Uganda. https://uganda.wcs.org/initiatives/oil-development.aspx. Accessed on April 14, 2021.

[51] NEMA. 2019. The East African Crude Oil Pipeline (EACOP) ESIA. National Environment Management Authority (NEMA). http://eacop.com/EACOP_ESIA.pdf.

[52] Nakayima, D. 2018. *Impacts of upstream oil and gas activities on environment, well-being and tourism in the Albertine Region of Uganda: Local community perspectives*. Doctoral dissertation, Makerere University.

[53] Uganda National Roads Authority (UNRA). 2017. Environmental and Social Impact Assessment for the proposed upgrade of Kabwoyabuhuka road (40km). Uganda National Roads Authority. Kampala, Uganda.

[54] Graham, E., Ovadia, J. S. (2019). Oil exploration and production in Sub-Saharan Africa, 1990-present: Trends and developments. The Extractive Industries and Society, *6*(2), 593-609.

[55] Total. 2019. Tilenga project Environmental and Social Impact Assessment Report: Volume III. Total in Uganda. https://ug.total.com/tilenga-project-environmental-and-social-impact-assessment-report. Accessed on April 16, 2021.

[56] Mugisa, S. 2016. Socio-economic effects of oil exploration among Hoima Municipality Communities, Uganda. (Masters dissertation, UTAMU, Kampala, Uganda).

[57] Dowhaniuk, N., Hartter, J., Ryan, S. J., Palace, M. W., & Congalton, R. G. 2017. The impact of industrial oil development on a protected area landscape: demographic and social change at Murchison Falls Conservation Area, Uganda. Population and Environment, *39*(3), 197-218.

[58] United Nations. 2015. Paris Agreement. https://unfccc.int/files/meetings/paris_nov_2015/application/pdf/paris_agreement_english_.pdf

Chapter 21

Climate Change in Ethiopia: Implication on Human Capital in Rural Community - Case Study of Bilate Basin Agro-Pastoral Livelihood Zone of Sidama

Firew Bekele Worana and Cheru Atsimegiorgis

Abstract

The objective of this study was to examine the trends of climate change and its subsequent impact on human capital development under the reference of human health and education of rural community of drought prone areas in Western Sidama (6.36°–7.14°N Latitude and 38.01°–38.56°E) of Ethiopia. Tropical Application of Metrology using SATellite (TAMSAT) data of both monthly rainfall and temperature was collected from Ethiopian National Meteorological Agency (NMA) for the period of 1987–2017. Data on perceived climate change; change driven impacts people experienced mainly on their health and education and copying or/and adaptation strategies affected community practiced was collected by employing both survey and participatory rural appraisal (PRA) techniques. In order to collect data from the household level, survey was employed for 400 households who were systematically sampled from 245,592 households of five drought prone administrative districts whereas PRA was employed to collect community level data. Mann–KENDALL TEST AND SEN'S SLOPE ESTIMATES (MAKESENS) and descriptive statistics were employed to analyze these data. The analysis result shows that there is increasing and decreasing trends of both temperature and rainfall, respectively. And increasing trend is statistically significant for temperature (α = 0.05; N = 31). Consequently, this change of climate variables has brought negative impacts on human capital mainly on health and education through various paths. Physiological inconvenience, prevalence of various diseases, and malnutrition were the main paths through which climate change impacts on human health were seen whereas students' failure in standard exam attributed to a roll over impacts of climate change since early child hood, increasing school dropout rate and decreasing demand of the households to family education mainly owing to diminish in agricultural yields were the education dimension impacts of change in climate variables. Though a temporary migration to less stress adjacent areas, receiving aids and use of health extension services were a household level copying mechanisms observed, the first two were seen to reproduce unintended negative effects such as interethnic conflicts, forcing children to drop the school and aid dependency syndrome among receivers that the household themselves, aid organizations and government should work in consortium on building resilience both at household and community levels.

IntechOpen

Keywords: bilate basin agro-pastoral livelihood zone, climate change, human capital, copying mechanism, resilience, Sidama

1. Introduction

Climate change is now more a reality [1] than a theory with a multiple implications on livelihood, health, and wealth of people across the globe [2]. According to the World Bank ([3], Xiii), as cited in Hameso [4], as the global climate is changed and the Earth is warmed, the rainfall pattern tends to shifts then by creating major climate extremes such as droughts, floods, and forest fires which collectively deny lives and livelihoods of millions. Poor people in Asia, Africa, and Latin America face prospects of tragic crop failure, agricultural productivity and, consequently, there happened increasing hunger, malnutrition, and disease [5]. Climate Change Impact [6], is any direct and/or indirect adversary impacts on one or more components of the small holders' livelihood (human lives, culture, ecosystem, economy, social well-fair, and infrastructure) brought about by the average value variations on precipitations and temperature throughout the time under consideration. It refers to [2, 7, 8], both consequences and outcome of direct outcomes such as human and animal morbidity and mortality, loss of biota including crop failure, destruction of materials, disturbance of life systems, hungry, drought, and other hazards in addition to shocks while indirect consequence is the derivative inertial adversaries of these outcome in thc form of residue. Adverse effects of climate change, on the other hand is, changes in the physical environment or biota resulting from climate change which has significant deleterious effects on the composition, resilience, or productivity of natural and managed ecosystems or on the operation of socio-economic systems or on human health and welfare [2]. There is high and increasing agreement among scientific society on the fact that [7–9] climate change affects all systems of the earth and its impact is greater in rain-fed agriculture of Africa.

Emerging literatures examine impacts of climate change on human capital development mainly on education by analyzing through different pathways [6, 10, 11], such as food and nutrition insecurity, infectious diseases (malaria, cholera, and diarrhea), and exposure to direct heat stress. The 2°C increase in temperature from normal is likely to reduce the schooling age of hotter environment children by 1.5 compared to the children of normal temperature environment. Hot days reduce performance on high stakes exams possibly by reducing the amount of learning achieved over the course of the school years and ultimately by reducing high school graduation rates [12]. On the other hand similar source indicates that above average temperature reduces education performance by up to 15% and lead to lasting impacts on educational attainment. Climate change affects food and nutrition security and subsequently this affects the cognitive development of the child even since the prenatal age, as pointed out by Dewey and Begum [10].

It is reported by the same authors that students drop the school at heat stress season. Families claim loss of farm owing to bad climatic conditions and thus, failed to send their children to school for financial constraints. This is reinforced by the report of UNDP-UNEP-UNCCD [13] as cited in IPCC [6] revealing that droughts can intensify the pressure to transfer children to the labor market in Ethiopia. With this path of climate change impacts in to health and education, although it is believed that formal education contributes to poverty reduction and economic development by fostering skills, intellectual ability, and employment opportunities, there is high and increasing confidence [7], among scientists that climate change affects education and earning from it then by keeping the poor in poverty trap.

DOI: http://dx.doi.org/10.5772/intechopen.98993

Further, although 57 million primary school-aged children worldwide (majority were in developing nations) remained out of school in 2015, in 2016, the United Nations released a set of more ambitious sustainable development goals (SDGs), one of which aims to achieve universal primary and secondary attainment by 2030 [14]. If climate change undermines educational attainment, this may have a compounding effect on underdevelopment that magnifies the direct impacts of climate change over subsequent times. On the other hand climate change affects human health and production capability through multiple and interactive paths. Climate change threatens [15], the health of people and communities by affecting their resilience and increasing their exposure, sensitivity, and adaptive capacity at all. On top of this, social determinants of health, such as those related to socioeconomic factors and health disparities, may amplify, or otherwise influence climate-related health effects, particularly when these factors simultaneously occur or close in time or space. All in all, climate change impact on human capital has been conceptualized by the interactive paths as depicted hereunder (**Figure 1**).

Although this theorized impacts of climate change on human capital stock and its capability, there is critical scientific studies' gap generally at a national level in Ethiopia and specifically at a local level of the study area. This study therefore investigated the impacts of precipitation and temperature anomalies on the health and education of rural community. By doing so, as it is the first insight by its content and substituents, in addition to indicating the point of support oriented interventions to various civic and private institutions, it informs policy makings at local, regional and national levels. In addition, it could serve as reference to academia.

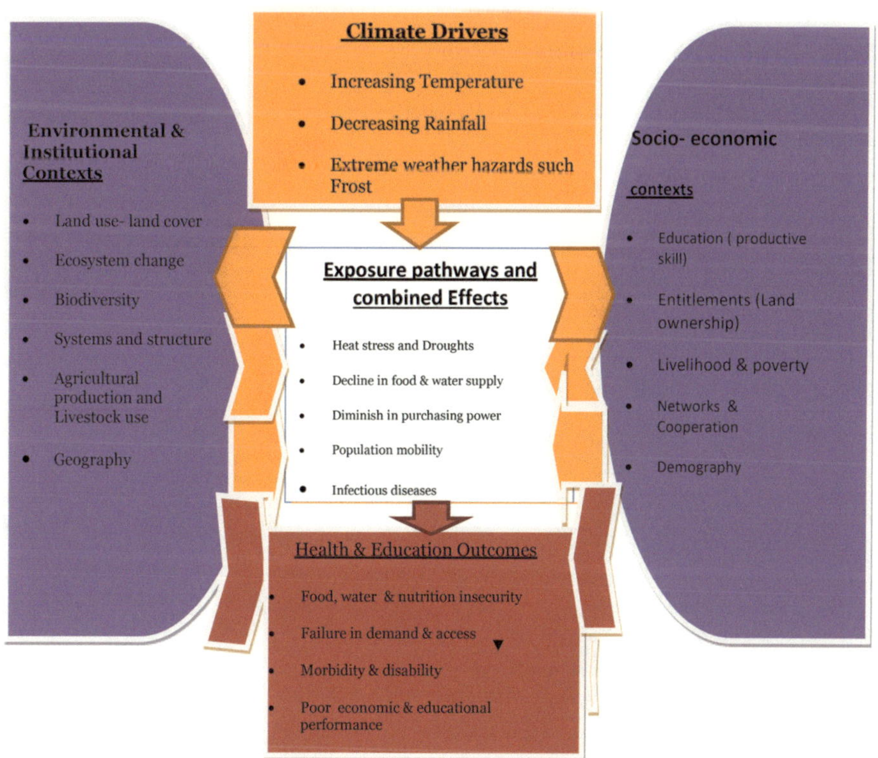

Figure 1.
Conceptualizing pathways of climate change impacts on human capital in rural community.

2. Study area

2.1 Description of the study area

This study was conducted within the 6.36°–7.14°N Latitude and 38.01°–38.56°E Longitude area in Ethiopia (**Figure 2**).

The study area is located in Sidama administrative region in Ethiopia. According to annual statistical report of the Sidama zone—now Sidama regional state—Finance and Economic Development Sector [16], Sidama Administration Zone was one of the 14 zones in the SNNPRS region. It is located in the north eastern part of the region and bounded by Oromia in the North, East, and South East, with Gedeo Zone in the South, and Wolayta Zone in the West. Its geographic location of Sidama regional state lies between 6°14′–7°18′ North Latitude and 37°92′–39°14′ East Longitude.

The region consists 30/Thirty/woredas (administrative districts), which are: Hawassa Zuria, Malga, Wondo-genet, Gorche, Wonsho, [Aleta] Chuko, Loka-abaya, Bursa, Bona-zuriya, Chire, Shebedino, Dalle, Aleta-wondo, Dara, Hula, Aroresa, Bensa, A[Ha]rbegona, Boricha, Hawela, Darara, Bilate-Zuria, Dara-Otilicho, Teticha, Shafamo, Chirone, Chabe-Gambeltu, Bura, Daela, and Hokko Woreda. Whereas, Aleta wondo, Aleta Chuko, Yirgalem, Leku, Chuko, and Daye are City Administrations. Total population of Sidama [17], is projected to be 4,271,739. The average population density of Sidama region in the 2017/18 is 635 person per/km^2. It is one of the densely populated regions in the country. The total area of the region is 6981.9 km^2. Its western part tilts towards southern part of Great East African Rift Valley in Ethiopia [18].

2.2 Agro-ecological description

Typically, Sidama is classified in to three distinctive ago-ecological zones [19, 20]:

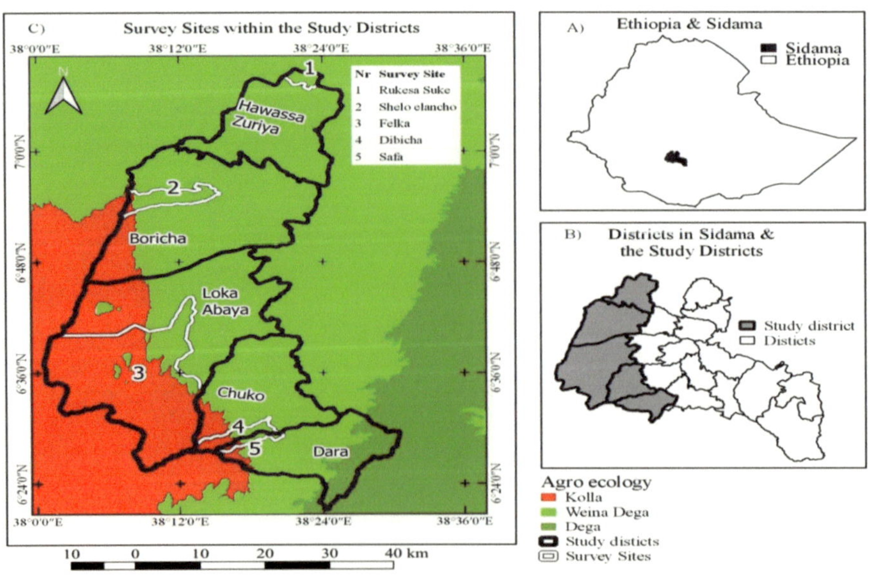

Figure 2.
Map of the study area.

2.2.1 'Qolla' (dry and hot tropical climatic zone)

It's is semi-arid agro-ecological zone. This is one of agro-ecological zones this study considered. It is the low land and lies between 500 and 1500 m above sea level. Receiving mean annual 400–1000 mm and 20–30°C rainfall and temperature, respectively, this agro-ecological zone constitutes 30% of total land of Sidama. Economically, settlers of this agro-ecological zone practice agriculture. Here, farming is dominated by annual crops such as maize (known as Maize livelihood belt), sorghum, and haricot bean whereas coffee and chat are also practiced as cash crops in higher altitude (transitional zone to Woina Dega) of the zone. Furthermore, this agro-ecological zone is mainly known for its livestock production in Sidama. On the other hand, in Sidama the area is known for extended and persistent droughts and hence most of the households had been in chronic food insecurity [8, 21].

2.2.2 'Woina Dega' (mid land, moist to humid, warm subtropical climate)

By constituting 54% of total land of Sidama, 'Woina Dega' agro-ecological zone lies within 1500–2500 masl. Climatically, it receives 1000–1800 mm mean annual rainfall and its mean annual temperature is known to be ranged within 15–20°C. The economic activity of this zone in Sidama is similar with that of the low land but here farming is dominated mainly by perennial crops such as coffee, 'enset', chat (Catha Adulis), tomato, and maize.

Although it is not uncommon to practice livestock economy in this zone, the farming dominance resulted in to shortage of grazing lands and made it problematic and hence it remained as only a supplementary sub-sector of the agriculture.

Though densely populated, this agro-ecological zone known to contribute higher share of cash crops from Sidama zone to local, regional, national, and international markets. Compared with the lower latitude zone, this agro-ecological zone, particularly to its higher altitude, is more food secured in terms of production and access dimensions thanks to 'enset', predominantly produced in this zone, and cash crops (mainly coffee, chat, and peen appeal) that are the sources of cash to access foods from market.

2.2.3 'Dega' (high lands, wet and cool temperate climate)

Constituting 15% of the total area of Sidama land, this agro-ecological zone has the elevation of 2500–3500 m above sea level. Climatically, this this agro-ecological zone receives 1200–1800 mm mean annual rainfall and has mean annual temperature of 10–15°C. This is the agro-ecological zone where most Sidama's rivers including the largest river, River Ganale, rises and tributes to Wabishabele River in Somalia.

2.2.3.1 Farming system and land use

The districts have bimodal rainfall pattern with two cropping seasons, short rainy season Spring (*Belg*) extending from February to May and the main rainy season *summar* (*Kiremt*) from June to September. The *Belg* rains are mainly used for planting short cycle crops such as maize production and seed bed preparation for *the Kiremt* crops.

The *Kiremt* rains are used for planting of cereal crops like some grains and vegetable crops and are also responsible for the growth and development of perennial crops such as enset, coffee, pineapple, and ch'at.

3. Material and methods

3.1 Data: source and type

This study employed data from secondary and primary sources. Secondary source dada include climate and agricultural data and primary source data include household survey and participatory rural appraisal (PRA; key informant, focused group discussion, and participant observation). Climate data for this study which encompassed the area of 6.36°–7.14°N Latitude and 38.01°–38.56°E Longitude was generated from National Meteorological Agency (NMA) on both monthly rainfall (mm) and daily maximum and minimum temperature for the period 1987–2017 which is 31 years' time. The NMA processes and documents a TAMSAT (**T**ropical **A**pplication of **M**etrology using **SAT**ellite data and ground-based observation) data in collaboration with World Meteorological Organization (WMO). This form of data is functional [22], in Matewos and Tefera [23], where there is no well-organized, well calibrated, validated, and reliable spatiotemporal weather data on station based observation (**Figure 3** and **Table 1**).

Agricultural data was collected from Bureau of Finance and Economic Development. Qualities of these data were controlled by identifying frontier outliers.

3.1.1 Household survey

This tool was employed to collect data from 400 respondents at a household level.

The sample size was determined based on Yemane [24]. Accordingly, the sample size will be as:

$$\boldsymbol{n} = \boldsymbol{N}/1 + \boldsymbol{N} * \boldsymbol{e}^2, \tag{1}$$

whereas
n stands for sample
N stands for population
E stands for error term (error margin). Thus,

$$\boldsymbol{n} = 7468/1 + 7468(0.05)^2 = >7468/18.665 = >400$$

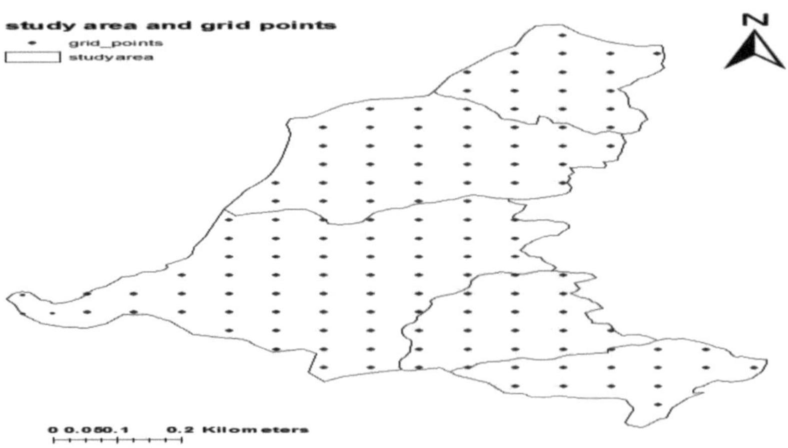

Figure 3.
Grid points of the study area.

No.	Study district	Total HHs (*N*)	Sampled farmers' association	Total HHs of the sample site	Sample size (*N*)
1.	Hawassa Zuria	30,841	Rukessa Suke	1195	64
2.	Boricha	65,957	Shello Elancho	1697	91
3.	Loka Abaya	27,085	Falka	1397	75
4.	Aleta Chuko	44,888	Dibbicha	1994	107
5.	Dara	46,453	Safa	1185	63
	Total	215,224	5	7468	400

Source: SAZFEDD (2017/18).

Table 1.
Population and samples.

Structured and semi-structured questions were employed to collect data on major research issues such as experiences of farmers on the trend of changes and variability in local climate (temperature and rainfall), major indicators of the climate change and its driven impacts on health and education of their households.

3.1.2 Key informant interview (KII)

This tool was employed to collect data from a community level. For this purpose, the study included key informant interview to gain in-depth information about their experience and observations on the climate change as well as impacts of change on the health and education at a community level.

3.1.3 Focus group discussion (FGD)

With the intention to collect data from household to the community ranges, as an augmentation on KII data, one FGD was designed for each of five administrative districts (five FGD's) under the study.

Normally, each of five groups had nine members who were from the composition of most elders, senior government representatives, the eldest household heading women, senior rural development agents, youth league leader in the structure of youth in local political arrangement, senior academician, religious society leader, senior health extension, and the representatives of civic societies who in one way or in other were expected to observe and/or feel the change and variability of climate and resulting impact on [their] farms and livestock within the society.

3.1.4 Participant observation

As another tool of PRA, this tool was employed to collect data on climate changes by observing various phenomena including bio-physical status of natural environments; personal physical condition and household assets (residences, farm, grazing fields and cattle on them and their physical statuses); and community's asset (social service institutions such as the institutions of health and education; ecosystems and their interactions). Aerial photographs were captured as evidence on what were observed and documented along with their relevant data.

Survey was conducted through trained enumerators under the supervision of the researcher under the assistance of moderators. There were one enumerator and one moderator for each of five districts and totally, there were eight individuals involved for overall data collection. PRA was carried out by researcher with the assistance of moderators.

3.2 Methods of data analysis

3.2.1 Climate data analysis

Climate data was analyzed based on time series analysis. It carries insightful importance in business and policy planning. It's applicable for a number of purposes including to study the past behavior of the phenomena under consideration; to compare the current trends with that in the past or the expected trends and to compare the performance of two different series of a different type for the same time duration. Mann–KENDALL TEST AND SEN'S SLOPE ESTIMATES (MAKESENS) was employed to test the trends the rainfall and temperature of the study area over last 31 years—1987–2017. Unlike a parametric analysis, in a regardless of the normal distribution of the population (*N*) of the study, the non-parametric regression (Mann–Kendall's (MK) test) fits for temporal scale changes of the units corresponding to each times. Thus, it was employed to test the rainfall and temperature trends of all 31 years (372 months) of Western Sidama drought prone districts.

The Mann–Kendall (MK) trend test tool [25–27] is commonly used, from among other alternative models of trend analysis, to determine if a trend exists, and can handle seasonal patterns within the data.

For the time series $x_1, \dots, x_n$, the MK test uses the following statistic equation:

$$S = \sum_{i=1}^{n-1} \sum_{j=k+1}^{n} \mathbf{Sign}(\mathbf{Xj} - \mathbf{Xi}) \quad (2)$$

If $S > 0$, then later observations in the time series tend to be larger than those that appear earlier in the time series and $s = 0$ implies no difference while the reverse is true if $S < 0$.

3.3 Standardized precipitation index (SPI)

The standardized precipitation index (SPI) is a widely used index to characterize meteorological drought on a range of timescales.

$$\text{SPI} = \text{X–Xm}/SD \quad (3)$$

Where,

X is actual precipitation; Xm, mean precipitation; SD, standard deviation.

Weather conditions per SPI value (**Table 2**).

3.3.1 Survey data analysis

Survey data was analyzed by using descriptive statistics.

3.3.2 PRA data analysis

Data from this source was analyzed based on issue by issue and case by case augmentation in such a way to ensure triangulation.

SPI value	Category
≥ +2.00	Extremely wet
+1.50 to +1.99	Very wet
+1.00 to +1.49	Moderately wet
−0.99 to +0.99	Mildly dry
−1.00 to −1.49	Moderately dry
−1.50 to −1.99	Severely dry
≤ −2.00	Extremely dry
Source: McKee et al. (1993).	

Table 2.
Standardized precipitation index.

4. Results and discussion

4.1 Rainfall and temperature trends

4.1.1 Rainfall trends

In terms of volume, as can be read from **Figure 4**, annual rainfall is seen decreasing with the magnitude of −3.8 mm/year which is not statistically significant (**Table 3**).

Similar finding in the region was reported by Belay et al. [28], Matewos and Tefera [23], and Misrak et al. [29] that both short and long period rainfall has decreased.

4.1.1.1 Standardized precipitation index (SPI)

Of all 31 years, 15 (48.4% of times considered) years, the area has been receiving below normal amount of annual rainfall (**Figure 5**). The year 2016 was the driest year followed by its successor while 2006 was the wettest one followed by 1997. In a decadal wise, most (10 out of 15 or 67%) droughts were recorded within 1998–2010. This implies that the years of 2000s were the years of frequent and continuous drought.

Of 15 droughts recorded within the time series (1987–2017), 11 (74%)—**Table 4**—were mild with SPI ranges of −0.02 to −0.96; 2 (13%) were moderate with the

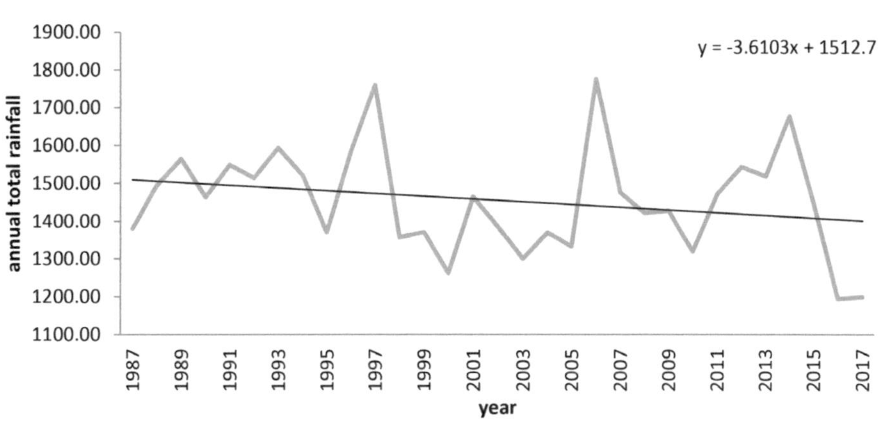

Figure 4.
Annual total rainfall trend of the study area.

Season	MK-test (z-test)	Trend	Significance	Sen's slope (mm/year)
Annual total rainfall	−1.19	Decreasing	Insignificant	−3.481
Belg total rainfall	−1.46	Decreasing	Insignificant	−1.974
Kiremt total rainfall	−0.87	Decreasing	Insignificant	−1.165

Source: Output of NMA data.

Table 3.
Annual and inter seasonal rainfall trends.

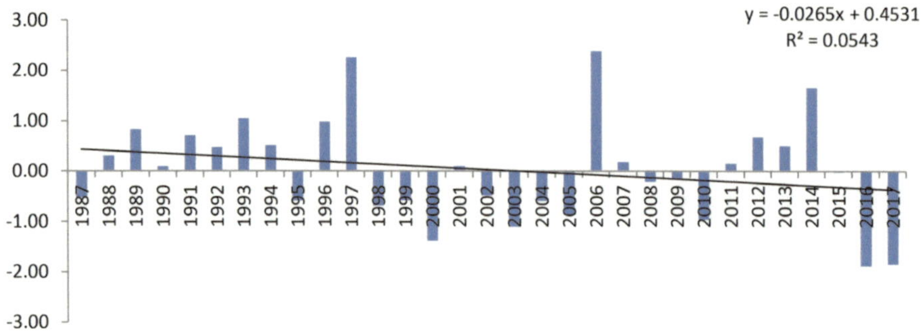

Figure 5.
Standardize annual mean rainfall anomalies of the study area. Source: MAKESENS output.

SPI values of −1.38 and −1.10 recorded in 2000 and 2003, respectively and the rest 2 (13%) were severe droughts with the SPI values of −1.88 and −1.85 (**Figure** 5), recorded in 2016 and 2017, respectively. Similar findings were reported by other researcher such as Matewos and Tefera [23] and Misrak et al. [29].

Though it is a stumbling block among scholars, definitions of the drought whatever it is climatological, atmospheric, meteorological, hydrological, agricultural, or be it is water resource management, commonly agree on that it is condition of insufficient moisture caused by insufficient (normal need) precipitation over a series of times.

Thus, according to McKee et al. [30] and Hayes et al. advices, precipitation (standardized) which is major factor for the sufficiency and insufficiency of water sources (ground water, stream flow, snowpack, and reservoir storage), time scale, and probability are vital variables to analyze climate born drought. While precipitation analysis enable to look at the supply side of water resource both for consumptive and usable demands, time scale points out to look the duration of

SPI values	SPI values (drought severity)	Number frequency (N = 31)	% of occurrence	Probability of occurrence of such drought
−2 and less	Extremely sever	0	0	Never ever happened
−1.99 to −1.5	Severe	2	13	1 in 15.5 years
−1.49 to −1.00	Moderate	2	13	1 in 15.5 years
−0.99 to 0.00	Mild	11	74	1 in 2.8 years
		15	100	

Source: Own computation based on McKee et al. (1993) classification.

Table 4.
Annual drought frequency probability analysis.

sufficiency or deficit for these demand. Similarly, analysis on probability of occurrence of drought enables to inform stakeholder for further preparedness on the likelihood of associated risks. On top of this, extremely severe drought has not ever recorded for the last 31 years as meteorological data has revealed (**Table 4**).

On the other hand, in each of 15 years and 6 months, the study area experiences two droughts, one severe, and another moderate, while a mild drought occurs in every 2.8 years. The implication is the area is definitely prone to drought varying from most frequently occurring mild drought to severe drought through the moderate one (**Table 4**).

4.1.2 Temperature

Annual average maximum temperature has shown the increasing trend which is statistically significant (at α = 0.05; P = 0.012; **Table 5**). From **Figure 6**, it can be learnt that annual average maximum temperature ranges 30.81–31.77°C for the last 31 years which implies the raise by 0.96°C.

Throughout the study time series, 1987–2017, annual average maximum temperature is characterized by increase in the end of 1980s, fall in the beginning of 1990s, raise again in the late of 1990s, fall onset of 2000s, raise then again in the late of 2000s, abrupt down again in the early 2010s and the sharp rise again at the momentum which touched the highest (31.77°C) level in 2016 which as **Figure 6** shows is, also, associated with SOI which is linked to strong El Niño event in 2015/16. With annual average maximum temperature of 31.77°C, the year 2016 is the hottest year where the 1996 is the wettest with annual average maximum temperature of 30.81°C of all 31 years. On the other hand, global land surface mean temperature rose in 1990s which is known as the hiatus or pause and then tended to

Maximum temperature	MK test (z-test)	Trend	Significance	Sen's slope (mm/year)
Annual average	2.01	Increasing	**	0.012
Belg average	2.58	Increasing	***	0.028
Kiremt average	1.90	increasing	***	0.014

** *Significant trend at α = 0.05.*
*** *Significant trend at α = 0.01.*

Table 5.
Annual, Belg, and Kiremt average maximum temperature.

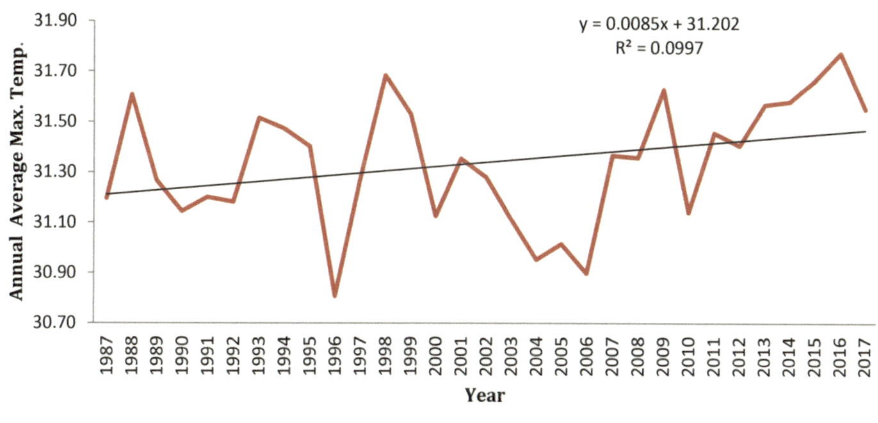

Figure 6.
Annual average maximum temperature of the study area.

decline then since 2000s, Parker, Wendy. Relatively, recent years have become warmer and warmer than the early years. This finding is similar with report by Ministry of Water Resources [31].

Thus, finding on local level temperature coincides with facts both at national and global levels.

4.2 Climate change impact on human health

Survey data revealed there is climate change impacts on human health recognized at a household and community level. The climate change impact on human health was explained through different paths.

The 77, 64, and 54% of the respondents believed that climate change has brought about negative impact on them and their family health through physiological inconveniences driven from extreme heat stress; occurrence of different diseases and increasing malnourishment rate (**Table 6**).

On the other hand, it was learnt from FGD and KIIs that climate change has brought about clear impact on the health of households and communities at large.

Intolerable heat stress and chilling frosts; diarrhea, typhoid, typhus and malaria; food and nutrition insecurities were commonly mentioned indicators of climate change impacts on the households and communities of the study area. This finding is in a line with Smith et al. [32] that reported multichannel impact of climate change on human health including heat stress, food insecurity, malnutrition and infectious diseases. Moreover, participant in four (80%; Safa of Dara, Falka of Lokka Abayya, Shello Ellancho of Borricha, and Rukessa Sukke of Hawassa Zuria) of five FGD held in the study area pointed out that impact of heat stress is exacerbated by acute water shortage in the districts.

In a spatial wise, the shortage was more resounded in Borricha, Lokka Abbayya, and Hawassa Zuria districts. Added, apart from health extension information, health services at health posts were neither easily accessible nor were they competent which signifies institutional malfunction, as participants claimed. Similar finding was reported by other scientific researchers including Hameso [33] and Matewos [34] that despite the fact that 97% coverage of health posts in the study area, the services were not sufficient and hence, as the poor members of the community could not afford for far distanced privately owned health services which, mostly, were found in the cities, they silently remain vulnerable to climate induced health shocks and further morbidities. These claims were partly found to be

Indicators of impact	Number of respondent (N = 400)	Proportion
Occurrence of different human diseases	256	64
Heat stress inconvenience on human physiology	254	77
Malnourishment rate increased	180	54
Couldn't afford education costs and thus cannot send children to the school	207	69
Children dropped the school for health attributed to heat stress and climate related health complications	194	66
Student failed standard test at the end b/c in one or another way of poor nutrition at the early age; poor attendance and follow at the lower grades	234	76

Table 6.
Impact of climate change on human capital.

acceptable by KIIs who were health extension and rural development agents in each of the districts.

Nutshell, data revealed that climate change has brought about negative impact on human health of the study area through the paths of generating physiological inconvenience (36%); mushrooming various climate driven human diseases (35%); and by exposing people to malnutrition (29%; **Figure 7**).

4.3 Climate change and responses from households and community at large

It was grasped from the participants of FGD and informants of KIIs that households and community at a large practice verities of copying mechanisms and adaptation strategies to withstand and to live with these climate shocks and reduce their impacts on their health. Receiving aids and transhumance were frequently mentioned practices to cope up climate shocks. Transhumance was being practiced to adjacent riverine localities of Bukito-Bura, Galade, Chiracha, Abaya Zuria, and Bilate Zuria. Despite participants and informants acknowledged the fact that this copying mechanism helped them to escape would be risks of droughts, at the same time, they also mentioned that they were exposed to get in to resource scarcity conflicts with pre-residing ethnic groups in the localities. According to their narrations, conflicts with Wolayita and Guji-Oromo ethnic groups were the typical ones. Thus, transhumance as a copying mechanism to family health impact of climate change offsets risks at hand while onsets other risks which is resource scarcity conflicts among rural community.

Furthermore, it was also raised by KIIs especially by school directors that, as long as households move with all the family members, students move with the transhumant family then by dropping out their school. Thus, children's right to education was being denied.

Receiving aid as another copying mechanism was also recognized to support the sustenance of lives exposed to climatic hazards.

On top of this, survey data revealed that 57% of the respondents responded receiving aid in terms of cash and kind. Though so, 64% of them responded that responded that receiving the said aid has encouraged the culture of dependency (**Figure 8**; **Table 7**).

Further, data revealed that the dependency syndrome among aid receiving rural household is manifested through poor saving (73%); negligence to adopting adaptation options (66%); skipping farming season (62%); and decreasing industriousness (52%).

Sum up, though copying mechanisms peoples practiced in the study areas in a response to negative impacts of climate changes were recognized to save human lives that could have not alive otherwise, data at the same time, pointed out that these mechanisms reproduced unintended negative impacts on community's peace

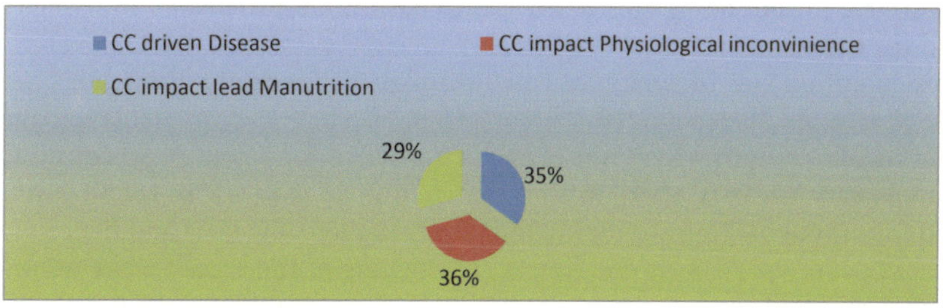

Figure 7.
Climate change impact on human health. Source: Survey 2020.

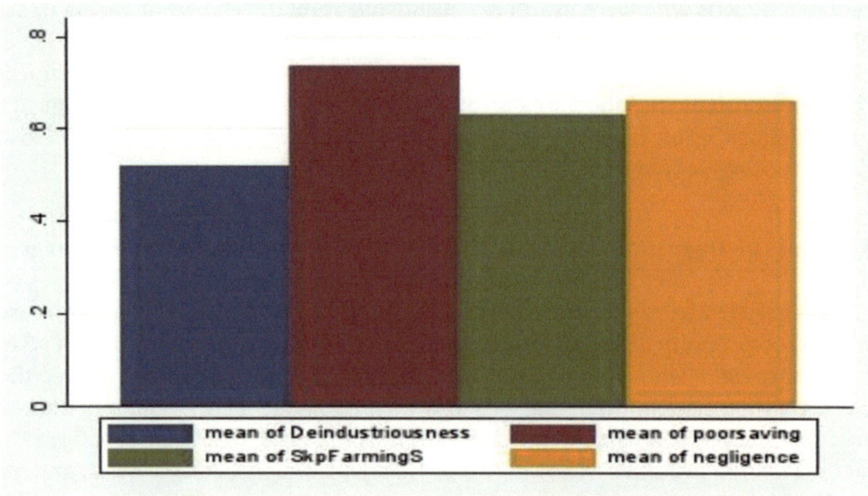

Figure 8.
Manifestations of aid dependency among rural households. Deindustriousness, decreasing industriousness; SkipFarmingS, skipping farming season; negligence, negligence to adopt adaptation strategies.

Issues	Obs	Mean	Std. Dev.	Min	Max
If a households receive aid	400	0.57	0.49	0	1
If receiving aid encouraged dependency culture	228	0.64	0.48	0	1

Table 7.
Aid receiving status of the households.

and stability, child's education, and individuals' commitment to economic growth and sustainable development.

Thus, the aid organizations and government should work on antecedently to climate crisis so that to reduce further multiplying impacts on the households and on the community at large.

4.4 Climate change impact on education

Result of survey pointed out that climate change has brought about negative impacts on the education of the people in the study area. And it was identified that paths of impacts are also different. Negative impact of the climate change was claimed to deny the child's right of access to education at rural settings. On top of this, 69% (**Table 6**) of the household believed that they had not been able to send their children to the school either at the onset of the school year or/and during the school course for not being able to afford for basic child provisions and facilities needed for the school.

This condition was further confirmed through all the group discussions held in the study area. As livestock was deteriorated in number; as crop yields remained only for hand to mouth which was even not sufficient for a household consumption, the groups narrated that, there was no source of finance to afford for all costs needed for school facilities of their children throughout the year and hence some household never enroll their children at the opening of the school while others' children drop amid the semesters. Per the FGD, the more the drier a year is, the less likely that a household send their children to the school; most likely the child, if was send perhaps, to drops out the school and follow the household's paths of copying mechanisms including temporary migration. This was affirmed through key informant interview with the school directors. Different adaptation strategies were being practiced at a community levels so as to reduce the impact of climate change on education including "School Feeding" program, aggressive afforestation at schools and digging boreholes in the schools so as to provide heat stressed and intolerably thirsted students with water at the school. On top of this, participants on focus group discussions and key informants (mainly school leaders) unambiguously disclosed that afforestation at the school has been contributing a lot so as to reduce heat irritations and inconveniences at schools. But at the same times, they did not report that as water tapes are running deeper following hydrological droughts in the areas, the boreholes they dug manually no more were being the sources of water for the schools; nor were there water pipelines nearby. This condition, therefore, requires the use of technology (which is beyond their affording capacities) such as water drilling machineries which dig up water from far deep and supply for the dehydrated students at the schools. This finding is alike with reports by Belay et al. [28] in the study of central rift valley of Ethiopia. On the other hand, according to key informant, particularly in three districts—Dara, Alata Chuko, and Boricha (**Figure 9**) there were some initiatives of "School Feeding" in some primary schools of the districts from the consortium of civic organizations and government to fill the food supply gaps of the households to their school children for which the FGD participants were very happy.

For its contribution to reduce school dropout rates and increasing better learning motivation of the children at the school, the "the School feeding" program was also recognized by the school directors and CEOs of the districts who were my key informants. Remarkably, from the report of year ended 2018 of ESDA, its program has contributed to 36% reduction of dropout rate while at the same time facilitated for better learning. However, both FGD participants and KIIs from the rest of two

Figure 9.
School feeding at Alawo Arfe Primary School under the supply of Ebenezer Supporting and Development Association (ESDA) in Boricha district in 2018. Source: ESDA.

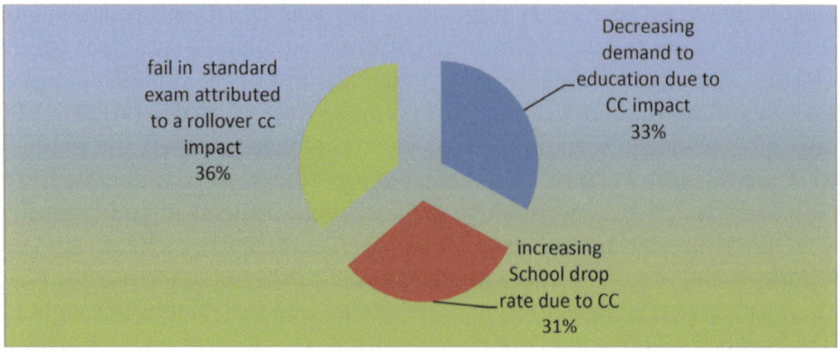

Figure 10.
Education dimension of climate change impacts on human capital. Source: Survey 2019.

districts—Falka of Lokka Abayya and Rukessa Sukke of Hawassa Zuria—claim that there was no such kind of "School Feeding" interventions over their districts.

Linked to this, it was learnt from other key informant from the administrative zone level that the said program was not launched formally at the zone level and what was observed at two district level was that a pilot works for further implementation to which the informant was not sure. On the other hand, 66% of the household believed that children drop the school for health impairment attributed to heat stress and climate related morbidities (**Table 6**). Further, as the same table portrays, 76% of the respondent believed that student failed standard test at the end because in one or another way of poor nutrition at the early age; poor attendance and follow at the lower grades. This finding is in a line with what Park [35] reported that heat stress has a profound negative impact on both short and long term educational attainments and the impact increases with lower adaptive capacity of economically poor community. Indeed, the Intergovernmental Panel on Climate Change [6] Fifth Assessment and report predicted that climate-induced health and income shocks could in turn negatively affect educational outcomes if, for example, children experience poorer health and nutrition in early life, thereby impairing cognitive and physical development; if households are unable to pay school fees; or if children must participate in income generation activities during school ages.

All in all, climate change impact on education as explained through three above said components can be weighted as 36, 33, and 31%, respectively for failing for standard examination attributed to a roll over impacts of climate change; decreasing demand to education due to deteriorating purchasing power which as a consequence of crop failure and declining livestock and increasing dropout rate as a function of extreme heat stress and other health related complications (**Figure 10**).

5. Conclusion

Data collected from secondary source and primary sources revealed that there is climate change that is defined through increasing temperature and decreasing rainfall over the study area. This change has brought about various impacts on rural community. Human capital (health and education) is one of livelihood dimensions that climate change has brought about negative impacts. To this end, the impact is 47% (**Figure 11**) for human health assessed through three sub-components (climate change impact driven diseases (physiological inconvenience (17%); prevalence of diseases (16%); and malnutrition (14%)) that are tracked in to in to one (health impact). On the other hand, the impact of climate change is 53% (**Figure 11**) for

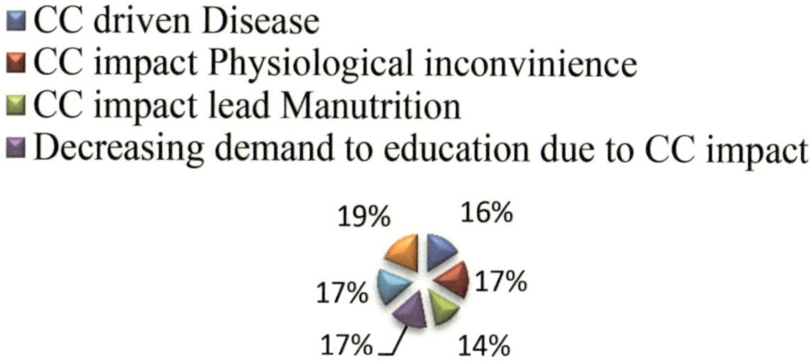

Figure 11.
Climate change impacts on health and education.

education assessed through other three sub-components (fail in standard exam attributed to a roll over impacts of climate change since early child hood (19%); increasing dropout rate (17%); and decreasing demand to education mainly owing to diminish in agricultural yield (17%)).

The latter is more linked to decrease in livestock and crop yields which are principal components and only the source of income for most of the households.

This diminution on the agricultural output resulted in to decline in food supply; diminish on their purchasing power (demand) of the households. Consequently, these households had not able to afford for education and education facilities of their children who either could not enroll to the school; or could not attend regularly; or drop it amid or fail to standard examinations of local and regional levels.

Thus, negative impact of climate change on human capital which is 53% for education and 47% for health implies, other things remain constant, that deteriorating future adaptive capacity of the rural community.

Data from survey and PRA also revealed that harmers in rural settings tried to cope up with and adapt climate change impacts on their health. It was also observed that there were some efforts of philanthropic and government organizations in supporting these endeavors the farmers. It was learnt these copying mechanism farmers practiced did contribute to save would be died lives. Nevertheless, it was also witnessed that some of these practices had led to unintended negative effects such as reinforcing interethnic; encouraging dependency syndrome and forcing children to drop the school.

Therefore, interventionists should plan to intervene before onset of climate change led crisis so as to reduce further mushrooming socio-economic problems during and crisis aftermaths.

Author details

Firew Bekele Worana[1*] and Cheru Atsimegiorgis[2]

1 Department of General Forestry, Wondo Genet College of Forestry and Natural Resources, Hawassa University, Ethiopia

2 Department of Statistics, Hawassa University, Ethiopia

*Address all correspondence to: firewbekele97@yahoo.com; heerihegi@gmail.com

References

[1] Sachs J. Common Wealth: Economics for Crowded Planet. Penguin Books; 2008

[2] UNFCCC. United Nations Framework on Climate Change Convention. New York: FCCC/INFORMAL/84; 1992

[3] World Bank. (2010). World Development Report 2010: Development and Climate Change. Washington, DC: The World Bank.

[4] Hameso S. Development challenges in the age of climate change: The case of Sidama. Workshop on the Economy of Southern Ethiopia. 01 March 2012. Department of Economics Hawassa, Hawassa University, Ethiopia. Ethiopia: Ethiopian Economics Association; 2012

[5] Adger WN, Dessai S, Goulden M, Hulme M, Lorenzon I, Nelson DR, Naess LO, Wolf J, Wreford A. Are there social limits to adaptation to climate change? Climatic Change. 2009;93:335–354

[6] IPCC. Climate change 2014: Synthesis report. In: Core Writing Team, Pachauri RK, Meyer LA, editors. Contribution of Working Groups I, II and III to the Fifth Assessment Report of the Intergovernmental Panel on Climate Change. Geneva, Switzerland: IPCC; 2014

[7] IPCC. Climate change 2007: Synthesis report. In: Contributions of Working Groups I, II, and III to the Fourth Assessment Report of the Intergovernmental Panel on Climate Change. Geneva: IPCC; 2007

[8] Tafesse, M. (2019). Climate Change - Induced Impacts on Smallholder Farmer in selected Districts of Sidama,Southern Ethiopia. Climate 2019, 7,70;DOI: 10.3390.cli7050070.

[9] Aryal M, Regmi PP, Thapa RB, Pande KR, Pant KP. Impact of climate variables to major food crops' yield in mid-hills of western development region, Nepal. Journal of Agriculture and Environment. 2016;17:1–8

[10] Dewey KG, Begum K. Long-term consequences of stunting in early life. Maternal & Child Nutrition 2011;7:5–18

[11] Graff ZJ, Shrader J. Temperature extremes, health, and human capital. The Future of Children. 2016;26:31–50

[12] Jisung, P. (2017). Hot Temperature, Human Capital and Adaptation to Climate Change. 1-38.

[13] UNDP-UNEP-UNCCD. Climate Change in African Dry Lands: Option and Opportunities for Adaptation and Mitigation. 2017. Available at: http://www.undp.org [Accessed: 20 April 2020]

[14] United Nations. GOAL 4: Ensure Inclusive and Equitable Quality Education and Promote Lifelong Learning Opportunities for All. 2016. Available at: https://sustainabledevelopment.un.org/sdg4 [Accessed: May 2019]

[15] Crimmins A. Impacts of climate change on human health in the United States: A scientific assessment. 2016. Available from: USGCRP resources: health2016globalchange.gov. EPA resources: www.epa.gov/climatechange/impacts.health (CAAAC-06/29/16)

[16] SZFEDS. Annual Statistical Report. Hawassa, Ethiopia: MA Printing Press; 2017/2018

[17] SZAFED. (2017). Sidama Administrative Zone Finance and Economic Development Department, 2017. Hawassa, Ethiopia: Artistic Press.

[18] Tafesse, M., & Tewodros, T. (2019). Local level rainfall and temperature variability in drought prone districts of rural Sidama, central rift valley region of Ethiopia. Physical Geography,DOI: 10.1080/02723646.2019.1625850.

[19] MoA. Agro Ecological Zonation of Ethiopia. Ethiopia: Addis Ababa; 2000

[20] Kumo WL. The Sidama Nations: History, Culture and Political Economy. Createspace Independent Publishing Platform; 2016

[21] Atara A, Tolossa D, Denu B. Assessment of food security situation of the rural households: The case of Boricha Woreda of Sidama Zone, Ethiopia. GeoJournal. 2019;86:711–727. DOI: 10.1007/s 10708-10094-2

[22] Dinku T, Hailemariam K, Ross M, Elena T, Stephen C. Combined use of satellite estimates and rain gauge observations to generate high-quality historical rainfall time series over Ethiopia. International Journal of Climatology. 2014;34:2489–2504

[23] Matewos T, Tefera T. Local level rainfall and temperature variability in drought-prone districts of rural Sidama, central rift valley region of Ethiopia. Physical Geography. 2020;41(1):36–53. DOI: 10.1080/02723646.2019.1625850

[24] Yemane T. Statistics: An Introductory Analysis. 2nd ed. New York: Harper and Row; 1967

[25] Kendall MG. Rank Correlation Methods. 4th ed. London: Charles Griffin; 1975

[26] Gilbert RO. Statistical Methods for Environmental Pollution Monitoring. NY: Wiley; 1987

[27] Gibbons JD, Chakraborti S. Nonparametric Statistical Inference. NY: Marcel Dekker; 2003

[28] Belay A, Recha JW, Woldeamanuel T, Morton JF. Smallholder farmers' adaptation to climate change and determinants for their adaptation decisions in the central Rift Valley of Ethiopia. Agriculture and Food Security. 2017;6:24. DOI: 10.1186/s40066-017-01100-1

[29] Misrak TH, Woldeamanuel T, Tadesse M. Spatial and temporal climate variability and change in the bilate catchment, central Rift Valley lakes region, Ethiopia. Physical Geography. 2021;42(3):199–225. DOI: 10.1080/02723646.2019.1698094

[30] Mckee, T. B., Doesken, N. J., & Kliest, J. (1993). THE RELATIONSHIP OF DROUGHT FREQUENCY AND DURATION TO TIME SCALES. Eighth Conference on Applied Climatology,17-22 January 1993, Anaheim, California (pp. 1-6). California: Department of Atmospheric Science,Colorado State University,Fort Collins, CO 80523.

[31] MoWR and NMSA. Initial National Communication of Ethiopia to the United Nations Framework Convention on Climate Change (UNFCCC). Ethiopia: Addis Ababa; 2001

[32] Smith KR, et al. Human health: Impacts, adaptation, and co-benefits. In: Field CB, et al., editors. Climate Change 2014: Impacts, Adaptation, and Vulnerability. Part A: Global and Sectorial Aspects. Contribution of Working Group II to the Fifth Assessment Report of the Intergovernmental Panel on Climate Change. Cambridge, UK: Cambridge University Press; 2014. pp. 709–754

[33] Hameso S. Farmers and policy-makers' perceptions of change in Ethiopia. Climate and Development. 2018;10(4):347–359. DOI: 10.1080/17565529.2017.1291408

[34] Matewos T. Climate change-induced impacts on smallholder farmers in selected districts of Sidama, Southern Ethiopia. Climate. 2019;7:70. DOI: 10.3390/cli7050070

[35] Park J. Hot Temperature, Human Capital and Adaptation to Climate Change. 2017

Chapter 22

Mitigating Climate Change Effects on Maternal and Prenatal Health in Nigeria

Adebanke L. Adebayo

Abstract

Generally, women and children have been considered at-risk populations, especially pregnant women, and their unborn babies. In the past decade, there has been overwhelming evidence linking climate change—extreme heat and air pollution—to adverse pregnancy, reproductive, and overall maternal health outcomes across the globe. This formative report highlights the effects of climate change. Using autoethnography and an adapted Delphi method—the combination of the different expert opinions—this report makes contextualized recommendations for women to mitigate the effects of climate change on maternal health outcomes in Nigeria.

Keywords: climate change, maternal health, prenatal health, heat wave, air pollution, flood, preterm birth, low birth weight, stillbirth, preeclampsia, Nigeria

1. Introduction

The discourse surrounding climate change has continuously emphasized the adverse effects on developing nations, especially among vulnerable populations—women and children. Globally, 700 newborn and 830 women die daily, with 99% of these deaths occurring in developing nations [1]. These alarming statistics that define maternal and neonatal health problems in developing nations will grow as the effect of climate change increases. According to the Intergovernmental Panel on Climate Change (IPCC), the negative influences of climate change on maternal and neonatal health in developing nations pose worse consequences if left unimpeded [2, 3]. Developing nations are already dealing with many public health issues such as malaria, diarrheal diseases, diabetes, and tuberculosis [4, 5]. Hence, a high confidence vote has been placed on African nations as being one of the most vulnerable continents to climate change and climate variability, a situation aggravated by the interaction of 'multiple stresses,' occurring at various levels, and low adaptive capacity [6].

Nigeria is listed among the top 10 countries of the world in the climate change vulnerability index [2]. Due to Nigeria's unique population, the consequences of climate change are likely to include increased water and food shortages, high exposure to heat stress, and dangerous levels of ultraviolet radiation. These climate change consequences have far more dire effects on at-risk populations—women and children. Also, extreme climate change influence on rainfall and temperature events can lead to increased incidence of diarrhea, cholera malnutrition, and even mortality in younger children.

Several reports have focused specifically on the direct and indirect negative effects of climate change on child and maternal health outcomes [7–11]. However, only a few studies have been conducted in Nigeria about climate-related health issues, especially related to climate change's direct and indirect effects on maternal and prenatal health outcomes [12–14]. Although most of the effects of climate change on health are yet to be foregrounded in climate research in Nigeria, it does not change the population's vulnerability and the current adverse health effects that are yet to be accounted for among at-risk populations.

2. Climate change framing in Nigeria

Nigeria is in West Africa—bordering Niger in the north, Cameroon in the west, Benin in the west, and Gulf of Guinea in the South. Nigeria is the most populous African country and seventh in the world with over 200 million people. She is often considered the giant of Africa because of her population and economy. Nigeria has the largest economy in Sub Saharan Africa and relies heavily on crude oil export. Aside from crude oil being the major source of Nigeria's economy, agriculture plays a pivotal role in the population's local economy and general sustainability. She has a diverse geography, ranging from arid to humid equatorial—equatorial in the south, tropical in the center, and arid in the north. Nigeria has a tropical climate with varying rainy and dry seasons, depending on location—more rainy days in the south, west, and east, and less precipitation in the north and northeast. With Nigeria having low precipitation in the north to high precipitation in the south, the climate change can lead to drought and desertification in the north, then flood and erosion in the south. Climate change would most likely impact sectors such as agriculture, fisheries, food security, water, forestry, health, economy, and energy.

Over the past couple of years, in the global context, there has been several media discourse and research around climate change impact on public health raised by the World Health Organization (WHO), United Nations (UN), Sustainable Development Goals (SDGs), and other governmental and private parastatal. The health concerns raised from the global discourse have led to the issuance of a global health concern [2, 15]. Other studies have been based on numerous studies of experts seeking to conceptualize climate change and reframe the perceived idea that climate change is a distant threat and highlight the anthropogenic nature [16–20].

Climate change is largely framed primarily as a public health issue within the global context, and the same is often projected to African nations [21, 22]. However, within the Nigerian context and Africa at large, there is a different discourse regarding climate change and its effects. For example, the media in Nigeria report climate change issues through the local and national lenses rather than the "expected" global context [23–25]. These media reports are centered on the negative effects of climate change on agricultural yields, economic strains, and infrastructural damage.

Only recently, after signing the Sustainable Development Goals (SDGs) and Millennium Development Goals (MDGs) in 2015, Nigeria began to report climate change effects on health. However, even with the SDGs and MDGs accentuation of climate change threats on health, the coverage and reports of climate change effects on health are still lagging. The media coverages are often centered around irregular rainfall, flood, drought, vulnerable farming populations, and desertification related to agricultural productivity—food quality and supply—and economic effects. Hence, it is through the lenses of agricultural sustainability and economic growth that climate change is mostly viewed.

3. Brief overview of climate change and health in Nigeria

Although Nigeria's coverage of climate change has been focused on agricultural and economic impact, there are growing reports and studies in the areas of health. Climate change studies in Nigeria have shown that increasing temperatures, heat stress, water quality, and air pollution can significantly have adverse direct and indirect effects on health outcomes [15, 26–28].

3.1 Direct effects of climate change on health outcome

Direct health impacts stem from extreme events such as heatwaves, floods, droughts, windstorms, and wildfires. Currently, there are numerous news reports of excruciating heat waves across the country. In the last 5 years, the whole country has been experiencing intense heat waves [29]. For example, in the first 4 months of 2019, the Nigerian Meteorology Agency (NiMet) reported an increase in the mercury of up to 5° higher than average—around 35°C—in the humid commercial capital [30]. Also, there were recorded incidences of deaths from heatwave in the northern part of Nigeria [30, 31].

It is pertinent to note that high temperatures can trigger tropical diseases, such as heat cramps, heat strokes, cerebrospinal meningitis [32], and malaria. Increasing temperatures and drought can also reduce freshwater quantity, increase heat stress, compromise hygiene, and raise the risk of water-borne diseases such as typhoid fever, cholera, dysentery, diarrhea, and even polio. Air pollution from climate change can increase the incidences of respiratory and cardiovascular diseases.

3.2 Indirect effects of climate change on health outcome

Indirect climate change effects can arise from malnutrition due to food shortages from floods and droughts. Rising temperatures, rising sea levels, floods, and rainfall patterns can affect freshwater supply, predisposing people to infections and water-borne illnesses. Increased rainfall intensity, flooding, stagnated water, and polluted groundwater will increase diseases like hepatitis and malaria commonly experienced in Southern Nigeria [32–34]. Heavy rainfall events can also lead to contaminated drinking water from sewage, industrial and chemical waste, leading to the outbreak of infections.

In coastal eco-zones, windstorms, and extreme rainfall, rising sea levels, and floods can cause injuries, drowning, death, severe physical and mental trauma, particularly for citizens who live along major river deltas, on islands, and in low-lying coastal areas [35, 36]. For example, in parts of Southern Nigeria, flooding from sea level rise has contaminated freshwater aquifers, rivers, and stock-watering points. This has increased salinity in these bodies of water and polluted them with sediment and sewage [37].

4. Maternal and prenatal health in Nigeria

Globally, 2.4 million children died in the first month of life in 2019 alone—approximately 7000 neonatal deaths every single day, most of which occurred in the first week of birth—about 1 million dying on the first day and close to 1 million dying within the next 6 days [38]. Most of these neonatal deaths happen in developing countries such as; India, Pakistan, *Nigeria*, China, Bangladesh, the Democratic Republic of the Congo, Ethiopia, Indonesia, Afghanistan, and the

United Republic of Tanzania [38, 39]. In Nigeria, the number of neonates that still die of preventable/treatable causes and diseases as of 2020 is as high as 38 per 1000 live births [40, 41].

Similarly, according to the World Health Organization (WHO), approximately 830 women die from pregnancy and child-related causes every day, and 99% of all maternal deaths occur in developing nations [42]. It is pertinent to note that nearly 20% of all global maternal deaths happen in Nigeria [43]. Compared to other developed countries, a Nigerian woman has a 1 in 22-lifetime risk of dying during pregnancy, childbirth, or postpartum/post-abortion. In contrast, in the most developed countries, the lifetime risk is 1 in 4900 [43].

With these alarming statistics and the prevalent climate change health effect on maternal and neonatal health outcomes, the mortality numbers might not decrease at the desired pace suggested in the Millennium Development Goal 5 (MDG) and Sustainable Development Goal 3 (SDG) nationally and globally, respectively.

5. Climate change and maternal/prenatal health outcomes

The alarming statistics that define maternal and neonatal health problems in Nigeria will most likely continue to grow as the effect of climate change increases. There is evidence that extreme temperatures and air pollution lead to adverse birth outcomes, including gestational age, birth weight, stillbirth, and neonatal stress from unusually hot temperature exposures [9, 44–46].

5.1 Increasing temperatures

According to Intergovernmental Panel on Climate Change (IPCC), the increasing global temperature and its detrimental influences on neonatal and maternal health have led them to strong concerns about the negative effects of climate change on the growth and development of babies both before (pre-natal) and after (postnatal) birth. However, climate change-related health issues will adversely affect developing nations—like Nigeria—already dealing with many public health issues such as malaria, diarrheal diseases, diabetes, and tuberculosis [4, 5].

Molina and Saldarriaga [47] have identified five pathways through which climate change temperature can affect the growth and development of a healthy fetus which includes exposure to extreme temperatures, maternal infection by a temperature-sensitive disease (e.g., respiratory infections) or by a biological vector-borne disease, maternal mental illnesses, and food insecurity brought about by less predictable growing conditions.

Temperature rises will extend the habitats of vectors diseases such as malaria that significantly harm children's health. Pregnant women are vulnerable to extreme temperatures and are especially susceptible to dehydration, which releases labor-inducing hormones [48]. Newborns are especially sensitive to ambient temperature extremes because their capacity for regulating body temperature is limited.

Extreme heat waves related to climate change may result in adverse pregnancy outcomes including but not limited to preterm births, still births, miscarriages, low birth weight, neonatal deaths, preeclampsia and eclampsia.

A study carried out to determine whether there is a seasonal variation in pre-eclamptic presentation undergoing cesarean delivery in a tropical rainforest belt in Nigeria showed seasonal variation in the cesarean delivery required for preeclampsia/eclampsia [12]. In an older retrospective report of eclampsia's meteorological relations in Lagos—the busiest city in Nigeria—the study showed that the incidence of this condition varies significantly with weather changes [14].

Droughts and flooding, where sanitary infrastructure is inadequate, will result in increased frequency of epidemics and enteric diseases, with maternal and neonatal health being the most vulnerable in developing nations [49, 50]. Furthermore, drought and flood reduce pregnant women's access to safe food and water, resulting in adverse psychological stress and disrupted health care, and increasing their vulnerability to dehydration (leading to pre-term labor), infectious agents and increased risk of exposure to environmental toxins.

5.2 Air pollution

During pregnancy, women have a unique immunological condition that causes modulation of the immune system at different stages of the pregnancy, which can lead to adverse reactions to environmental toxins and [51]. Also, pregnant women's ability to thermoregulate is compromised, making them particularly susceptible to heatwaves and air pollutants [52, 53].

Studies have shown links between prenatal exposure to environmental chemicals and adverse health outcomes that that cuts across life include impacts on fertility and pregnancy, neurodevelopment, and cancer. Warmer/higher temperatures can allow certain chemicals to break down into toxic elements, vaporize and easily enter the air we breathe, leading to adverse effects on fertility and pregnancy outcomes [11, 54, 55]. Climate change affects how chemicals deposited into the environment by factories, car emissions, and congestion in urban areas interacting with the air we breathe. Therefore, prenatal and early childhood exposure to air pollution, especially PM2.5, contributes to respiratory illness such as asthma, pneumonia and chronic obstructive pulmonary disease [56].

Because of the dangers these environmental exposures will have on maternal health outcomes, both the American College of Obstetricians and Gynecologists (ACOG) and the International Federation of Gynecology and Obstetrics (FIGO) have alerted the members of the reproductive and maternal health community about the far-reaching effects of climate change on maternal and neonatal health outcomes [57, 58].

As stated above, both increasing temperatures and air pollution are important factors impacting pregnancy and birth outcomes. Women who are exposed to hot temperatures/extreme heat, heat stress, or air pollution are more likely to have negative birth outcomes such as, but not limited to, miscarriages and stillbirth, preterm and low birth weight, and other preventable pregnancy and birth complications such as preeclampsia.

6. Autoethnography and Delphi technique

The recommendations discussed below are contextualized and targeted towards Nigerian women to help them actively engage in behaviors to mitigate the effects of climate change on maternal health outcomes. The recommendation from this formative report builds upon two key approaches: autoethnography and an adaptation of the Delphi technique.

The autoethnography conducted explored the major sociocultural perspectives of women in Nigeria and the researcher's perspective about climate change and health [59]. Through the ethnography, (a) there was an examination of the women's ways of speaking and relating about the changes in their immediate environment; (b) there was a core evaluation of the investigative use of space and place as it relates to climate events; (c) the researcher being part of the culture and possessing the same cultural identity was able to use personal experience as data to describe, analyze, and understand the women's sociocultural nuanced experiences [59].

Building on the existing autoethnography, this report employed the Delphi technique—a group facilitating technique that seeks to obtain consensus on the opinions of experts or individuals knowledgeable on the subject through a series of interviews [60]. The Delphi technique is a multistage process designed to combine [61] the different expert opinions into a consensus to best understand a problem, seek solutions, educate, and make informed recommendations to improve health outcomes. When there is limited or conflicting evidence, it is often used to develop evidence-based design recommendations and considerations for healthcare-built environments [62]. It has also been extensively used to generate many indicators relevant to health behaviors in communicating public health issues [63].

For this formative report, experts were purposefully selected—convenience sampling—to apply their knowledge to the existing research discourse on climate change and its effects on maternal/prenatal health outcomes (topical interview). These experts included obstetrician-gynecologists in Nigeria.

Experts were interviewed to explore their various opinions. They were asked three core questions; (a) what do you think women, especially pregnant women, need to know about the changing climate? (b) how do you think women protect themselves and their developing babies from the effects of climate change—rising temperatures; (c) how do you think women protect themselves and their developing babies from the effects of climate change—air pollution.

The experts provided nuanced contextual information related to the sociocultural and economic effects and how it is intertwined with climate effects on health outcomes. These experts' communication adequately informed the recommendations in this report.

7. Recommendations to mitigate climate change effects on maternal and prenatal health outcomes in Nigeria

The recommendations below are contextualized and targeted towards Nigeria women with active consideration for the sociocultural and socioeconomic factors that typify an average Nigerian woman. However, it is crucial to note that communicating the negative effects of climate change might be challenging. For example, it is easier to describe the need for nutrition than the consequences of being exposed to extreme temperatures, especially to women who are busy co-bread winners of their families and go about their daily hustle to make ends meet for their families. The extreme temperature is least considered when these women have hungry mouths to feed. However, numerous sociocultural and socioeconomic factors vary across ethnic groups with implications for engaging in some of these protective recommendations. The recommendation from this formative review builds upon three key factors: existing literature, autoethnography [59], and findings from the adaption of the Delphi technique conducted.

7.1 Extreme heat (hotter days)

Limit the amount of time you spend outdoors:

a. Stay indoors or in the shade—out of the Sun—especially during mid-day.

b. Stay hydrated: Drink a lot of water.

c. Wear light clothing to avoid excessive sweat.

When outdoors:

a. Use headcovers or umbrellas as shade. Wearing headgears is almost typical for women in northern Nigeria. Hence, wearing headgears can also serve as protection from the Sun.

7.2 Air pollution

Spend as little time as possible in places:

a. Where outdoor air pollution is worst, such as busy roadways and industrial plants.

b. Where indoor air pollution is worst because of generator fumes or poorly vented indoor gas/biofuel cooking.

c. During pregnancy, most women become very sensitive to unpleasant smells—this attribute can also be used to navigate spaces with certain forms of air pollutants in and around your homes and workspaces.

d. Download an air quality app on your smartphone so that you will know when outdoor air pollution is dangerous.

7.3 Information seeking

Ask your doctor about other protective measures for your baby's health from the risks associated with climate change.

8. Conclusion

Irrespective of the recommendations made, it is important to highlight that these protective actions would differ from one region of the country to another because of the climate variations. Also, self-efficacy is not just in the individual; it is also in the social and physical environmental barriers preventing the individual from engaging in certain behaviors or making health decisions. Climate change remains a monster staring Nigerian women's reproductive health in the face through the adverse effects of extreme temperatures and air pollution.

Acknowledgements

Dr. Bruce Bekkar, Dr. Edward Maibach, Dr. Gary Kreps, Dr. Oluseyi Asaolu and Dr. Oluwaseun Fadare.

Conflict of interest

The authors declare no conflict of interest.

Author details

Adebanke L. Adebayo
George Mason University, Fairfax, United States

*Address all correspondence to: aadebay@gmu.edu

References

[1] Maternal Health [Internet]. WHO|Regional Office for Africa. Available from: https://www.afro.who.int/health-topics/maternal-health [cited: 04 Sep 2021]

[2] IPCC—Intergovernmental Panel on Climate Change [Internet]. 2018. Available from: http://www.ipcc.ch/ipccreports/tar/wg2/index.php?idp=378 [cited: 15 Nov 2018]

[3] Intergovernmental Panel of Climate Change (IPCC). Current and Historic National Climate Change Impacts, Adaptation and Vulnerability. Contribution of the Working Group to the Fourth Assessment Report of the Intergovernmental Panel on Climate Change. 7701 Telegraph Road, Alexandria, Virginia 22315-3868: Institute for Water Resources, U.S. Army Corps of Engineers; 2007, p. 65

[4] Boutayeb A. The double burden of communicable and non-communicable diseases in developing countries. Transactions of the Royal Society of Tropical Medicine and Hygiene. 2006; **100**(3):191-199

[5] Stevens P. Diseases of poverty and the 10/90 gap [Internet]. 2004. Available from: http://who.int/intellectualproperty/submissions/InternationalPolicyNetwork.pdf [Accessed: 01 December 2016]

[6] Boston 677 Huntington Avenue, Ma 02115 +1495-1000. Africa CDC head is driving a new public health agenda on the continent [Internet]. News. 2020. Available from: https://www.hsph.harvard.edu/news/features/africa-cdc-head-is-driving-a-new-public-health-agenda-on-the-continent/ [cited: 04 Sep 2021]

[7] Senegal I, Namibia O, Abdrabo M, Ama E, Lennard C, et al. Chapter 22 Africa. In: Climate Change 2014: Impacts, Adaptation, and Vulnerability. Part B: Regional Aspects. Contribution of Working Group II to the Fifth Assessment Report of the Intergovernmental Panel on Climate Change. pp. 1199-1265

[8] Adebayo AL. The effects of climate change on maternal and child health [Internet]. Medium. 2019. Available from: https://medium.com/@bankiesworld/the-effects-of-climate-change-on-maternal-and-child-health-d4b2efc45521 [cited: 05 Jul 2020]

[9] Bakhtsiyarava M, Grace K, Nawrotzki RJ. Climate, birth weight, and agricultural livelihoods in Kenya and Mali. American Journal of Public Health. 2018;**108**(S2):S144-S150

[10] Grace K. Considering climate in studies of fertility and reproductive health in poor countries. Nature Climate Change. 2017;**7**(7):479-485

[11] Rylander C, Øyvind Odland J, Manning ST. Climate change and the potential effects on maternal and pregnancy outcomes: An assessment of the most vulnerable—The mother, fetus, and newborn child. Global Health Action. 2013;**6**(1):19538

[12] Sorensen C, Murray V, Lemery J, Balbus J. Climate change and women's health: Impacts and policy directions. PLOS Medicine. 2018;**15**(7):e1002603

[13] Okafor UV, Ezegwui HU. Cesarean delivery in preeclampsia and seasonal variation in a tropical rainforest belt. Journal of Postgraduate Medicine. 2010;**56**(1):21

[14] Omoruyi EP, Onafalujo AK. Effects of climate change on health risks in Nigeria. Asian Journal of Business and Management Sciences. 2011;**1**(1): 204-215

[15] Agobe JT, Good W, Hancock KW. Meteorological relations of eclampsia in Lagos, Nigeria. British Journal of Obstetrics and Gynaecology. 1981; **88**(7):706-710

[16] Brügger A, Dessai S, Devine-Wright P, Morton TA, Pidgeon NF. Psychological responses to the proximity of climate change. Nature Climate Change. 2015;**5**(12):1031-1037

[17] O'Connor RE, Bord RJ, Fisher A. Rating threat mitigators: Faith in experts, governments, and individuals themselves to create a safer world. Risk Analysis. 1998;**18**(5):547-556

[18] Leiserowitz A, De Franca DM, Poortinga W, Pidgeon NF. Cross-national comparisons of image associations with "Global Warming" and "Climate Change" among laypeople in the United States of America and Great Britain. Journal of Risk Research. 2006;**9**(3):265-281

[19] Nisbet MC. Communicating climate change: Why frames matter for public engagement. Environment: Science and Policy for Sustainable Development. 2009;**51**(2):12-23

[20] Swim JK, Clayton S, Howard GS. Human behavioral contributions to climate change: Psychological and contextual drivers. American Psychologist. 2011;**66**(4):251-264

[21] Whitmarsh L. What's in a name? Commonalities and differences in public understanding of "climate change" and "global warming". Public Understanding of Science. 2009;**18**(4):401-420

[22] Annual report 2020|UNFCCC [Internet]. Available from: https://unfccc.int/annualreport [cited: 04 Sep 2021]

[23] Tagbo E. Media coverage of climate change in Africa: A study of Nigeria and South Africa. Oxford, UK: Reuters Foundation; 2010

[24] Abdulhamid Y. The Impact of Climate Change in Nigeria [Internet]. 2011. Available from: https://www.semanticscholar.org/paper/The-Impact-of-Climate-Change-in-Nigeria.-Abdulhamid/6bafa71383ea445c603e593c864c9876be189ad4 [cited: 24 May 2020]

[25] Batta HE, Ashong AC, Bashir AS. Press coverage of climate change issues in Nigeria and implications for public participation opportunities. JSD. 2013;**6**(2):56

[26] Nwabueze C, Egbra S. Newspaper framing of climate change in Nigeria and Ghana. Applied Environmental Education & Communication. 2016; **15**(2):111-124

[27] Akpodiogaga-a P, Odjugo O. General overview of climate change impacts in Nigeria. Journal of Human Ecology. 2010;**29**(1):47-55

[28] Idowu AA, Ayoola SO, Opele AI, Ikenweiwe NB. Impact of Climate Change in Nigeria. Iranica Journal of Energy & Environment. 2011;**2**(2):145-152

[29] Haider H. Climate Change in Nigeria: Impacts and Responses. 2019. Available from: https://opendocs.ids.ac.uk/opendocs/handle/20.500.12413/14761 [cited: 04 Sep 2021]

[30] NiMet [Internet]. Available from: https://www.nimet.gov.ng/ [cited: 08 Sep 2021]

[31] Deadly heatwave rocks Nigeria. 2002. Available from: http://news.bbc.co.uk/2/hi/africa/2038164.stm [cited: 04 Sep 2021]

[32] Nigeria suffers severe heatwave with no relief in sight|Climate News|Al Jazeera [Internet]. Available from: https://www.aljazeera.com/

news/2019/4/6/nigeria-suffers-severe-heatwave-with-no-relief-in-sight [cited: 04 Sep 2021]

[33] Abdulkadir A, Lawal AM, Muhammad TI. Climate change and its implications on human existence in Nigeria: A review. Bayero Journal of Pure and Applied Sciences. 2017; **10**(2):152-158

[34] Building Nigeria's Response to Climate Change (BNRCC). National Adaptation Strategy and Plan of Action on Climate Change for Nigeria. Prepared for the Special Climate Change Unit of Nigeria's Federal Ministry of Environment. 2011

[35] Osuafor AM, Nnorom NR. Impact of Climate Change on Food Security in Nigeria. AFRREV STECH: An International Journal of Science and Technology. 2014;**3**(1):208-219

[36] Abdussalam AF, Monaghan AJ, Steinhoff DF, Dukic VM, Hayden MH, Hopson TM, et al. The impact of climate change on meningitis in Northwest Nigeria: An assessment using CMIP5 climate model simulations. Weather, Climate, and Society. 2014;**6**(3):371-379

[37] Ozor N, Madukwe MC, Enete AA, Amaechina EC, Onokala P. Barriers to climate change adaptation among farming households of Southern Nigeria. Journal of Agricultural Extension. 2010;**14**(1):114-124

[38] Newborns: Improving survival and well-being [Internet]. Available from: https://www.who.int/news-room/fact-sheets/detail/newborns-reducing-mortality [cited: 04 Sep 2021]

[39] WHO|PMNCH Fact sheet: Stillbirths [Internet]. WHO—World Health Organization. Available from: https://www.who.int/pmnch/media/press_materials/fs/201106_stillbirths/en/ [cited: 04 Sep 2021]

[40] Maternal and newborn health [Internet]. Available from: https://www.unicef.org/health/maternal-and-newborn-health [cited:04 Sep 2021]

[41] IGME: Levels and Trends in Child Mortality [Internet]. UNICEF DATA. 2020. Available from: https://data.unicef.org/resources/levels-and-trends-in-child-mortality/ [cited: 04 Sep 2021]

[42] Neonatal mortality [Internet]. UNICEF DATA. Available from: https://data.unicef.org/topic/child-survival/neonatal-mortality/ [cited:04 Sep 2021]

[43] WHO|Maternal health in Nigeria: Generating information for action [Internet]. WHO—World Health Organization. Available from: http://www.who.int/reproductivehealth/maternal-health-nigeria/en/ [cited: 04 Sep 2021]

[44] MacVicar S, Berrang-Ford L, Harper S, Huang Y, Bambaiha DN, Yang S. Whether weather matters: Evidence of association between in utero meteorological exposures and foetal growth among Indigenous and non-Indigenous mothers in rural Uganda. PLoS One. 2017;**12**(6):e0179010

[45] Kuehn L, McCormick S. Heat Exposure and Maternal Health in the Face of Climate Change. Int J Environ Res Public Health. 2017;**14**(8):853

[46] Zhang Y, Yu C, Wang L. Temperature exposure during pregnancy and birth outcomes: An updated systematic review of epidemiological evidence. Environmental Pollution. 2017;**225**:700-712

[47] Molina O, Saldarriaga V. The perils of climate change: In utero exposure to temperature variability and birth outcomes in the Andean region. Economics and Human Biology. 2017;**24**:111-124

[48] Balbus J, Crimmins A, Gamble JL, Easterling DR, Kunkel KE, Saha S, et al.

Ch. 1: Introduction: Climate Change and Human Health. The Impacts of Climate Change on Human Health in the United States: A Scientific Assessment [Internet]. U.S. Global Change Research Program; 2016. Available from: https://health2016.globalchange.gov/downloads#climate-change-and-human-health [cited: 20 Jun 2020]

[49] PMNCH|PMNCH Knowledge Summary #32. Protecting women's and children's health from a changing climate [Internet]. WHO—World Health Organization. Available from: https://www.who.int/pmnch/knowledge/publications/summaries/ks32/en/ [cited: 04 Sep 2021]

[50] Mor G, Cardenas I, Abrahams V, Guller S. Inflammation and pregnancy: The role of the immune system at the implantation site. Annals of the New York Academy of Sciences. 2011; **1221**(1):80-87

[51] USGCRP. The Impacts of Climate Change on Human Health in the United States: A Scientific Assessment [Internet]. Washington, DC: U.S. Global Change Research Program; 2016. pp. 1-312. Available from: https://health2016.globalchange.gov/executive-summary.html [cited: 04 Sep 2021]

[52] Lundgren K, Kuklane K, Gao C, Holmér I. Effects of heat stress on working populations when facing climate change. Industrial Health. 2013;**51**(1):3-15

[53] Di Renzo GC, Conry JA, Blake J, DeFrancesco MS, DeNicola N, Martin JN, et al. International federation of gynecology and obstetrics opinion on reproductive health impacts of exposure to toxic environmental chemicals. International Journal of Gynecology & Obstetrics. 2015; **131**(3):219-225

[54] The Reality of Maternal Mortality in Central African Republic [Internet]. The Borgen Project. 2017. Available from: https://borgenproject.org/maternal-mortality-in-central-african-republic/ [cited: 03 Dec 2020]

[55] Nathaniel DeNicola MD. Toxic environmental exposures in maternal, fetal, and reproductive health [Internet]. Contemporary OB/GYN. 2018. Available from: https://www.contemporaryobgyn.net/article/toxic-environmental-exposures-maternal-fetal-and-reproductive-health [cited: 21 Apr 2020]

[56] Exposure to toxic environmental agents. Fertility and Sterility. 2013; **100**(4):931-934

[57] International Federation of Gynecology and Obstetrics (FIGO) opinion on reproductive health impacts of exposure to toxic environmental chemicals [Internet]. Program on Reproductive Health and the Environment. Available from: https://prhe.ucsf.edu/international-federation-gynecology-and-obstetrics-figo-opinion-reproductive-health-impacts-exposure [cited: 04 Sep 2021]

[58] Climate Change and Women's Health [Internet]. Available from: https://www.acog.org/en/clinical-information/policy-and-position-statements/position-statements/2018/climate-change-and-womens-health [cited: 04 Sep 2021]

[59] Adebayo AL. Sociocultural perspective of climate change and maternal health outcome in West Africa. In: Proceedings of the Central States Communication Association: Global Inspiration: A Discussion on Healthcare Barriers and Concerns for Underserved Populations [Internet]. 2021. Available from: https://www.csca-net.org/aws/CSCA/asset_manager/get_file/548924?ver=13

[60] Hasson F, Keeney S, McKenna H. Research guidelines for the Delphi

survey technique. Journal of Advanced Nursing. 2000;**32**(4):1008-1015

[61] Lynn MR, Layman EL, Englebardt SP. Nursing administration research priorities: A national Delphi study. JONA: The Journal of Nursing Administration. 1998;**28**(5):7-11

[62] Taylor E. We agree, don't we? The Delphi method for health environments research. HERD. 2020;**13**(1):11-23

[63] Vantamay N. Using the delphi technique to develop effectiveness indicators for social marketing communication to reduce health-risk behaviors among youth. The Southeast Asian Journal of Tropical Medicine and Public Health. 2015;**46**(5):9

Chapter 23

Climate Change and Its Potential Impacts on Insect-Plant Interactions

Somala Karthik, M.S. Sai Reddy and Gummudala Yashaswini

Abstract

The most dynamic and global environmental issue to date is climate change. The consequences of greenhouse effect and climate change from rising temperatures, frequent droughts, irregular rainfall, etc. are already evident. Insects and plants are affected by climate change and extreme weather events and the direct impact of anthropogenic climate change has been reported on every continent, in every ocean and in most major taxonomic groups. In the modern period, as a result of natural cycles and anthropogenic activities and their effects on the global climate, plants are typically susceptible to new environmental factors, *i.e.* higher levels solar radiation, rise in temperatures, greenhouse effect and changes in rainfall patterns over the seasons. Increased temperatures, CO_2 and rapid changes in rainfall patterns can dramatically alter the biochemistry of plants and thus plant defence responses. This can have important implications in insect fertility, feeding rates, survival, population size, and dispersal. The relationships between plants and insects are thus changed with significant consequences for food security and natural ecosystems. Similarly, mismatches between plants and insect pollinators are caused by the acceleration of plant phenology by warming. Human nutrition which depends on insect pollination can be affected with reduction in plant reproduction and fitness. Thus, understanding abiotic stress reactions in plants and insects is relevant and challenging in agriculture. In the preparation and implementation of effective strategies for future insect pest management programmes, the impact of climate change on crop production, mediated by changes in the populations of extreme insect pests should be carefully considered.

Keywords: Climate change, temperature, CO_2, insects, plants

1. Introduction

Climate change is a worldwide threat that is unavoidable and immediate which encompasses a combination of natural and anthropogenic changes in the environment. Worldwide attention has been attracted by recent changes in global climate phenomena and consequent losses. Climate change, according to the Intergovernmental Panel on Climate Change (IPCC), is described as "any change in climate over time, whether due to natural variability or as a result of human activity". Human activities are responsible for much of the warming that has been observed over the last 50 years. From 1990 to 2100, the global mean surface temperature is expected to rise by 1.4 to 5.8°C. In the next 100 years, if temperatures increase by around 2°C, the detrimental global warming effects will begin to spread in much of the world's region [1]. In addition, CO_2 levels rose from 280 ppm to 401 ppm in 2015 (Mauna Loa Observatory: Hawaii).

Insects constitute over half of the estimated 1.5 million organism species of the biodiversity identified so far on the planet and are fundamental to the structure and function of ecosystems. Insects are among the most susceptible groups of organisms to climate change as they are ectothermic, so thermal changes have strong direct effect upon their growth, reproduction and existence [2]. The effects of climate change on insect pests are of greater significance because insects are involved in many biotic interactions, such as plants, natural enemies, pollinators and other organisms, which play a major role in the ecological functioning of insect pests [3]. The impact of climate change on arthropod extinction rates is 100 to 1000 times greater than what has occurred previously, with about 45 to 275 species becoming extinct on a daily basis. An increase in a temperature rise of 6°C would result in the mass extinction of species, including humans. For example, due to hot temperatures (like heat waves) related to climate change, have resulted in a decrease in bumblebee populations by 46 per cent in North America and by 17 per cent across Europe compared to the base period of 1901 to 1974. In India Basavarajappa S, has observed a 2 per cent decline in rock bee, *Apis dorsata* colonies every year in Mysore due to increase in temperature, altering its local climate.

Climate change and extreme weather events affect insects and plants, and the direct effect of anthropogenic climate change has been recorded on every continent, in every ocean and in the majority of major taxonomic groups. In the modern era, plants are habitually vulnerable to new environmental factors *i.e.,* solar radiation, high temperatures, rise in CO_2 levels and shifts in pattern of rainfall over the seasons, as a result of natural cycles and anthropogenic activities and their impact on the global environment. Because of the close relationship between insects and host plants, through the changes undergone by their host plants, insect herbivores are likely to experience direct and indirect consequences of climate change. Global climatic changes are also expected to influence interactions between insects and plants in many ways. They may directly affect insects through changes in physiology, behaviour and life history parameters, as well as indirectly through changes in their own life history experienced by host plants.

2. Factors governing the climate change

Over thousands or millions of years, global climate change typically occurred very slowly. But today, by contrast, our environment is changing fast. There are many factors that govern the climate change around the world. The most important factors are discussed below:

2.1 The sun and the cosmic rays

Climate change is influenced by natural changes like the amount of solar energy reaching the Earth. The rate of energy emitted by the Sun varies slightly from day to day. Over many millennia the relationship between Earth and Sun can change the geographic distribution of the energy of the sun throughout the earth's surface. The orbit of Earth around the Sun is an ellipse and when it changes in shapes, the Earth moves nearer to the Sun which makes our climate much warmer.

The orientation of earth's axis can also affect the amount of sunlight reaching the earth's surface [4]. The angle of rotation of the earth's axis varies over time and it shifts from 22.1° to 24.5° and back again for around 41,000 years. With increase in the angle the summers become warmer and the winters turn colder. The Sun also emits particle radiation, primarily protons and electrons, which comprise the solar wind. These particles come near to the earth, but the earth's magnetic field averts them from reaching the surface. The earth's atmosphere reaches more intense executions, known as solar cosmic rays. Cosmic solar rays cannot be reaching the earth's surface, but are extremely energetic, collide with atoms at the top of the atmosphere, causing major magnetic field perturbations to disrupt power lines and electrical equipment [5]. It has been suggested that changes in solar output might affect our climate-both directly, by changing the rate of solar heating of the Earth and atmosphere, and indirectly, by changing cloud forming processes. The increase in absorption of solar radiation results in rise in temperatures which in turn results in upsurge of CO_2 levels. Shrivastava [6] suggested that rise of 1°C will result in the release of 30 petagrams of carbon from the soils, which is almost twice the amount emitted due to human activities annually.

2.2 The greenhouse effect

Greenhouse gases are the molecules that are capable of absorbing infrared radiation released from the surface of the Earth and re-radiating it back, thereby leading to the greenhouse effect phenomenon. During the history of the Earth, greenhouse gases concentrations such as water vapour, carbon dioxide, methane, nitrous oxide, ozone and certain artificial chemicals like Chloro Fluoro Carbons (CFCs) have varied considerably, and these fluctuations have triggered major climate changes at a wide range of timescales. Human activities, particularly the combustion of fossil fuels (coal, oil and natural gas), agriculture and land clearing, are responsible for rising concentrations of greenhouse gases. This has intensified the greenhouse effect, leading to earth's warming.

2.3 Human influence

The factors above mentioned affect the climate naturally. However, we could not forget the effect of human activities on the changes in climate. Early in history, influence of human on the climate would have been quite small. Since, the beginning of the Industrial Revolution, at the end of 19th Century in the atmosphere there was a rise in the emission of the amount of greenhouse gases. The number of trees being cut down by humans has also increased, resulting in reduced uptake of carbon dioxide by the forests. Black carbon (BC), a solid particle or aerosol that is not a gas, leads to atmospheric warming. Unlike GHGs, in addition to absorbing infrared radiation, BC can also directly absorb incoming and reflected sunlight. It may also settle on the snow and ice, darkening the surface and thereby increasing the snow's absorption of sunlight and accelerating the melting process. Sulphates, organic carbon, and other aerosols might cause cooling by reflecting sunlight. Clouds can interact with warming and cooling aerosols, alters a number of properties of cloud like the rate of formation, dissipation, reflectivity, and precipitation. They may contribute to cooling, by reflecting sunlight and warmth and by trapping the outgoing heat.

True insights about climate change can be provided by factors such as temperature, precipitation (amount, frequency and timing), humidity, wind (velocity, timing), gaseous concentration etc.

Factors	Insects	Plants
a. Temperature	• Evolutionary changes • Reproductivity • Life period • Metabolism • Activity • Migration	• Photosynthesis • Respiration • Phytochemicals • Germination • Flowering
b. Humidity	• Development • Survival • Behaviour • Physiology • Reproduction	• Transpiration • Nutrients from the soil • Photosynthesis • Pollination • Incidence of diseases
c. Precipitation	• Survival • Development • Reproduction • Distribution	• Photosynthesis • Spread of diseases • Development • Transpiration • Pollination • Competitive Suppression • Life period
d. Wind	• Distribution • Behaviour • Abundance • Reproduction • Survival rate • Pollinating	• Photosynthesis • Transpiration • Lodging • Chilling injuries • Pollination
e. Greenhouse gases	• Reproductive capabilities • Distributional ranges • Physiology • Behaviour • Population dynamics	• Photosynthesis • Stomatal conductance • Oxidative stress • Carbon to Nitrogen (C:N)

3. Effect of different climate change factors on insect pest, plants and their interactions

In agriculture, climate change can interfere in normal plant physiologies such as photosynthesis, respiration, transpiration, nutrient absorption, balance of minerals and exchange of ions etc. It may also intervene with the production of crops by altering the population and function of insect pests. Climate variables such as temperature, humidity, precipitation etc. are accountable for the growth, development and multiplication of organisms like insects, fungi, bacteria, virus etc. As with the changing climate, populations of pest are also expected to change. In addition, climate change is expected to fetch modifications in host plant resistance against insect pests. The resistance can be overcome by faster disease cycles and altered physiologies of insect pest. As global warming is caused by climate change, several insect species are affected in terms of their distribution, demography, and life history parameters. The response of an insect population to a swiftly changing climate will be inconsistent when insects interact with different competitors, predators and parasitoids. This also affects overall food production systems that can be at critical risk due to the consequences of climate change [7]. These changes inflict consequences on human livelihood, including the rapid spread of pest and diseases of important crops. This has brought new challenges to agricultural sustainability.

3.1 Effect of temperature on insect pest and plants

The global average temperature is expected to increase by at least 4°C by the end of the 21st century, due to the increased frequency and intensity of drought and heat waves [8]. Temperature has a strong effect on insect growth, survival and reproduction and enrols a major role in controlling the development and growth of their host plants. In addition, the development of plant secondary chemicals as well as the structural characteristics used to protect against herbivores are influenced by temperature. Thus, for both insects and plants, temperature has potentially significant consequences (**Figure 1**). Phytochemical and morphological changes in host plants are caused by changes in temperature. For example, at night temperatures of 17°C, the concentration of catecholic phenolics (chlorogenic acid and rutin) in tomatoes was significantly higher than at other temperatures [9]. Also, Rivero et al. [10] reported low polyphenol oxidase (PPO) activity of peroxidase (POX) at 35°C in tomatoes; it has been also reported that there is a substantial decrease in protease inhibitor activity in tomato at temperatures below 22°C [11]. At elevated temperatures, the thickness of leaf trichomes normally rises [12].

In alfalfa (*Medicago sativa*), the concentrations of plant secondary metabolites (sapogenins and saponins) were elevated at increased temperatures, suppressing the growth of caterpillar (*Spodoptera exigua*). By contrast, the Green-veined

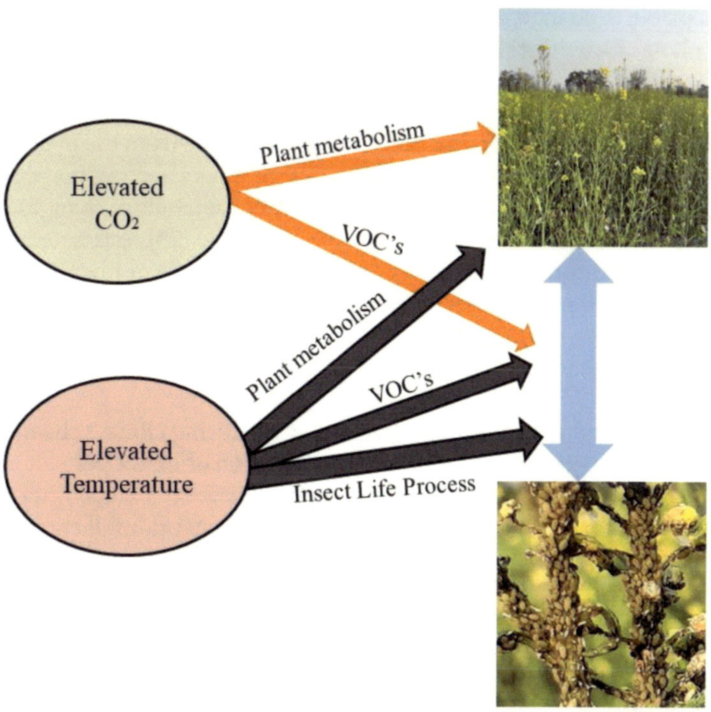

Figure 1.
Effects of elevated CO_2 and temperature on plant, insect and their interaction.

butterfly, *Pieris napi* reacted to warming-mediated poor-quality foliage in Brassicaceae, by consuming significantly higher amounts of plant tissue [13]. However, when fed on oilseed rape plants subjected to different temperatures with nutritional quality variations, the production of aphids (*Myzus persicae* and *Brevicoryne brassicae*) was not affected [14]. Moreover, temperature-induced tobacco shifts (*Nicotiana tabacum*) have an impact on the tobacco hornworm, *Manduca sexta* that the normally accepted law of temperature size, which predicts an improved final mass of ectotherms (e.g. insects) at lower temperature, has been reversed [15].

3.2 Effect of carbon dioxide (CO_2) on insect pest and plants

Higher concentrations of CO_2 with the rise in temperatures in the atmosphere have direct effects on plant metabolism and affect the distribution, abundance and productivity of insects that feed on plants (**Figure 1**). The behaviour of phloem-feeding insects, when supplied with plants grown under increased CO_2, increases compared to leaf chewing insects [16]. When leaf chewing insects like grasshoppers and caterpillar larvae feed on plants that are grown under higher CO_2 levels, more leaf area is eaten than they actually eat [17]. *Spodoptera litura* has been reported to grow under higher levels of CO_2 as a serious pest [18]. The larvae of *Helicoverpa*, grown under high CO_2 ate much more leaf tissue than those under ambient CO_2. However, under elevated CO_2, adult moths increased and lived longer and laid considerably few eggs [19].

The change in CO_2 concentration also influences the plant biochemistry, along with the synthesis of secondary metabolites [20]. The higher concentration of CO_2 is subjected to increased ratio of carbon to nitrogen in plants. Insects are allowed to consume more in order to achieve sufficient dietary nitrogen, resulting in slower larval growth and increased mortality. Phytophagous insects can become more susceptible to changes in atmospheric CO_2 concentration by CO_2 cascading effects on plant biochemistry, as certain plant feeding insect species produce their pheromone molecules on the basis of compounds taken from the host plants [21]. Example: Bark beetles use the mevalonate pathway to generate pheromones, where certain components of aggregation pheromones originate from the hydroxylation of secondary metabolites derived from tree [22]. Besides affecting the plant biochemistry, along with the synthesis of secondary metabolites changes in CO_2 concentration could also affect the plant yield. Example: [23], estimated a yield loss in wheat, maize and cotton of 36 to 40 per cent in a scenario of low CO_2 emissions, and between 63 to 70 percent in a scenario of high CO_2 emissions.

3.3 Effect of precipitation on insect pest and plants

More frequent and extreme precipitation events during climate change are expected to have detrimental effects on the population of insect pests. It is one of the weather factor that acts upon the activities of several insects by means of soil moisture or directly when exposed. Increased summer rainfall encourages a rapid rise in the soil dwelling wireworms, *Agriotes lineatus* population and larvae of root chewing insects, *Agriotes lineatus* [24]. Soil moisture kills insects by means of submerging in water, or affects the soil texture by preventing the emergence of insects. It is also harmful mainly to the insects that are free living in the soil as eggs or as newly-hatched larvae or nymphs.

The effect of the intense raindrops or water in the leaf axils will dislodge or drown small insects such as aphids, or newly-hatched larvae or nymphs from the plants. High proportions of cabbageworm young larvae, *Pieris rapae* and diamondback moth, *Plutella xylostella* on cabbage, are killed by high precipitation. Intense precipitation also has a catastrophic effect on the boring insect eggs and newly-hatched larvae such as the European corn borer, *Ostrinia nubilalis*, before boring into the plants. It also destroys aestivating adults of the black cutworm larva, *Agrotis ipsilon* and results in drowning of larvae in low-lying areas. Changes in pattern of rainfall are tracked by desert locust, *Schistocerca gregaria* migratory patterns in Sub-Saharan Africa [25]. Precipitation also has a positive association with plant height, total area of the leaves, number of plants and number of leaves, nitrogen and chlorophyll content of the leaves, which has a direct or indirect impact on the population of insect pests.

3.4 Effect of climate change on interaction between insect pests and plants

Climate change can directly affect insect-plant interactions and alter the functioning of both insect pests and plants. The development of secondary metabolites of the plants and other phytochemicals is also affected (**Table 1**). Both plant and herbivore structures can be modified by increasing temperature, CO_2, precipitation, etc. Rise in global temperature, atmospheric CO_2, and the duration of dry season are all likely to have consequences for tropical plant/herbivore interactions, with significant implications on food security and natural habitats. It will increase

GEC driver	Effects	Plant–herbivore	Plant–pollinator
Temperature	+ (Positive)	Flavonoids, condensed tannins, total phenolics, alkaloids, lignin, saponins, volatile terpenes	GLVs, volatile aromatic hydrocarbons, volatile terpenes
	-(Negative)	Condensed tannins, lignin	Volatile aromatic hydrocarbons, volatile terpenes, GLVs, volatile benzenoids, volatile nitrogen containing compounds, nectar sugar concentration
	0 (Neutral)	Cardenolides, lignin, condensed tannins, phenolic glycosides	Volatile terpenes, GLVs, nectar sugar concentration
Carbon dioxide	+ (Positive)	Glucosinolates, alkaloids, flavonoids, saponins	—
	-(Negative)	Phenolics, condensed tannins	Pollen protein
	0 (Neutral)	Saponins	Nectar sucrose concentration
Precipitation	+ (Positive)	Cardenolides, glucosinolates, iridoid glycosides, volatile terpenes, GLV	Volatile terpenes, GLVs, volatile benzenoids, and volatile aliphatics
	-(Negative)	Volatile terpenes, GLVs, volatile aliphatic ketones, volatile nitriles, phenolics, monoterpenes and sesquiterpenes	Volatile terpenes, GLVs, volatile benzenoids
	0 (Neutral)	Lignin, glucosinolates, GLVs, volatile terpenes, polyphenolics, condensed tannins, terpenes	Volatile terpenes, GLVs, volatile benzenoids, volatile nitrogen containing compounds

Table 1.
Effects (+, −, 0) of GEC drivers on plant chemical traits that mediate plant–herbivore and plant–pollinator interactions [26].

the effect of pests benefiting from reduced host defences due to stress resulted from the lack of adaptation to sub-optimal conditions of climate. Climate change could support non-resistant crops or cultivars, contributing to greater insect pest's infestation [27]. But, plants grown under increased temperatures or CO_2 would be less nutritious, as indicated by many researchers and longer larval period and increased mortality of insects is observed upon the insects feeding on them [28]. The defence mechanism of plants against insect pests is diminished by climate change, thereby rendering them susceptible to attack. For example: Early initiation of *H. armigera* infestation in cotton and pulses in Northern India [29]. It has also been found that CO_2 decreases the plant defences towards insect pests. For example, under increased levels of CO_2 in soybeans, the plant defence pathway signalling mediated by jasmonic acid (JA) does not work [30]. Plants become susceptible to insect pests like Japanese beetle, *Popillia japonica* and western corn rootworm, *Diabrotica virgifera* due to reduced production of defensive cysteine proteinase inhibitors (CystPIs). Additionally, the herbivore-induced plant volatiles (HIPVs) are influenced by higher temperatures and CO_2 [31].

3.5 Effect of climate change on plant volatile compounds

The production and release of plant volatile organic compounds (VOCs) can be influenced by changes in abiotic factors and are expected to influence how insects recognise and make use of plant VOCs in intra- and inter-specific interactions [32]. VOCs involved in a number of insect-plant interactions, ranging from positive (e.g., pollination and seed dispersal) to negative (e.g., herbivore defences). The atmosphere could be made more fragrant by global climate, due to release of higher levels of fragrant chemicals in a changing environment by plants. This, in turn, would affect how plants communicate with each other through competitive and allelopathic processes and how they protect themselves from pests, like insects, viruses and pathogens. Few major studies have been conducted to address the effect of changing temperature and gas concentration on VOCs metabolism and expression. Plants are required to develop increased concentrations of VOCs for extended time periods under higher temperatures, thus altering their ecological role in interactions of insects and plants. For example, monoterpene emissions are highly temperature-sensitive exhibiting a 3 fold increase for every 10°C increase in temperature [33]. Therefore, future herbivorous rates are reduced by the development and emission of higher concentrations of VOCs like methyl jasmonate or methyl salicylate that act as plant signalling molecules against insect attack. On the other hand, if a more fragrant atmosphere, confuses pollinators and seed dispersers, beneficial relations may also be interrupted, causing plant reproduction and fitness to be reduced.

VOCs are expected to increase at high CO_2 concentrations because of the positive relationship in between the carbon supply and VOCs production. On the basis of the resource allocation hypothesis, increased CO_2 concentrations are hypothesised to increase emissions of monoterpenes and sesquiterpenes into the atmosphere [34]. As per this theory, when there is an abundance of carbon availability relative to what is required for plant growth, increases the production of C-based plant secondary compounds. In conifers and cultivated plants, the development of certain C-based VOCs increases under high CO_2 conditions [35]. Higher temperature and CO_2 affects the emission of herbivorous mediated plant volatile organic compound (HIPVs) [36]. Any changes made to HIPVs would have a direct impact on the effectiveness of biological control. The olfactory perception of the volatiles will be diminished by change in temperatures, thereby affecting the host position capacity of the natural enemies. Higher CO_2 concentrations would also modify the levels of oxalic and malic acids in chickpea, affecting its herbivorous resistance [37].

3.6 Effect of climate change on plant: Pollinator interactions

For the health of natural habitats, plant-pollinator interactions are important, and most of the human diet is dependent on pollination by insects. By altering the phenology, morphology, and distribution of plants and insects, components of Global Environmental Change (GEC), including higher temperatures, increased CO_2 levels, and modified patterns of precipitation, can directly impact the interactions between plants and pollinators. Another important way where GEC factors can influence plant-pollinator interactions is the modification of phytochemicals (nectar and volatile chemistry) necessary for pollinator attraction (**Table 1**). Floral biogenic volatile organic compounds (BVOCs), that have a major function in attraction of the pollinators and plant-pollinator mutualisms, can be transformed by the components of global climate change. Most of the effects of temperature on floral BVOCs have been shown, with a consistent positive influence on global warming BVOC emissions. BVOCs are actively carried by a protein through the plasma membrane and expelled from the *Petunia hybrida* flowers [38], where the temperature and protein behaviour is always positively associated.

However, anthropogenic airborne pollutants such as ozone and diesel exhaust can destroy floral VOCs once released and increase the foraging times of pollinator. For instance, Farré-Armengol et al. [39] found that appropriate ozone levels in compound-specific ways degraded *Brassica nigra* floral BVOCs, altering the ratio of bouquet compounds that strongly inhibited the attraction of the generalist bumble bee pollinator, *Bombus terrestris*. It is apparent that airborne contaminants have major adverse effects often in unpredictable ways on the pollinator attraction towards flowers (eg. by changing BVOC ratios).

4. Impact of climate change on insects, plants and their interactions

Climate change has significant consequences in every field of agriculture. Climatic changes like temperature, precipitation, humidity and other meteorological components influence the relationship between insect pests and plants. Climate change has enhanced the pest population and their damage potential by increasing the distribution, improving survival rates and developing the adaptability of insect pests. The change in population, mobility, and insect pest behaviour is caused by increasing temperatures, changed precipitation patterns and disrupted gaseous composition of the atmosphere etc. A number of variables that decide how much plants can grow are influenced by climate change. At the same time, incidence of higher temperatures, decline in the supply of water and changes in soil conditions would actually make it harder for plants to flourish. The relationships between plants and insects are altered by increased CO_2 and temperature, with important consequences for food security. Via warming acceleration of plant phenology creates mismatches between plants and insect pollinators. Likewise, changing the development rate of plant in relation to the development of insect can intensify/mitigate the effects of herbivore.

4.1 Impact of climate change on insect pests

The insect pests are seriously affected by overall rise in global average temperatures, weather pattern changes and severe climatic events. With these seasonal and long term changes the population dynamics of many insect pests would be influenced. Different climate patterns primarily affect insect ecosystems and their survival strategies. Significant climate change drivers like higher temperatures and CO_2 levels and lower soil humidity, have an effect on the nature of population of insect pests and results in subsequent crop losses. Abiotic parameters impose direct effects on the rate of distribution and abundance of insect pest populations by adjusting their growth, survival, reproductivity, dispersal and number of generations per season. Because of the rapid climate change, insect pests are developing increased overwintering stages and number of generations with rapid population growth. Temperature is said to cause direct effects among the abiotic factors. For example, increasing temperatures, from 1.5 to 2.5°C, will surely increase the winter survival and prolong the range of pink bollworm, *Pectinophora gossypiella* [40]. During extended periods of drought, followed by heavy rainfall oriental armyworm, *Mythimna separata*, the populations raises due to the undesirable effects of drought on the activity and abundance of natural enemies of this insect pest [41].

4.2 Impact of climate change on beneficial insects

Climate change impacts the insect pest's natural enemies in a wide variety of ways. Plants grown under higher temperatures and CO_2 and lower precipitation provides various nutritional opportunities for different insect pests, eventually affecting the fitness of insect pest- feeding predators and parasitoids [42]. Despite of a wide variety of host and parasitoid species, variability in precipitation is the key cause for differences in caterpillar parasitism. Parasitism of mealy bug is reduced under conditions of water stress combined with dry conditions in cassava, *Manihot esculenta* [43]. In relation to herbivore hosts and their movement, natural enemies locate their hosts based on their tolerance to environmental extremes. Predatory bugs, *Oechalia schellenbergii* were found to be more effective in destroying the cotton bollworm larvae when pea plants are cultivated at high CO_2 levels [44]. Similarly, in feeding upon the aphid, *Aphis gossypii*, the coccinellid predator, *Leis axyridis*, was found to be more successful at higher CO_2 levels [45].

In hot summers rather than in moderate summers, ladybird beetles (*Coccinella septempunctata*) reduce aphid populations (*Sitobion avenae*) more effectively [46]. Rise in temperature affects the production and release of volatile compounds and extra floral nectar by plants. These secretions help the insects to avoid the attack from natural enemies. Natural enemies need to undergo climate change for breeding purposes, after overcoming temperature extremes; they need to find hosts efficiently through a broad spectrum of temperature and humidity environments. *Trichogramma carverae*, the egg parasitoid fails to recognise hosts at temperature above 35°C [47] and reduces fertility at 30°C [48]. Some parasitoids evolve earlier than hosts in rapid response to temperature and often engage in the extinction of the parasitoid population in absence of the hosts. At elevated temperatures, the rate of insect parasitism will be reduced as host species emerge and move through the susceptible stages quickly before the appearance of parasitoids. Mild winters in temperate regions enhance the survival of parasitoids. Ex: Aphid parasitoids from cereal crops become active during winter and reduce spring aphid populations [49]. The foraging behaviour of ants is often affected by temperature. In general, chemically recruited ants prefer to eat at temperatures lower than those that do not [50]. As a consequence, increased temperature results in pheromone decay changing the trail following action which is disadvantageous to the activity of ant feeding [51]. Hymenopteran parasitoids and small predators sometimes have a negative impact on rising temperatures. Ex: At 40°C BPH is 17 times more tolerant than its natural enemies *Cyrtorhinus lividipennis* and spider, *Pardosa pseudoannulata* [52].

4.3 Impact of climate change on invasive insect species

Climate change is altering important aspects of the environment such as temperature and precipitation, the occurrence of extreme weather events, as well as air composition and land cover. The main factors driving the survival of organisms are temperature, atmospheric CO_2 concentration and available nutrients. It is most likely that changes in these variables might stress the ecosystems and facilitate the chances of invasions. According to the Convention on Biological Diversity (CBD) invasive alien species are considered to be the greatest threat to biodiversity loss worldwide and by altering their geographical structure, function and diversity, inflicts high costs on agriculture, forestry and aquatic ecosystems. Climate change imposes direct effects on insect physiology and their behaviour and indirectly effect through biotic interactions. The introduction, establishment, distribution, impact and changes in the effectiveness of mitigation strategies of invasive insect species are expected to be the significant drivers of anthropogenic and global climate change. Global warming is expected to increase the ecological consequences such as new pests introduction, by changing phenological events such as flowering times mainly in plants of temperate species as many tropical plants can tolerate the phenological changes. The key issue favouring the introduction of insect susceptible cultivars or crops is the invasion of new insect-pests. For example, during 2018 and 2019, fall armyworm, *Spodoptera frugiperda* which is a recent invasive insect from Africa has spread to several countries like India, Thailand, Myanmar, China, Republic of Korea, Japan, Philippines, Indonesia and Australia. The relationship between temperature and the rate of development primarily affects its biology, distribution and abundance. As insect development occurs within a defined temperature range, a change in temperature will consequently affect the developmental rate, life-cycle duration and finally affects the survival. Rise in ambient temperature to near the thermal optimum of insects causes an increase in their metabolism and activity.

From the end of 2019 to early 2020, a desert locust (*Schistocerca gregaria*) outbreak has posed a significant risk to food security and livelihoods across many East African nations. Changes in climate such as increasing temperatures and precipitation over desert areas, and heavy winds combined with tropical cyclones can provide a new environment for reproduction, growth and migration of pest. This means that global warming played a role in establishing the conditions needed for the growth, outbreak and survival of the locust. Oceans absorb around 90 per cent of anthropogenic heat [53] and in the western part of the Indian Ocean in the tropical Ocean system, the most rapid warming occurs with a summer average rise of 1.2°C [54]. In neighbouring areas, this warming has increased the frequency and intensity of extreme climate events and thus favoured the movement of locust plague to various countries like Pakistan, India etc.

4.4 Impact of climate change on plant-pollinators interactions

Climate change is directly linked to the loss of habitat, nutritional deficiencies and lack of various diets, as the abnormal climate affects the growth of plants and flowers. Flowers are forced by climate change to bloom half a day earlier each year, meaning plants are now flowering a month earlier than 45 years ago. Finally, plants that flower earlier mean that they are not pollinated and the bees and butterflies do not have any food left. A study conducted in Spain between 1952 and 2014 found that from the mid-1970s, (*Apis mellifera*) populations appeared early in the spring, as they have adapted quickly to warmer temperatures [55]. Climate change however, has the ability to disrupt the mutualism between plants and pollinators

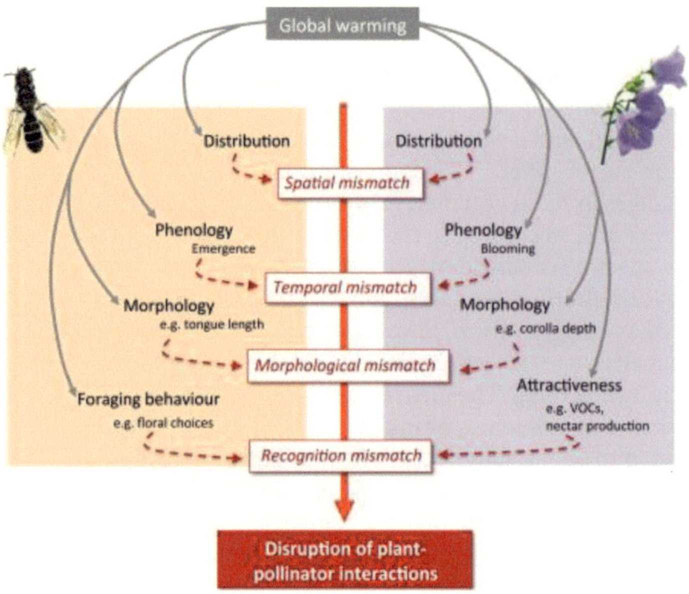

Figure 2.
Potential impacts of global warming on plant-pollinator interactions [56].

and thus lead to potential mismatches, placing plant and pollinator species at risk of extinction (**Figure 2**).

The reduced co-occurrence of interacting partners, the mismatches in plant-pollinator interactions may occur in a shared habitat; this decrease can be temporal or spatial. Increasing attention has been given to such types of temporal mismatches between plants and pollinating insects. A modification of the flowering period of the plant and/or the phenology of the pollinator either of which can be advanced or delayed can drive these mismatches. The co-occurrence of plants and pollinators, needed for interaction to occur, may also be spatially disrupted. The geographical overlap between interacting partners may decrease or increase during global warming, depending on the plasticity, adaptability and life history features of the species in question. In addition to temporal or spatial mismatches, climate change also has the ability to affect the interactions between plant-pollinators that are mediated by physiological or morphological characteristics. The mechanical fit of the interaction can be affected in order to have access to plant resources, in addition to plant morphology, because success of pollination depends on morphological characteristics like length of tongue or overall size of the body. For example, in many species, average rise in temperature has been shown to adversely affect the size of body. In addition, temperature rises will affect the pollinator's foraging behaviour, plant's attractiveness, together with the quality and quantity of plant resources.

4.5 Impact of climate change on plants

Whether it is heat waves, increased flooding or droughts, climate change has many impacts on plants. In addition to these global warming knock-on effects, rising concentrations of carbon dioxide and temperatures has a direct effect on the growth of plant, reproduction and resilience. Rise in local and global temperatures pose a major challenge to the growth and development of plants [57]. The Intergovernmental Panel on Climate Change (IPCC) has suggested that global temperatures would persist to rise by another 1.5°C by 2030 and 2052, if the present global warming patterns remain the same. Heat stress can damage all plant growth phases from the time of germination to reproduction, resulting in restricted production of important staple food crops [58]. The effect of heat stress on wheat yields, for instance is negative. For every 1°C increase in global mean temperature, a 4–6 per cent decrease in average global wheat yields is expected [59]. Climate change enforces plants to change their dates for leaves and blooming. It is suspected that warmer temperatures potentially destroy tropical forests resulting in more gases causing atmospheric warming and with increase in temperature; cold regions have become increasingly adaptable to growth of plants.

Necessary processes like photosynthesis, respiration, metabolism, and behaviour of stomata are regulated by CO_2. CO_2 concentrations have been rising, from around 350 ppm in 1986 to over 415 ppm in 2019 [60] and are expected to rise to 550 ppm by 2050 as reported by the IPCC. Elevated CO_2 improves the efficacy of photosynthetics, and thereby improves crop growth and yield. Rubisco's improved carboxylation ability that is comparatively poor at present-day CO_2 concentrations in the atmosphere has become the main reason for this improved photosynthesis. However, with increase in CO_2 concentration, at the CO_2 fixation site will raise the CO_2/O_2 ratio, contributing to the effectiveness of Rubisco's carboxylation by reducing the photorespiration rate (**Figure 3**). Under conditions of elevated CO_2, an increase in root to shoot ratio was observed, in this condition plants synthesise a great number of chloroplasts, mesophyll cells, longer stems and extended diameter, length and number of large roots, more lateral root development with changes in branching patterns [62].

4.6 Impact of climate change on insect-plant interactions

Insects and plants are affected by climate change and severe weather actions and the direct impact of anthropogenic climate change has been reported on each and every continent, ocean, and in many main taxonomic groups. Plants experience new environmental problems like higher CO_2 and O_3 levels, increased temperature and UV radiation, and changes in rainfall pattern across the seasons as a result of recent activities of human and their influence on global climate. Insects constitute

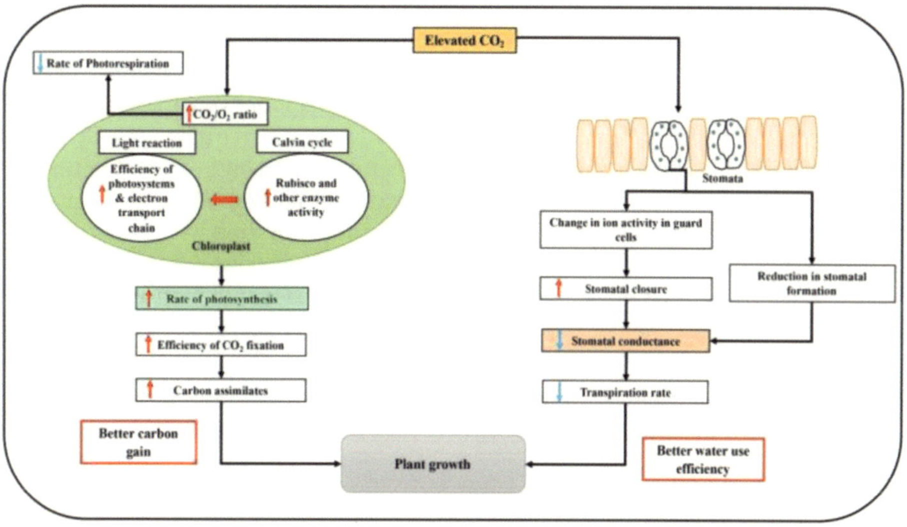

Figure 3.
Effects of CO2 induced photosynthesis and stomata conductance on plant growth responses [61].

nearly half of the biodiversity and are vital for the structure and function of ecosystem. Because of their close relationship with host plants, through the changes undergone by their host plants, herbivorous insects are likely to experience climatic change direct and indirect consequences. In many ways, global climate changes are reported to influence the interactions between insects and plants. They could directly influence insects, through changes in parameters of physiology, behavioural and life history, as well as indirectly, by means of change in their morphology, biochemistry, physiology and patterns of richness, diversity and abundance experienced by host plants [63]. By functioning as herbivores, pollinators, predators and parasitoids, insects play major roles in ecosystem services and by altering their abundance and diversity, have attained the capacity to modify the services they offer [64]. Over past 20 years, the studies documenting the impacts of climate change on insects have risen exponentially.

4.6.1 Increased temperature

In many global change scenarios meant for plants and insect herbivores, the ecological-niche models use revealed a definite spatial mismatch among the monophagous butterfly, *Boloria titania*, and its larval host plant, *Polygonum bistorta* due to each species expressing differential range expansion in response to changes in climate and land use [65]. These findings indicate that, because of species-specific responses to climate change problems, temperature increase and other altered factors by humans have the capacity to disturb the insect- plant interactions at trophic level. Another example of the impacts of rising temperatures on the generation of asynchrony between insects and their food sources is the winter moth, *Operophtera brumata*. The outbreaks of climate-dependent psyllid, *Cardiaspina* sp. and their effects on the *Eucalyptus* dieback across thousands of hectares of Western Sydney's seriously endangered Cumberland Plain Woodlands (CPW) are due to the effect of change in temperatures. Summer heat waves (maximum above 46°C) combined with resource shortages due to defoliation triggered the *Cardiaspina* sp. outbreak in 2013 and in the CPW it became unnoticeable [66]. Conversely, by mid-2015, population levels grew and large parts of the CPW were defoliated again until a heat wave led to extreme decline in populations of psyllid in early 2017 (up to 46°C maximum).

While most of the studies, have concentrated on negative interactions involving insects on the effect of global warming on trophic interactions. Memmott et al. [67] have discussed that how climate change can interrupt or even eradicate mutual interactions like pollination and dispersion of seed in between organisms. By means of simulations based on a real network of interactions between 1,419 pollinating insect species with 429 species of plants, they showed that 17 to 50 per cent of all pollinators studied would suffer a decrease in the supply of food with phenological progress of their floral resources by two weeks. For specialist pollinators, this reduction would be even more extreme. Data on the impact of climate change on the synchrony of host-parasitoid interactions are not as widespread as interactions between plant-herbivores and predator–prey [68] but recent studies have proven that parasitoid and host asynchrony affects of climate change can be direct or indirect through changes in host plant.

4.6.2 Enriched atmospheric CO_2

It affects the physiology of plants, with significant implications on plant growth and biochemical composition. Plant chemical composition influences both positive and negative trophic interactions and decomposition, which will then react to atmospheric CO_2 concentrations [69]. Even though the impacts of increased CO_2 on plants are erratic and not uniform, increased activity of photosynthetic, production and leaf area/biomass are often exhibited by plants grown under high CO_2 conditions. Higher CO_2 levels could change the primary and secondary metabolism of plants as well. The increase in the supply of carbon for tissues of plant and the subsequent C/N ratio changes influence the amount of nitrogen in plant tissues, triggering a “nitrogen dilution effect”. This lower nitrogen concentration, combined with higher C/N ratio with possible influence on the plants secondary metabolism, suggests lower leaf protein concentration and thus reduces the nutritional value of herbivores. Increased CO_2 usually raises the concentration of leaf carbohydrates and reduces the amount of nitrogen (N) in combination with elevated temperatures. Higher CO_2 exposure depresses the jasmonic acid (JA), a plant defence hormone while stimulating salicylic acid (SA) production. This results in increased vulnerability to chewing insects and increased tolerance to pathogens.

In addition to higher CO_2, elevated ozone (O_3) concentrations in the troposphere also affect plants and insects indirectly. In North America and Europe, tropospheric ozone layer is known as main hazardous and well-known pollutant affecting the ecosystems of agriculture and forests. Since the pre-industrial period, O_3 concentrations have increased by almost 40 per cent and are reported to affect directly the plant species and affect herbivorous insects indirectly. O_3 in plants triggers a cascade of adverse physiological effects, disrupting the process of photosynthesis and reducing the carbohydrates supply in the plant [69]. While higher CO_2 concentrations stimulate the productivity and development of plants, O_3 tends to have detrimental impacts on plants, usually leading to reduced growth and lower quality of nutrition in the leaves. This modification in plants quality resulted in the increased rate of herbivory due to overcompensation by insects because of lower nutritional features of tissues. Plants grown under increased O_3 conditions generally display lower photosynthetic rates, reduced leaf area, premature leaf abscission and damaged branch and root growth. Increased O_3 concentrations are expected to have indirect effects on insect and would depend on the extent of change in the condition of host plant (bottom-up factors) or the influence of natural enemies (top-down factors). Elevated O_3 may alter the population of natural enemies by making changes in their diversity, number and prey quality or by changing the behaviour of natural enemies [64].

5. Impact of climate change on the insect pest management strategies

Dramatic changes in the geographical distribution and population development of insect pests, interactions between insect-host plants, the behaviour and abundance of natural enemies, and the efficacy of crop defence technologies may be caused by global warming and climate change. As a consequence of global warming, the distribution and relative abundance of some insect species susceptible to increase in temperatures in the temperate regions may decrease, while insect pests currently confined to the tropical and subtropical regions may migrate to the temperate regions along with a shift in the production areas of their host plants. As a consequence of global warming and climate change, the relative effectiveness of pest control strategies is likely to change. There is an immediate need to evaluate, under varying environmental conditions, the efficiency of different IPM technologies and develop suitable strategies for mitigating the adverse effects of climate change [70].

Although some impacts of climate change may be optimistic, evidence indicates that pest issues are likely to become more volatile and greater in amplitude overall.

However, due to the complex interacting factors of increasing CO_2 levels, shifting climate regimes and altered frequency/intensity of extreme weather events, predicting the impact of climate change on insect pests is not simple [71]. In addition, differences in the thermal preferences of insects and their natural enemies may result in a lack of cooperation between the two and an increased risk of host outbreaks [72]. Changes in the effectiveness of methods of insect pest control as well as changes in policies of land use and crop management are the result of other indirect responses to insect pests, which can also have a higher impact on the pressure of insect pests than the direct effects of climate change alone. A few examples of direct and indirect effects are the following impacts on insect pests if changing climate conditions are studied in isolation:

5.1 Increases in temperature

The severity of damage caused by insect pests may be increased by increases in temperature. In USA, where increasing temperatures leading to greater insect populations in southern regions have bring about in higher use of insecticides compared to colder, higher latitude provinces, such growing insect populations and pressures will lead to more frequent insecticide applications. Such upsurges of toxic chemical applications may have serious adverse effects on human and environmental health. Temperature changes can also decrease the efficacy of some insecticides, such as a decrease in the toxicity of lambda-cyhalothrin, bifenthrin and spinosad to *Ostrinia nubilalis* as a result of elevated temperature after exposure [73]. The effectiveness of parasitoids in the control of pest species and the expression of defensive characteristics used by insect pests against their larval parasitoids has been found to affect even with minor variations in thermal conditions [74].

5.2 Altering precipitation

Extreme or insufficient precipitation can have a major impact on crop and pest interactions, as hot and humid conditions favour many species that are highly susceptible to moisture and rainfall. Also, as found during floods in Iowa in 1993, water-stressed crops are more likely to be affected by pests [75]. Changes in precipitation events are compounded by outbreaks of desert locusts, as demonstrated by their incursion of greater than 10 countries in northern and western Africa in 2004 after heftier than usual rainfall, resulting in severe crop injuries and food scarcities. Locust epidemics are only expected to become more common as the frequency and severity of precipitation events are predicted to increase in the future.

5.3 Increasing CO_2 levels

Increased CO_2 levels can directly lead to increased crop harvests, but any increase in yields can be partially or fully offset by losses caused by insects, pathogens and weeds. For example, in North America, cabbage loopers, *Trichoplusia ni* are observed to ingest a higher amount of leaves under higher CO_2 levels, which is believed to be due to the decreased levels of nitrogen observed in cabbage leaves that grown under these conditions [76].

5.4 Extreme weather actions

Extreme weather conditions can unpredictably affect interactions between crops, pests and diseases, likely leading to the failure of some crop protection strategies and subsequent reductions in yields. For example, *Trichogramma evanescens* populations were so reduced in May 1993 by exceptionally dry and warm weather conditions in Slovakia that no record of active parasitism of European corn borer eggs was reported that year [77]. In hurricanes, intense air streams can also move fungal spores or insects from overwintering sites to places where additional problems can be caused. Winds associated with Hurricane Wilma thus spread citrus canker widely in Florida, killing 170,000 acres of fruit trees grown commercially [78]. Ecosystems affected by extreme climate events are also automatically more fragile and vulnerable to invasions of space by aliens and indigenous organisms.

6. Climate change impact and risk analysis

Changes in species abundance and diversity due to climate alteration will lead to a decrease in the effectiveness of insect pest management systems, so current monitoring methods need to be strengthened and new ones need to be created to recognise possible changes in pest distribution, population ecology, risk assessment, yield loss and impact assessment. Potential enhancements in pest endurance strategies require wider and deeper inter-centre collaborations to create new IPM options or to disseminate existing ones to new areas where farmers can find them suitable. Excessive use of synthetic insecticides results from existing sensitivities to environmental contamination, human health threats and the return of pests. Numerous botanically and biologically based products are currently used as eco-friendly products. However, both of these pest management methods are extremely environmentally sensitive. Due to rising temperatures and UV radiation and decreasing relative humidity, many of these control tactics may be ineffective [79]. Appropriate pest management techniques, which will be successful in global warming situations in the future, must therefore be created. The resistance of host plants, natural plant products, bio-pesticides, natural enemies and agricultural practises provide a potentially viable alternative to integrated pest control. But, as a result of global warming, the relative effectiveness of many of those control mechanisms is likely to change. Climate change is greatly influenced by biological regulation, which is considered to be an important and successful aspect of IPM programmes, as the relationship between natural enemies and host pests is affected.

The troubling aspect in the absence of natural enemies is the transfer of insect species to new terrains, as it can lead to outbursts of pests. The biggest challenge in the future is to establish efficient model forecasting that would cover the approaches for their management. It is urgent to establish and incorporate modelling methods for predicting changes in the topographical distribution and population development of insect pests and adapting approaches to minimise crop losses. Weather-based pest management systems are valuable decision-making tools that help farmers recognise the risk of outbreaks of pests under different climatic conditions. For alert systems, weather, plant-insect relationship information is very important to take appropriate action to avoid outbreaks of pests and to avoid economic losses. For sustainable agriculture and the mitigation of the effects of climate change on agriculture, assessing the impacts of climate change on crop yield and climate-smart crop growth is significant.

7. Conclusion

In modern era climate change is globally acknowledged fact. It has a serious effect on the diversity, distribution, occurrence, reproduction, development, growth, voltisim and phenology of insect pests and plant species. It also affects the activity of plant defence and resistance system, invasive insect species, natural enemies, pollinators and insect pest management strategies. Food protection in the 21st century will be the greatest challenge for humanity in the years to come, considering the declining efficiency of production due to the depletion of the natural resource base, the drastic effects of climate change on the diversity and abundance of insect pests, and the scale of crop losses. Coping with climate change is very tedious, due to its uncertainty, ambiguity, unpredictability and differential effects over time and place. It is important and challenging in agriculture to understand abiotic stress reactions in plants, insect pests, invasive insect species, natural enemies and pollinators. The effects of climate change on crop production, mediated by changes in populations of extreme insect pests, should be carefully considered in the planning and implementation of adaptation and mitigation strategies for future pest management programmes. It is then vital to look at the possible impacts of climate change on crop safety in a concerted manner and to establish effective actions to mitigate the impacts of climate change on food security.

Author details

Somala Karthik*, M.S. Sai Reddy and Gummudala Yashaswini
Department of Entomology, P.G. College of Agriculture, Dr. Rajendra Prasad Central Agricultural University, Pusa, Bihar, India

*Address all correspondence to: somalakarthik1995@gmail.com

References

[1] IPCC. Climate change 2001: scientific basis, Contribution of working group I to the third assessment report of the intergovernmental panel climate change (IPCC). Cambridge University Press. 2001; Cambridge http:/ /www.grida.no/ climate/ipcc_tar/

[2] Bale JSB, Masters GJ, Hodkinson ID, Awmack C, Bezemer TM. Herbivory in global climate change research: Direct effects of rising temperature on insect herbivores. Global Change Biology. 2002; 8: 1-16.

[3] Moore BA, Allard GB. Climate change impacts on forest health. Forest Health and Biosecurity Working Papers, Forestry Department Food and Agriculture Organization of the United Nations. Working Paper FBS/9E, FAO, Rome, Italy. 2008; pp 35.

[4] Melillo JM, Richmond TT, Yohe G. Climate change impacts in the United States. Third National Climate Assessment. 2014; 52.

[5] Florides LA, Christodoulides P, Messaritis V. Global warming and CO_2. Vs sun. In: Harris, S. A. (Ed.). Global Warming. Croatia, IntechOpen. 2011: 23-62.

[6] Shrivastava A. Rising temperature causes soil to release more carbon dioxide. Down to Earth. 2016.

[7] IPCC. Climate change 2014: synthesis report. Contribution of Working Groups I, II and III to the Fifth Assessment Report of the Intergovernmental Panel on Climate Change [Pachauri RK, Meyer LA. (eds.)]. IPCC, Geneva, Switzerland. 2014; 151 pp.

[8] Brown PT, Caldeira K. Greater future global warming inferred from Earth's recent energy budget. Nature. 2017; 552(7683):45-50.

[9] Bradfield M, Stamp N. Effect of night time temperature on tomato plant defensive chemistry. Journal of Chemical Ecology. 2004; 30(9): 1713-1721.

[10] Rivero RM, Sánchez E, Ruiz JM, Romero L. Influence of temperature on biomass, iron metabolism and some related bioindicators in tomato and watermelon plants. Journal Of Plant Physiology. 2003; 160(9):1065-1071.

[11] Green TR, Ryan CA. Wound-induced proteinase inhibitor in tomato leaves: Some effects of light and temperature on the wound response. Plant Physiology. 1973; 51(1):19-21.

[12] Bickford CP. Ecophysiology of leaf trichomes. *Functional Plant Biology*. 2016; 1;43(9):807-14.

[13] Bauerfeind SS, Fischer K. Increased temperature reduces herbivore host-plant quality. Global Change Biology. 2013; 19(11):3272-3282.

[14] Himanen SJ, Nissinen A, Dong WX, Nerg AM, Stewart CN, Poppy GM, Holopainen JK. Interactions of elevated carbon dioxide and temperature with aphid feeding on transgenic oilseed rape. Global Change Biology. 2008; 14(6):1437-1454.

[15] Diamond SE, Kingsolver JG. Fitness consequences of host plant choice: A field experiment. Oikos 2010;119(3): 542-550.

[16] Bezemer TM, Jones TH. Plant-insect herbivore interactions in elevated atmospheric CO_2: Quantitative analyses and guild effects. Oikos 1998; 212-222.

[17] Lindroth RL, Arteel GE, Kinney KK. Responses of three saturniid species to paper birch grown under enriched CO_2 atmospheres. Functional Ecology. 1995; 306-311.

[18] Kranthi KR, Kranthi S, Gopalakrishnan N, Asokan R, Mayee CD. *Bt* resistance-Its management and prospects in the current context of climate change. IPM Strategies to Combat Emerging Pests in the Current Scenario of Climate Change. 2009; pp. 237-261.

[19] Chen FJ, Wu G, Lu J, Ge F. Effects of elevated CO_2 on the foraging behaviour of cotton bollworm, Helicoverpa armigera. Insect science. 2005b; 12(5): 359-365.

[20] Ode PJ, Johnson SN, Moore BD. Atmospheric change and induced plant secondary metabolites are we reshaping the building blocks of multi-trophic interactions? Current Opinion in Insect Science. 2014; 5: 57-65.

[21] Boullis A, Francis F, Verheggen FJ. Climate change and tritrophic interactions: Will modifications to greenhouse gas emissions increase the vulnerability of herbivorous insects to natural enemies? Environmental Entomology. 2015; 44(2): 277-286.

[22] Blomquist GJ, Figueroa-Teran R, Chang E, Tittiger C. Pheromone production in bark beetles. Insect Biochemistry and Molecular Biology. 2010; 40(10):699-712.

[23] Schlenker W, Roberts MJ. Nonlinear temperature effects indicate severe damages to US crop yields under climate change. Proceedings of the National Academy of sciences. 2009; 106(37): 15594-15598.

[24] Staley JT, Hodgson CJ, Mortimer SR, Morecroft MD, Masters GJ, Brown VK, Taylor ME. Effects of summer rainfall manipulations on the abundance and vertical distribution of herbivorous soil macro-invertebrates. European Journal of Soil Biology. 2007; 43(3): 189-198.

[25] Hulme M, Doherty R, Ngara T, New M, Lister D. African climate change: 1900-2100. Climate Research. 2001; 17(2): 145-168.

[26] Jamieson MA, Burkle LA, Manson JS, Runyon JB, Trowbridge AM, Zientek J. Global change effects on plant–insect interactions: The role of phytochemistry. Current opinion in insect science. 2017; 23: 70-80.

[27] Lobell DB, Gourdji SM. The influence of climate change on global crop productivity. Plant Physiology. 2012; 160: 1686-1697.

[28] Coviella CE, Trumble JT. Effects of elevated atmospheric CO_2 on insect-plant interactions. Conservation Biology 1999; 13(4): 700-712.

[29] Sharma HC. Climate change effects on insects: Implications for crop protection and food security. Journal of Crop Improvement. 2014; 28: 229-259.

[30] Zavala JA, Casteel CL, Delucia EH, Berenbaum MR. Anthropogenic increase in carbon dioxide compromises plant defence against invasive insects. Proceedings of the National Academy of Sciences of the United States of America. 2008; 105: 5129-5133.

[31] Gouinguené SP, Turlings TCJ. The effects of abiotic factors on induced volatile emissions in corn plants. Plant Physiology. 2002; 129: 1296-1307.

[32] Peñuelas J, Staudt M. BVOCs and global change. Trends in Plant Science. 2010; 15(3):133-144.

[33] Constable JV, Litvak ME, Greenberg JP, Monson RK. Monoterpene emission from coniferous trees in response to elevated CO_2 concentration and climate warming. Global Change Biology 1999; 5(3):252-267.

[34] Lerdau M, Litvak M, Monson R. Plant chemical defense: Monoterpenes and the growth-differentiation balance

hypothesis. Trends in Ecology & Evolution. 1994; 9(2):58-61.

[35] Jasoni R, Kane C, Green C, Peffley E, Tissue D, Thompson L, Payton P, Paré PW. Altered leaf and root emissions from onion (*Allium cepa* L.) grown under elevated CO_2 conditions. Environmental and Experimental Botany. 2004; 51(3):273-280.

[36] Arneth A, Sitch S, Bondeau A, Butterbach-Bahl K, Foster P. From biota to chemistry and climate: Towards a comprehensive description of trace gas exchange between the biosphere and atmosphere. Biogeosciences. 2010; 7: 121-149.

[37] Sharma HC, War AR, Pathania M, Sharma SP, Akbar SM, Munghate RS. Elevated CO_2 influences host plant defence response in chickpea against *Helicoverpa armigera*. Arthropod-Plant Interactions. 2016; 10(2): 171-181.

[38] Adebesin F, Widhalm JR, Boachon B, Lefèvre F, Pierman B, Lynch JH, Porter JA. Emission of volatile organic compounds from petunia flowers is facilitated by an ABC transporter. Science. 2017; 356(6345): 1386-1388.

[39] Farré-Armengol G, Peñuelas J, Li T, Yli-Pirilä P, Filella I, Llusia J, Blande JD. Ozone degrades floral scent and reduces pollinator attraction to flowers. New Phytologist. 2016; 209(1): 152-160.

[40] Gutierrez AP, Ellis CK, Ghezelbash R. Climatic limits of pink bollworm in Arizona and California: Effects of climate warming. Acta Oecologica. 2006; 30(3):353-364.

[41] Sharma HC, Sullivan DJ, Bhatnagar VS. Population dynamics and natural mortality factors of the oriental armyworm, *Mythimna separata* (Walker) (Lepidoptera: Noctuidae) in south-Central India. Crop Protection. 2002; 21:721-732.

[42] Coviella CE, Stipanovic RD, Trumble JT. Plant allocation to defensive compounds: Interactions between elevated CO_2 and nitrogen in transgenic cotton plants. Journal of Experimental Botany. 2002; 53(367): 323-331.

[43] Calatayud PA, Polania MA, Seligmann CD, Bellotti AC. Influence of water-stressed cassava on *Phenacoccus herreni* and three associated parasitoids. Entomologia Experimentalis et Applicata. 2002; 102(2): 163-175.

[44] Coll M, Hughes L. Effects of elevated CO_2 on an insect omnivore: A test for nutritional effects mediated by host plants and prey. Agriculture, Ecosystems & Environment. 2008; 123(4): 271-279.

[45] Chen F, Ge F, Parajulee MN. Impact of elevated CO_2 on tri-trophic interaction of *Gossypium hirsutum*, *Aphis gossypii* and *Leis axyridis*. Environmental Entomology. 2005a; 34(1): 37-46.

[46] Skirvin DJ, Perry JN, Harrington R. The effect of climate change on an aphid–coccinellid interaction. Global Change Biology. 1997; 3(1): 1-11.

[47] Thomson LJ, Robinson M, Hoffmann AA. Field and laboratory evidence for acclimation without costs in an egg parasitoid. Functional Ecology. 2001; 15(2): 217-221.

[48] Scott M, Berrigan D, Hoffmann AA. Costs and benefits of acclimation to elevated temperature in *Trichogramma carverae*. Entomologia Experimentalis et Applicata. 1997; 85(3): 211-219.

[49] Legrand MA, Colinet H, Vernon P, Hance T. Autumn, winter and spring dynamics of aphid *Sitobion avenae* and parasitoid *Aphidius rhopalosiphi* interactions. Annals of Applied Biology. 2004; 145(2): 139-144.

[50] Ruano F, Tinaut A, Soler AJJ. High surface temperatures select for individual foraging in ants. Behavioral Ecology. 2000; 11(4): 396-404.

[51] Van Oudenhove L, Billoir E, Boulay R, Bernstein C, Cerdá X. Temperature limits trail following behaviour through pheromone decay in ants. Naturwissenschaften. 2011; 98(12): 1009-1017.

[52] Kambrekar DN, Guledgudda SS, Anand K. Impact of climate change on insect pests and their natural enemies. Karnataka Journal of Agricultural Sciences. 2015; 28(5):814-816.

[53] Barnett TP, Pierce DW, Santer BD, Gregory JM, Washington, WM. Penetration of human-induced warming into the world's oceans. Science. 2005; 309(5732): 284-287.

[54] Roxy MK, Ritika K, Terray P, Masson S. The curious case of Indian Ocean warming. Journal of Climate. 2014; 27(22): 8501-8509.

[55] Gordo O, Sanz J. Temporal trends in phenology of the honey bee *Apis mellifera* (L.) and the small white *Pieris rapae* (L.) in the Iberian peninsula (1952-2004). Ecological Entomology. 2006; 31: 261– 268.

[56] Gérard M, Vanderplanck M, Wood T, Michez D. Global warming and plant–pollinator mismatches. Emerging topics in life sciences. 2020; 4: 77-86.

[57] Priya M, Sharma L, Kaur R, Bindumadhava H, Nair RM, Siddique KHM, Nayyar H. GABA (γ-aminobutyric acid), as a thermo-protectant, to improve the reproductive function of heat-stressed mungbean plants. Scientific Reports. 2019; 9:77-88.

[58] Hussain HA, Men S, Hussain S. Interactive effects of drought and heat stresses on morpho-physiological attributes, yield, nutrient uptake and oxidative status in maize hybrids. Scientific Reports. 2019; 9: 3890.

[59] Asseng S, Ewert F, Martre P. Rising temperatures reduce global wheat production. Nature Climate Change. 2015; 5: 143-147.

[60] Sengupta R, Pandey K. World breached safe atmospheric CO_2 levels 33 years ago. Down to Earth. 2019.

[61] Gamage D, Thompson M, Sutherland M, Hirotsu N, Makino A, Seneweera S. New insights into the cellular mechanisms of plant growth at elevated atmospheric carbon dioxide concentrations. Plant, cell & environment. 2018; 41:1233-1246.

[62] Qaderi MM, Reid, DM. "Crop responses to elevated carbon dioxide and temperature, In Singh SN, (ed.), Climate Change and Crops". *Environmental Science and Engineering.* 2009.

[63] Kazakis G, Ghosn D, Vogiatzakis IN, Papanastasis VP. Vascular plant diversity and climate change in the alpine zone of the Lefka Ori, Crete. *Biodiversity and Conservation*. 2007; 16(6):1603-1615.

[64] Hillstrom ML, Lindroth RL. Elevated atmospheric carbon dioxide and ozone alter forest insect abundance and community composition. Insect Conservation and Diversity. 2008; 1(4):233-241.

[65] Schweiger O, Settele J, Kudrna O, Klotz S, Kühn I. Climate change can cause spatial mismatch of trophically interacting species. Ecology. 2008; 89(12):3472-3479.

[66] Hall AA, Gherlenda AN, Cook JM, Riegler M. Anatomy of an outbreak: The biology and population dynamics of a Cardiaspina psyllid species in an endangered woodland ecosystem. Agricultural and Forest Entomology. 2015; 17: 292-301.

[67] Memmott J, Craze PG, Waser NM, Price MV. Global warming and the disruption of plant–pollinator interactions. Ecology Letters. 2007; 10(8):710-717.

[68] Klapwijk MJ, Gröbler BC, Lewis OT. Influence of experimental warming and shading on host–parasitoid synchrony. Global Change Biology. 2010; 16(1):102-112.

[69] Lindroth R, Roth S, Kruger E, Volin J, Koss P. CO_2-mediated changes in aspen chemistry: Effects on gypsy moth performance and susceptibility to virus. Global Change Biology. 1997; 3(3): 279-289.

[70] ICRISAT. Pest management strategies in the face of climate change. 2015.

[71] Bebber DP, Ramotowski MA, Gurr SJ. Crop pests and pathogens move pole wards in a warming world. Nature Climate Change. 2013; 3(11):985-988.

[72] Furlong MJ, Zalucki MP. Climate change and biological control: The consequences of increasing temperatures on host–parasitoid interactions. Current Opinion in Insect Science. 2017; 20:39-44.

[73] Musser FR, Shelton AM. The influence of post-exposure temperature on the toxicity of insecticides to *Ostrinia nubilalis* (Lepidoptera: Crambidae). Pesticide Science. 2005; 61(5): 508-510.

[74] Iltis C, Martel G, Thiéry D, Moreau J, Louâpre P. When warmer means weaker: High temperatures reduce behavioural and immune defences of the larvae of a major grapevine pest. Journal of Pest Science. 2018; 91(4):1315-1326.

[75] Munkvold GP, Yang XB. Crop damage and epidemics associated with 1993 floods in Iowa. Plant Disease. 1995; 79(1):95-101.

[76] Trumble J, Butler C. Climate change will exacerbate California's insect pest problems. California Agriculture. 2009; 63(2):73-78.

[77] Cagan L, Tancik J, Hassan S. Natural parasitism of the European corn borer eggs *Ostrinia nubilalis* (Lep., Pyralidae) by *Trichogramma* in Slovakia—Need for field releases of the natural enemy. Journal of Applied Entomology. 1998; 122(1-5):315-318.

[78] Sutherst RW, Constable F, Finlay KJ, Harrington R, Luck J, Zalucki MP. Adapting to crop pest and pathogen risks under a changing climate. Wiley Interdisciplinary Reviews: *Climate Change*. 2011; 2(2):220-237.

[79] Niziolek OK, Berenbaum MR, Delucia EH. Impact of elevated CO_2 and temperature on Japanese beetle herbivory. Insect Science. 2012; 20(4): 513-523.